Localization and Metal–Insulator Transitions

Institute for Amorphous Studies Series

Series editors

David Adler
Massachusetts Institute of Technology
Cambridge, Massachusetts

and

Brian B. Schwartz
Institute for Amorphous Studies
Bloomfield Hills, Michigan
and Brooklyn College of the City University of New York
Brooklyn, New York

PHYSICAL PROPERTIES OF AMORPHOUS MATERIALS
Edited by David Adler, Brian B. Schwartz, and Martin C. Steele

PHYSICS OF DISORDERED MATERIALS
Edited by David Adler, Hellmut Fritzsche, and Stanford R. Ovshinsky

TETRAHEDRALLY-BONDED AMORPHOUS SEMICONDUCTORS
Edited by David Adler and Hellmut Fritzsche

LOCALIZATION AND METAL–INSULATOR TRANSITIONS
Edited by Hellmut Fritzsche and David Adler

Localization and Metal–Insulator Transitions

Edited by

Hellmut Fritzsche
University of Chicago
Chicago, Illinois

and

David Adler
Massachusetts Institute of Technology
Cambridge, Massachusetts

Plenum Press • New York and London

Library of Congress Cataloging in Publication Data

Main entry under title:

Localization and metal–insulator transitions.

(Institute for Amorphous Studies series)
Part of 3 vol. festschrift in honor of Sir Nevill Mott on his eightieth birthday.
Companion vols.: Tetrahedrally-bonded amorphous semiconductors/edited by David Adler and Hellmut Fritzsche. Physics of disordered materials/edited by David Adler, Hellmut Fritzsche, and Stanford R. Ovshinsky.
Includes bibliographies and indexes.
1. Metal–insulator transitions—Addresses, essays, lectures. 2. Anderson model—Addresses, essays, lectures. 3. Amorphous semiconductors—Addresses, essays, lectures. 4. Order–disorder models—Addresses, essays, lectures. 5. Mott, N. F. (Nevill Francis), Sir, 1905– . I. Fritzsche, Hellmut. II. Adler, David. III. Mott, N. F. (Nevill Francis), Sir, 1905– . IV. Series.
QC176.8.E4L629 1985 530.4′1 85-12161
ISBN 0-306-42077-5

A Division of Plenum Publishing Corporation
233 Spring Street, New York, N.Y. 10013

Printed in the United States of America

To
Sir Nevill Mott
with respect and affection on the occasion
of his eightieth birthday

PREFACE

This volume and its two companion volumes, entitled Tetrahedrally-Bonded Amorphous Semiconductors and Physics of Disordered Materials, are our way of paying special tribute to Sir Nevill Mott and to express our heartfelt wishes to him on the occasion of his eightieth birthday. Sir Nevill has set the highest standards as a physicist, teacher, and scientific leader. Our feelings for him include not only the respect and admiration due a great scientist, but also a deep affection for a great human being, who possesses a rare combination of outstanding personal qualities. We thank him for enriching our lives, and we shall forever carry cherished memories of this noble man.

Scientists best express their thanks by contributing their thoughts and observations to a Festschrift. This one honoring Sir Nevill fills three volumes, with literally hundreds of authors meeting a strict deadline. The fact that contributions poured in from all parts of the world attests to the international cohesion of our scientific community. It is a tribute to Sir Nevill's stand for peace and understanding, transcending national borders.

The editors wish to express their gratitude to Ghazaleh Koefod for her diligence and expertise in deciphering and typing many of the papers, as well as helping in numerous other ways. The blame for the errors that remain belongs to the editors.

David Adler
Massachusetts Institute
of Technology
Cambridge, Massachusetts

Hellmut Fritzsche
The University of Chicago
Chicago, Illinois

Stanford R. Ovshinsky
Energy Conversion Devices
Troy, Michigan

CONTENTS

Contents of Companion Volumes:

INTRODUCTION

Hellmut Fritzsche and David Adler

This volume, one of three in this series presented as a Festschrift in honor of the eightieth birthday of Sir Nevill Mott, is concerned with the subjects of free-carrier localization and metal-insulator transitions. These represent two strongly related areas in which the giant contributions of Sir Nevill are ubiquitous. From his earliest work on Mott transitions and Mott insulators almost 50 years ago to his current analysis of the relationships between electronic correlations, disorder, dimensionality, and localization, Sir Nevill's publications have dominated the entire field. In this Festschrift, 72 authors from 11 countries throughout the world have contributed 39 papers in his honor. The volume is divided into three parts.

Part One is devoted to experimental studies of metal-insulator transitions. It begins with Rosenbaum's fascinating analysis of the approach of disordered insulators towards the metallic state, Si:P exemplifying the electron glass and $Hg_{1-x}Cd_xTe$ the electron crystal. Castner and Sharfman then discuss the effect of an applied magnetic field on the metal-insulator transition in Si:P, followed by reviews of metal-insulator transitions in IV-VI compounds by Allgaier and in VO_2 by Villenueve and Hagenmuller. Next, Rao and Ganguly consider the effects of composition on the transition in perovskite oxides, while Minomura describes the transitions induced by pressure in crystalline and amorphous materials. Metal-insulator transitions in amorphous Mo-Ge alloys, including the appearance of superconductivity, are described by Yoshizumi et al., Micklitz analyzes the transitions in metal/rare-gas mixtures, and Adkins does the same in discontinuous metal films. The following three papers deal with metal-insulator transitions in liquids, in cesium by Hensel et al., in Se-Te alloys by Cutler and Rasolodramanitra, and in a wide array of materials by Warren. Beaglehole and Clarkson then discuss the transition in microemulsions, Goodenough et al. analyze the role of interactions in perovskite and pyrochlore oxides, and von Molnar and Penney conclude the section with a review of transitions in disordered magnetic semiconductors.

Theoretical studies of metal-insulator transitions are the focus of Part Two, which opens with an analysis of the relevant thermodynamics by Yonezawa and Ogawa, and discussions of the interconnections between correlations and disorder by Castellani and DiCastro and of the competition between long-range interelectronic repulsion and chemical bonding by March. These are followed by reviews of Mott transitions in disordered semiconductors by Andreyev and Shlimak, the effects of exciton condensation on the transitions by Turkevich, and transitions in heavily doped semiconductors by Economou and Fertis. Next, Carini et al. analyze some surprising flux quantization in normal metals, and Cyrot suggests the possibility that heavy fermions arise from nearly localized f electrons. The subject of Anderson localization in general is then reviewed by Kramer and MacKinnon, while Kaveh investigates the effects of phase correlations. Following these, Emin discusses localization due to electron-phonon coupling, Wegner analyzes the density correlations near a mobility edge, and Pollak and Friedman derive an expression for the thermoelectric power of a material characterized by phonon-assisted hopping conduction. Finally, Halperin et al. review transport behavior in the vicinity of the percolation threshold of continuum systems.

The final section deals with systems with lower effective dimensionalities. These include polyacetylene, a quasi-one-dimensional semiconductor discussed by Chen et al., as well as grain boundaries in polycrystalline germanium, metal-ammonia solutions intercalated in graphite, and layered solids such as La_2NiO_4, all quasi-two-dimensional systems, reviewed by Landwehr and Uchids, by Solin, and by Honig and Buttrey, respectively. Friedel and Noguera then analyze the band structure of a collection of resonant states, followed by two contrasting views of localization in two dimensions, an approach using scaling theory by Abrahams and one analyzing the results of a numerical calculation of the outward diffusion of an electron by Scher and Adler. The next two papers are primarily concerned with transport in two-dimensional channels in field-effect transistors. Azbel analyzes Lee oscillations in the regime in which Mott variable-range hopping predominates, while Long et al. describe some results on GaAs and InP devices, concentrating on the existence of the Mott minimum metallic conductivity. Finally, Firsov reviews localization in quasi-one-dimensional and anisotropic quasi-two-dimensional systems.

As even a brief perusal of this volume indicates, not only has the subject of metal-insulator transitions remained exciting throughout the past 50 years, but the work of even the past two years has opened up fruitful new areas for both theorists and experimentalists. We look forward to guidance from Sir Nevill on unraveling some of the recently uncovered mysteries.

THE DISORDERED INSULATOR: ELECTRON GLASSES AND CRYSTALS

T. F. Rosenbaum

The James Franck Institute and Department of Physics
The University of Chicago, Chicago, IL 60637

The behavior of electrons in a random potential has been a particularly rich problem for both theorists and experimentalists, as attested to by this collection of articles. In three-dimensional disordered systems there is the extra twist of a possible metal-insulator transition, going back to the classic papers by Mott[1] and by Anderson[2], which emphasize, respectively, the role of electron correlation and disorder in the phenomenon of electron localization. Most recent work[3] has concentrated on the unusual properties of the disordered metal and the approach to the metal-insulator transition critical point from above. In fact, the nature of the disordered insulator near the critical point not only provides an equally good test of theories of the transition, but it is a fascinating entity unto itself. Under the proper circumstances, barely localized electrons can form glasses or even crystals, depending upon the relative importance of the disorder and the interactions. In this paper, I will briefly review results on the approach to the metal-insulator critical point from below, and summarize recent experiments on the electron glass in Si:P and the electron crystal in $Hg_{1-x}Cd_xTe$.

The Insulator-Metal Transition

On the insulating side the approach to the critical point is characterized by a diverging polarizability (dielectric susceptibility). Simple dimensional considerations[4] suggest and calculations[5] confirm a relationship between the critical behavior of the conductivity σ in the metal and the donor dielectric susceptibility $4\pi\chi$ in the insulator. For large wavevectors q, the disordered insulator behaves like a metal, so $4\pi\chi(q) \sim 1/q^2$. This behavior is cut off at $q_c \sim 1/\xi$, where ξ is the localization length, and the dielectric susceptibility diverges as $4\pi\chi(o) \sim 1/q_c^2 \sim \xi^2 \sim (n_c/n-1)^{-2\nu}$. This is just twice the characteristic exponent in the metal, where $\sigma(o) \sim 1/\xi \sim (n/n_c-1)^{\nu}$. I plot in Figure 1 the critical divergence of the low frequency donor dielectric susceptibility in Si:P, culled from capacitance measurements[6], a Kramers-Kronig analysis of the far-infrared absorption[7], and a resonant transmission microwave cavity technique[8].

The best fit $\zeta \equiv -2\nu = 1.07 \pm 0.1$ is, as expected, twice that determined in the metal, $\nu = 0.48 \pm .07$. This correspondence is illustrated in Figure 2, where we have extended measurements of both the dielectric susceptibility[9] and the conductivity[10] deep into the critical region by uniaxially stressing Si:P at milliKelvin temperatures.

It would be valuable to compare the critical behavior of $4\pi\chi$ and σ in those systems where measurements[3] of σ give $\nu \approx 1$.

The Electron Glass

The manifestations[3,11] of electron interactions in the disordered metal, originally confirmed by both tunnelling[12,13] and transport[13,14] measurements, suggest an even greater role for electron correlations in the disordered insulator, where metallic screening is absent. Historically, a number of experiments have been explained on the basis of the non-interacting Anderson model or "Fermi glass"[15,16]. The most widely applied consequences of this model have been the Mott "$T^{1/4}$" variable range hopping law[16] and the

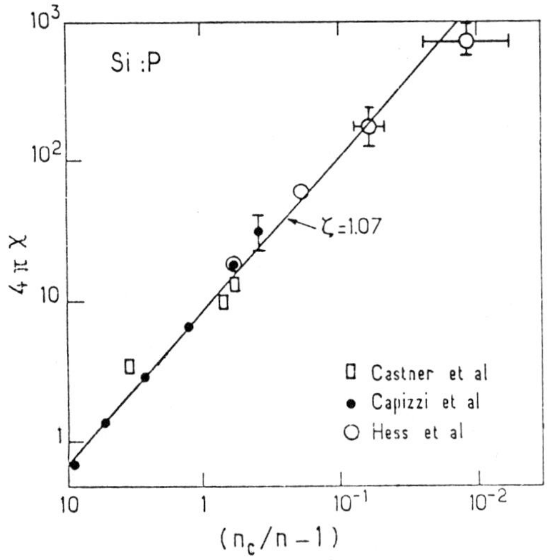

Fig. 1. Divergence of the dielectric susceptibility at the insulator-metal transition (From Refs. 6, 7, 8).

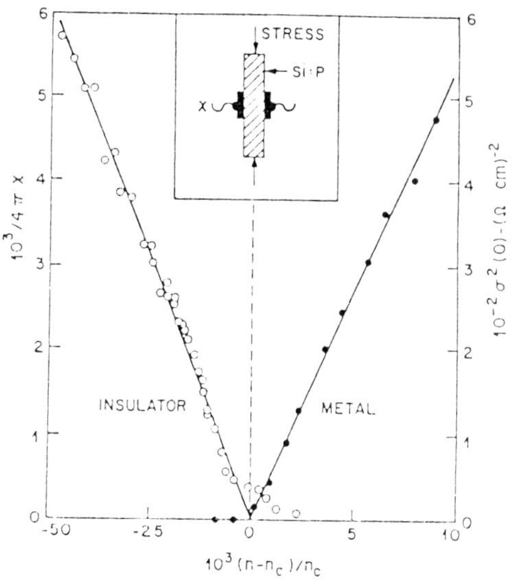

Fig. 2. The dielectric susceptibility diverges with twice the exponent of the conductivity (From Refs. 9, 10).

Austin-Mott[17] low frequency ac conductivity. These calculations have been modified by Efros and Shklovskii[18] who show that long-range (1/r) Coulomb interactions give rise to a soft Coulomb gap[19] where the density of states goes continuously to zero at the Fermi level. The hopping conductivity, in contrast, is enhanced[18,19] giving a "$T^{\frac{1}{2}}$" law in the interacting "electron glass". Physically, a large number of low-lying states remain unfilled because of the Coulomb repulsion of nearby occupied sites. These unfilled sites can contribute to the conductivity, however, as electrons are allowed to hop from filled to unfilled sites.

It is difficult to discriminate between the $T^{\frac{1}{4}}$ and $T^{\frac{1}{2}}$ laws experimentally. Although σ changes rapidly in the insulator, the fits are usually constrained to small intervals in temperature[16]. Better evidence for the electron glass follows from the critical behavior[9,20] of the ac conductivity $\sigma(\omega)$. Bhatt and Ramakrishnan[20] demonstrated the need to include the large density of low-lying hopping modes in the Coulomb insulator even when describing the scaling form of $\sigma(\omega)$ near the insulator-metal transition. The different properties of the interacting and non-interacting insulators are summarized in Table I.

The critical behavior of $\sigma(\omega)$ not only provides a clear test of whether interactions are essential to the physics of the disordered insulator, but it allows a direct measurement of a fundamental quantity, the localization

Table 1. Summary of properties where σ = conductivity, ε = dielectric constant, ξ = localization length, ω = frequency, and T = temperature.

Non-Interacting Fermi Glass	Interacting Electron Glass
$\sigma(T) \propto \exp(T/T_o)^{1/4}$	$\sigma(T) \propto \exp(T/T_o)^{1/2}$
$\sigma(\omega) \propto \xi^5$	$\sigma(\omega) \propto \xi^2$
$\varepsilon \propto \xi^2$	$\varepsilon \propto \xi^2$
$\sigma(\omega) \propto \varepsilon^{5/2}$	$\sigma(\omega) \propto \varepsilon$

length. Figure 3 demonstrates that Si:P is indeed an electron glass, with $\sigma(\omega)$ and ε diverging in the same fashion (slope = 1 on a log-log plot).

Why are these disordered insulator called glasses? As first found by Pollak and Geballe[21] in 1961, deep in the insulator $\sigma(\omega) \sim \omega^s$ with $s \approx 0.8$. The sublinear dependence of $\sigma(\omega)$ implies, via a Kramers-Kronig analysis, a corresponding infrared divergence of the dielectric constant and glass-like behavior. Si:P retains these qualities near the insulator-metal transition, as shown by the frequency dependence of the real and imaginary parts of the conductivity in Figure 4.

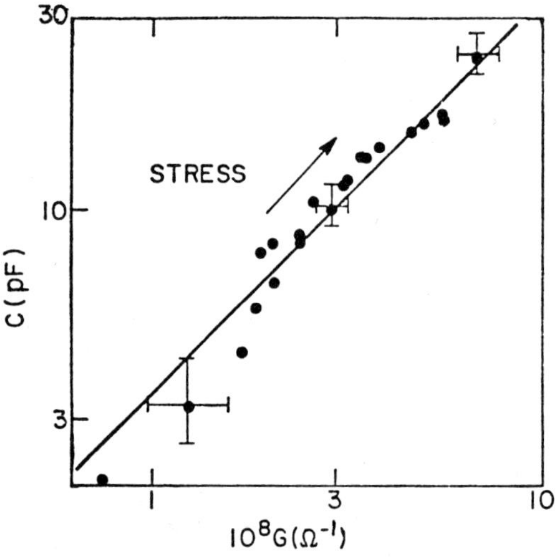

Fig. 3. Divergence of the real and imaginary parts of the low frequency (31 kHz) conductivity in Si:P at the approach to the insulator-metal transition. The slope of one indicates an interacting electron glass (see text).

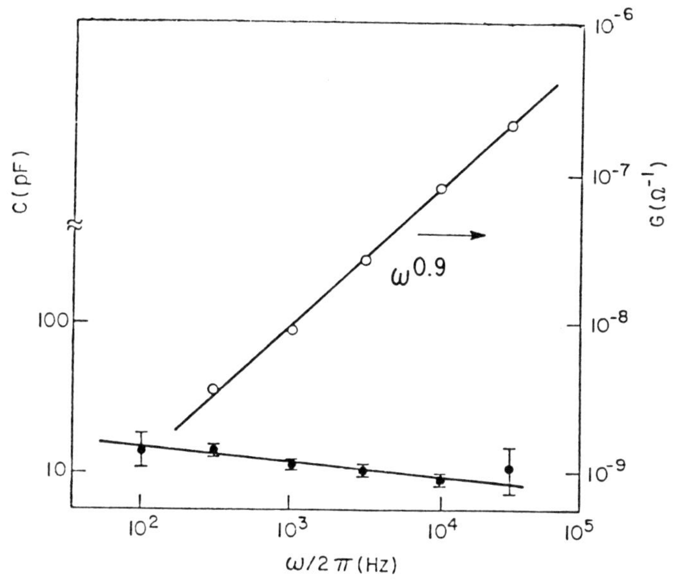

Fig. 4. Glassy properties of Si:P indicated by sublinear frequency dependence of the conductance and corresponding capacitive behavior at T = 13 mK.

The glassy properties[22,23] of the disordered insulator merit more attention, particularly from experimentalists. In analogy to other glasses, there should be unusual hysteretic and relaxational effects. Moreover, the long-range Coulomb potential may give rise to new types of glassy behavior.

The Electron Crystal

If the concentration of electrons is sufficiently dilute then the lowest energy ground state may be an electron or Wigner crystal[24]. Assuming that the electron gas moves in a uniform background of positive charge, then the repulsive Coulomb energy e^2/r dominates the kinetic energy $\hbar^2/2mr^2$ at large r. It takes too much energy for the electrons to move near each other; they would rather sit in a lattice!

In practice, a magnetic field H can be used to help nucleate[25] the Wigner crystal. For sufficiently large H the electrons are confined to the lowest spin-polarized Landau level, removing two degrees of freedom and leaving dispersion only along the field direction. The high mobility of the $Hg_{1-x}Cd_xTe$ compounds at a low carrier density allows this extreme quantum limit to be reached at H of only a few kiloOersteds and makes them attractive candidates for studying the electron crystal.

For the samples we studied[26,27], x = 0.24, the effective mass ratio $m/m^* = 80$ and the dielectric constant $\varepsilon_\infty = 20$, resulting in an effective Bohr radius $a^* = \varepsilon_\infty(m/m^*)a_o = 900\text{Å}$ and a metal with an electron density

$n \simeq 1.40 \times 10^{14}\ cm^{-3}$. The large size of the electron wavepackets helps average over the disorder, approximating a positive jellium background, and the low density allows the Coulomb energy to dominate all other energies in the problem (in particular, the Fermi energy). The metal-insulator transition with H is essentially a Mott transition[28] where the lowest energy state in the insulator is the electron crystal. If the sample is sufficiently inhomogeneous, then the correlation length over which electrons are ordered can be very short. If it approaches a*, then magnetic freeze-out of the electrons onto individual impurity sites is a more accurate description of the insulating ground state.

A perfect electron lattice at zero-temperature would have infinite conductivity as it would be free to slide without dissipation. However, even weak disorder will be sufficient to pin the electron lattice to the host crystal by introducing fluctuations on a length scale $\gg n^{-1/3}$. The pinning of the lattice should lead to nonlinear I-V characteristics from which an electron crystal correlation length can be deduced[27].

The inset to Figure 5 shows the non-linear I-V characteristics of $Hg_{0.76}Cd_{0.24}Te$ at T = 10 mK and H = 64 kOe. Deviation from Ohmic behavior begins at $V_d = 8.6 \times 10^{-5}$ V, which corresponds to an electric field $E_d = 4.9 \times 10^{-4}$ V/cm. The non-linear onset can be seen more clearly in the main part of Figure 5, where we have subtracted out the Ohmic portion V/R_0 such that deviations from zero represent non-Ohmic character. The small value of E_d implies a large correlation length for the electron crystal, which we estimate[27] of order 0.1 mm. In addition, the non-linear region is accompanied by a large increase in the noise, similar to depinning results in charge density wave compounds[29].

The low density of electrons makes experimental probes of the Wigner crystal (especially structural ones) difficult. Promising directions include ac depinning characteristics, including narrow-band noise, and the melting transition from electron solid to correlated fluid. As the electron crystal is better characterized, we should learn more about the fundamentals of what makes a solid.

Acknowledgments

I would like to thank my collaborators at both Bell Laboratories and The University of Chicago for making this work possible. The work at The University of Chicago was supported by the National Science Foundation under Grant No. DMR 83-05065.

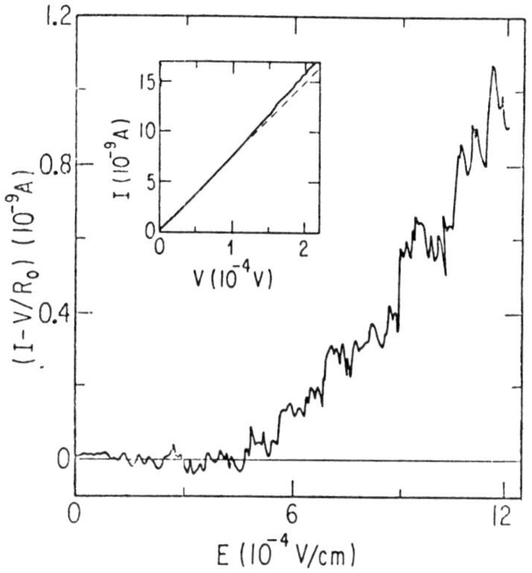

Fig. 5. Non-Ohmic effects indicating depinning of the electron crystal at T = 10 mK, H = 64 kOe.

References

1. N. F. Mott, Proc. Camb. Phil. Soc. 32, 281 (1949); Proc. Phys. Soc. Lond. A62, 416 (1956).
2. P. W. Anderson, Phys. Rev. 109, 1492 (1958).
3. Recent reviews include P. A. Lee and T. V. Ramakrishnan, Rev. Mod. Phys., in press; P. A. Lee and R. C. Dynes, Science, 1984; "Electron-Electron Interactions in Disordered Systems," A. L. Efros and M. Pollak, eds. (Amsterdam, Elsevier Science, 1985).
4. T. F. Rosenbaum, K. Andres, G. A. Thomas, and R. N. Bhatt, Phys. Rev. Lett. 45, 1723 (1980).
5. D. Belitz, A. Gold, and W. Goetze, Z. Phys. B44, 273 (1981); Y. Imry, Y. Gefen, and D. J. Bergmann, Phys. Rev. B26, 3436 (1982).
6. T. G. Castner, Phil. Mag. B42, 873 (1980).
7. M. Capizzi, G. A. Thomas, F. DeRosa, R. N. Bhatt, and T. M. Rice, Phys. Rev. Lett. 44, 1019 (1980).
8. H. F. Hess, K. Deconde, T. F. Rosenbaum, and G. A Thomas, Phys. Rev. B25, 5578 (1982).
9. M. A. Paalanen, T. F. Rosenbaum, G. A. Thomas, and R. N. Bhatt, Phys. Rev. Lett. 51, 1896 (1983).
10. M. A. Paalanen, T. F. Rosenbaum, G. A. Thomas, and R. N. Bhatt, Phys. Rev. Lett. 48, 1284 (1982).
11. B. L. Altshuler and A. G. Aronov, Solid State Commun. 30, 115 (1979); Sov. Phys. JETP 50, 968 (1979).
12. R. C. Dynes and J. P. Garno, Phys. Rev. Lett. 46, 137 (1981); W. L. McMillan and J. Mochel, Phys. Rev. Lett. 46, 556 (1981).

13. G. Hertel, D. J. Bishop, E. J. Spencer, J. M. Rowell and R. C. Dynes, Phys. Rev. Lett. 50, 743 (1983).

14. T. F. Rosenbaum, K. Andres, G. A. Thomas, and P. A. Lee, Phys. Rev. Lett. 46, 568 (1981).

15. P. W. Anderson, Comm. on Solid State Phys. 2, 193 (1970); L. Fleishman, D. C. Licciardello, and P. W. Anderson, Phys. Rev. Lett. 40, 1340 (1978).

16. N. F. Mott, Phil. Mag. 19, 835 (1969); N. F. Mott and E. A. Davis, "Electronic Processes in Non-Crystalline Materials" (Oxford U. Press, 1979).

17. I. G. Austin and N. F. Mott, Advances in Physics 18, 41 (1969).

18. A. L. Efros and B. I. Shklovskii, J. Phys. C8, L49 (1975); A. L. Efros, J. Phys. C9, 2021 (1976); Phil. Mag. 43, 829 (1981).

19. M. Pollak, Discuss. Faraday Soc. 50, 13 (1970); Proc. Roy. Soc. London A325, 383 (1971).

20. R. N. Bhatt and T. V. Ramakrishnan, J. Phys. C17, L639 (1984); R. N. Bhatt, Phil. Mag. B50, 189 (1984).

21. M. Pollak and T. H. Geballe, Phys. Rev. 122, 1742 (1961).

22. See articles by H. Kamimura and by M. Pollak, in Proceedings of the International Conference on Heavy Doping and the Metal-Insulator Trans. in Semicond, P. T. Landsberg, ed., Santa Cruz, CA, 1984, in press as a special issue of Solid State Electronics.

23. J. H. Davies, P. A. Lee, and T. M. Rice, Phys. Rev. Lett. 49, 758 (1982).

24. E. P. Wigner, Phys. Rev. 46, 1002 (1934).

25. W. G. Kleppmann and R. J. Elliott, J. Phys. C8, 2729 (1975).

26. T. F. Rosenbaum, S. B. Field, D. A. Nelson, and P. B. Littlewood, Phys. Rev. Lett. 54, 241 (1985).

27. S. B. Field, D. H. Reich, B. S. Shivaram, T. F. Rosenbaum, D. A. Nelson, and P. B. Littlewood, to be published.

28. N. F. Mott, "Metal-Insulator Transitions" (London, Taylor and Francis, Ltd., 1974).

29. See, for example, the review by G. Gruner, Physica D8, 1 (1983).

TUNING THE METAL-INSULATOR TRANSITION

IN N-TYPE SILICON WITH A MAGNETIC FIELD

Theodore G. Castner and William N. Shafarman
Department of Physics and Astronomy
University of Rochester
Rochester, N.Y.

INTRODUCTION

Since the classic calculation of Yafet, Keyes, and Adams[1] demonstrating the shrinkage of shallow donor wave functions in large static magnetic fields, it has been widely recognized that a magnetic field could be used to tune a metallic sample through the Metal-Insulator (MI) transition. Several experimental groups[2-5] have successfully tuned n-type InSb through the MI transition with modest magnetic fields. For InSb it is easy to bring the magnetic length $\lambda=(\hbar c/eH)^{1/2}$ to a value much less than the donor Bohr radius ($a_D^*\sim600$Å) with reasonable fields since $\lambda=81$ Å $(10/H(T.))^{1/2}$. One thereby readily achieves the strong field limit ($\lambda<<a_D^*$) for InSb. For n-type Si and Ge the donor Bohr radii are very much smaller ($a_D^*<20$Å for Si and $a_D^*<47$Å for Ge) and one remains in the weak field limit (intermediate regime for Ge) for the largest static laboratory fields (H~30T.) available. As a result there have been no reported successful efforts in tuning the MI transition in Si with a magnetic field to compare with the remarkably successful tuning of the critical density N_c for Si:P by Paalanen et al.[6] utilizing uniaxial stress. Below some new experimental evidence is presented, for both barely insulating and barely metallic samples, which can be interpreted in terms of tuning of N_c by a large static magnetic field. On the insulating side both low-temperature magnetocapacitance measurements and the magnetic field-dependence of Mott variable range hopping for Si:As samples yield evidence for a decrease in the localization length with increasing magnetic field. Barely metallic samples indicate a new type of magnetoresistance behavior in addition to exhibiting low-field behavior similar to that of Si:P[7]. A new type of low-temperature behavior at higher fields again suggests N_c is increasing with magnetic field. These results both support in part the type of zero-temperature magnetic phase diagram for the MI transition proposed by Shapiro[8].

A convenient way of characterizing the magnetic behavior of the transport properties is in terms of the various lengths in the problem. These are the magnetic length λ, the mean free path ℓ for metallic samples, the average spacing between donors given by d, and the Bohr radii in the dilute limit. In the dilute concentration range, where the magnetoresistance behavior is of the percolation type, Shklovskii[9] characterizes the behavior as the weak-field case when $\lambda>>a_D^*$ and the strong-field case when $\lambda<<a_D^*$. Since $\lambda_{min}\sim48$Å it is not possible with

static magnetic fields to achieve the strong field case for n-type Si. In the immediate vicinity of the critical density N_c ($N_c^{-1/3} = d_c \sim 4a_D^*$ from the Mott criterion) the characteristic scaling length is the localization (correlation) length $\xi(N)$ on the insulating (metallic) side of the transition. $\xi(N)$ is given by

$$\xi(N) = \xi_o |1-N/N_c|^{-\nu} \qquad (1)$$

where the prefactor ξ_o is of order a_D^* and the critical exponent has been determined accurately for Si:P[6] to be $\nu = 0.50 \pm .05$. For $|1-N/N_c| << 1$ $\xi(N)$ becomes large and can become much larger than $d_c \sim 4a_D^*$ and ℓ for $|1-N/N_c|$ sufficiently small. In this case the decreasing magnetic length with increasing field will first approach $\xi(N)$ and it may also be possible to achieve $\lambda < \xi(N)$. But in analogy to the percolation case for dilute samples one would expect $\xi(N)$ itself to be affected by the magnetic field. Thus $\xi(N,H)$ will be characterized by the field-dependent parameters $\xi_o(H)$, $\nu(H)$, and $N_c(H)$. We will return to the field dependence of ξ shortly. In Fig. 1 the characteristic lengths $\lambda(H), d(N)$, and $\xi(N)$ are shown versus magnetic field and donor density respectively for Si:P ($N_c = 3.74 \times 10^{18}/cm^3$) and Si:As ($N_c = 8.25 \pm 0.1 \times 10^{18}/cm^3$). The mean-free-path ℓ for $N > N_c$ is also shown in a cross hatched band indicating some uncertainty in the magnitude of this quantity. $\ell(N) \sim d(N)$ and varies slowly in the barely metallic region. Fig. 1 shows for $\nu = 1/2$ and $\xi_o = a_D^*$ that one requires $|1-N/N_c| < 0.05$ to obtain $\xi(N) > d(N)$ and $\ell(N)$ [this condition is obviously filled more readily if ξ_o is considerably larger than a_D^*]. For our purposes the critical region here will be that with $|N/N_c - 1| < 0.1$. Fig. 1 shows it is possible to have the magnetic length approach ξ, the zero-field value of the localization (correlation) length for barely insulating (metallic) samples for fields H~20 tesla. In fact it is just possible to have λ approach d_c for Si:P and Si:As. At these fields it would not be surprising to have magnetic field-induced changes in the localization (correlation) length.

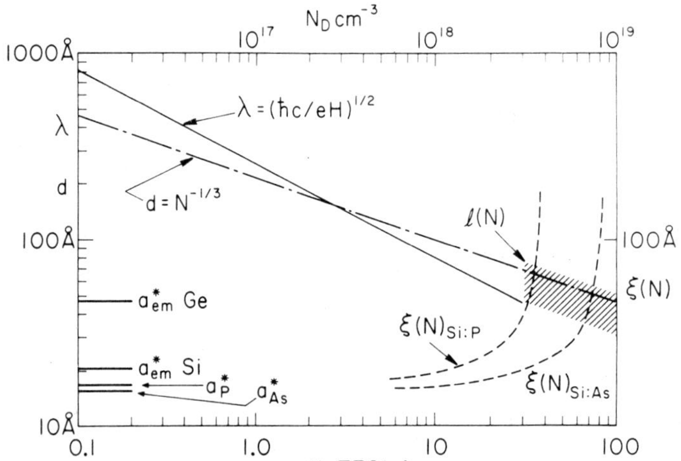

Fig. 1. The magnetic length λ vs. H and the average donor spacing d and localization length $\xi(N)$ versus donor density.

The effect of the magnetic field on the three parameters ξ_o, N_c and ν in Eq. (1) is of direct relevance to the magnetic phase diagram of the MI transition and has not been determined for n-type Si. It has been suggested by Castellani et al.[10] that $\nu(H)\rightarrow 1$ at large fields (recalling $\nu(0)=0.5$ for Si:P and Si:As). Since $\xi_o=a_D^*$ and $\lambda_{min}>>a_D^*$ one might expect the effect of a field on ξ_o to be a relatively small effect. The effect of H on N_c(the tuning of the transition) is clearly a dominant effect in InSb and would be expected to be very much smaller in n-type Si. Using a field-dependent Mott criterion

$$N_c(H)^{1/3}a_D^*(H) = 0.26 \tag{2}$$

with $a_D^*(H)=a_D^*(0)(1-1/2\gamma_D^{\ 2})$ where $\gamma_D^{\ 2}=\varepsilon_h a_D^{*3}H^2/m^*c^2$ one obtains $N_c(H)=N_c(0)(1+\eta H^2)$ with $\eta=1.8\times10^{-5}$ T.$^{-2}$ for Si:P. This yields less than a 1% increase in $N_c(H)$ for H=20T. However, the experimental evidence presented in this paper suggests the increase in $N_c(H)$ with H is several orders of magnitude larger than predicted by the above estimate. Suppose we consider the behavior of $\xi(N,H)$ in two extreme cases, namely 1) N_c= constant, $\nu(0)=1/2,\nu(H)\rightarrow 1$; and 2) ν=constant=1/2 and $N_c(H)=N_c(0)[1+\eta H^p]$. Case 1) predicts a symmetrical (about $N_c(0)$) increase in $\xi(N,H)$ for both barely insulating and barely metallic samples. On the other hand, case 2) predicts an antisymmetrical behavior about $N_c(0)$ predicting a decrease in $\xi(N,H)$ for $N<N_c(0)$ and an initial increase in $\xi(N,H)$ for $N>N_c(H)$. The magnetotransport measurements to be discussed below yield experimental evidence that can answer which case best approximates the experimental results. These two cases for $\xi(N,H)$ along with the zero-field concentration dependence of $\xi(N)$ are shown in Fig. 2. The figure shows that case 1) corresponds to an increase in $\xi(N,H)$ on both sides of the transition while case 2) (with $\eta>0$) corresponds to a decrease in the localization length on the insulating side and an initial increase in the correlation length on the metallic side of the transition.

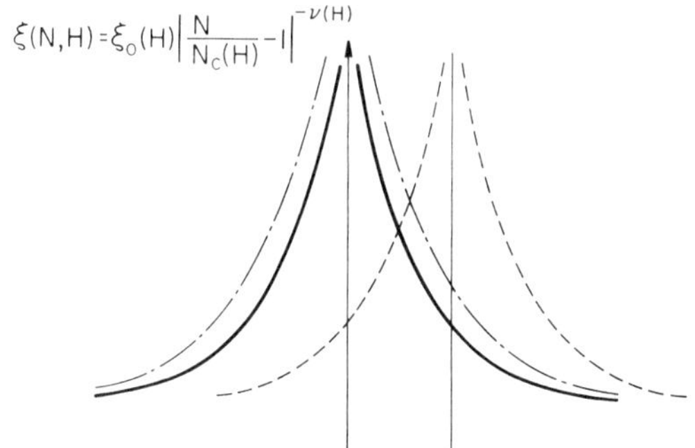

Fig. 2. The localization (correlation) length $\xi(N,H)$ vs. N for zero field (——); case 1) N_c=const., $\nu(0)=\frac{1}{2}$, and $\nu(H)\rightarrow 1$ (— - —); and case 2) $\nu=\frac{1}{2}$=const., $N_c(H)$ increasing with field (- - -).

MAGNETOCAPACITANCE MEASUREMENTS

The standard experimental approach for monitoring the shrinkage of donor impurity wave functions has been through magnetoresistance measurements in the hopping regime. This area has recently been reviewed by Chroboczek[11] and has been considered in detail by Shklovskii[12]. At sufficiently low temperatures the hopping contribution to the low frequency dielectric response is frozen out and the dielectric 'constant' $\varepsilon'(N)$ is given by the host contribution ε_h plus a contribution from the donors. Because the donors are coupled by Heisenberg exchange interactions into clusters with various low lying spin states, the form of the donor contribution to the polarization is best described in terms of an effective donor polarizability $\alpha_{eff}(N,T,H)$ and $\varepsilon(N,T,H)$ is given by

$$\varepsilon'(N,T,H) = \varepsilon_h(T,H) + 4\pi N\alpha_{eff}(N,T,H) \tag{3}$$

neglecting the Lorentz local field correction. Eq. (3) is certainly valid in the intermediate concentration regime for $N<N_c(0)/2$. Closer to the critical regime as $N\to N_c$ and as $T\to 0K$ $\varepsilon(N)$ is given by[13]

$$\varepsilon(N) = \varepsilon_h + 4\pi e^2 N(E_F)\xi^2 \tag{4}$$

from a Thomas-Fermi model with a $q^2+\xi^{-2}$ denominator. Empirically Paalanen et al.[6] have demonstated at $T\sim 3mK$ that $\varepsilon'(N)-\varepsilon_h=\chi_0(1-N/N_c)^{-\zeta}$ with $\zeta=1.0$. Thus these results show $\zeta=2\nu$ and from Eq. (1) that $\varepsilon'(N)-\varepsilon_h\propto\xi^2$. Magnetocapacitance measurements determining $\varepsilon'(N,T\to 0K,H)$ will permit a direct measure of the magnetic field dependence of the localization length $\xi(N,H)$.

Magnetocapacitance measurements reported by New et al.[14,15] on Si:P and Si:As samples in the intermediate concentration regime have shown a strongly temperature-dependent decrease of $\varepsilon(N,T,H)$ until $g\mu_B H>10kT$. For $H/T<<5T./K$ the magnetocapacitance decrease is proportional to $(H/T)^2$. For larger fields the behavior is approaching the diamagnetic shrinkage behavior associated with the lowest energy Zeeman state of the high spin state. The high field decrease in $\varepsilon'(N,T,H)$ is weakly temperature dependent and shows a field dependence considerably less than the quadratic field dependence expected for isolated donors in the weak-field limit. Typical data for a Si:P sample is shown in Fig. 3 for the quantity $\{\varepsilon'(N,T,0)-\varepsilon'(N,T,H)\}/\varepsilon'(N,T,0)$. Although this data shows a substantial temperature-dependent decrease in $\varepsilon'(N,T,H)$ with H, it cannot be used to say anything significant about the change in $\xi(N,T\to 0,H)$ with H. The data in Fig. 3 must be interpreted in terms of the different polarizabilities of the different spin states of donor clusters. Furthermore, there is a broad distribution of exchange energies present. Accounting for these details, in addition to the more complex diamagnetic behavior of donor clusters, must be done to explain the data.

The most relevant available magnetocapacitance data at present is for a Si:As sample ($N=7.3\times 10^{18}/cm^3$, $N/N_c=.88$) shown in Fig. 4. This shows how strong the temperature dependence of $\varepsilon'(N,T)$ is for a sample in the critical regime. However, most of the hopping contribution to $\varepsilon'(N,T)$ is frozen out at 60 mK. The 60 mK field dependence of $\varepsilon'(N,H)-\varepsilon_h$ can qualitatively be explained in terms of a decrease in $\xi(N,H)$ with field, although one cannot rule out an additional field-dependent term leading to

$$\varepsilon'(N,T,H) = \varepsilon_h + 4\pi e^2 N(E_F,H)\xi(N,H)^2 + f(H,T) \tag{5}$$

Although the existence of $f(H,T)$ cannot be ruled out, the data can be explained using case 2) discussed above with N_c increasing quadratically with H. The rapid decrease of $\varepsilon'(N,H)-\varepsilon_h$ with H is not consistent with an increasing value of $\nu(H)$ with H because increases in $\nu(H)$ will increase

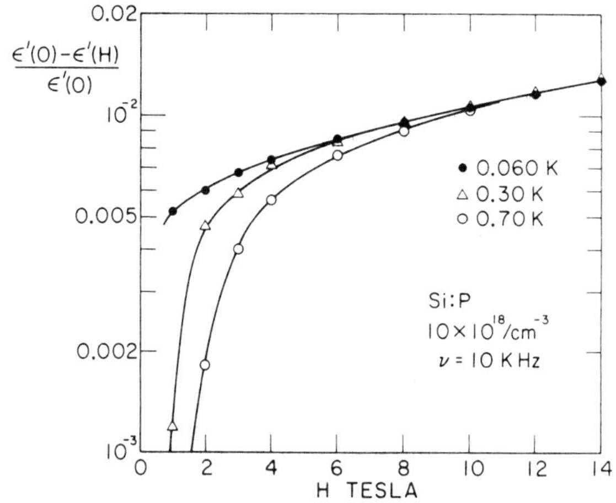

Fig. 3. $\{\varepsilon'(N,T,0)-\varepsilon'(N,T,H)\}/\varepsilon'(N,T,0)$ vs. H for a $1.0\times10^{18}/cm^3$ Si:P sample for T=0.060K, 0.30K, and 0.70K.

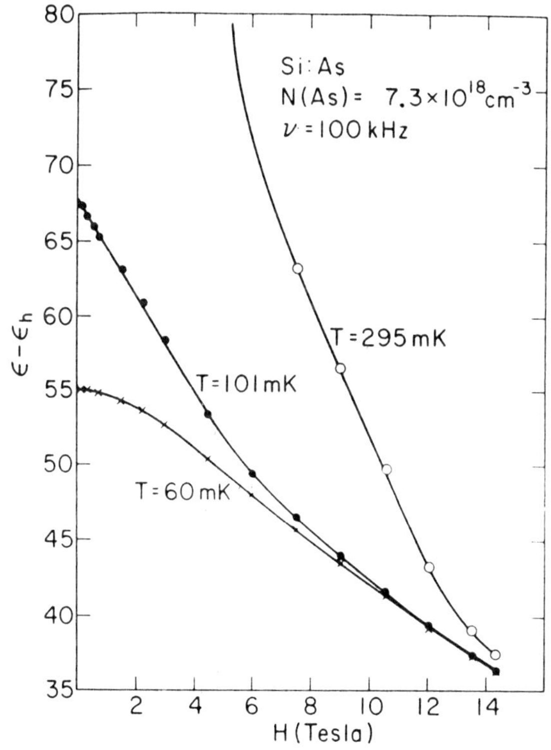

Fig. 4. $\varepsilon'(N,T,H)-\varepsilon_h$ vs. H for a $7.3\times10^{18}/cm^3$ Si:As sample at T=60mK, 101mK, and 295mK measured at ν=100 kHz.

$\varepsilon'(N,H)$ with H. It would be highly desirable to have magnetocapacitance data for samples with $N/N_c(0)$ closer to unity. Such experiments would require temperatures $T \ll 60$ mK, which argues against the use of Bitter magnets as the source of the large magnetic fields needed in magnetocapacitance studies on barely insulating samples in the milliKelvin temperature range. An alternative method of obtaining experimental information on the magnetic field dependence of the localization length $\xi(N,H)$ with less restrictive requirements on the temperature has been found from the field dependence of Mott VRH conduction[16].

THE MAGNETIC FIELD DEPENDENCE OF MOTT VARIABLE RANGE HOPPING NEAR THE MI TRANSITION

The observation of Mott VRH conduction[17] in disordered systems has frequently been reported in amorphous semiconductors and randomly doped semiconductors. The first definitive report of this behavior for a crystalline doped semiconductor was for n-type Ge by Allen and Adkins[18] The form of Mott VRH conduction is given by

$$\sigma_{DC} = \sigma_o e^{-(T_o/T)^m} , \tag{6}$$

where T_o is a characteristic temperature and the exponent $m=1/4$. Recently Hess et al.[19] have inferred Mott VRH for barely insulating Si:P from 400 MHz data taken in the temperature range 20 mK to 120 mK. Shafarman and Castner[16] have reported four terminal σ_{DC} measurements on barely insulating Si:As samples in the concentration range 7.17×10^{18} to $8.17\times10^{18}/cm^3$. Using the logarithmic derivative approach suggested by Hill[20] and taking points equally spaced in intervals of $1/T$, it has been possible to show $0.18<m<0.28$ for all the samples studied at zero magnetic field. Dilution refrigerator measurements at FBNML have shown (but with less accuracy) that the Mott $m=1/4$ law holds down to 50 mK while recent higher temperature measurements confirm the Mott law behavior to about 15 K, above which the behavior changes to simple exponential ($e^{-\Delta\varepsilon/KT}$) behavior.

Fig. 5 shows the variation of $T_o(N)$ with N as N approaches N_c. $T_o(N)$ varies over many orders of magnitude and seems to approach zero at $N_c=8.25\pm0.1\times10^{18}/cm^3$, a value slightly smaller than that obtained by Newman and Holcomb[21] but within their stated experimental errors. The data presented here is based on the room temperature $\rho(N)$ versus N calibration determined by Newman and Holcomb[21]. Mott's VRH calculation leads to $T_o(N)$ given by

$$T_o(N) = \text{constant}/kN(E_F)\xi(N)^3 \tag{7}$$

where the constant is of order 20. With $\xi(N)$ given by Eq. (1) with $\nu=1/2$ and the density of states $N(E_F)$ falling approximately as found for Si:P by Thomas et al.[22] Eq. (7) does not adequately describe the very steep variation of $T_o(N)$ with N shown in Fig. 5. The situation can be improved by incorporating the static dielectric constant into the energy difference $\Delta\varepsilon_{ij}$ leading to $T_o \propto \{k\,N(E_F)^2\xi(N)^5\}^{-1}$ for $4\pi N(E_F)e^2\xi^2 \gg \varepsilon_h$ near the transition.

A study of Mott VRH in a magnetic field leads to the possibility of studying the effect of the field on $\xi(N,H)$ through the effect of H on the characteristic temperature $T_o(N,H)$. Such a study of InSb by Tokumoto, Mansfield, and Lea[23] has shown a change in m from 1/4 to 1/2 leading to $\rho=\rho_o\exp(T_o/T)^{1/2}$ for fixed magnetic fields. A theoretical explanation for this behavior ($m\simeq1/2$) has been given by Shklovskii[24] based on nonresonant subbarrier tunneling in a transverse magnetic field. The extremely large

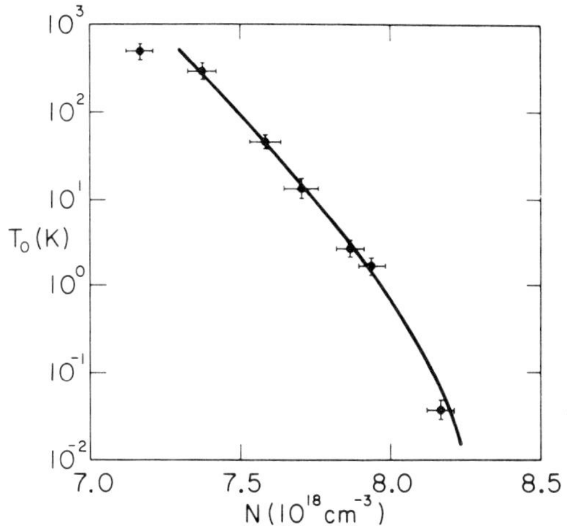

Fig. 5. Variable range hopping characteristic temperature $T_o(N)$ vs. N for barely insulating Si:As samples.

donor Bohr radius (a_D^*~600Å) in InSb argues these results are in the strong field limit. For the results discussed below for Si:As (a_D^*=15.5Å) one remains in the weak field limit over the entire field range from 0 to 20 tesla.

Fig. 6 shows $\ell n\sigma_{DC}$ vs. $T^{-1/4}$ for a $7.72\times10^{18}/cm^3$ Si:As sample for zero field and fields of 4.36, 6.0, 10.9 and 19 tesla. The exponent m remains approximately ¼ and independent of magnetic field from zero to 19 tesla. The 19T. data shows some slight curvature at each end, but the best overall fit is still m≃¼ . However, the slope (slope = $T_o(H)^{1/4}$) increases steadily with magnetic field showing that the principal effect of the magnetic field is to increase the characteristic temperature $T_o(N,H)$. Data for a $8.06\times10^{18}/cm^3$ Si:As sample at the same fields also shows m=¼ for all the field values and shows correspondingly larger relative increases in $T_o(N,H)$ with H. On the other hand, a more dilute $7.38\times10^{18}/cm^3$ sample shows deviations of m to larger values at H=10.9T. and at H=19T. shows m~1. It is worth noting that Shklovskii[25] has shown in the weak field limit of VRH that $\ell n\rho(H)/\rho(0)\propto H^2(T_o/T)^{3/4}$. However, here we are considering the weak field behavior in the barely insulating critical regime where the principal effect of the magnetic field is on the localization length $\xi(N,H)$ which is large as N approaches N_c. In Fig. 7 the ratios $T_o(N,H)/T_o(N,0)$ for these three samples are shown versus H. Perhaps the most important feature is that for a given field H the ratio $T_o(N,H)/T_o(N,0)$ increases rapidly with N as $N\to N_c(0)$, namely the ratio seems to show critical behavior. For a given sample the best fit over the entire field range is given by $\ell nT_o(N,H)\propto H$. From Eqs. (7) and (1) the ratio will be given by

$$\frac{T_o(N,H)}{T_o(N,0)} = \frac{N(E_F,0)}{N(E_F,H)}\left[\frac{\xi_o(0)}{\xi_o(H)}\right]^3 \frac{\{1-N/N_c(H)\}^{3\nu(H)}}{\{1-N/N_c(0)\}^{3\nu(0)}} \qquad (8)$$

If the principal effect were due to case 1) ($\nu(H)$ increasing), then the ratio $T_o(H)/T_o(0)$ would actually decrease to values less than one. In fact case 2) with $N_c(H)=N_c(0)\{1+\eta H^p\}$ gives a good account of the data in the low field range where the field dependence of $N(E_F)$ and ξ_o can be neglected. For this case Eq. (8) reduces to

$$\frac{T_o(N,H)}{T_o(N,0)} \approx \left[\frac{1+\eta H^p/(1-N/N_c(0))}{1+\eta H^p}\right]^{3\nu} \tag{9}$$

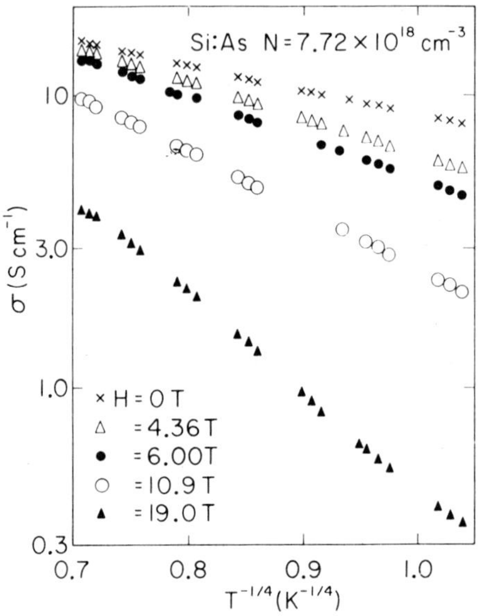

Fig. 6. $\ell n\ \sigma_{DC}$ vs. $T^{-1/4}$ for a $7.72\times10^{18}/cm^3$ Si:As sample for magnetic fields H=0, 4.36, 6.0, 10.9, and 19.0 tesla.

For a fixed field H the ratio increases rapidly as $\{1-N/N_c(0)\}\rightarrow 0$. The low-field increase in the ratio for the three samples can be quantitatively explained with p=2 and η in the range 0.0035 to 0.0085 T^{-2}. Eq. (9) also predicts a saturation of $T_o(N,H)/T_o(N,0)$ for $\eta H^2 \gg 1$ which is not observed. Presumably the decrease in $N(E_F)$ and ξ_o with H in the high field region keep the ratio increasing with H as predicted by Eq. (8). On the insulating side of $N_c(0)$, where there are spin states of large clusters, qualitatively the effect of the field is to produce large Zeeman splittings of cluster states, which has the effect of reducing $N(E_F,H)$ with H. However, no quantitative account of this can be given. Using the Miller-Abrahams[26] result $a_D^*(H)=a_D(0)\{1-\frac{1}{2}\gamma_D^2\}$ one estimates only a 1% decrease in $\xi_o(H)$ for H=20T. suggesting that most of the extra change in the ratio $T_o(N,H)/T_o(N,0)$ originates from the ratio $N(E_F,0)/N(E_F,H)$. The large

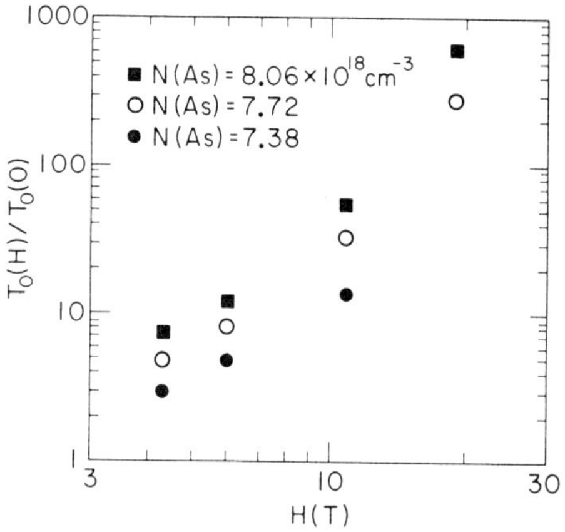

Fig. 7. $T_o(N,H)/T_o(N,0)$ vs. H for three Si:As samples.

density dependence of the ratio $T_o(N,H)/T_o(N,0)$ strongly supports the behavior characteristic of case 2) arising from a $N_c(H)$ increasing with field (a decrease in $N_c(H)$ with increasing H would give the wrong sign of the effect). Although $\nu(H)\to 1$ of case 1) is ruled out by the large ratios in $T_o(N,H)/T_o(N,0)$, one observes that a strong decrease of $\nu(H)$ toward zero with increasing field can also lead to large increases in the ratio. However, this alternative does not appear to be consistent with both the large density and field variations of the ratio, since as $\nu(H)\to 0$ there could no longer be much field dependence from the factor $\{1-N/N_c(H)\}$. The simplest explanation of the data is given by case 2).

MAGNETORESISTANCE OF BARELY METALLIC SAMPLES

There have been numerous magnetoresistance investigations of n-type Ge, n-type Si and certain III - V semiconductors. These studies were concerned with the anisotropy of $\rho(H)/\rho(0)$ and with the interesting question of the negative magnetoresistance[27,28] observed at low fields in metallic samples. The emphasis in the work discussed below is on the low temperature behavior (T<<1K) in the field range 0-20T. for Si:As samples in the concentration range $8.4\times10^{18}/cm^3$ to $10.2\times10^{18}/cm^3$. A relevant previous low-temperature study of barely metallic Si:P by Rosenbaum et al.[7] concentrated on the low-field behavior at very low temperatures.

These authors found a positive magnetoresistance of the form

$$\Delta\rho(H)/\rho(0) = \{\rho(H)-\rho(0)\}/\rho(0) = AH^{\frac{1}{2}} \tag{10}$$

independent of temperature for $g\mu_B H >> k_B T$. A in general consists of two contributions, namely 1) a positive contribution A_i due to electron interactions, as calculated by Lee and Ramakrishnan[29], and a negative localization contribution A_ℓ which has been calculated by Kawabata[28]. In the Si:P results[7] and the Si:As results discussed below, $A=A_i+A_\ell$ is positive and $\rho(H)/\rho(0)$ is dominated by electron-electron interactions for $N_c(0)<N<1.25\ N_c(0)$. The new feature not considered previously is the behavior shown at higher fields as shown in Fig. 8 for a $8.4\times10^{18}/cm^3$ Si:As sample. At the lowest temperatures (T<0.15K) $\rho(H)/\rho(0)$ first exhibits $H^{\frac{1}{2}}$ behavior with increasing field, but then beyond a density-dependent inflection field $H_i(N)$ starts to increase more rapidly and approaches a new high-field behavior characterized by

$$\ell n\{\rho(H)/\rho(0)\} = f(N,T)H^2 \tag{11}$$

At high fields the behavior is definitely more accurately fit by $\ell n\rho(H)/\rho(0)\propto H^2$ than $\Delta\rho(H)/\rho(0)\propto H^2$. This may be significant in that the electron wave functions are localized on a length scale of order $\lambda\propto H^{-\frac{1}{2}}$ and the magnetoresistance behavior $\ell n\rho(H)/\rho(0)\propto H^2$ is analagous to that in the insulating side in the percolation regime[30]. The prefactor f(N,T) decreases very rapidly with increasing N and increases very slowly

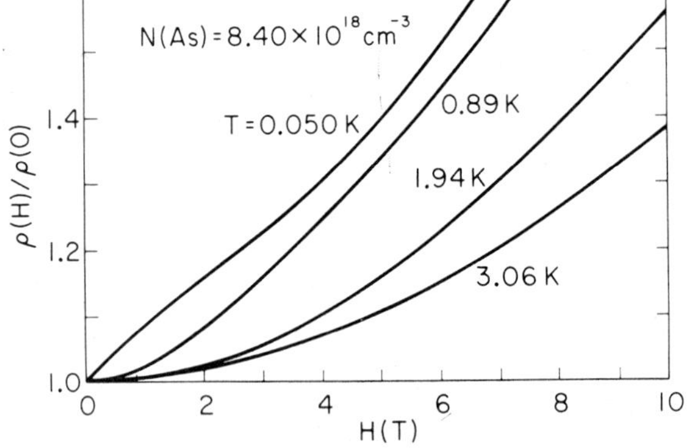

Fig. 8. $\rho(H)/\rho(0)$ vs. H for a $8.40\times10^{18}/cm^3$ Si:As sample for T=0.050K, 0.89K, 1.94K, and 3.06K.

with decreasing temperature for 50mK<T<0.5K. Values for the inflection fields for the five barely metallic samples studied are shown in Table 1 along with the values of $\lambda(H_i)$ and $A(N)$ and $f(N,T\to 0K)$.

Our interpretation of the inflection field $H_i(N)$ in the magnetoresistance is that it represents the crossover between $AH^{\frac{1}{2}}$ behavior and $\ln\rho(H)/\rho(0)=fH^2$behavior and occurs for the magnetic length λ of order the correlation length $\xi(N,H)$, namely

$$(\hbar c/eH_i)^{\frac{1}{2}} = \xi_o(H)|N/N_c(H_i)-1|^{-\nu(H_i)} \qquad (12)$$

If no field dependence of $\xi(N)$ were allowed one would obtain $H_i=(ea_D^{*2}/\hbar c)^{-1}(N/N_c(0)-1)$ which has the wrong functional dependence and the wrong order of magnitude. Case 1) is also ruled out because it leads to $H_i\propto\{N/N_c(0)-1\}^2$ which is clearly inconsistent with the data. However, case 2) leads to, for $N_c(H)=N_c(0)\ \{1+\eta H^2\}$ and $a_D^{*2}H_i e/\hbar c<<1$, the result

$$H_i = (1/\eta)^{\frac{1}{2}}\ \{N/N_c(0)-1\}^{\frac{1}{2}} \qquad (13)$$

This is consistent with the data for $\eta=0.008$ T.$^{-2}$, a value that is virtually the same as that required to explain the $T_o(N,H)/T_o(N,0)$ ratios for the field-dependent Mott VRH conduction. Thus, magnetic tuning of $N_c(H)$ as given in case 2) correctly describes the behavior of both barely insulating and barely metallic samples.

Table 1. Inflection Fields, Magnetic Lengths, and Magnetoresistance Parameters for Metallic Si:As Samples

$N\times 10^{-18}/cm^3$	H_i(T.)	$\lambda(H_i)$(Å)	$A(T.^{-\frac{1}{2}})$	$f(T\to 0)(T.^{-2})$
8.40	1.6±0.3	203	0.13	5.1×10^{-3}
8.80	3.1±0.3	145	0.12	1.7×10^{-3}
9.38	3.5±0.3	137	0.11	1.1×10^{-3}
9.74	5.2±0.5	112	0.084	7.7×10^{-4}
10.19	5.6±0.4	108	0.076	4.6×10^{-4}

In normal metals the magnetoresistance is characterized by the parameter $\omega_c\tau=k_F\ell/(\lambda^2k_F^2)$. One might compare the magnetic length λ with ℓ or with $(\ell/k_F)^{\frac{1}{2}}$. However, these quantities are very slowly varying for $N_c(0)<N<1.2\ N_c(0)$ and could not possibly lead to an inflection field H_i changing by a factor of three (see Table 1) in this narrow concentration range. In this concentration range the correlation length $\xi(N,0)>\ell$ and the first length reached by λ with increasing field is the field-dependent correlation length $\xi(N,H)$. For small fields the effect of the field for $N_c(0)<N_c(H)<N$ is to increase the correlation length substantially. This explains the relatively large values of $\lambda(H_i)$ given in Table 1.

The form of $\rho(H)/\rho(0)$ over the entire field range as $T\to 0$ (or for $g\mu_BH>>kT$) can be written accurately as

$$\rho(H)/\rho(0) = \{1+AH^{\frac{1}{2}}\}e^{f(N,T)H^2} \qquad (14)$$

Differentiation of Eq. (14) leads to the inflection field $H_i(N)=(A(N)/8f(N,T\to 0))^{2/3}(1+\text{corrections})$. The results in Table 1 show $A(N)$ decreasing slowly with N while $f(N,T\to 0)$ decreases very rapidly with N, thus accounting for the strong increase in $H_i(N)$ with N. The functional form $H_i(N)\propto(N/N_c(0)-1)^{1/2}$ given in Eq. (13) from setting $\lambda(H_i)=\xi(N,H_i)$ should be considered an approximate form. The magnitude depends critically on η.

Although the temperature dependence of $f(N,T)$ is very small for $T<0.5K$ for the metallic samples ($N_c(0)>8.4\times10^{18}/cm^3$) it is not zero. The data suggests the temperature dependence of $f(N,T)$ is decreasing with increasing N and is zero within experimental error for $T<1.0K$ for the $10.2\times10^{18}/cm^3$) sample. Since this high-field data is taken in the regime $N_c(0)<N<N_c(H)$, namely in the "magnetic insulator" regime, it is not yet clear why the temperature dependence of $\rho(H)/\rho(0)$ has become negligible.

DISCUSSION AND CONCLUSIONS

From both the analysis of the effect of a fixed magnetic field on Mott VRH for barely insulating samples and also from the analysis of the inflection field $H_i(N)$ observed in the magnetoresistance of barely metallic samples, one concludes that the critical density $N_c(H)$ is increasing quadratically with magnetic field. This magnetic tuning of N_c with H can be represented on a T=0K phase diagram of H vs. $N_c(H)$ as shown in Fig. 9 after Shapiro[8]. Shapiro's phase diagram (solid line) initially shows a decrease in $N_c(H)$ with H resulting from the anticipated negative magnetoresistance. However, the results for Si:P and Si:As for $N_c(0)<N<1.24N_c(0)$ do not show this negative magnetoresistance. For $N/N_c(0)-1<0.2$ the tuning of $N_c(H)$ is of the form $N_c(H)=N_c(0)\{1+\eta H^2\}$. The magnitude of η is more than 100 times larger than one would obtain from a field-dependent Mott criterion $N_c(H)^{1/3}a_D^*(H)=0.26$ and the Miller-Abrahams[26] calculation of $a_D^*(H)=a_D^*(0)\ \{1-1/2\ \varepsilon_h a_D^{*3}H^2/m^*c^2\}$. This one-hundred-fold increase is approximately the ratio $(d/a_D^*(0))^3$. One would expect a much larger effect on $N_c(H)$ since the size of the wave functions is of order $\xi(N,H)$ rather than $a_D^*(H)$. Nevertheless no satisfactory theory of the magnitude of $\{N_c(H)-N_c(0)\}$ exists.

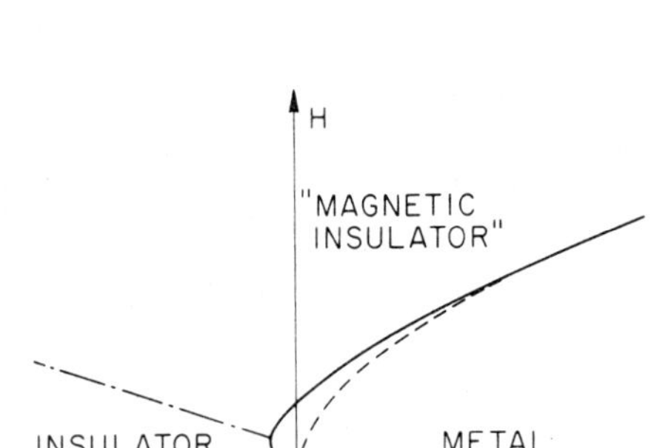

Fig. 9. T=0K magnetic field metal-insulator phase diagram after Shapiro (Ref. 8). The "magnetic-insulator" phase occurs for $N_c(0)<N<N_c(H)$.

Shapiro has speculated on the behavior in the area designated the "magnetic insulator" regime. In our case (dashed line) this would correspond to the region above the dashed line, which corresponds to $H > H_i(N) = (1/\gamma)^{1/2} (N/N_c(0)-1)^{1/2}$. The transport properties in this regime for $H >> H_i(N)$ and $N > N_c(0)$ seem to be characterized by a negligible temperature dependence of $\rho(H,T)/\rho(0,T)$. The reasons for this are not clear and additional work is required. In the "magnetic insulator" regime $N_c(0) < N < N_c(H)$ one might expect to observe Mott VRH with a characteristic $T_o(N,H)$.

The behavior discussed above for barely insulating and barely metallic Si:As samples is interpreted in terms of the magnetic tuning of N_c with magnetic field. It is instructive to compare these results with the results for InSb obtained by Robert et al.[4] and more recently by Mansfield et al.[5] For InSb the quantity $\gamma_D = 8H(T.)$ is a factor of about 900 times larger than the effective mass value for n-type Si where $\gamma_{em} = 0.0088H(T.)$. Thus for InSb one is already in the strong field limit at H=1T. Robert et al.[4] and Mansfield et al.[5] have tuned metallic samples with $N/N_c(0) > 100$ through the MI transition to insulating behavior with high fields that depend on the excess donor density $(N-N_c(0))$ {for a compensated sample $N \to (N_D - N_A)$; for InSb $N_c(0) \simeq 7\times10^{13}/cm^3$}. Robert et al.[4] show the threshold field H_c required to induce the transition is given by $H_c \propto (N-N_c(0))^{0.84}$. This translates into the relation $N_c(H) = N_c(0) + \sigma H^{1.2}$, which is a different field dependence of the tuning than found for Si:As. This difference may result from the fact that the InSb experiments were done almost exclusively in the high field limit. Nevertheless, the qualitative features are similar and the magnetic phase would be the same as in Fig. 9. Mansfield et al.[5] find the T=0K conductivity $\sigma(0,H)$ for a metallic sample obeys the scaling form

$$\sigma(0,H) = \sigma_c \left[\frac{a_\perp{}^2 a_\parallel}{a_c{}^3} - 1\right]^{\nu(H)} , \qquad (15)$$

where $a_\perp(H)$ and $a_\parallel(H)$ are given in Ref. 1 and $a_c \sim (\tfrac{1}{4}) N^{-1/3}$. They report a critical exponent $\nu(H) \simeq 1.2$ for two different samples which are strongly metallic in zero field. This value of $\nu(H)$ would seem to support the proposal of Castellani et al.[10] that $\nu(H) \to 1$ at high fields and may suggest that $\nu(H)$ only approaches one in the high field limit with $\lambda << a_D{}^*$. It would be instructive to study barely metallic InSb with a threshold field H_c sufficiently small that one could remain in the low field regime with $\lambda >> a_D{}^*$. The Si:As results are in the low field regime. Although the Si:As results can best be explained with case 2) $\nu(H) = \nu(0) = \tfrac{1}{2}$ and N_c increasing with H we cannot rule out small changes in $\nu(H)$ in the higher field range (10T.<H<20T.). The Mansfield et al.[5] results described by Eq. (15) actually satisfy the magnetic field-dependent Mott criterion rather well using the field-dependent parameters $a_\perp(H)$ and $a_\parallel(H)$ given by Yafet, Keyes, and Adams[1]. On the other hand, the low field results for Si:As are not well described by a low-field field-dependent Mott criterion using the parameters $a_\perp(H)$ and $a_\parallel(H)$ in the low field limit. The reason for this is not understood but is probably related to the large value of localization (correlation) length very near the MI transition.

The lack of any shift in the localization (correlation) critical exponent $\nu(H)$ from the value $\nu(0) = \tfrac{1}{2}$ with magnetic field can be argued to support the strong spin-flip scattering universality class discussed by Castellani and DiCastro[31].

ACKNOWLEDGEMENTS

The authors are grateful for the collaboration with J. S. Brooks, M. J. Naughton, and K. P. Martin of Boston University on dilution refrigerator measurements at the Francis Bitter National Magnet Laboratory. The magnetocapacitance results and the magnetoresistance results showing inflection fields would not have been possible without the DR capability. Technical assistance from L. Rubin, P. Tedrow, and B. Brandt at FBNML is greatly appreciated as is the direct help with the experiments from D. Koon, M. Migliuolo, and V. Zarifis. This work was supported in part by NSF Grant DMR-8306106.

REFERENCES

1. Y. Yafet, R. W. Keyes, and E. N. Adams, J. Phys. Chem. Solids 1:137 (1956).
2. R. W. Keyes and R. J. Sladek, J. Phys. Chem. Solids 1:143 (1956).
3. S. Ishida and E. Otsuka, J. Phys. Soc. Japan 42:542 (1977).
4. J. L. Robert, A. Raymond, R. L. Aulombard, and C. Bousquet, Phil. Mag. B 42:1003 (1980).
5. R. Mansfield, M. Abdul-Gader and P. Rozooni, Proc. of Int. Conf. on Heavy Doping and the Metal-Insulator Transition in Semiconductors, Santa Cruz, 1984, ed. by P. Landsberg, Solid State Electronics (in press).
6. M. Paalanen, T. F. Rosenbaum, G. A. Thomas, and R. N. Bhatt, Phys. Rev. Lett. 51:1896 (1983).
7. T. F. Rosenbaum, R. G. Milligan, G. A. Thomas, P. A. Lee, T. V. Ramakrishnan, and R. N. Bhatt, Phys. Rev. Lett. 47:1758 (1981).
8. B. Shapiro, Phil. Mag. B50:241 (1984).
9. B. I. Shklovskii, Zh. Eksp. Tear. Fiz.61:2033 (1971) {Sov. Phys. JETP 34:1084 (1972)}.
10. C. Castellani, C. DiCastro, P. A. Lee, and M. Ma, Phys. Rev. B30:527 (1984).
11. J. A. Chroboczek, Phil. Mag. B42:933 (1980).
12. B. I. Shklovskii, Fix. Tekh. Poluprovodn, 6:1197 (1972).
13. Y. Imry, "Anderson Localization", ed. by Y. Nagaoka and H. Fukuyama, Springer-Verlag, Berlin (1982) p. 140.
14. D. New, N. K. Lee, H. S. Tan, and T. G. Castner, Phys. Rev. Lett. 48:1208 (1982).
15. D. New, T. G. Castner, M. J. Naughton and J. S. Brooks, Conf. Proc. The Application of High Magnetic Fields in Semiconductor Physics, Grenoble 1982, Lecture Notes in Physics 177, Springer-Verlag, Berlin (1983) p. 475.
16. W. N. Shafarman and T. G. Castner, "Proc. of the 17th Int. Conf. on the Physics of Semiconductors", ed. by Chadi and Harrison, Springer-Verlag (in press).
17. N. F. Mott, J. Non-Cryst. Solids 1:1 (1968).
18. F. R. Allen and C. J. Adkins, Phil. Mag. 26:1027 (1972).
19. H. F. Hess, K. DeConde, T. F. Rosenbaum, and G. A. Thomas, Phys. Rev. B25:5578 (1982).
20. R. M. Hill, phys. stat. sol. (a)35:K29 (1976).
21. P. F. Newman and D. F. Holcomb, Phys. Rev. B28:628 (1983).
22. G. A. Thomas, Y. Ootuka, S. Kobayashi, and W. Sasaki, Phys. Rev. B24, 4886 (1981).
23. H. Tokumoto, R. Mansfield, and M. J. Lea, Sol. St. Commun., 35:961 (1980).
24. B. I. Shklovskii, Pis'ma Sh. Eksp. Fiz. 36:43 (1982) {JETP Lett. 36:51 (1982)}.

25. B. I. Shklovskii, and A. L. Efros, "Electronic Properties of Doped Semiconductors", Sol. St. Sciences 45, Springer-Berlin (1984) p. 211.
26. A. Miller and E. Abrahams, Phys. Rev. 120:745 (1960).
27. Y. Toyozawa, J. Phys. Soc. Japan 17:986 (1962).
28. A. Kawabata, Sol. St. Commun. 34:431 (1980); J. Phys. Soc. Japan 49:628 (1980).
29. P. A. Lee and T. V. Ramakrishnan, Phys. Rev. B26:4009 (1982).
30. N. Mikoshiba, Phys. Rev. 127:1962 (1962).
31. C. Castellani and C. DiCastro, Proc. of the VII Sityer Conf., June 1984, Lecture Notes, Springer-Verlag (in press).

METAL-SEMICONDUCTOR TRANSITIONS IN DOPED IV-VI SEMICONDUCTORS

R. S. Allgaier

Theodore Associates, Inc.
10510 Streamview Court
Potomac, Maryland 20854

INTRODUCTION

The largest segment of postwar semiconductor research deals with the column-IV elements and their offspring, the III-V compounds. Just one column to the right, there is another closeknit family of nonmetals, the column-V semimetals As, Sb, and Bi, and the IV-VI semiconductor compounds. This review will focus on PbTe, SnTe, GeTe, and some of their alloys.[1] Up until a few years ago, there would have been very little to say about the metal-semiconductor (M-SC) transition in these materials, since one of their distinctive properties has been the lack of extrinsic carrier freezeout. The discovery of deep-level defects and impurities in the IV-VI compounds has changed this situation completely, and has had a strong impact on many of their other basic properties. Currently, deep-level research generates about a third of all the papers published on the IV-VI family.

A broad goal of this review is to convince the reader that deep-level effects in IV-VI semiconductors have made basic and applied research on these materials more interesting and useful, and much more challenging. A more specific purpose is to honor Sir Nevill Mott on his eightieth birthday. It was nearly fifty years ago that he first discussed the metal-nonmetal transformation of bismuth, a member of the V/IV-VI family.[2]

PREHISTORY: BEFORE 1940

Descriptions of some nonmetallic properties of PbS were published as early as 1865 and 1874. Probably the most extensive research during this early period was carried out on Bi. For example, during the 25 years following the 1879 discovery of the Hall effect, about half of all the published reports of the effect dealt with Bi.[3] In 1936, Mott and Jones[2] noted that the relative displacement of the two sublattices in Bi is favored by the "tendency of the total energy to diminish as far as possible." This displacement opens up gaps in the allowed energy bands, changing Bi from a good metal with 2 1/2 filled bands to a nonmetal with five filled bands, and lowering the energy of the highest occupied electronic states.

FIRST PHASE: 1945-1965

It is convenient to divide postwar research on the IV-VI compounds into two 20-year periods (for clarity, I will transfer a few events from one phase to the other). At the beginning of the first phase, the scientific community became more generally aware of the wartime use of PbS as a sensitive infrared detector. But this application involved very poorly defined polycrystalline films. Over the next 20 years, research therefore focussed on bulk single crystals of the IV-VI semiconductors, and many of their fundamental properties were determined. Optical studies revealed that the forbidden energy gaps in the lead "salts", PbS, PbSe, and PbTe, are small and direct, and increase with increasing temperature. In 1960, it was surprising to find that this direct gap is not at the zone center. Their magnetoresistance anisotropy suggested that near their extrema, both the conduction and valence bands consist of a <111>-oriented multivalley structure. A comparison of Shubnikov-de Haas and Hall coefficient data established that there are four valleys in each band, so that the band edges are located at the zone-face L points.

Transport measurements revealed that modest room-temperature mobilities in the lead salts increase very rapidly with decreasing temperature T (as $T^{-5/2}$ in the best samples), level off at values between 1×10^5 and 4×10^6 cm^2/V-sec, and do not decrease again, as would be expected from the usual kind of ionized impurity scattering. These very high mobilities did not generate much excitement, since they were always accompanied by carrier densities in excess of 10^{17} cm^{-3}. Such values suggested that the available samples were very impure, and could not be used to determine the basic properties of the pure compounds, nor to evaluate their potential for high-quality devices. Actually, almost all of the extrinsic carriers in those early samples of IV-VI compounds were associated with deviations from stoichiometry. The lead salts grow in the NaCl structure, and were viewed as doubly ionized versions of that ionic lattice; thus each missing chalcogen or Pb atom produces two electrons or two holes, respectively. In fact, in the presence of so many carriers from lattice vacancies, it was difficult to establish that any deviations from the usual characteristics were due to the presence of foreign impurities, unless they had been introduced deliberately, and in large numbers.

The very high dielectric constants needed to account for the high carrier mobilities at 4.2 K were also invoked to explain another very characteristic feature of these semiconductors, the complete lack of carrier freezeout at low temperatures, even in magnetic fields as high as 20 T. Application of the simple hydrogenic model of a shallow impurity to the lead salts predicts ionization energies in the microvolt range, some two orders of magnitude less than kT at 1 K. And in any event, it seemed likely that at the high extrinsic carrier densities typical of the lead salts, the shallow levels will broaden into an impurity band (IB) which will overlap the adjacent conduction band or valence band (CB, VB), even if the ionization energy is not quite so small. However, it was puzzling (and never really explained) that this merged band seems to have no effect on the free-carrier properties in the overlap region. Thus, undoped lead salts, and SnTe and GeTe as well, were always found to be on the metallic side of the M-SC transition commonly observed in other semiconductors as a function of extrinsic carrier density.

At one time it seemed possible that a M-SC transition would take place in the lead salts, if a technique could be devised to lower the carrier density associated with stoichiometry deviation in a precisely controlled fashion. But even if a low average carrier density can be achieved, there is a basic problem: the product of vacancy concentrations of the two sublattices is fixed by the Schottky constant corresponding to some high T at which the vacancies were frozen in, these vacancy concentrations are statistically distributed, and the net carrier density is the difference between them. Hence, fractional deviations from the average difference will grow uncontrollably as the average value decreases, leading to a grossly inhomogeneous conducting medium, and ultimately to a material with randomly distributed n- and p-type regions. It has been found on a number of occasions that carrier mobilities at all T begin to drop rapidly when carrier densities below 10^{16} cm^{-3} are reached. At low T, this could be due to the extra defect scattering from an increasing degree of compensation. But at higher T, it is more likely that the inhomogeneous state just described is developing.

During the first 20-year period, a few studies were carried out on the effects of incorporating foreign impurities into IV-VI compounds. It was found, for example, that adding Na increased p, while adding Cl increased n. Such changes seem reasonable, assuming that Na substitutes for Pb, and Cl for Te. But carriers associated with these and many other foreign impurities in the IV-VI semiconductors, like those from lattice vacancies, do not freeze out at low T. At the end of this period, Rosenberg and Wald[4] studied the lattice and electrical properties of $Pb_{1-x}Cd_xTe$, $Pb_{1-x}In_xTe$, and the corresponding selenide alloys, with x as large as 0.1. They found that Cd and In added electrons to each system, but the changes were as much as two orders of magnitude smaller than the dopant concentrations. They were very close to discovering deep levels in the IV-VI family, but unfortunately, they did not investigate the dependence of the carrier densities on T or on the dopant concentration within a given system.

During the first phase, it was also recognized that the high dielectric constants in the lead salts and SnTe reflect a general tendency towards instability in the NaCl-type lattice of these compounds. To varying degrees, the lattice is approaching the unstable simple cubic form, and is on the verge of doing the same thing that Bi does to avoid the problem. If the lattice <u>were</u> simple cubic, the {111} faces of the Brillouin zone would vanish. With hindsight, it is natural to find that the lead salts, SnTe, and GeTe have small energy gaps on these faces. The most "symmetrical" of the IV-VI compounds, GeTe, actually becomes rhombohedral below 700 K. The cubic-rhombohedral transition was found to decrease gradually in $Ge_{1-x}Sn_xTe$ alloys with increasing x, extrapolating to roughly 50 K at the SnTe limit. Later it was discovered that SnTe itself does become rhombohedral at low T, with a transition temperature which drops to zero as p increases. More recently, it has become clear that the major factor which stabilizes the cubic phase is structural defects, rather than carriers per se. The lead salts themselves remain cubic, but transitions have been seen in $Pb_{1-x}Sn_xTe$ down to x = 0.45, and it takes less than 1% GeTe in $Pb_{1-x}Ge_xTe$ to make this alloy become rhombohedral. The latter result was surprising, and has been ascribed to the instability of the small Ge atom at the center of a much larger Pb vacancy. The cubic-rhombohedral transitions in the column-V semimetals and in the IV-VI compounds are distinguishable from each other because the latter case leads to a ferroelectric phase, although most of the usual ferroelectric properties are masked by the high carrier densities normally present in these compounds.

The electronic and lattice properties mentioned thus far imply that a strong electron-phonon interaction is present in IV-VI compounds, so that it no longer seems quite so remarkable that superconductivity was discovered in GeTe and SnTe, in both cases when $p > 5 \times 10^{20}$ cm^{-3}. This is the point near which the Fermi level enters a high density-of-states region associated with the 12-valley subsidiary maxima in the valence bands of these two compounds. From time to time, superconductivity in the lead salts has been reported, but never confirmed by further tests. It seems likely that a Pb precipitate was involved. As x increases, the superconducting transition temperature increases in $Sn_{.97-x}Ag_xTe$, while in $Sn_{1-x}In_xTe$ it decreases, vanishes near x = 0.08, and reappears and increases again at higher x. The latter result was interpreted on the basis of a simple ionic model, which predicts that adding In introduces electrons to the system. Thus the vanishing of the superconducting transition corresponds to a low-carrier-density, $p \rightarrow n$ transition region.

It was only towards the end of this first research period that the close connection between the band structures of the group-V semimetals and the IV-VI semiconductors became clear. It took about 30 years, ending about 1965, to establish firmly that there were three, tilted "Shoenberg" pockets in the Bi CB, and one VB pocket along its trigonal axis. The situation seemed much more complex in Sb and As; this was mostly due to the mistake of associating a prominent feature in the quantum-oscillatory data with the Shoenberg CB valleys, whereas it was actually due to the multivalley VB structure. Starting from the simpler four-valley, direct-gap, L-point band model for the cubic IV-VI compounds suddenly made the band structure relationships very obvious. It was already known (from theory and experiment) that when those cubic semiconductors are subjected to a Bi-like strain, the energy in the three valleys inclined to the strain axis decreases relative to that of the valley along the axis, in both bands. Sufficient strain will lead to CB-VB overlap, thus creating the three-electron/one-hole pocket Bi model. Moreover, the multivalley VB structure of Sb and As turned out to correspond to the secondary VB maxima of the lead salts. In Bi-Sb alloys, the single VB valley must drop and be replaced by a rising multivalley structure; hence it is not surprising to find a range of compositions in which those alloys are semiconducting.

SECOND PHASE: 1965-1985

The review of IV-VI semiconductor research during the second phase will focus almost entirely on deep-level dopants. But an important and relevant event which must first be mentioned was the discovery of the band-crossing phenomenon in $Pb_{1-x}Sn_xTe$ and the corresponding selenide alloys in 1966 and 1967.[5,6] It is important to note that this crossover begins at the L point, but spreads into a volume of momentum space which grows with increasing x, decreasing T, and increasing pressure P. This evolution actually conspired to conceal the phenomenon. For example, the susceptibility effective mass in SnTe was found to increase with increasing T, as it does in PbTe.[7] In the latter case, this simply reflects the increasing band gap E_g; in the former, it is due to the fact that the Fermi level E_F lies outside the inverted region in the samples studied. The band-gap behavior has also been studied, much less thoroughly, than the SnTe-GeTe and PbTe-GeTe alloy systems. Almost 20 years have passed since the crossover was discovered, but so far as I know, the alloy composition at which the bands cross back again has yet to be identified.

Deep Levels in Undoped, Imperfect Samples

In 1969, Pratt and Parada first reported a calculation which provided a radically different basis for the lack of freezeout in the IV-VI compounds.[8] They found that a Pb vacancy and a Te vacancy in PbTe remove one and four doubly degenerate levels from the VB and transfer them into the CB. Since four and six electrons are removed when the vacancies are created, two holes appear in the VB and two electrons in the CB, respectively. Later, Pratt showed that interstitial Pb and Te produce two CB electrons and no carriers, respectively.[9] These results provide a convenient, plausible explanation for the lack of extrinsic carrier freezeout and for the lack of any effects on free-carrier properties from shallow levels which had broadened and merged with the adjacent CB or VB. But in a broader sense, this deep-level explanation marks the end of an old era and the beginning of a new and very different one, the nature of which will emerge in the sections to follow.

The Complexities of Doping IV-VI Semiconductors

To describe and assess what has already been published about the effects of doping IV-VI compounds would surely fill a good-sized book. Singly doped compounds and their alloys already constitute ternary and quaternary systems, and multiply-doped systems are now being studied. Moreover, the ever-present, electrically active sublattice vacancies might as well be counted as two additional "chemical" components. My starting assumption is that all dopants and defects will introduce deep levels somewhere in the energy spectrum of the host. The energy level for a given impurity in a given host will depend on whether the impurity occupies either sublattice or an intersitial position, or is slightly displaced from one of these "regular" locations. Some of the obvious assumptions about position may turn out to be erroneous.[10] The impurity energy level may change with its concentration, either because its character is evolving, or because it is altering the band properties of the host medium. In general, different results can be expected according to whether, for example, In is introduced into PbTe as In, InTe, In_2Te_3, or even as InPb. And is the described doping effect the net result of replacing Pb by In, or of replacing the Pb vacancy by In? Published results are not always clear on such alternatives. And on top of all this, there is the large and very difficult question of the consequences of lattice relaxation around a defect or impurity.

The First Deep Impurity Level: Indium in PbTe

In 1971, Averkin et al. found that in n-type $Pb_{1-x}In_xTe$, the Hall coefficient R in the extrinsic T range exhibited a broad maximum, instead of the usual T-independent behavior.[11] They also discovered that n saturated in the mid-10^{18} cm^{-3} range, although as much as 6% InTe had been added. Rosenberg and Wald had found the same result at x = 0.055 in this system.[4] These results were interpreted in terms of the model sketched in Fig. 1(a). The In level lies in the CB, about 70 meV above its minimum, is doubly degenerate, and contributes one electron per In atom. As more In is added, E_F rises to E_{In}, and is thereafter pinned at the In level. The carriers which remain in this quasi-local level, or narrow IB, contribute little or nothing to the Hall effect. As T increases, E_F and E_{In} fall towards the CB edge, thus explaining the low-T side of the Hall maximum. At still higher T, E_{In} drops into the forbidden energy gap, and the decrease in R reflects a gradual changeover from a conductivity σ which is dominated by free carriers in the CB to one in which IB conduction predominates. The notion that IB conduction could become important on the high-temperature side of a Hall maximum may seem improbable, but is supported by the observation of some drastic changes in transport behavior for $x > 0.15$.[12] Actually, the decline in σ_{CB} at high T is inevitable in this model: the usual exponential increase of free-carrier density with increasing T becomes an exponential decrease, since the activation energy (measured downward from the CB edge) has the form $E = -A + BkT$. Hence $\exp(-E)$ becomes $\exp(-B)\exp(+A/kT)$. The model of Fig. 1(a) therefore predicts a M-SC transition, but with the T ranges reversed from those of the usual transition seen in extrinsic semiconductors.

The Indium Level in PbSnTe Alloys

Striking new features in the electronic properties appear when In is added to $Pb_{1-x}Sn_xTe$, including the occurrence of M-SC transitions as a function of x and P.[13] They may be described in terms of Figs. 1(b) and 1(c). As x increases, E_{In} drops very rapidly, entering the gap at x = 0.20 and the VB at x = 0.30. Between these two values of x, low-T experiments showed very clearly that with increasing P, E_{In} leaves the gap region, enters the CB or VB (depending on x), and returns to the gap. With E_{In} near mid-gap energy, R and σ rise by four and drop by seven orders of magnitude from their usual values. After decades of living with always-metallic extrinsic IV-VI compounds, it is intriguing to consider the nature of the conductivity at 4.2 K in such a low-σ sample. There are, say, 10^{19} available carriers, one for each added In atom. Less than 10^3 will be excited into the CB. Assuming carrier mobilities of 10^5 and 10^{-5} cm^2/V-sec for carriers in the CB and IB leads to the conductivity ratio $\sigma_{IB}/\sigma_{CB} = 10^6$, so that the IB contribution completely dominates the conductivity at this low T. However, for the Hall effect, in which the square of the mobility enters each term, the CB contribution should still dominate, by four orders of magnitude, and should be very sensitive to the magnitude of the magnetic field B. This interesting prediction does not seem to have been explored experimentally.

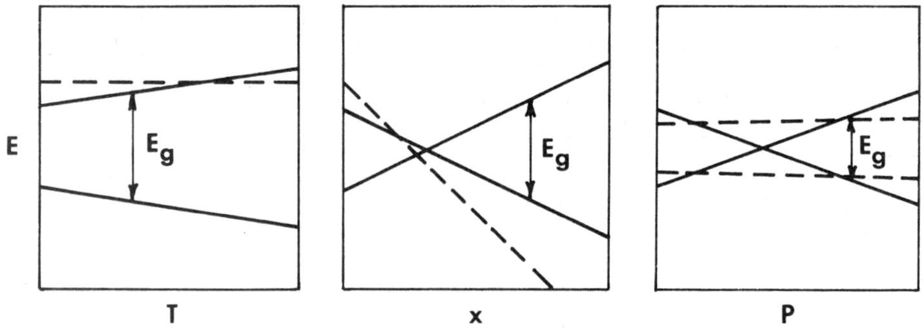

Fig. 1. The indium impurity level (dashed lines). Dependence (a) on temperature T in PbTe, (b) on composition x in $Pb_{1-x}Sn_xTe$, and (c) on pressure P in $Pb_{1-x}Sn_xTe$ at $x \approx 0.22$ (upper line) and at $x \approx 0.28$ (lower line).

Some structure in transport properties at $T \approx 20$ K was first noticed in In-doped $Pb_{1-x}Sn_xTe$ near $x = 0.20$.[14] There is a downward cusp near 20 K, and with decreasing T, a sharp drop in R and rise in σ occurs. One obvious possibility, based on Figs. 1(a) and 1(b), is that these features correspond to a sharpened M-SC transition: E_{In} has been lowered by increasing x, and now crosses the CB edge near 20 K. It was also noticed that the transport properties near this composition are sensitive to the In content, which is not surprising when E_{In} lies close to the CB edge.[15] It was also discovered that if the sample is shielded from higher-T radiation, σ continues to drop below 20 K, and that the high-σ state seen in unshielded samples decays towards the low-σ one (over a period of hours) when the radiation is removed.[16] Numerous studies eventually made it clear that below 20 K, the cubic sample has become rhombohedral. In the dark, it is clearly ferroelectric as well, since spontaneous voltages as large as 200 E_g/e were detected, and kinks in I-V curves corresponding to domain rotations were also seen.[17] But the metastable, metallic phase below 20 K was also found to be noncubic,[18] and it is not yet clear how these two low-T phases differ. It has also been suggested that above the transition, Jahn-Teller distortions are already present around the In atoms.[17] As noted earlier in this review, the rhombohedral phase has not been detected in undoped $Pb_{1-x}Sn_xTe$ for $x < 0.45$. Doping with In may precipitate a phase change at lower x because of its small size, as in the case of $Pb_{1-x}Ge_xTe$ at still lower x. Adding In also introduces a reservoir of electrons at an energy at or above the CB minimum, and a Bi-like distortion will lower three of the four CB valleys, permitting some of those electrons to drop to lower energy states.

The Need for Alternative Models for Indium-Doped IV-VI Compounds

The one-electron model sketched in Fig. 1 provides a simple picture for describing many, and perhaps most, of the observed effects of In doping. But this model predicts that In-doped samples should be paramagnetic and should exhibit a low-T Schottky anomaly in the specific heat for $B \neq 0$. These characteristics have never been observed.[19] In 1975 (the same year that Anderson published his negative-U paper), Andreev et al. proposed a variable valence model for In-doped $Pb_{1-x}Sn_xTe$.[14] They suggested that the reaction $2In^{2+} \longrightarrow In^{+} + In^{3+}$ occurs spontaneously, because the Coulomb repulsion of two electrons on the same In atom is more than compensated for by the lattice distortion surrounding this state. Others have proposed variations on this theme, using such words or phrases as "autocompensation," "tunnel impurity autolocalization," "Jahn-Teller instability,", "multiple-charge state," "ionic-host/negative-U," and "Peierls instability."[20] Which of these theories has the most merit is by no means clear, nor is it obvious that the simple model of Fig. 1 is completely without merit.

Other Effects of Indium Doping

Another effect of a quasi-local In level in the CB of PbTe is to electrically homogenize the material. This is evident from the high-B saturation of the magnetoresistance (a rare phenomenon in the lead salts),[21] and from an equally rare agreement between the Landau-level broadening deduced from the Dingle temperature, and from the scattering time deduced from the low-T mobility.[22] The pinning of the Fermi level due to the presence of the In level in the CB or VB in $Pb_{1-x}Sn_xTe$ also predicts that oscillations in R vs B should occur in quantizing magnetic fields, and could also lead to a M-SC transition in the extreme-quantum-limit region. When E_{In} lies in the CB, the long-term metastability of the low-T metallic state prevents these phenomena from being observed. But the relaxation is much faster for $x > 0.30$, when E_{In} has dropped into the VB. The oscillations are seen in this case, but since $g > 2$ in this regime, p increases with increasing B in the extreme quantum limit, so that a M-SC transition is not expected.[23] Switching effects and avalanche breakdown (with initiation delays up to one hour) have also been observed.[24]

Characteristics of Other Column-III Dopants

There is some experimental evidence to support the notion that as the column-III dopant becomes lighter and the lattice perturbation becomes stronger, the upward shift in E_{III} will grow. For example, optical and electrical experiments suggest that E_{Ga} is about 200 meV, and E_B more than 400 meV, above the CB minimum.[25] There is one very anomalous result concerning the P dependence of E_{Ga} in $Pb_{.8}Sn_{.2}Te$.[26] In this alloy, E_{Ga} is still above the CB minimum at atmospheric pressure. But as P increases, E_{Ga} drops rapidly through the gap and then levels off in the VB, thereafter resembling the behavior of the lower dashed line in Fig. 1(c). Murase has suggested that a larger energy difference between the Ga^{+} and the Ga^{3+} states, due to the smaller sizes of the ions (compared to In), is responsible for this unusual behavior.[23]

The least perturbation of the lead salts should be expected from Tl, which is immediately adjacent to Pb in the periodic table. This dopant does exhibit acceptor behavior, as expected if E_{Tl} remains in the VB, is doubly degenerate, and is occupied by only one of its own electrons. But there are some distinctive features of the Tl IB and its impact on the properties of PbTe.[27,28] The Tl IB is much broader (up to about 0.1 eV) than in the case of In. This may reflect a stronger interaction between the IB and the 12-pocket, high density-of-states subsidiary valence band in PbTe. This interaction may also explain the sharp drop in hole mobility which occurs when p is large enough to lower E_F to E_{Tl}, an effect not observed when E_F reaches the In level in the CB of PbTe. Some have argued that Tl is too large to generate the lattice distortions which lead to a negative-U system, so that Tl exhibits ordinary deep-level acceptor behavior.[27] A magnetic susceptibility study should settle this question.

But the most striking effect of Tl is the occurrence of superconductivity above 1 K in PbTe doped with just 1% Tl.[28] This is especially intriguing after so many decades of controversy about the possibility of superconductivity in the lead salts. The effect might be explained in this case as it has been for superconducting GeTe and SnTe: at high enough p, the Fermi level lies below the subsidiary VB maxima, greatly enhancing the electron-phonon interaction. But the specific impact of the Tl IB has been established by multiple-doping experiments which make it possible to sweep E_F across the IB. It was found that the superconducting transition rises and falls again as p increases. Unexpectedly, it was also discovered that the addition of less than 1% GeTe suppresses the superconductivity.[29] An increase in the transition temperature had been expected, since the addition of only 1/2 % of GeTe to PbTe brings about a soft-mode phase change.

Deep Levels Associated with Other Dopants and Defects

Although it is in Column II of the periodic table, the effects of Cd doping are similar in several ways to those of In.[30] In PbTe, Cd exhibits donor action, saturating near 1×10^{18} cm^{-3}. For $Pb_{.8}Sn_{.2}Te$, the saturation value has dropped to 1×10^{17} cm^{-3}. But when 2% CdTe was added to PbTe, strong freezeout effects were seen, with n dropping to 1×10^{15} cm^{-3} at 77 K. This is evidently due to the fact that E_g in PbCdTe alloys increases by more than 20 meV for each 1% of added CdTe. Thus the dopant itself causes E_{Cd} to drop from the CB into the gap, bringing about a M-SC transition. The same phenomenon clearly takes place in PbCdS alloys: for lightly doped alloys, a donor level some 20 meV below the CB edge was detected, but all samples were p-type when more than 8% CdS was added.[31]

A very clearcut interpretation has been developed to explain the behavior of ion-bombarded $PbSe_{1-x}Te_x$ and $Pb_{1-x}Sn_xTe$ alloys, for $0 \leq x \leq 1$.[32] First of all, the Pratt-Parada predictions for vacancies and interstitials in[8,9] PbTe have been confirmed, except that the measurements suggest that interstitial Pb donates one, not two, electrons to the CB. The net effect of roughly equal numbers of the four types of defects created by ion bombardment is to add electrons to the system, for all x in $PbSe_{1-x}Te_x$, and for $x < 0.2$ in $Pb_{1-x}Sn_xTe$. The formula, per "set" of four defects, is $\Delta n = +1 = 2V_{Te} - 2V_{Pb} + 1I_{Pb} + 0I_{Te}$ (V = vacancy, I = interstitial). But the electron density saturates above a certain damage level. The energy corresponding to the saturation level in PbTe is 55 − 0.17T meV, very similar to the behavior of the In level in PbTe. This energy dependence changes very little with x in the $PbSe_{1-x}Te_x$ alloys, but drops very rapidly with increasing x in $Pb_{1-x}Sn_xTe$, entering the VB when $x > 0.3$. This composition dependence is also similar to that of In in the same alloy systems. It appears that this level is the lowest of the four which are shifted into the CB from the VB when a Te vacancy is created. At low T, this level is filled, except when it lies in the CB. Thus when it is in the VB, as in SnTe or in Sn-rich PbSnTe alloys, a Te vacancy will be neutral. Then $\Delta n = -1$ in the above formula, since the first term no longer contributes, thus explaining why p always increases, without saturation, as the damage in these materials increases. This result also provides a new way to explain why ordinary undoped and undamaged SnTe and GeTe are always p-type.

Multiple Doping and Compensation Effects

Most studies of doped IV-VI compounds have added only a single kind of impurity. But in one of the earliest papers on In in PbTe, it was found that E_F will rise above E_{In} if the host is simultaneously doped with iodine, and $N_I > N_{In}$.[21] Much more recently, the studies of superconductivity in Tl-doped PbTe used Na to sweep E_F across the Tl IB.[28] These two results demonstrate, first of all, that $E_I > E_{In}$ and (measuring downward from the

VB edge) $E_{Na} > E_{Tl}$, so that Na and I are also deep-level impurities, even though their energy levels in the CB and VB have not been identified.

Multiple-doping experiments have attracted more attention recently.[33] An interesting general feature of the results is that they are not additive. For example, changes in deep-impurity energy-level structure occur which do not seem to be related in any obvious fashion to the levels associated with the individual impurities. Another new effect arises when a second impurity is used to vary the Fermi energy within the IB formed by the first impurity. Transport measurements reveal a basic change in the character of the IB (assuming it is in the CB) from an acceptor-like behavior when nearly empty, to a donor-like behavior when nearly full.

Another recent series of experiments deal with an effect generally referred to as self-compensation.[34] They involve a deep-level dopant and an unusually large density of lattice vacancies (larger than the allowable maximum in the undoped host) which generate carriers of opposite signs. The results can sometimes be interpreted by a simple thermodynamic model which minimizes the free energy, but in other cases, vacancy-impurity complexes must be invoked to explain the results.

CONCLUSIONS AND FUTURE OPPORTUNITIES

In my opinion, the picture which has been emerging from a decade of exploration of deep-level dopants in the IV-VI semiconductor family may be described as follows: Many different impurities may be incorporated in rather large amounts into the soft, deformable lattices of the IV-VI semiconductors. All of these dopants, as well as intrinsic lattice vacancies, form deep levels which cover a large energy spectrum, relative to the size of the energy gaps. This last statement is in accord with the observation that the levels have been found in or near the forbidden energy gap for only a small fraction of all of the dopants studied. As the host is changed, or for a given host, as the temperature, pressure, or magnetic field are varied, the band gap will sweep across this spectrum of impurity levels, carrying with it a set of evolving band-edge carrier properties. The band-gap and nearby energy regions in the IV-VI compounds are very sensitive to these parameters, because of varying relativistic contributions to the band structure, and because of the basic instability of the lattice structure. By their very nature, by contrast, the deep levels are relatively stable with respect to one another and with respect to the vacuum level. Thus it is not surprising to find that an In impurity level and a Te vacancy level exhibit similar dependences on temperature and alloy composition.

As in any rapidly growing field, the number of unanswered questions is increasing at least as fast as those which have been answered. A number of important, basic problems have not been solved, or even addressed as yet. The very first is to develop a proper multistate theory to account for the behavior of the Fermi level. In this review, I have discussed the impurity data in terms of a simple, single-level picture--not because I believe it is right, but because it is so easy to describe various phenomena in terms of such a model. A second broad problem is to acquire more concrete, detailed information about the two different, low-temperature non-cubic phases seen in several IV-VI alloy systems. A third basic question involves the location of the various impurities in the IV-VI lattice. In most of the past research on doped IV-VI semiconductors, this problem has been "solved" by assumption, not by experiment. Dow has observed that the obvious choice may be less justified in the case of the IV-VI family. His preliminary calculations of a large number of deep-impurity levels sometimes produced results in accord with experimental findings by placing the impurity on the "wrong" sublattice.[10] And one more, very challenging task will be to develop a means to assess the impact of lattice relaxation around an impurity or defect on the deep-level spectrum and other relevant properties.

There are several currently-fashionable areas of solid-state research which ought to lead to some interesting results, when extended to include studies of deep-level IV-VI semiconductors. One example would be the investigation of order-disorder transitions in stable or metastable V/IV-VI alloys, in analogy to the work that Dow and his associates have carried out on the IV/III-V systems.[35] Here, the new feature is that one of the constituents is semimetallic. A second example would be a broad spectrum of experiments involving superlattices. New features here would include, again, the use of a semimetallic component, the juxtaposition of alloy compositions with direct and inverted band gaps, the application of deep-impurity Fermi level pinning to bring about desired band-edge energy shifts between the superlattice components, and the ability to fabricate Döhler-style doping superlattices by modulating the stoichiometry deviation, rather than by introducing actual foreign impurities. A general feature of all of these studies will be the greatly diminished effect of Coulomb scattering from any kind of defect in the IV-VI semiconductors, relative to that expected in most other semiconductors.

Papers on IV-VI and V/IV-VI superlattices have begun to appear.[36] One report on the PbTe-Bi system reveals that as the Bi/PbTe layer thickness ratio decreases, the Bi-atom positions approach those of a simple cubic structure. I began this review with a reference to Mott and Jones's 1936 explanation for the lattice distortion which takes Bi from a simple cubic metal to a double-layer rhombohedral nonmetal. In 1985, it appears that in a suitable environment, the reverse transformation in Bi may be on the verge of taking place.

ACKNOWLEDGMENTS

I am deeply indebted to George and Athanasia Lalos for the facilities and support which made it possible to prepare this review.

REFERENCES

Most of the general background information for the first three sections of this paper was taken from:

1. W. W. Scanlon, in "Solid State Physics, Advances in Research and Applications," F. Seitz and D. Turnbull, ed., Academic, New York (1959), Vol. 9, pp. 83-137.
   Yu. I. Ravich, B. A. Efimova, and I. A. Smirnov, "Semiconducting Lead Chalcogenides," Plenum, New York (1970).
   G. Nimtz and B. Schlicht, in "Narrow-Gap Semiconductors (Springer Tracts in Modern Physics)," G. Höhler and E. A. Niekisch, ed., Springer-Verlag, Berlin (1983), Vol. 98, pp. 1-117.
   H. Bilz, A. Bussman-Holder, W. Jantsch, and P. Vogl, "Dynamical Properties of IV-VI Compounds (Springer Tracts in Modern Physics)," G. Höhler and E. A. Niekisch, ed., Springer-Verlag, Berlin (1983), Vol. 99.
   J. K. Hulm, M. Ashkin, D. W. Deis, and C. K. Jones, in "Progress in Low Temperature Physics," C. J. Gorter, ed., North-Holland, Amsterdam (1970), Vol. 6, pp. 205-242.
   M. H. Cohen, L. M. Falicov, and S. Golin, <u>IBM J. Res. Dev.</u> 8: 215 (1964).
2. N. F. Mott and H. Jones, "The Theory of the Properties of Metals

and Alloys," Clarendon, Oxford (1936) [Reprint, Dover, New York (1958)], p. 167.
3. T. C. McKay, Proc. Am. Acad. 41: 385 (1906).
4. A. J. Rosenberg and F. Wald, J. Phys. Chem. Solids 26: 1079 (1965).
5. J. O. Dimmock, I. Melngailis, and A. J. Strauss, Phys. Rev. Lett. 16: 1193 (1966).
6. A. J. Strauss, Phys. Rev. 157: 608 (1967).
7. R. F. Bis and J. R. Dixon, Phys. Rev. B 2: 1004 (1970).
8. N. J. Parada and G. W. Pratt, Jr., Phys. Rev. Lett. 22: 180 (1969); N. J. Parada, Phys. Rev. B 3: 2042 (1971).
9. G. W. Pratt, in "Physics of IV-VI Compounds and Alloys," Gordon and Breach, London (1974), pp. 85-91.
10. J. D. Dow, private communication.
11. A. A. Averkin, V. I. Kaidanov, and R. B. Mel'nik, Sov. Phys. Semicond. 5: 75 (1971).
12. V. G. Golubev, N. I. Grechko, S. N. Lykov, E. P. Sabo, and I. A. Chernik, Sov. Phys. Semicond. 11: 1001 (1977); S. N. Lykov, Yu. I. Ravich, and I. A. Chernik, Sov. Phys. Semicond. 11: 1016 (1977).
13. B. A. Akimov, R. S. Vadkhva, V. P. Zlomanov, L. I. Ryabova, and S. M. Chudinov, Sov. Phys. Semicond. 11: 637 (1977); B. A. Akimov, L. I. Ryabova, O. B. Yatsenko, and S. M. Chudinov, Sov. Phys. Semicond. 13: 441 (1979); B. A. Akimov, V. P. Zlomanov, L. I. Ryabova, S. M. Chudinov, and O. B. Yatsenko, Sov. Phys. Semicond. 13: 759 (1979); D. Eger, A. Zemel, H. Shtrikman, and N. Tamari, Mat. Res. Bull. 15: 1333 (1980); A. Zemel, D. Eger, H. Shtrikman, and N. Tamari, J. Electron. Mater. 10: 301 (1981).
14. Yu. V. Andreev, K. I. Geiman, I. A. Drabkin, A. V. Matveenko, E. A. Mozhaev, and B. Ya. Moizhes, Sov. Phys. Semicond. 9: 1235 (1975); K. I. Geiman, I. A. Drabkin, A. V. Matveenko, E. A. Mozhaev, and R. V. Parfen'ev, Sov. Phys. Semicond. 11: 499 (1977).
15. O. V. Aleksandrov, G. A. Kalyuzhnaya, K. V. Kiseleva, and N. I. Strogankova, Inorg. Mater. 14: 998 (1978); G. A. Kaljuzhnaya, T. S. Mamedov, K. H. Herrmann, and M. Wendt, Kristall und Technik 14: 849 (1979); B. A. Akimov, N. B. Brandt, A. A. Zhukov, L. I. Ryabova, and D. R. Khokhlov, Sov. Phys. Semicond. 15: 1294 (1981).
16. B. A. Akimov, N. B. Brandt, S. A. Bogoslovskii, L. I. Ryabova, and S. M. Chudinov, JETP Lett. 29: 9 (1979); B. M. Vul, I. D. Voronova, G. A. Kalyuzhnaya, T. S. Mamedov, and T. Sh. Ragimova, JETP Lett. 29: 18 (1979); B. A. Akimov, N. B. Brandt, and L. I. Ryabova, Moscow Univ. Phys. Bull. 37: 12 (1982); B. A. Akimov, N. B. Brandt, S. O. Klimonskiy, L. I. Ryabova, and D. R. Khokhlov, Phys. Lett. 88A: 483 (1982); B. M. Vul, S. P. Grishechkina, and T. Sh. Ragimova, Sov. Phys. Semicond. 16: 928 (1982); B. A. Akimov, N. B. Brandt, L. I. Ryabova, V. V. Sokovishin, and S. M. Chudinov, J. Low Temp. Phys. 51: 9 (1983).
17. K. H. Herrmann, G. A. Kalyuzhnaya, K.-P. Möllmann, and M. Wendt, Phys. Status Solidi A 71: K21 (1982); K. H. Herrmann and K.-P. Möllmann, Phys. Status Solidi A 80: K101 (1983); K.-P. Möllmann, K. H. Herrmann, and R. Enderlein, Physica 117B & 118B: 582 (1983); K. H. Herrmann and K.-P. Möllmann, Phys. Status Solidi B 76: K67 (1983).
18. T. Ichiguchi and H. D. Drew, in "Proceedings of the 17th International Conference on the Physics of Semiconductors," San Francisco, 1984, to be published.
19. I. A. Drabkin, M. A. Kvantov, and V. V. Kompaniets, Sov. Phys. Semicond. 13: 1206 (1979); S. N. Lykov and I. A. Chernik, Sov. Phys. Semicond. 14: 1112 (1980); I. A. Drabkin, M. A. Kvantov, V. V. Kompaniets, and Yu. P. Kostikov, Sov. Phys. Semicond. 16: 815 ((1982).

20. K. Weiser, A. Klein, and M. Ainhorn, Appl. Phys. Lett. 34: 607 (1979); Yu. Kagan and K. A. Kikoin, JETP Lett. 31: 335 (1980); B. A. Volkov and O. A. Pankratov, Sov. Phys. Dokl. 25: 922 (1980); V. I. Litvinov and K. D. Tovstyuk, Sov. Phys. Semicond. 24: 508 (1982); I. A. Drabkin and B. Ya. Moizhes, Sov. Phys. Semicond. 15: 357 (1981); K. Weiser, Phys. Rev. B 25: 1408 (1982); A. Bardasis and S. Das Sarma, Phys. Rev. B 29: 780 (1984).
21. V. I Kaidanov, R. B. Mel'nik, and I. A. Chernik, Sov. Phys. Semicond. 7: 522 (1973).
22. S. N. Lykov and I. A. Chernik, Sov. Phys. Semicond. 14: 25 (1980).
23. B. A. Akimov, L. I. Ryabova, and S. M. Chudinov, Sov. Phys. Solid State 21: 416 (1979); K. Murase, S. Takaoka, T. Itoga, and S. Ishida, in "Applications of High Magnetic Fields in Semiconductor Physics (Lecture Notes in Physics)," G. Landwehr, ed., Springer-Verlag, Berlin (1983), Vol. 177, pp. 374-377; S. Takaoka, T. Itoga, and K. Murase, Solid State Commun. 46: 287 (1983).
24. B. A. Akimov, N. B. Brandt, V. I. Stafeev, V. N. Nikiforov, and O. B. Yatsenko, JETP Lett. 32: 127 (1980); B. A. Akimov, N. B. Brandt, B. S. Kerner, V. N. Nikiforov, and S. M. Chudinov, Solid State Commun. 43: 31 (1982).
25. A. N. Veis, V. I. Kaidanov, N. A. Kostyleva, R. B. Mel'nik, and Yu. I. Ukhanov, Sov. Phys. Semicond. 7: 630 (1973); A. N. Veis and R.R. Yafaev, Sov. Phys. Semicond. 18: 295 (1984).
26. B. A. Akimov, N. B. Brandt, L. I. Ryabova, D. R. Khokhlov, S. M. Chudinov, and O. B. Yatsenko, JETP Lett. 31: 279 (1980).
27. B. F. Gruzinov, I. A. Drabkin, Yu. Ya. Eliseeva, E. Ya. Lev, and I. V. Nel'son, Sov. Phys. Semicond. 13: 767 (1979); A. N. Veis, V. I Kaidanov, S. A. Nemov, S. N. Emelin, A. Ya. Ksendzov, and Yu. K. Shalabutov, Sov. Phys. Semicond. 13: 106 (1979); V. I. Kaidanov and S. A. Nemov, Sov. Phys. Semicond. 15: 306 (1981).
28. I. A. Chernik and S. N. Lykov, Sov. Phys. Solid State 23: 817 (1981); I. A. Chernik and S. N. Lykov, Sov. Phys. Solid State 23: 1724 (1981); I. A. Chernik and S. N. Lykov, Sov. Phys. Solid State 23: 2062 (1981); I. A. Chernik, S. N. Lykov, and N. I. Grechko, Sov. Phys. Solid State 24: 1661 (1982); S. A. Kaz'min, S. N. Lykov, R. V. Parfen'ev, I. A. Chernik, and D. V. Shamshur, Sov. Phys. Solid State 24: 832 (1982).
29. N. A. Erasova, S. N. Lykov, and I. A. Chernik, Sov. Phys. Solid State 25: 150 (1983).
30. L. M. Rogers and A. J. Crocker, J. Phys. D 4: 1006 (1971); E. Silberg and A. Zemel, Appl. Phys. Lett. 31: 807 (1977); E. Silberg and A. Zemel, J. Electron. Mater. 8: 99 (1979); A. Silberg and A. Zemel, J. Phys. D 15: 275 (1982); Y. Sternberg, N. Yellin, and L. Ben-dor, Mat. Res. Bull. 19: 201 (1984).
31. A. K. Sood, K. Wu, and J. N. Zemel, Thin Solid Films 48: 87 (1978).
32. L. Palmetshofer, Appl. Phys. A 34: 139 (1984).
33. B. A. Akimov, A. I. Elizarov, K. R. Kurbanov, L. I. Ryabova, V.V. Sokovishin, and A. V. Fedorov, Sov. Phys. Semicond. 17: 632 (1983); V. I. Kaidanov, S. A. Nemov, Yu. I. Ravich, and A. M. Zaitsev, Sov. Phys. Semicond. 17: 1027 (1983).
34. L. I. Bytenskii, V. I. Kaidanov, R. B. Mel'nik, S. A. Nemov, and Yu. I. Ravich, Sov. Phys. Semicond. 14: 40 (1980); L. I. Bytenskii, V. I. Kaidanov, R. F. Kuteinikov, R. B. Mel'nik, S. A. Nemov, and Yu. I. Ravich, Sov. Phys. Semicond. 15: 563 (1981); L. I. Bytenskii, V. I. Kaidanov, V. P. Mateenko, R. B. Mel'nik, and S. A. Nemov, Sov. Phys. Semicond. 18: 303 (1984).
35. K. E. Newman, A. Lastras-Martinez, B. Kramer, S. A. Barnett, M. A. Ray, J. D. Dow, and J. E. Greene, Phys. Rev. Lett. 50: 1466 (1983); K. E. Newman and J. D. Dow, Phys. Rev. B 27: 7495 (1983); K. E. Newman and J. D. Dow, Appl. Phys. Lett. 42: 1033

(1983).

36. S. V. Gaponov, B. M. Luskin, and N. N. Salashchenko, Sov. Phys. Tech. Phys. Lett. 5: 210 (1979); M. Kinoshita and H. Fujiyasu, J. Appl. Phys. 51: 5845 (1980); H. Kinoshita, T. Sakashita, and H. Fujiyasu, J. Appl. Phys. 52: 2869 (1981); M. Kinoshita, S. Takaoka, K. Murase, and H. Fujiyasu, in "Proceedings of the 2nd Symposium on Molecular Beam Epitaxy, and Related Clean Surface Technology," Tokyo (1982), p. 81; H. Clemens, E. J. Fantner, and G. Bauer, Rev. Sci. Instrum. 54: 685 (1983); K. E. Ambrosch, H. Clemens, E. J. Fantner, G. Bauer, M. Kriechbaum, P. Kocevar, and R. J. Nicholas, Surf. Sci. 142: 571 (1984); H. Fujiyasu, A. Ishida, H. Kuwabara, S. Shimomura, S. Takaoka, and K. Murase, Surf. Sci. 142: 579 (1984); M. Kriechbaum, K. E. Ambrosch, E. J. Fantner, H. Clemens, and G. Bauer, Phys. Rev. B 30: 3394 (1984); E. J. Fantner and G. Bauer, in "Two-Dimensional Systems, Heterostructures, and Superlattices (Springer Series in Solid-State Sciences)," G. Bauer, F. Kuchar, and H. Heinrich, ed., Springer, Berlin (1984), Vol. 53, pp. 207-217; B. Y. Jin, H. K. Wong, G. K. Wong, J. E. Hilliard, and J. B. Ketterson, J. Appl. Phys. 55: 920 (1984); S. C. Shin, J. E. Hilliard, and J. B. Ketterson, Thin Solid Films 111: 323 (1984).

METAL-INSULATOR TRANSITIONS

IN PURE AND DOPED VO_2

Gérard Villeneuve and Paul Hagenmuller

Laboratoire de Chimie du Solide du CNRS
Ecole Nationale Supérieure de Chimie
et de Physique de Bordeaux
Talence, France

INTRODUCTION

Since its discovery by Morin in 1959 (1) the metal-insulator transition in the vanadium dioxide VO_2 has been the matter of numerous experimental as well as theoretical investigations. Up to now more than six hundred papers have been devoted to the subject, most of them published in the seventies, as a consequence of better knowledge on the narrow band materials, following a series of articles of Mott (2) and their quantitative treatment by Hubbard (3)

In its original paper, Morin explained the transition on hand of Slater's model (4) i.e. as a metal-antiferromagnetic insulator transition: magnetic ordering opens a gap in the middle of the Brillouin zone, provided that the exchange interaction is large enough and (or) the metallic non magnetic band sufficiently narrow. However, experiments showing the absence of any magnetic ordering in the insulating phase ruled out this interpretation.

In the following period there was a strong tendency to class the metal-insulator transition of VO_2 as a Mott transition, particularly from those people which did not really understand the meaning of a "Mott transition".

Structural investigations of the high temperature metallic phase, more symmetric than the insulating one brought serious arguments to the theory of a phonon induced metal-insulator transition in VO_2. A phenomenological picture established from chemical bonding considerations describes the behavior of VO_2 above and below the transition (5).

After the International Conference on "Metal-Insulator Transitions" in 1968 (6), great interest has been taken in the properties of VO_2 and their interpretation. In order to obtain more information, doping by small amounts of tri-, tetra- or pentavalent cationic impurities in the 0.1 %-5 % range and anionic ones has been undertaken in different laboratories. It appeared that the dopants fall into two classes. The first one, i.e. Nb^{5+}, W^{6+}, F^-, induces for formation of V^{3+} ions and gives rise to decreasing metal-insulator transition temperature at low concentrations but does not stabilize other crystallographic phases. The second one, i.e. trivalent impurities as Cr^{3+}, Al^{3+} or Fe^{3+}, is compensated for V^{5+} ions and leads to a complex phase diagram with at least two new insulating phases, many of which properties are significantly different from those of the insulating phase of pure VO_2. Since these phases involve large structural changes and since they are stabilized by impurity concentrations as low as 0.2 %, they must be interpreted as alternative phases of pure VO_2 whose free energy is only slightly larger than that of the low temperature variety of pure VO_2.

In this article we shall describe the main results obtained in the Laboratoire de Chimie du Solide du CNRS at Bordeaux, in the Laboratoire de Physique des Solides at Orsay, in the Bell Telephone Laboratory at Murray Hill as well as in the Laboratoire Central de Rercherches Thomson-CSF at Orsay. After some controversies in the beginning of the seventies, this collaboration proved to be quite fruitful even if each group kept its own interpretation as we shall see later.

We shall successively discuss the structural, electric and magnetic properties of the high temperature metallic phase and of the low temperature insulating phases of pure and doped VO_2. It will appear that the ideas developed by Mott since 1949 are still realistic.

STRUCTURAL PROPERTIES

The high temperature metallic phase has a rutile-type structure called R in the following as determined by Westman (7), it has been refined by Mc Whan et al. (8). The V^{4+} ions are located at the center of quasi-regular octahedra and form chains along the rutile c_R-axis; the vanadium-vanadium distance inside a chain is d_{V-V} = 2.851 Å at 360K and the interchain V-V distance is 3,528 Å (Fig. 1-a). The insulating low temperature phase (called M_1 hereafter) results from a monoclinic distorsion of the rutile structure (Fig. 1-b). The V-atoms form pairs which are slightly tilted with respect to the |001| direction of the undistorted rutile cell (9). The V-V distances are alternatively 2.62 and 3.12 Å (10) . As a consequence of the tilting the vanadium atoms remain no longer at the center of the octahedron and the V-O distances vary between 1.76 and 2.06 Å.

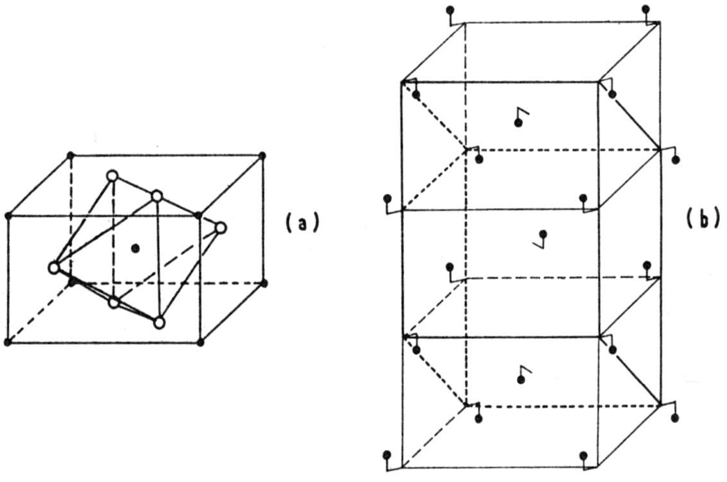

Fig. 1. Tetragonal R (a) and monoclinic M_1 (b) structures of vanadium dioxide.

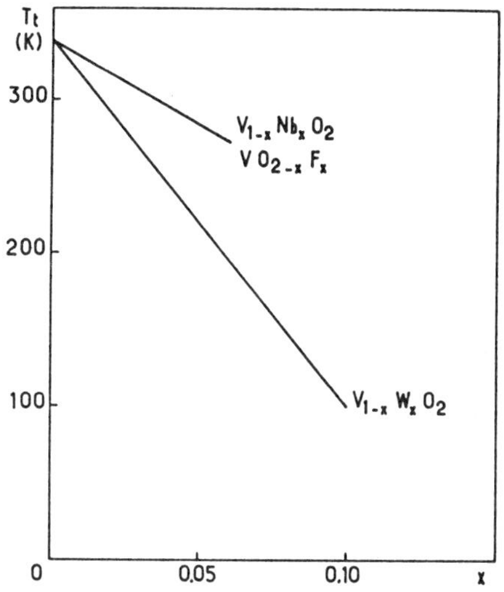

Fig. 2. Variation of the transition temperature of VO_2 doped alloys containing V^{3+} ions.

Doping with impurities of class I (i.e. Nb, W, F) only lead to lowering of the transition temperature (11-13) ; the decrease is about 11K per at. % of niobium or fluorine, and 24 K per at. % of tungsten (Fig.2).

Impurities of class II (i.e. Cr, Al, Fe) give rise to a complex phase diagram with two more intermediate insulating phases called M_2 and T (14-17). In the M_2 phase the V atoms form two types of chains parallel to the rutile c-axis (Fig. 3). In one type the V atoms are paired without tilting of the pairs ; in the other one they are tilted but not paired forming a zig-zag chain (14). The T phase has a triclinic symmetry (15) with a structure intermediate between those of M_1 and M_2 as shown by NMR experiments (18, 41), (Fig. 4).

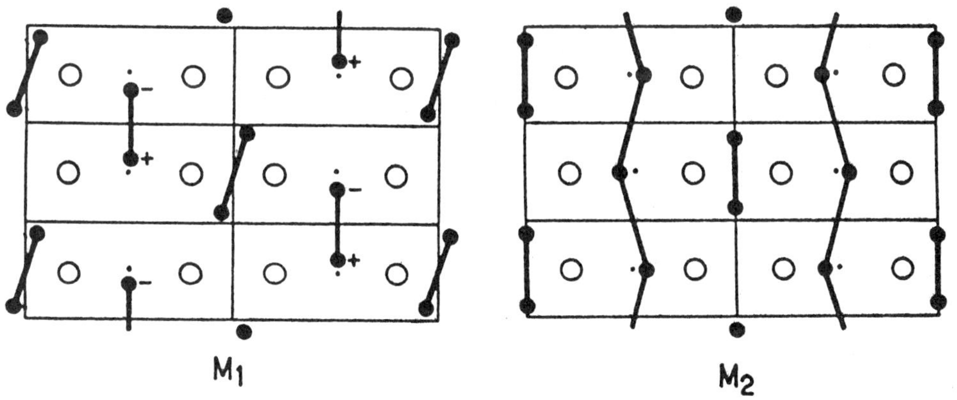

Fig. 3. The $(110)_R$ plane in the M_1 and M_2 phases.

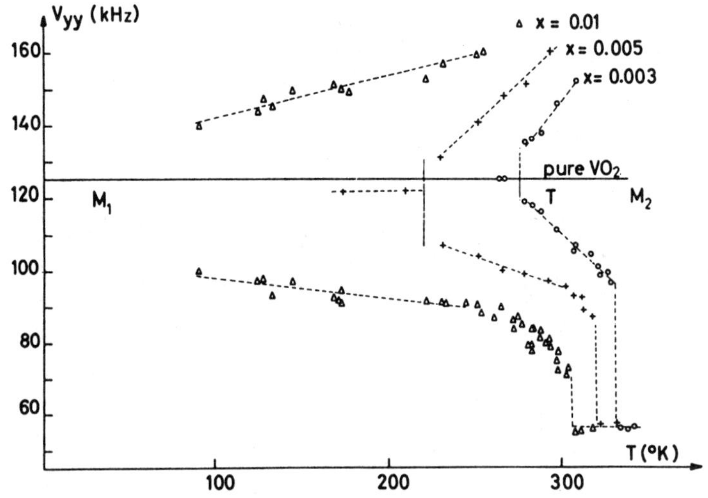

Fig. 4. Electric field gradient V_{yy} in M_1, T and M_2 phases in $V_{1-x}Cr_xO_2$.

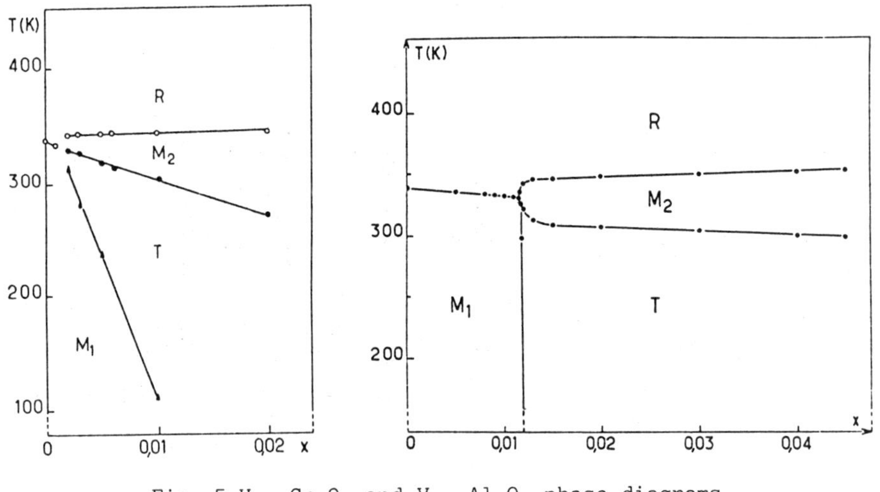

Fig. 5. $V_{1-x}Cr_xO_2$ and $V_{1-x}Al_xO_2$ phase diagrams.

Examples of phase diagrams are given in Fig. 5. M_2 phase also appears in pure VO_2 under uniaxial stress along the $|110|_R$ direction (19).

ELECTRICAL PROPERTIES

Above the transition temperature T_t = 340 K, VO_2 is a metallic conductor with a conductivity rather low ($\sigma \simeq 10^4\ \Omega^{-1}cm^{-1}$) compared to that of classical metals. A sudden discontinuity is observed at the transition when the phase becomes non metallic. The conductivity does not vary exponentially with 1/T except just a few degrees before the insulator-metal transition. The activation energy observed in this small temperature range would correspond to an intrinsic electrical gap of 0.8-0.9 eV (Fig. 6) (20-22).

Low Temperature Insulating Phase

Conductivity and thermoelectric power measurements were carried out on Nb doped VO_2 single crystals $V_{1-x}Nb_xO_2$ (23). At low impurity concentrations (x $\lesssim$ 1.4 %) the conductivity of the insulating M_1 phase is thermally activated up to 200 K with a ($\log\sigma$, T^{-1}) slope quasi identical to that of the α vs T^{-1} line (Fig. 7 and 8). Moreover above 200 K the conductivity tends to a saturation value proportional to the doping concentration. The observed activation energy has been ascribed to the energy needed to release the carriers trapped into the donors (V^{3+}-Nb^{5+}) from a localized impurity level into the conducting (i.e. excited) state of the host lattice (23, 24). Thus the saturation of the conductivity is due to saturation of the number of carriers, and one can obtain a rough estimation of their mobility in the conducting state: $\mu_D \simeq 0.25\ cm^2/V.s$ at 300 K (23), a value low enough to rule out any independant electron model.

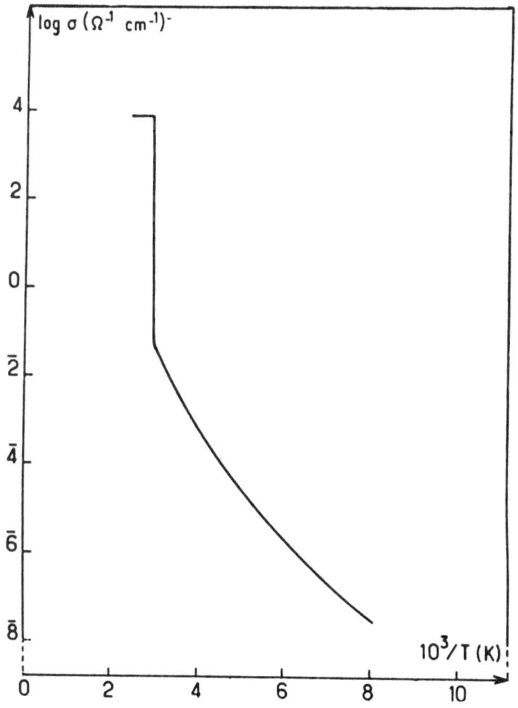

Fig. 6. Electrical conductivity of VO_2 vs temperature.

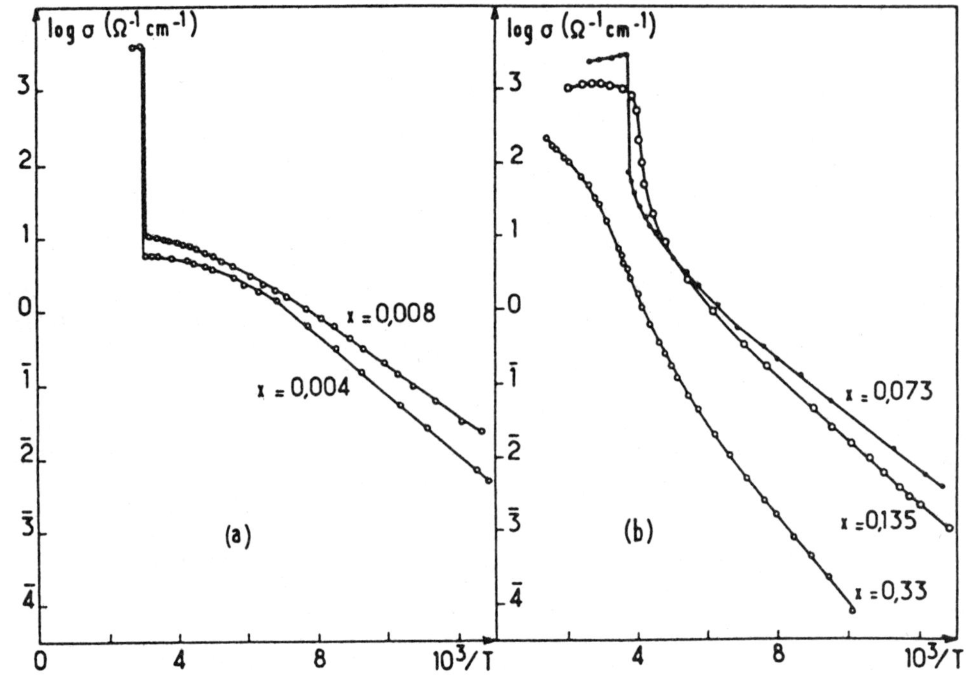

Fig. 7. Variation of log σ vs reciprocal temperature for the $V_{1-x}Nb_xO_2$ system.

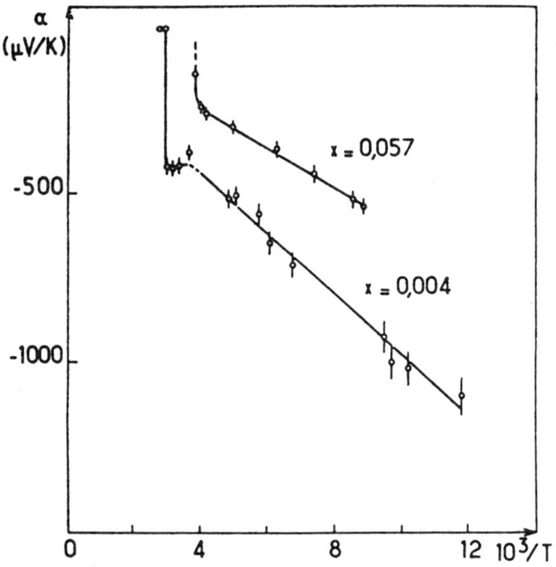

Fig. 8. Variation of the thermoelectric power vs reciprocal temperature for the $V_{1-x}Nb_xO_2$ system.

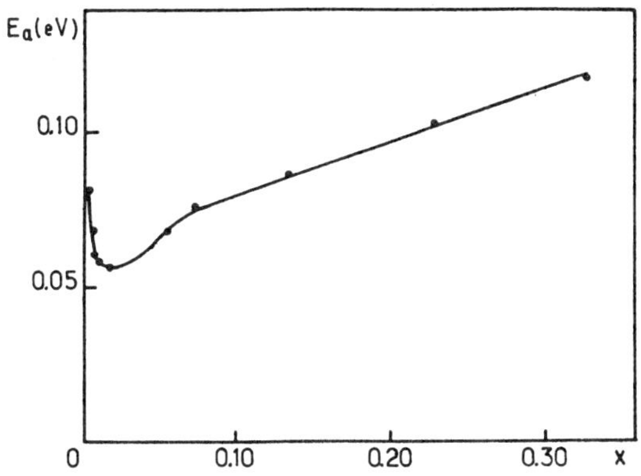

Fig. 9. Variation of the activation energy with x in $V_{1-x}Nb_xO_2$.

For x > 1.6 % in $V_{1-x}Nb_xO_2$ one can observe a regular increase of the activation energy up to x = 33 %, which is the highest concentration investigated on single crystals (Fig. 9). Usually, when the doping concentration is large enough, impurity levels turn into (metallic) impurity bands but disordering due to random distribution of impurities localizes the states in the tails of the bands. As a consequence, a mobility edge appears whose energies is shifted towards high energy at increasing concentration. A possible picture of the density of states is given in Fig. 10, in order to account for the behavior of the insulating Nb-doped VO_2 phase. An alternative explanation is that an activated hopping regime is established when x > 0.05 (24). Both interpretations are based to the onset of Anderson localization resulting from disorder.

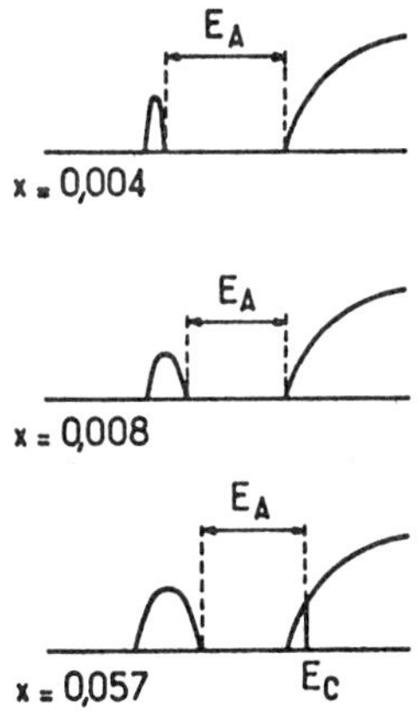

Fig. 10. Density of states vs Nb concentration for $V_{1-x}Nb_xO_2$ (schematic). E_c represents the mobility edge.

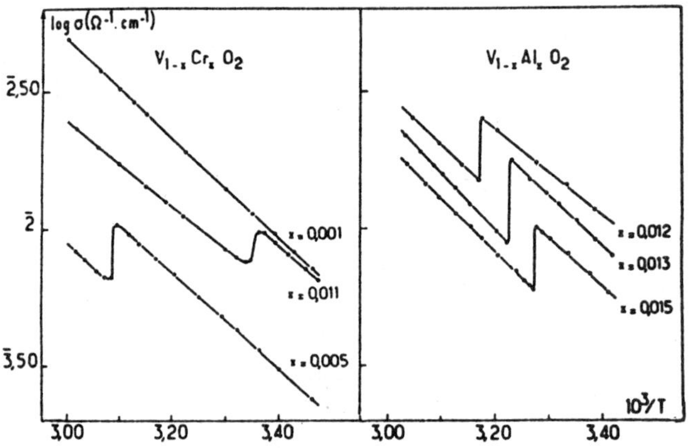

Fig. 11. Evolution of the electrical conductivity in M_1, T and M_2 phases of the $V_{1-x}Cr_xO_2$ and $V_{1-x}Al_xO_2$ systems.

The activation energy in the insulating M_1 phase of W-doped VO_2 is of the same order of magnitude : 0.096 eV for x = 0.006 (12). Bayard et al. found 0.05 eV in $VO_{2-x}F_x$ (13).

Doping with trivalent impurities leads to a quite different behavior of the insulating phases whose activation energies do not differ from that of the M_1 phase in pure VO_2, as shown in Fig. 11 (25, 26). At the T-M_2 transition the conductivity drops abruptly by a factor 2. Table 1 gives the activation energies values obtained by Villeneuve et al. (25) and Buchy and Merenda (26) in M_1, T and M_2 phases of VO_2 doped with Cr and Al. Since the three insulating phases are structurally very different, Zylbersztejn and Mott concluded that the electrical gap is not a structural gap (i.e. it does not arise from pairing), but it is mainly a Mott-Hubbard one (24). The same interpretation was proposed by Villeneuve et al. (25).

The non-linearity of the log σ vs T^{-1} curve at temperature lower than 250 K has been explained from a.c conductivity measurements at different frequencies (27-29). Below 200 K, a frequency dependent conductivity following a Pollak-Geballe law (30) : $\sigma(\omega) \alpha\ \omega^{s}$, is associated with a hopping conduction which becomes more and more important when the temperature decreases. This hopping conduction occurs between localized defects which are present even in nominally "pure" VO_2 and randomly distributed.

Table 1. M_1, T and M_2 activation energies.

	M_1	T	M_2
$V_{0.999}Cr_{0.001}O_2$*	0.36 eV		
$V_{0.995}Cr_{0.005}O_2$*		0.35 eV	0.37 eV
$V_{0.989}Cr_{0.011}O_2$*		0.33 eV	0.32 eV
$V_{0.988}Al_{0.012}O_2$*		0.32 eV	0.38 eV
$V_{0.987}Al_{0.013}O_2$*		0.38 eV	0.42 eV
$V_{0.985}Al_{0.015}O_2$*		0.36 eV	0.38 eV
$V_{0.996}Cr_{0.004}O_2$**		0.40 eV	0.40 eV

* : from ref.(25) ; ** : from ref.(26).

High Temperature Phase

In their high temperature rutile-phase, both VO_2 and NbO_2 are metallic conductors, but the solid solution $V_{1-x}Nb_xO_2$ becomes insulating when $x \simeq 0.20$ (estimated from single crystal experiments). This result could be explained by the expansion of the VO_2 lattice with increasing Nb contend leading to a band narrowing such that Hubbard's criterion for localization is satisfied above $x \simeq 0.20$. Goodenough gave an empirical formulation of Hubbard's criterion in terms of critical interatomic distance (31). The metal-non metal transition occurs for x = 0.20 when the lattice constant c_R, identical to the V-V distance, reaches the value c_R = 2.94 Å which corresponds to the critical distance calculated from Goodenough's formula for V^{4+} ions. Nevertheless Lederer et al. considered that the disorder due to Nb^{4+} ions distributed at random in the cationic sites of the lattice must be taken into account in order to give a complementary explanation of this surprising metal-non metal transition (32). This idea came from a paper published by Mott (33), who extended Anderson's localization in an impurity band (34) to a degenerate electron gas, introducing the concept of "Fermi glass". Lederer et al. pointed out that the presence of both electron correlations and disorder, strongly lowers Anderson's criterion for localization as well as Mott- Hubbard's one. They concluded that the combination of randomness and of strong electron correlations in a metallic phase should lead to complete localization in all states of the band, and to the formation, for $x > 0.20$, of a "strongly correlated Fermi glass".

Extrapolating this analysis, it is tempting to emphasize the influence of electron correlations in the metallic phase of pure VO_2 which could be described as a highly correlated Brinkman-Rice metal (35, 39). We shall see later that this description is still much debated.

MAGNETIC PROPERTIES

Low Temperature Phases

In its low temperature insulating phase, pure VO_2 exhibits a small constant paramagnetic susceptibility down to 20 K (36), $\chi_M \simeq 65.10^{-6}$ emu/mole due to the contributions of both diamagnetism of the ions and Van Vleck temperature independant paramagnetism (Fig. 12).

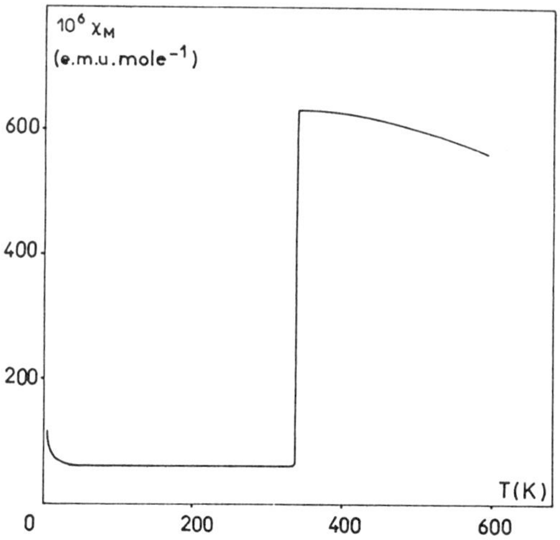

Fig. 12. Magnetic susceptibility of "pure" VO_2.

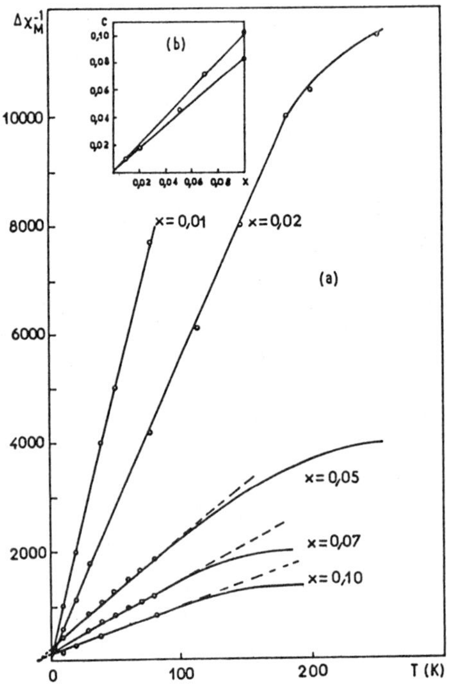

Fig. 13. Reciprocal impurity susceptibility of $V_{1-x}Nb_xO_2$.

The V-V pairs are in a singlet spin state. Below 20 K a susceptibility increase in observed, due to impurities or point defects present even in so-called "pure" VO_2. The impurity susceptibility defined by $\Delta\chi = \chi - \chi_0$ (where χ_0 is the low temperature constant susceptibility) fits well with a Curie law with a Curie constant corresponding to the presence of 2×10^{-3} unpaired V^{4+} ions (S = 1/2) or 7.6×10^{-4} unpaired V^{3+} ions (S = 1). Such defects could be responsible for the hopping conduction mechanism (see above).

The presence of isolated V^{4+} ions in "pure" VO_2 is confirmed by EPR studies at low temperature (37). Grunin et al. observed a quasiaxial hyperfine interaction ; they also pointed out that the spin hamiltonian parameters are practically independent of the structural distortion from a detailed analysis of the V^{4+} spectra in monoclinic VO_2 and quadratic TiO_2.

Magnetic susceptibility data relative to doped VO_2 clearly indicate an electron localization in the insulating phases. In the $V_{1-x}Nb_xO_2$ system ($x < 0.07$), the impurity susceptibility defined by $\Delta\chi = \chi - \chi_0$, where χ_0 is the low temperature constant susceptibility of pure VO_2, follows a Curie-Weiss law $\Delta\chi = \frac{C}{T-\theta}$, with $\theta \simeq -1$ K (36), as shown in Fig. 13.
The experimental Curie constant corresponds to a spin S = 1 per impurity due to the formation of V^{3+} (d^2) ions associated to Nb^{5+} (d^0) ions. This hypothesis was confirmed by EPR spectra observed at 112 GHz and 1.8 K by D'Haenens et al. for Nb-doped VO_2 (38).

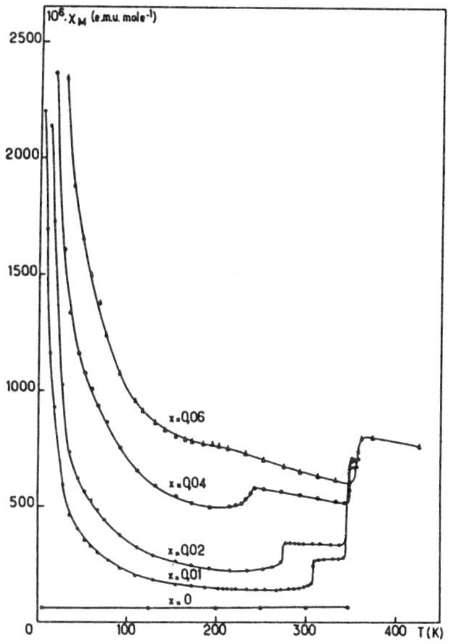

Fig. 14. Magnetic susceptibility of $V_{1-x}Cr_xO_2$.

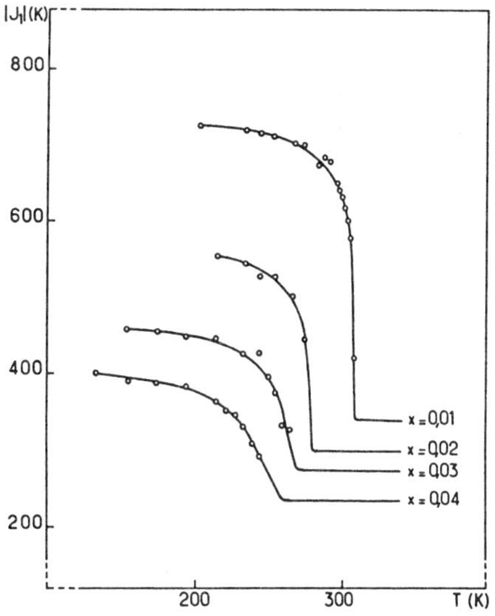

Fig. 15. Thermal evolution of the exchange interaction at the T-M_2 transition.

Magnetic susceptibility data on VO_2 doped with Ti, W and F are also well interpreted in terms of electon localization (40, 12, 13), which rules out any independent electron description of the insulating M_1 phase.

Magnetic properties have been investigated for the M_1, T and M_2 insulating phases of trivalent Cr and Al-doped VO_2. In the M_1 phase, the effective moment per Cr corresponds to the presence of Cr^{3+} ions associated to V^{5+} ; no change is observed in the behavior of $V_{1-x}Al_xO_2$ (M_1) phase compared to that of undoped VO_2, confirming the association of Al^{3+} and V^{5+} diamagnetic ions (41) (Fig. 14).

At the T-M_2 phase transition, the susceptibility jumps abruptly for the low concentrations of impurities. Since the chains of paired vanadium in the M_2 phase are non magnetic, the jump can only result from the magnetic moments associated with the equidistant atoms in the zigzag chains (18). The susceptibility of the M_2 phase is that of non interacting antiferromagnetic Heisenberg chains with S = 1/2. The M_2-T transition can be viewed as a dimerization of those chains as it is suggested by the evolution of the coupling constant J at the transition (Fig. 15)

High Temperature Phase

Pure VO_2 exhibits a very large susceptibility in the metallic phase, along with an appreciable temperature dependence (although not as rapid as in V_2O_3). The question still open is whether the enhanced Pauli paramagnetism is due to the Stoner exchange enhancement factor (42) :

$$\frac{\chi}{\chi_p} = \frac{1}{1-U^*N}$$

where χ is the experimental d-spin susceptibility, χ_p the expected Pauli susceptibility when the density of states at the Fermi level is N and U* the effective intra-atomic Coulomb repulsion which takes into account screening as well as covalency effects.

An alternative interpretation has been proposed by Mc Whan et al. (39): the large magnetic susceptibility of VO_2 results mainly from spin fluctuations in a strongly correlated electron gas following the theory of Brinkman and Rice (35). This results in a strong enhancement of the effective mass ; as a consequence the enhancement of the magnetic susceptibility and that of the electronic heat capacity should be the same. Magnetic data reported on doped-VO_2 are unable to give a definitive answer so far one cannot conclude from them to an obvious evidence of either hypothesis.

The Brinkman - Rice highly correlated electron gas interpretation is supported by the following experimental facts:

(i) in the $V_{1-x}W_xO_2$ system, for x = 0.14, the metallic phase is stable down to 0 K. Thus its electronic heat capacity can be measured directly at low temperature. The effective mass enhancement is found to be the same as that of the magnetic susceptibility (39):

$$\frac{m^*\gamma}{m} = \frac{m^*\chi}{m}$$

(ii) a metal-non metal transition vs composition is observed in the high temperature rutile phase of the $V_{1-x}Nb_xO_2$ system. In the non-metallic region, the susceptibility becomes of Curie-Weis type with well localized V^{4+} and V^{3+} ions. One can include that the high temperature metallic phase of pure VO_2 is close to the M-I transition, i.e. it could be described as a highly correlated electron gas

(iii) Knight shift determinations made on the metallic phase of pure and doped VO_2 by Pouget led this author to the same conclusion (43).

Let us consider now the arguments favorable to a Stoner exchange enhanced susceptibility :

(i) the first one given by Zylbersztein and Mott (24) is that the Brinkman-Rice model is not necessary to explain some properties observed for the metallic phase : this would be true for $V_{0.86}W_{0.14}O_2$, where an impurity concentration as large as 14 % may strongly modify the density of states at the Fermi level with respect to that of pure VO_2

(ii) the metallic variety of pure VO_2 can be described with an enhancement of the electron specific heat three times lower than that of the magnetic susceptibility (24)

(iii) NMR spin-lattice relaxation measurements in the metallic phase of pure VO_2 and $V_{1-x}W_xO_2$ have been recently reported by Takanashi et al. (44). These authors consider the relaxation rate as an exchange enhanced Korringa relaxation process.

TEMPORARY CONCLUSION

Vanadium dioxide has been the subject of numerous investigations, as its complex behavior gives rise to many questions for both solid state physicists and chemists. On the other hand it is the symbol of the progress that can be realized when both theoreticians and experimentalists work together.

In 1973, Hearn and Hyland claimed that : "Whilst there exists divided opinions as to the nature of the semiconducting phases in such oxides as VO_2, V_2O_3 and Ti_2O_3, the existence of a metallic phase as a thermally excited state of the system is, in its own right, no less enigmatic" (42).

In 1985, Villeneuve and Hagenmuller tell : Whilst there exist divided opinions as to the nature of the metallic phase in VO_2 (exchange enhanced or highly correlated electron gas), the nature of its semiconducting ground state is now well understood on hand of Mott-Hubbard correlations.

What about in 1995 ?

ACKNOWLEDGEMENTS

Les auteurs remercient en premier lieu Sir Nevill Mott qui n'a pas son égal pour expliquer simplement des concepts physiques compliqués. Depuis 1972, date de son premier séjour dans notre laboratoire, chacune de ses visites fut pour nous un enrichissement intellectuel. Nous tenons également à remercier toutes les personnes avec qui nous avons collaboré, que ce soit sous la forme d'échange d'idées ou de travail programmé : M. Bayard, A. Bordet, F. Buchy, F. Carmona, A. Casalot, R. Comes, P. Delhaes, M. Drillon, J.B. Goodenough, J.C. Launay, H. Launois, P. Lederer, M. Marezio, N. Nygren, M. Pouchard, J.P. Pouget, T.M. Rice et A. Zylberzstejn.

REFERENCES

1. F.J. Morin, Phys. Rev. Lett. 3:34 (1959).
2. See for example : N.F. Mott, Proc. Phys. Soc. (London) A62:416 (1949); N.F. Mott, Phil. Mag. 6:287 (1961) ; N.F. Mott, Nuevo Cim. Suppl. 7:318 (1958).
3. J. Hubbard, Proc. Roy. Soc. A276:238 (1963).
J. Hubbard, Proc. Roy. Soc. A277:401 (1964).
4. J.C. Slater, Phys. Rev. 82:538 (1951).
5. J.B. Goodenough, J. Solid State Chem. 3:490 (1971).
6. Special issue of Rev. Modern Phys. 40:677 (1968).
7. S. Westman, Acta Chem. Scand. 15:217 (1961).
8. D.B. Mc Whan, M. Marezio, J.P. Remeika and P.D. Dernier, Phys. Rev. B10:490 (1974).
9. G. Andersson, Acta Chem. Scand. 10:623 (1956).
10. J.M. Longo and P. Kierkegaard, Acta chem. Scand. 24:426 (1976).
11. G. Villeneuve, A. Bordet, A. Casalot, J.P. Pouget, H. Launois and P. Lederer, J. Phys. Chem. Solids, 33:1953 (1972).
12. T. Horlin, T. Niklewski and M. Nygren, Mat. Res. Bull., 7:12 (1972)
13. M. Bayard, M. Pouchard, P. Hagenmuller and A. Wold, J. Solid. State Chem. (1975).

14. M. Marezio, D.B. Mc Whan, J.P. Remeika and P.D. Dernier, Phys. Rev. B5:2541 (1972).
15. G. Villeneuve, M. Drillon and P. Hagenmuller, Mat. Res. Bull. 8:1111 (1973).
16. M. Drillon and G. Villeneuve, Mat. Res. Bull. 9:1199 (1974).
17. E. Pollert, G. Villeneuve, F. Ménil and P. Hagenmuller, Mat. Res. Bull. 11:159 (1976).
18. J.P. Pouget, H. Launois, T.M. Rice, J.P. Dernier, A. Gossard, G. Villeneuve and P. Hagenmuller, Phys. Rev. B10:1801 (1974).
19. J.P. Pouget, J. Launois, J.P. D'Haenens, P. Merenda and T.M. Rice, Phys. Rev. Lett. 35:773 (1975).
20. C.R. Everaert and J.B. Mac Chesney, J. Appl. Phys. 30:2872 (1968).
21. L. Ladd and W. Paul, Solid State Comm. 7:425 (1979).
22. J.C. Launay, G. Villeneuve and M. Pouchard, Mat. Res. Bull. 8:997 (1973).
23. G. Villeneuve, J.C. Launay and P. Hagenmuller, Solid State Comm. 15:1683 (1974).
24. A. Zylbersztejn and N.F. Mott, Phys. Rev. B11:4383 (1975).
25 G. Villeneuve, M. Drillon, J.C. Launay, E. Marquestaut and P. Hagenmuller, Solid State Comm. 17:657 (1975).
26. F. Buchy and P. Merenda, quoted in ref. 24.
27. J.F. Palmier, Y. Ballini and P. Merenda, Solid State State Comm. 14:575 (1974).
28. S. Kabashima, T. Goto, N. Nishimura and T. Kawakubo, J. Phys. Soc. Japan 32:158 (1972).
29. A. Mansingh, R. Singh and M. Sayer, J. Phys. Chem. Solids 45:79 (1984).
30. M. Pollak and T.H. Geballe, Phys. Rev. 122:1742 (1961).
31. J.B. Goodenough, J. Appl. Phys. 37:1415 (1966)
32. P. Lederer, H. Launois, J.P. Pouget, A. Casalot and G. Villeneuve, J. Phys. Chem. Solids, 33:1969 (1972).
33. N.F. Mott, J. Non Cryst. Solids 1:1 (1968)
N.F. Mott, Adv. Phys. 16:49 (1967).
34. P.W. Anderson, Phys. Rev. 109:1492 (1958).
35. W.F. Brinkman and T.M. Rice, Phys. Rev. B2:4302 (1970).
36. J.P. Pouget, P. Lederer, D.S. Schreiber, H. Launois, D. Wohlleben, A. Casalot and G. Villeneuve, J. Phys. Chem. Solids 33:1961 (1972).
37. W.S. Grunin, V.A. Ioffe and I.B. Patrina, Phys. Stat. Sol. (b)63:629 (1974).
38. J.P. D'Haenens, D. Kaplan, P. Merenda and J. Tuchendler, in : "Proceedings of the 12th International Conference on the Physics of Semiconductors", Stuttgart (1974).
39. D.B. Mc Whan, J.P. Remeika, J.P. Maita, H. Okinaka, K. Kosuge and S. Kachi, Phys. Rev. B7:326 (1973).
40. T.M. Rice, D.B. Mc Whan and W.F. Brinkman, in "Proceedings of the 10th International Conference on the Physics of Semiconductors", Boston (1970).
41. G. Villeneuve, M. Drillon, P. Hagenmuller, M. Nygren, J.P. Pouget, F. Carmona and P. Delhaes, J. Phys. C10:3621 (1977).
42. C.J. Hearn and G.J. Hyland, Phys. Letters, 43A:87 (1973)
43. J.P. Pouget, Thesis, Orsay (1975).
44. K. Takanashi, H. Yasuoka, Y. Ueda and K. Kosuge, J. Phys. Soc. Japan, 52:3953 (1983).

COMPOSITION-CONTROLLED METAL-INSULATOR TRANSITIONS IN METAL OXIDES§

C.N.R. Rao* and P. Ganguly

Solid State and Structural Chemistry Unit
Indian Institute of Science
Bangalore 560 012, INDIA

1. INTRODUCTION

The family of perovskite-related oxides of the 3d transition metals such as $LaBO_3$ ($B=Ti^{3+}$, V^{3+}, Cr^{3+}, Mn^{3+}, Fe^{3+}, Co^{3+} and Ni^{3+}) are interesting from the point of view of metal-insulator (MI) transitions in strongly correlated systems[1-3]. These oxides exhibit enormous range of interesting properties. Thus, $LaTiO_3$ and $LaNiO_3$ have low resistivities at room temperature ($\sim 10^{-2}$ and $\sim 10^{-3}$ Ω cm respectively). $LaNiO_3$ has a positive temperature coefficient of resistivity (TCR) over a wide temperature range while $LaTiO_3$ shows a change in sign of TCR around 130K with metallic behaviour in the high temperature region[3,4]. $LaVO_3$, $LaMnO_3$, $LaCrO_3$ and $LaFeO_3$ are magnetic insulators with well defined magnetic ordering temperatures ($T_N \approx 140$, 180, 280 and 750K, respectively) and high resistivities[1-3] (> 10^3 Ωcm at 300K). A fascinating member of this series of oxides is $LaCoO_3$ which shows an intricate temperature-dependence of the electronic configuration of trivalent Co ions including changes in the spin-state and a probable disproportionation to di- and tetra-valent Co ions[2,5,6]. Around 1200K, $LaCoO_3$ undergoes a first-order MI transition. There is an intimate relationship between the electronic configurational changes and the changes in the electrical and magnetic properties preceding the MI transition. In a sense, $LaCoO_3$ may be looked upon as showing a MI transition arising from thermally induced compositional changes.

Insulating $LaVO_3$, $LaCoO_3$ and $LaMnO_3$ can be rendered metallic by replacing a part of the La ions by Sr ions[5,7-9]. In such oxides, $La_{1-x}Sr_xBO_3$, a gradual MI transition occurs when $x \gtrsim x_c \approx 0.25$-$0.3$. The V and Co systems were discussed

§Communication No.293 from the Solid State and Structural Chemistry Unit.
*To whom all correspondence should be addressed.

by Mott[10] and Rao et al[11] in the light of localization and the concept of minimum metallic conductivity (σ_{min}). The system $Ln_{1-x}Sr_xCoO_3$ (Ln=La, Nd, Pr etc.) has been studied extensively employing Mossbauer spectroscopy[7,9,12]. In these oxides, electron transfer from B^{3+} to B^{4+} ions renders them metallic when $x > x_c$. Another type of MI transition studied by us recently[13,14] is in oxides of the type $LaNi_{1-x}B_xO_3$ (B=Cr, Mn, Fe or Co) where the $3d^n$ levels of the M ions are expected to be separated energetically from those of the Ni^{3+} ions. The common feature in all these oxides is that irrespective of the details of the MI transition, the TCR changes sign at a critical value of x when the value of the conductivity also crosses a critical value (500 Ω^{-1} cm^{-1}) which corresponds well with the value of σ_{min} in such systems[10,15].

In this article, we shall examine the important features of composition-controlled MI transitions in oxides of the type $La_{1-x}Sr_xBO_3$ and $LaNi_{1-x}B_xO_3$. In the former, disorder is created due to a random distribution of Sr^{2+} (at $x < 0.5$) and the disordering potential is the charge potential and the Co^{4+} or V^{4+} holes form an impurity band. In the latter, the Ni^{3+} and B^{3+} ions have the same charge but have different $3d^n$ energies and the disorder arises mainly due to changes in the exchange potential. One anticipates a correlation between the magnetic and the electrical properties across the MI transition in such oxides. An interesting feature in oxides discussed here is that conduction electrons as well as the localized magnetic electrons have predominantly d character unlike in systems with localized moment impurities in metals (e.g. Mn in Cu).

2. $La_{1-x}Sr_xCoO_3$ and $La_{1-x}Sr_xVO_3$

The $La_{1-x}Sr_xBO_3$ system (B=V, Mn or Co) is interesting in as much as when $x \gtrsim x_c \approx 0.25$-$0.30$, the insulating $LaBO_3$ becomes a metal[7-9]. A similar MI transition is found in $Pr_{1-x}Sr_xCoO_3$ and $Nd_{1-x}Sr_xCoO_3$ when $x \gtrsim 0.30$. The MI transition in these systems is considered to be due to delocalization of the B^{4+} holes which are bound to the Sr^{2+} ions for small values of x. ^{57}Co Mossbauer studies[9] of $La_{1-x}Sr_xCoO_3$ show that when $x < 0.125$ the different oxidation states and spin configurations of cobalt can be distinguished; when $x \gtrsim 0.125$, a single Mossbauer resonance with an isomer shift intermediate between that of the tri- and tetra-valent states is found indicating that the two states cannot be distinguished within the time scale involved in Mossbauer studies ($\sim 10^{-7}$ sec.). This observation suggests the presence of an impurity band for $x \gtrsim 0.125$. Close to the MI transition, the states at the Fermi energy are localized in the Anderson sense[10,11]. Predictions based on the Anderson model were substantiated by the observation of a linear log ρ vs $T^{-1/4}$ plot at low temperature[9,16]. The value of the conductivity at which there is a change in the sign of TCR occurs is around 500 Ω^{-1} cm^{-1} (Fig.1).

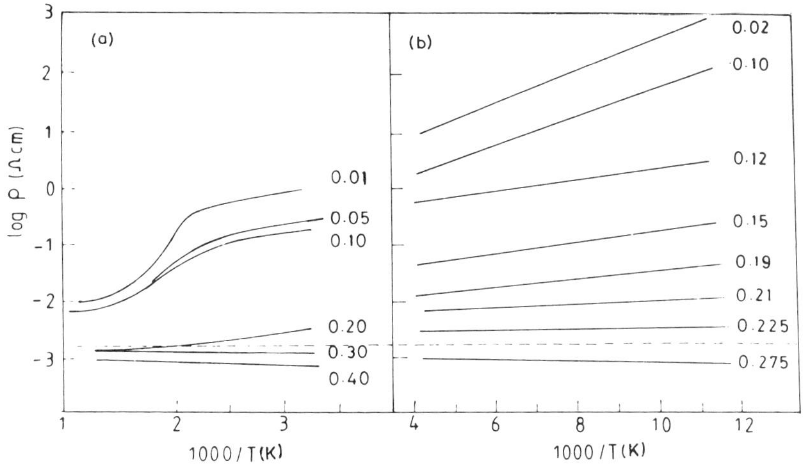

Fig.1. Log ρ vs 1/T plots of (a) $Pr_{1-x}Sr_xCoO_3$; (b) $La_{1-x}Sr_xVO_3$. (from Refs.8 and 12). Dotted line shows values of ρ at which TCR changes sign.

$La_{1-x}Sr_xCoO_3$ is an itinerant electron ferromagnet when $x \gtrsim 0.3$. Mossbauer studies[9] have shown the appearance of six-finger spectra below the Curie temperature (~ 232K for the x=0.5). A μ_{eff} of 2.8 μ_B is obtained in the paramagnetic region. Interestingly, the layered perovskite with the K_2NiF_4 structure $La_{0.5}Sr_{1.5}CoO_4$ (same as $SrO.La_{0.5}Sr_{0.5}CoO_3$) is a ferromagnetic insulator[17] at low temperatures (< 200K) with a μ_{eff} value close to 2.8 μ_B in the paramagnetic region. The presence of localized electron moments in $La_{1-x}Sr_xCoO_3$ is in contrast to $La_{1-x}Sr_xVO_3$ in the metallic phase ($x \gtrsim 0.25$) which shows a nearly temperature-independent susceptibility[18] at high temperatures.

Other points of interest regarding these systems are the following. Both $La_{1-x}Sr_{1+x}VO_4$ and $La_{1-x}Sr_{1+x}CoO_4$ which have the layered perovskite structure have fairly high resistivities[19,20] being several orders of magnitude higher than the corresponding members of the three-dimensional $La_{1-x}Sr_xCoO_3$ system and are never metallic. The dimensionality of the system seems to be important for the occurrence of a MI transition as a function of composition.

There has been no explanation for the nearly universal value of x_c (~ 0.28 ± 0.03) at which the MI transition occurs in the $La_{1-x}Sr_xBO_3$ systems. In $La_{1-x}K_xVO_3$, x_c is around 0.13 which indicates that the critical parameter[21] is the percentage of B^{4+} ions and not the value of x. This would suggest that the Bohr radius, a_H, of B^{4+} is nearly the same for the different B ions and that the criterion[22] $n_c^{1/3} a_H \sim 0.26$ is satisfied. An $x_c \approx 0.25$-0.30 is reminiscent of Pauling's criterion[23] that the metallic state is attained when the number of sites with metallic orbitals is greater than 25%. The value of x_c is also close to the value for the percolation threshold[24] for cubic lattices (~0.30). The activation energy $E_a = (x_c - x)^{1.6 \pm 0.3}$ as x approaches x_c from the insulating side[5,12,16]. The uncertainties in the exponent is too large to distinguish between mobility-edge-controlled or percolation-edge-controlled Mi transitions. Interestingly, in the layered perovskite oxide system, $La_{1-x}Sr_{1+x}VO_4$, a decrease in resistivity is observed[19] when $x > 0.6$ which is the site percolation threshold for a square-planar array[24].

3. The $LaNi_{1-x}B_xO_3$ SYSTEM

In Fig.2 we show the log ρ vs T plots of a few series of oxides of the $LaNi_{1-x}B_xO_3$ system[13,14]. A common feature in all these oxides is that TCR changes sign around a resistivity $\sim 2\times10^{-3}$ Ω cm. shown by dotted line. The critical concentration, x_c, below which TCR is positive is different for the different B^{3+} ions ($M=Cr\ 0.05<x_c<0.10$; $M=Mn\ 0.03<x_c<0.05$; $M=Fe\ 0.25<x_c<0.35$ and $M=Co\ 0.35<x_c<0.50$). The marked localization effects of Cr and Mn compared to Fe and Co may be explained by considerations of the $3d^n$ energy levels E^n. E^n is expected to decrease with increasing atomic number and such a trend is seen in the energy levels of B^{3+} impurities in single crystals of TiO_2 obtained by Mizushima et al[25,26] and also in the estimates of Wilson (see Fig.7 of Ref.27). E^n of Cr^{3+} is expected to be farthest away from that of Ni^{3+}. The Cr^{3+}-O-Ni^{3+} and Mn^{3+}-O-Ni^{3+} exchange integral is thus expected to be smaller than the Fe^{3+}-O-Ni^{3+} exchange integral. Localisation effects may therefore be expected to be strongest when B=Cr or Mn.

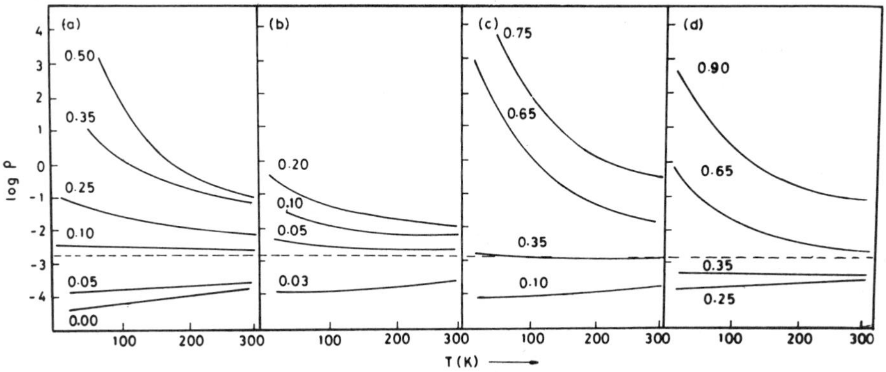

Fig.2. Log ρ vs T plots of $LaNi_{1-x}B_xO_3$: (a) B=Cr; (b) B=Mn; (c) B=Fe; (d) B=Co. (from Refs. 13 and 14). Dotted line shows value of ρ at which TCR changes sign.

The magnetic susceptibility behaviour of these oxides[14,28] are shown in Fig.3. When B=Cr, Mn or Co, the susceptibility diverges below a temperature T for some values of x and T decreases with increasing x. When B=Fe, there is evidence for antiferromagnetic ordering at low temperatures when $x \leq 0.65$. In $LaNiO_3$, a σ* band of mainly e_g character is assumed to be quarter-filled[1,2]. Ferromagnetic ordering is predicted in the insulating narrow band regime when $n_\ell \leq 1/2$. Cr^{3+}, Mn^{3+} and low- or intermediate spin Co^{3+} ions have one or no e_g electron and the solid solutions would satisfy the condition that $n_\ell \leq 1/2$ for

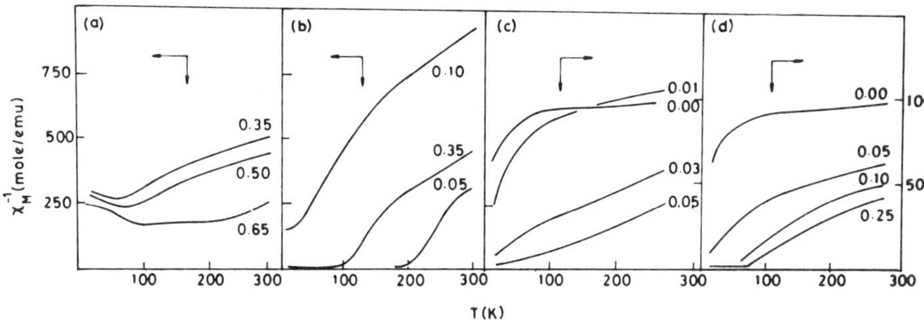

Fig.3. χ_M^{-1} vs T plots of $LaNi_{1-x}B_xO_3$ (T<300K): (a) B=Fe; (b) B=Cr; (c) B=Mn; (d) B=Co (from Refs.14 and 18).

ferromagnetic ordering. Since Fe^{3+} has two e_g electrons n_ℓ would be greater than 1/2 in the solid solutions and antiferromagnetic ordering is anticipated. XPS studies on these systems have, however, shown that in the solid solutions there is no overlap of d bands which are to be treated as rigid bands[29]. We find that the magnetization of La_2CrNiO_6 or La_2CoNiO_6 at 4.2K does not saturate even at 20KG and at this field the magnetic moment is less than 1 μ_B per mole. The non-integral value of the magnetic moment could suggest that as in the ferromagnetic metals, a broad s band overlaps the d band. It is worth noting that in the energy level diagram of Mizushima et al[25,26] the Ni^{3+} energy level is located within the valence band of TiO_2. The low value of the magnetization may also be accounted for by a ferrimagnetic type of ordering (as suggested by the nature of the χ^{-1} vs T plots in Fig.3) or by only a fraction of the Ni^{3+} ions neighbouring the M^{3+} ions being polarized ferromagnetically to form super-paramagnetic or giant magnetic moments.

It is of interest to estimate the bandwidth of $LaNiO_3$ and compare it with the difference in energy between the B^{3+} and Ni^{3+} ions so that one may know whether the B^{3+} energy level is within the itinerant electron d band of $LaNiO_3$ or not. An estimate of the bandwith of $LaNiO_3$ may be obtained from the temperature-independent Pauli-paramagnetic susceptibility of $LaNiO_3$ which is estimated to be $\sim$600-700$\times 10^{-6}$ emu/mol. The Fermi temperature in this case would be of the order of 1000K and bandwidths would only be of the order of 0.1-0.2 eV, much smaller than earlier estimates of d bandwidths (see for example Fig.5 of Ref.27). Since the $3d^n$ level of Ni^{3+} is expected to be energetically lower than the other B^{3+} ions by about 0.5 to 2.0 eV (see for example Fig.1 of Ref.26), a rigid band model would imply that in the metallic phase ($x<x_c$) the B^{3+} electrons would be at a higher energy than that of the itinerant electron band associated with Ni^{3+}.

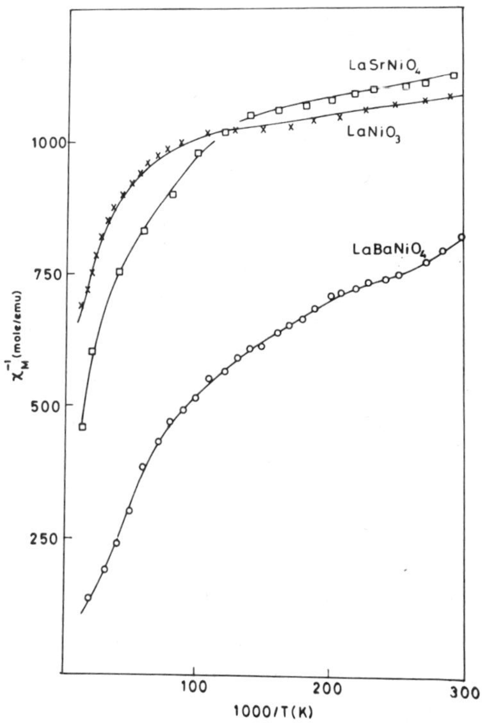

Fig.4. χ^{-1} plots of $LaNiO_3$, $LaSrNiO_4$ and $LaBaNiO_4$ (from Refs.13, 20 and 32).

Goodenough et al[30] and Mott[10] interpret the high-temperature susceptibility of $LaNiO_3$ in terms of the Brinkman-Rice model[31]. The enhanced susceptibility of $LaNiO_3$ with respect to the Pauli value is due to an increase in the effective mass m*. In Mott's interpretation[10] of the Brinkmann-Rice model, m* is related inversely to the fraction of charge carriers, ξ ($m^* = 1/\xi$). In $LaNiO_3$, the high value of the susceptibility would imply an $m^* \approx 10m_0$ or $\xi \approx 0.10$. The rapid increase in susceptibility at low temperatures (Fig.4) has been attributed[14] to impurities in $LaNiO_3$. We note, however, that the nature of the χ^{-1} vs T plot of $LaNiO_3$ (Fig.4) resembles the predicted variation of the susceptibility in the spin-polaron model of Mott (see Fig.7 of Ref.10). The magnetic susceptibility behaviour of $LaNiO_3$ below 300K is similar to that of the layered perovskites $LaSrNiO_4$ or $LaBaNiO_4$ (see Fig.4) although their electrical resistivities are considerably different ($\sim 10^{-1}$ and 10^3 for $LaSrNiO_4$ and $LaBaNiO_4$, respectively)[21,32,33]. It seems therefore that just as in $La_{1-x}Sr_xBO_3$ systems, dimensionality does not affect the magnetic properties of $LaNiO_3$ but strongly influences the resistivity.

Other points that have emerged from our studies of the MI transition in $LaNi_{1-x}B_xO_3$ are the following: (i) the conductivity when $x > x_c$ can be fitted to an empirical equation of the type,

$$\sigma = A \exp(-E_a/k(T+\theta')) \qquad (1)$$

so that there is an effective temperature, T_{eff}, which is greater than T, θ being positive. We have interpreted θ to be an entropy generating factor associated with the width of a partially filled band so that there is a finite conductivity at 0K. Furthermore, $E_a \rightarrow 0$ and $\theta \rightarrow 600$-700K.

(ii) A spin-glass behaviour is observed[14] around 40K in the x=0.10 composition in $LaNi_{1-x}Mn_xO_3$, while the x=0.05 composition shows evidence for the formation of giant magnetic moments. In the more insulating $LaCo_{1-x}Mn_xO_3$ system (which also shows ferromagnetic interaction), there is no evidence for spin-glass behaviour although giant moments are formed at low temperatures.

(iii) The Seebeck coefficient, α, is always negative in the metallic phase. Plots of α vs x for the various B ions show near universal behaviour (Fig.5).

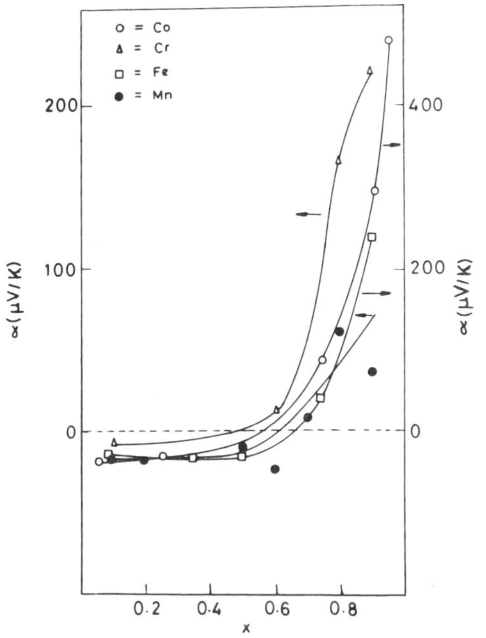

Fig.5. Seebeck coefficient vs x plots of $LaNi_{1-x}B_xO_3$.

4. GENERAL CONSIDERATIONS

In all the composition-controlled metal-insulator transitions investigated by us, TCR changes sign when $\rho \sim 2\text{-}3\times10^{-3}\ \Omega$ cm (or $\sigma \approx 500\ \Omega^{-1}cm^{-1}$) and is independent of the details of the transition. The TCR in several of the oxide systems seems to be determined by the value of the resistivity at 300K. Thus, all the oxides which have $\rho_{300K} > 2\times10^{-3}\ \Omega$ cm show a negative TCR while those with $\rho_{300K} < 2\times10^{-3}\ \Omega$ cm show a positive TCR. Na_xWO_3 bronzes also obey this criterion. Oxides which are metallic such as CrO_2, $SrCrO_3$, $LaNiO_3$ and $SrVO_3$ have resistivities less than $2\times10^{-3}\ \Omega$ cm as do the metallic phases of oxides such VO_2, V_2O_3, V_4O_7, V_3O_5, Ti_2O_5 and Ti_2O_3 which show MI transitions as a function of temperature[34]. This value is close to the value of σ_{min} predicted by Mott[10,15] for concentrated systems ($\sigma_{min} \sim 0.05\ e^2/ha$; $\underline{a} = 3\text{-}4$ Å) and the applicability of σ_{min} at high temperature has been discussed by Mott[35]. Mott's derivation is based on the premise that the mean free path cannot be less than interatomic spacing and σ_{min} is derived from the metallic phase when $k_F L = 1$. We may also calculate a maximum value of the conductivity for activated diffusion, σ_{max} (diff), as the conductivity when the activation energy goes to zero. Zener[36] has shown that in systems such as $La_{1-x}Sr_xMnO_3$, $\sigma = xE_{ex}/kT \cdot e^2/ha$. For $\underline{x}=0.3$ and $E_{ex} = kT$, the value of $\sigma_{max}(\text{diff}) \simeq \sigma_{min}$.

Zener's derivation implies that as the metallic phase is approached from the insulating side the transfer integral and hence the bandwidth W increases. The residence time, τ, of an electron at a lattice site therefore decreases ($\tau \sim h/W$ from the uncertainty principle). When $\tau \ll \tau_{vib}$, the vibration period of a lattice, the electron may be considered to be itinerant since the electron can diffuse from its initial location before the atom can respond to its presence and trap it. Since $\tau_{vib} \approx 10^{-13}$ sec, we may define a τ_{min} ($\sim 10^{-14}$ sec) which would define the minimum relaxation time in the metallic phase[5]. Conductivity for this value of τ_{min} can be written as,

$$\sigma_{\tau_{min}} = ne^2 \frac{\tau_{min}}{m^*} \qquad (2)$$

From σ_{min} to be equal to $\sigma_{\tau_{min}}$, m* should be $\sim 10\ m_0$. This is roughly the value of m* seen from a consideration of the magnitude of the enhanced susceptibility of $LaNiO_3$ and $La_{1-x}Sr_xVO_3$ ($x \sim 0.275$) using the Brinkmann-Rice model[10] or from photoelectron spectral studies of $La_{1-x}Sr_xVO_3$ in the metallic phase[37]. The experimentally observed criterion[22], $n_c^{1/3}\ a_H \sim 0.26$ (where n_c is the density of charge carriers) implies that the $4\pi n_c a_H^3$ is around 10% of the volume available from close-packing considerations[5]. In concentrated systems the number of potential sites $\underline{n}$ will thus be greater than n_c and only 10% need be ionized for the MI transition to occur.

The change of sign of TCR in oxide systems around $\rho \sim 2000\ \mu\Omega$ cm can be related to the Mooij criterion[38,39] which states that for amorphous metals alloy system TCR changes sign when $\rho \sim 200\ \mu\Omega$ cm. Along with the change in the sign of TCR, there is often a change in the sign of the Seebeck coefficient which becomes negative when TCR is positive. The Mooïj criterion has been reviewed by Naugle[39]. The concept of a τ_{min} suggested by us makes closest contact with the model proposed by Jonson and Girvin[40] who showed that in the high-resistivity regime the adiabatic phonon approximation breaks down. This leads to negative TCR because of phonon-associated hopping. Indeed, if we assume $m^* = m_0$ and $\tau_{min} = 10^{-14}$ sec Eqn.(2), the value of the resistivity corresponding to τ_{min} becomes ~200 μΩ cm. σ_{min}[~(1/3)(e^2/ha)] expected from the Ioffe-Regel condition is also close to that predicted by the Mooij criterion.

ACKNOWLEDGEMENT

The authors thank the University Grants Commission for support of this research.

REFERENCES

1. J.B. Goodenough, Prog. Solid State Chemistry, 5:145 (1972) and references therein.
2. (a) J.B. Goodenough, In "Solid State Chemistry" (Ed. C.N.R. Rao) Marcel Dekker, N.Y. (1974) and references therein.
   (b) C.N.R. Rao, Indian J. Chem., 51:979 (1974).
3. P. Ganguly, Om Parkash and C.N.R. Rao, Phys. Stat. Solidii, 26:569 (1976) and references therein.
4. D.A. Maclean and J.E. Greedan, Inorg. Chem., 20:1025 (1981).
5. C.N.R. Rao and P. Ganguly in "Metallic and Non-Metallic States of Matter" (Eds. P. Edwards and C.N.R. Rao), Taylor & Francis, London (1985).
6. V.G. Bhide, D.S. Rajoria, G. Rama Rao and C.N.R. Rao, Phys. Rev., B6: 1021 (1972) and references therein.
7. G.H. Jonker and J.H. Van Santen, Physica, 16:337, 599 (1950); Physica, 19:120 (1953).
8. P. Dougier and A. Casalot, J. Solid State Chem., 2:396 (1970).
9. (a) V.G. Bhide, D.S. Rajoria, C.N.R. Rao, G. Rama Rao and V.G. Jadhao, Phys. Rev.., B12:2832 (1975).
   (b) C.N.R. Rao, Om Parkash, D. Bahadur, P. Ganguly and S. Nagabhushana, J. Solid State Chem., 22:353 (1977).
10. N.F. Mott, Adv. Phys., 21:785 (1972).
11. C.N.R. Rao, V.G. Bhide and N.F. Mott, Phil. Mag., 32:1277 (1975).
12. C.N.R. Rao and Om Parkash, Phil. Mag., 35:1111 (1977).
13. P. Ganguly, N.Y. Vasanthacharya, C.N.R. Rao and P.P. Edwards, J. Solid State Chem., 54:400 (1984).
14. N.Y. Vasanthacharya, P. Ganguly, J.B. Goodenough and C.N.R. Rao, J. Phys. C., 17:2745 (1984).
15. N.F. Mott, Adv. Phys., 16:49 (1967); J. Non-Crystalline Solids, 1:1 (1968); Repts. Prog. Phys., 47:309 (1984).
16. M. Sayer, R. Chen, R. Fletcher and A. Mansingh, J. Phys. C., 8:2059 (1975).
17. P. Ganguly and S. Ramasesha, Mag. Letts., 1:131 (1980).
18. P. Dougier and P. Hagenmuller, J. Solid State Chem., 15:158 (1975).

19. T. Shin-ike, T. Sakai, G. Adachi and J. Shiokawa, Mater. Res. Bull., 12:831 (1977).
20. K.K. Singh, Ph.D. Thesis, Indian Institute of Science, Bangalore, 1982.
21. G.V. Bazuev, O.V. Makarova and G.P. Shveikin, Inorg. Materials, 18:720 (1982).
22. P.P. Edwards and M.J. Sienko, Phys. Rev., B17:2573 (1978).
23. L. Pauling, Phys. Rev. Letters, 47:277 (1981) reference therein.
24. See for example B.K. Shante and S. Kilpatrick, Adv. Phys., 20:225 (1971).
25. K. Mizushima, M. Tanaka and S. Iida, J. Phys. Soc. Japan, 32:1519 (1972); K. Mizushima, M. Tanaka, K. Asai and S. Iida, AIP Conf. Proc., 18:1044 (1973).
26. K. Mizushima, M. Tanaka, A. Asai, S. Iida and J.B. Goodenough, J. Phys. Chem. Solids, 40:1129 (1979).
27. J.A. Wilson, Adv. Phys., 21:143 (1972).
28. N.Y. Vasanthacharya, P. Ganguly and C.N.R. Rao, J. Solid State Chem., 53:140 (1984).
29. W.H. Madhusudhan, S. Kollali, P.R. Sarode, P. Ganguly, M.S. Hegde and C.N.R. Rao, Pramana, 12:317 (1979).
30. J.B. Goodenough, N.F. Mott, G. Demazeau, M. Pouchard and P. Hagenmuller, Mater. Res. Bull., 8:647 (1973).
31. W.F. Brinkman and T.M. Rice, Phys. Rev., B2:4302 (1970).
32. R.A. Mohan Ram, K.K. Singh, W.H. Madhusudan, P. Ganguly and C.N.R. Rao, Mater. Res. Bull., 18:703 (1983).
33. G. Demazeau, M. Pouchard and P. Hagenmuller, J. Solid State Chem., 18:159 (1976); G. Demazeau, J.L. Marty, B. Buffat, J.M. Dance, N. Pouchard, P. Dordor and B. Chevalier, Mater. Res. Bull., 17:37 (1982).
34. C.N.R. Rao and G.V. Subba Rao, Transition Metal Oxides: Crystal Chemistry, Phase Transition and Related Aspects, NSRDS-NBS Monograph 49 (National Bureau of Standards, Washington DC) 1974; Phys. Stat. Solidii, 1a:597 (1970).
35. N.F. Mott, Phil. Mag., B44:265 (1981) and references therein.
36. C. Zener, Phys. Rev., 82:403 (1951).
37. R.G. Edgell, M.R. Harrison, M.D. Hill, L. Porte and G. Wall, J. Phys. C., 17:2889 (1984).
38. J.H. Mooij, Phys. Status Solidii., 17a:521 (1973).
39. D.G. Naugle, J. Phys. Chem. Solids, 45:367 (1984).
40. M. Jonson and S. Girvin, Phys. Rev. Lett., 43:1447 (1979).

PRESSURE-INDUCED INSULATOR-METAL TRANSITION

Shigeru Minomura

Department of Physics
Hokkaido University
Sapporo 060, Japan

INTRODUCTION

Research on insulator-metal transitions has been a subject of fundamental interest ever since Wilson (1931) introduced a model for the electronic band structure of solids. A insulator or semiconductor is defined as a material which shows at the absolute zero of temperature a energy gap between the full valence band and the empty conduction band. A metal, on the other hand, is a material with a partially occupied conduction band. As a consequence, the insulator-metal transitions are interpreted as indicating a band-crossing transition.

Mott (1949, 1968) first suggested that a crystalline array of hydrogen-like atoms at the absolute zero of temperature should show a sharp transition from insulator to metal, if the interatomic distance a decreases to a certain critical value a_0. If a_0 is large, the material should be an insulator with Hubbard intra-atomic energy U. If a_0 is small, the material should be a metal with overlapping Hubbard bands. Such a transition is known as the Mott transition, and is believed to occur in transition-metal compounds.

In the early 1960s, Drickamer and his co-workers developed the ultra-high pressure supported anvil devices for measurements of optical absorption, electrical resistance, x-ray diffraction and Mössbauer effect, and observed the insulator-metal transitions in a variety of materials at high pressure to 500 kbar. They demonstrated the pressure-induced insulator-metal transitions in iodine, Se, Si, Ge, zinc-blende crystals, thallous halides, etc. (Drickamer 1965). In the early 1970, McWhan and Remeika (1970), and Jayaraman et al. (1970) determined the pressure-temperature phase diagram for Cr or Ti doped V_2O_3, which shows three phases; paramagnetic insulator and metal, and antiferromagnetic insulator. Jayaraman et al. (1975) observed the insulator-metal transitions in rare-earth compounds at high pressure. These transitions are interpreted as indicating a 4f-5d electronic transition.

This review describes experimental aspects of the pressure-induced insulator-metal transitions in covalent crystalline and amorphous semiconductors. In the first section we discuss the behavior of iodine which is associated with metallization and molecular dissociation at high pressure. In the second section we deal with the behavior of Se and Te which is accompanied by a change of structure to puckered layers at high pressure. In the third section we discuss the behavior of Si, Ge and some group III-V compounds which are associated with a change of structure to 6-fold coordinates at high pressure.

IODINE

Molecular iodine crystallizes into a base-centered orthorhombic structure with a space group D_{2h}^{18}-Cmca at atmospheric pressure. Drickamer and his co-workers first demonstrated the insulator-metal transition in iodine at 180 kbar and room temperature which is associated with the metallization in the molecular phase (Suchan et al., 1959; Balchan and Drickamer 1961; Riggleman and Drickamer, 1962, 1963). The transition is accompanied by a continuous decrease to zero in optical energy gap and activation energy. From high-pressure x-ray studies, Lynch and Drickamer (1966) proposed two different models for the insulator-metal transition, either the metallization of molecular phase or the dissociation into monatomic phase.

Recently the x-ray analysis of iodine at high pressure has been extensively performed by Shimomura et al. (1978) and Takemura et al. (1979, 1980, 1982). The x-ray diffraction patterns of iodine at high pressure to 206 kbar are identified as a base-centered orthorhombic structure which is the same as that at atmospheric pressure. However, at 210 kbar they reveal the growth and development of a new phase which is identified as a body-centered orthorhombic structure with a space group D_{2h}^{25}-Immm. The diffraction profiles observed and calculated for iodine at 206 and 300 kbar are shown in Figs. 1 and 2. The projection of atomic positions on the basal plane for iodine at 1 bar, 200 kbar and 300 kbar are shown in Fig. 3. The crystal structure of the low-pressure phase below 206 kbar consists of diatomic molecules very weakly bonded together to form the layers parallel to the bc plane, whereas that of the high-pressure phase above 210 kbar shows the molecular dissociation. In the high-pressure monatomic phase each atom has 12 neighbors as in Fig. 4.

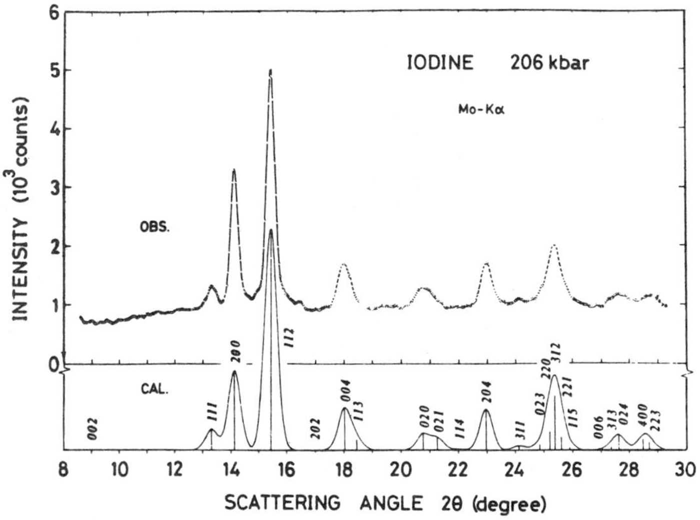

Fig. 1. X-ray diffraction patterns observed and calculated for the molecular phase of iodine at 206 kbar.

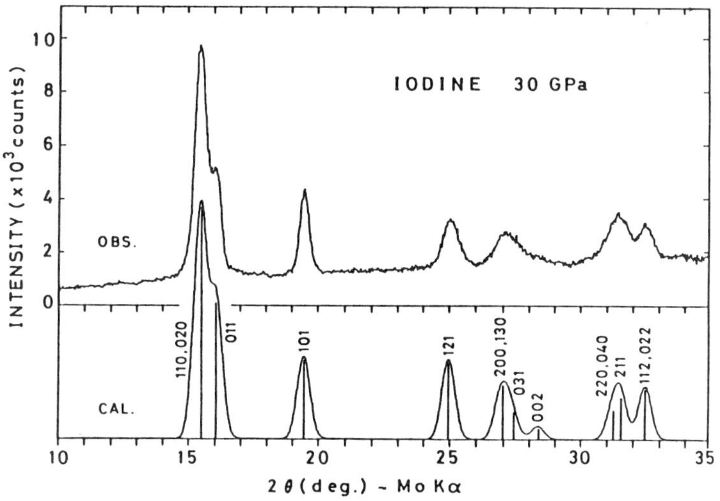

Fig. 2. X-ray diffraction patterns observed and calculated for the monatomic phase of iodine at 300 kbar.

At 300 kbar the second and third coordinate distances (r_2 and r_3) are longer than the first coordinate distance (r_3) by a factor of 1.04 and 1.16, respectively. The variation of interatomic distances with pressure is shown in Fig. 5. In the low-pressure phase the molecular tilt angle measured from the C anxis increases from 32.2° to 38.9° with increasing pressure to 198 kbar, while the bond length remains unchanged. The volume compression arises from the decrease in intermolecular atomic distances.

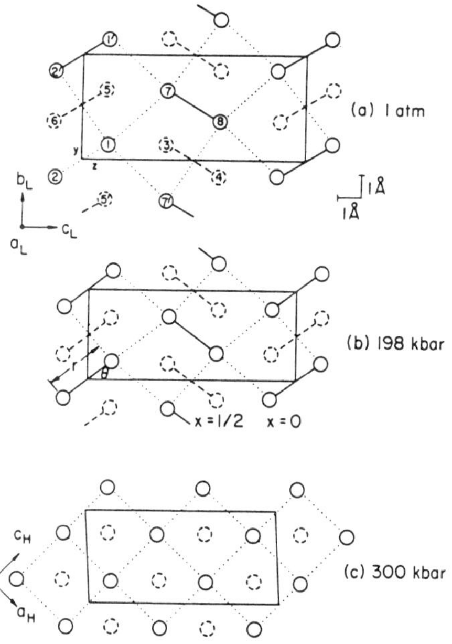

Fig. 3. Projection of atomic positions on the basal plane for iodine at (a) 1 atm, (b) 198 kbar, and (c) 300 kbar. The solid rectangle represents the basal plane of the orthorhombic unit cell.

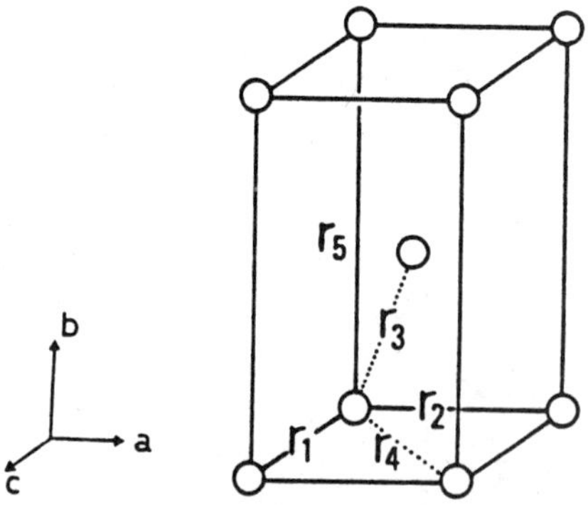

	r_i(Å)	n	r_i/r_1
r_1	2.899	2	1.00
r_2	3.050	2	1.05
r_3	3.373	8	1.16
r_4	4.208	4	1.45
r_5	5.273	2	1.82

r_i : inter-atomic distance

n : coordination number

Fig. 4. Interatomic distances for the monatomic phase of iodine at 300 kbar.

The molecular dissociation at 210 kbar is accompanied by a decrease in volume of 4%. From the change in axial ratios with pressure as in Fig. 6, the structure of monatomic phase will change to a body-centered tetragonal lattice at about 450 kbar, and to a face-centered cubic lattice at extremely high pressure.

The high-pressure experiments on iodine demonstrate the insulator-metal transition which is associated with the metallization in the molecular phase at 180 kbar and the molecular dissociation into the monatomic phase at 210 kbar. It is the most interesting feature that the transition pressure of metallization is lower than that of dissociation. Such a behavior of iodine is used as a prototype for the insulator-metal transition in diatomic molecular crystals, in particular for hydrogen.

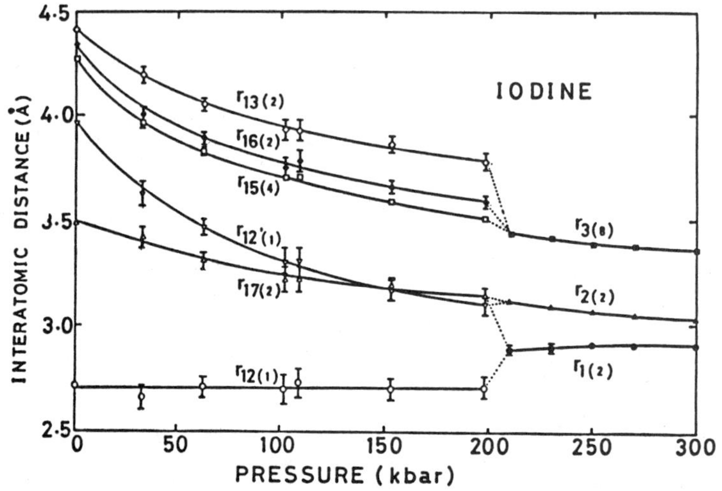

Fig. 5. Variation of interatomic distances with pressure for the molecular and monatomic phases of iodine.

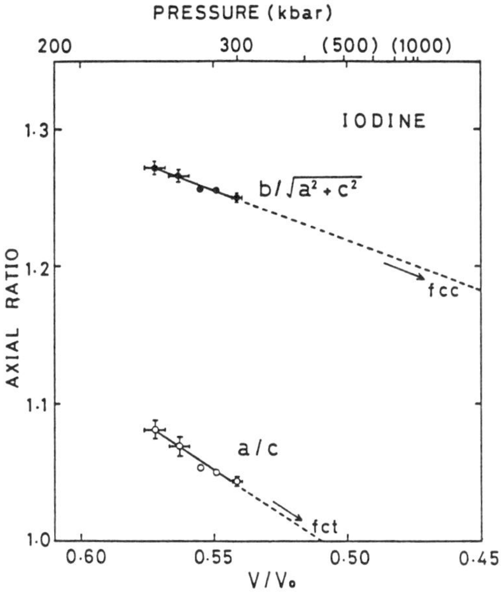

Fig. 6. Variation of axial ratios with pressure for the monatomic phase of iodine.

Metallic hydrogen has been a topic of considerable interest, both theoretical and experimental. In particular, the metallization and dissociation of molecular hydrogen have attracted much recent attention (Friedli and Aschcroft 1977; Chakravarty et al., 1981; Van Straaten et al., 1982; Ross and McMahan, 1982). It is predicted from theoretical calculations that hydrogen will transform to a molecular metallic phase at about 2.0∿2.5 Mbar and to a monatomic metallic phase at about 2.5∿5.5 Mbar. The behavior of condensed hydrogen at high pressure has been investigated by the static and shock-wave methods. The resistance measurements at high pressure and low temperature by Vereshchagin et al. (1975) show that solid hydrogen becomes a metal with a sharp drop in resistance which seems to be the first order transition. This feature is apparently inconsistant with the metallization in the molecular phase.

SELENIUM AND TELLURIUM

The.crystal structure of trigonal Se and Te consists of infinite helical chains which spiral around the c axis with three atoms per turn. Each atom within the chains is tigthly bonded to two neighbors with covalent character. The interchain bonding is much weaker. It has been known for many years that Se and Te show the insulator-metal transition at about 130 and 40 kbar, respectively(Bridgman, 1952; Sachan et al., 1959; Balchan and Drickamer, 1960; Riggleman and Drickamer, 1962, 1963). The ealier high-pressure x-ray studies by Jamieson and McWhan (1965) for Te and by McCann and Cartz (1972) for Se were unable to determine the structure of high-pressure phases.

Recently there has been extensive high-pressure studies on Se and Te (Minomura et al., 1975, 1979; Minomura, 1978; Aoki et al., 1980). The lattice parameter a decreases rapidly with increasing pressure while the parameter c increases by a small amount. These features are interpreted as indicating a distortion which arises from a rapid decrease in interchain atomic distance and a small increase in bond angle. The insulator-metal transition occur at the c/a ratio of 1.42 both in trigonal Se and Te. The variation of electrical resistivity with pressure for trigonal Se is shown in Fig. 7. With increasing pressure trigonal Se shows the decrease in resistivity by a factor of more than 10 orders of magnitude, and at 180 kbar the insulator-metal transition accompanied by a sharp drop in resistivity. On the other hand, α-monoclinic and amorphous Se show the insulator-metal transitions at 100 and 105 kbar, respectively.

In trigonal Se and Te five of six optical modes are Raman active; A_1 mode and two doubly-degenerated E modes. The A_1 mode is the chain-expansion type. The E modes are separated into predominantly angle-bending and bond-stretching type. The variation of phonon frequencies with pressure for A_1, E' and E" modes in trigonal Se and Te is shown in Fig. 8. The phonon frequencies of A_1 mode decrease rapidly with increasing pressure while those of E modes remain unchanged. The softening of A_1 mode with pressure is interpreted as indicating an interference of interchain interactions with the intrachain bonding.

The x-ray diffraction pattern for the high-pressure phase of Te at 45 kbar is shown in Fig. 9. This phase is identified as a monoclinic structure with a space group C_2^2, which consists of puckered layers with 4 atoms per unit cell as in Fig. 10. The lattice constants are a=3.104 Å, b=7.513 Å, c=4.766 Å, and β=92.709°. Each atom has 4 neighbors in the same layer, and 4 next neighbors within the same double layers. The layer structure is regarded as an intermediate step in the course of compression from a helical chain to a β-P_0 structure (Jamieson and McWhan, 1965).

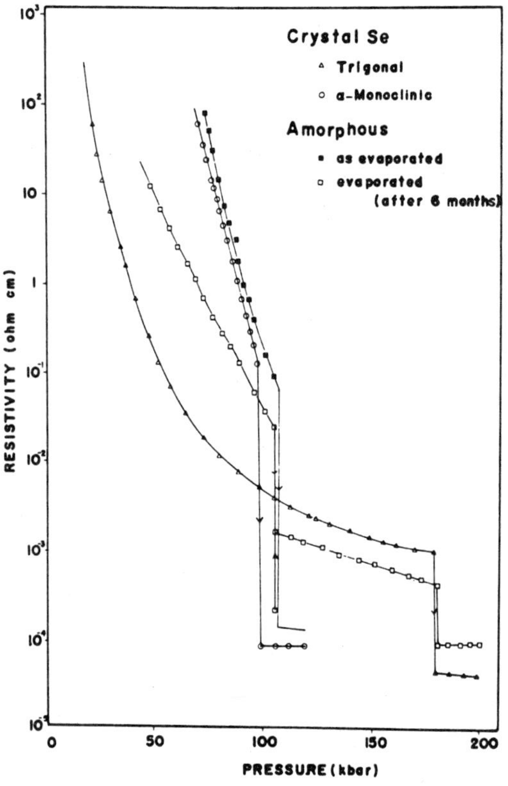

Fig. 7. Variation of resistivity with pressure for trigonal, α-monoclinic and amorphous Se.

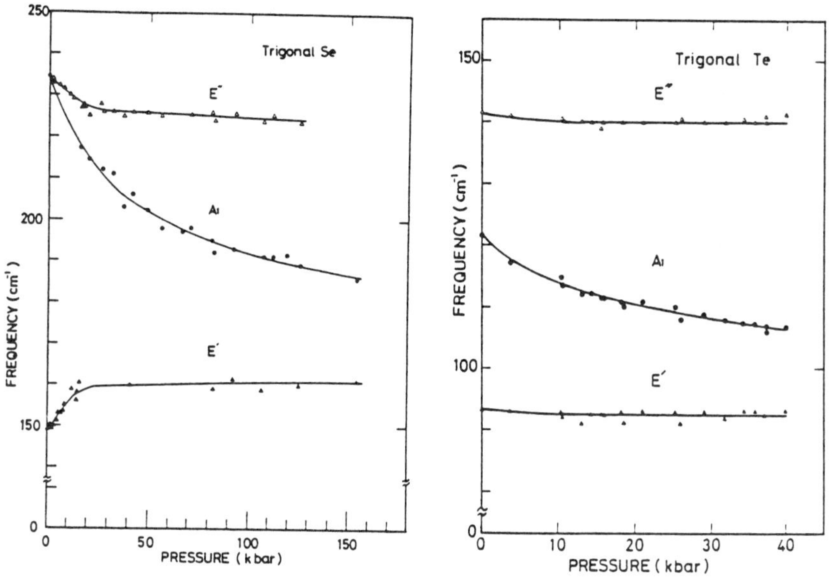

Fig. 8. Variation of Raman-active frequencies with pressure for trigonal Se and Te.

The electronic states of Se and Te have been studies by measurements of the XPS and UPS (Shevchik et al., 1973) and by calculations of the band structure and density of states (Joannopoulos et al., 1974, Doerre and Joannopoulos, 1979). In trigonal Se and Te, the uppermost valence band arises from the non-bonding orbitals. The upper and lower p-like valence bands contain states predominantly involved in interchain and intrachain bonding, respectively. It is predicted that the compression gives rise to an interference of interchain interactions with the intrachain bonding. The metallic character for the high-pressure monoclinic phase of Te is demonstrated by the unfilled p-like untibonding states.

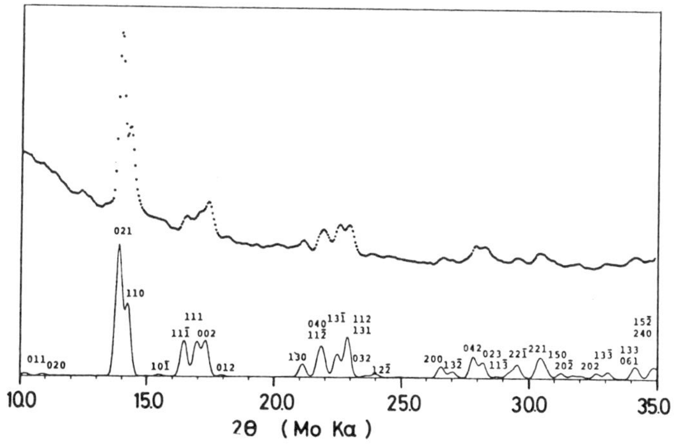

Fig. 9. X-ray diffraction patterns observed and calculated for monoclinic Te at 45 kbar.

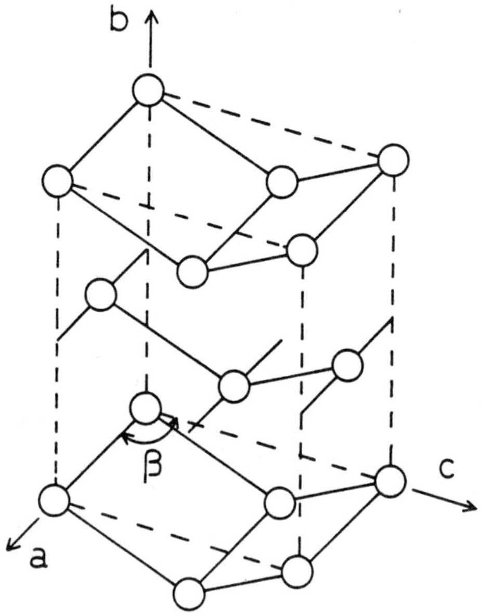

Fig. 10. Puckered layer structure of monoclinic Te.

TETRAHEDRALLY BONDED SEMICONDUCTORS

Our knowledge of tetrahedrally bonded crystalline and amorphous semiconductors has been advanced by several existing developments during the past three decades. In particular, the high-pressure studies have served for understanding the fundamental properties of these materials as a function of volume or interatomic distance. The high-pressure optical studies provided an empirical rule governing the shift of band energy extreme (Paul and Warschauer, 1963; Drickamer, 1965). The high-pressure electrical measurements provided a common phenomena of the insulator-metal transitions (Drickamer, 1965). The theoretical studies predicted the electronic band structure and insulator-metal transition pressures (Van Vechten, 1973; Yin and Cohen 1980).

The earliest studies on the insulator-metal transitions in tetrahedrally bonded crystalline semiconductors were measurements of the optical absorption edge (Slykhouse and Drickamer, 1958; Edwards et al., 1959; Edwards and Drickamer, 1961), and the electrical resistance (Minomura and Drickamer, 1962; Samara and Drickamer, 1962). The insulator-metal transitions are associated with the discontinuous changes in optical energy gap and activation energy. The high-pressure x-ray studies by Jamieson (1963) has shown that the high-pressure phases are identified as a β-Sn or rocksalt structure. Recently the transition pressure in tetrahedrally bonded semiconductors have been predicted reasonably by quantum dielectric theory (Van Vechten,1973) and pseudopotential calculations (Morita et al., 1972; Yin and Cohen, 1980).

The insulator-metal transitions in tetrahedrally bonded amorphous semiconductors have been also studied by measurements of the optical absorption edge, electrical resistance, superconductivity, and x-ray diffraction (Shimomura et al., 1974, 1976; Asaumi et al., 1976; Minomura et al., 1977, 1980; Minomura, 1978, 1981, 1984). The variation of optical energy gap with pressure for glow-dischange (GD) and reactively sputtered

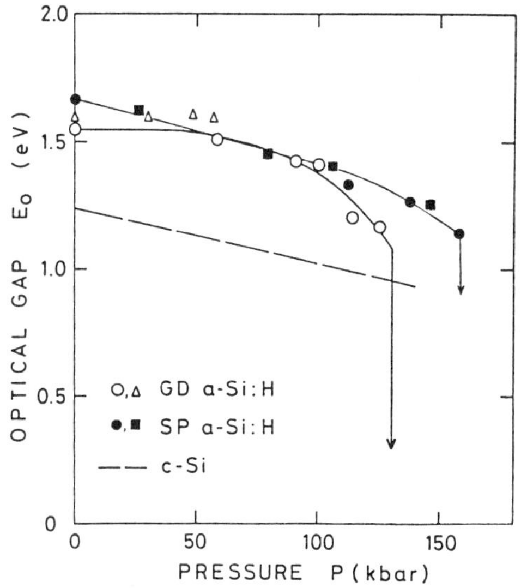

Fig. 11. Variation of optical energy gap with pressure for GD and SP a-Si:H, and c-Si.

(SP) a-Si:H, and c-Si is shown in Fig. 11. With increasing pressure the optical gap decreases nonlinearly, and drops to zero at 130∿150 kbar. The variation of electrical resistance with pressure is shown in Fig. 12. The insulator-metal transitions accompanied by a discontinuous drop in resistance occur in evaporated a-Si at 100 kbar, GD a-Si:H at 130 kbar, and c-Si at 150 kbar. On the other hand, the SP a-Si:H becomes metallic with a continuous decrease in resistance at about 150 kbar. As in Fig. 13, the high-pressure phases become superconducting with the transition temperature of 6.8∿7.4 K. The variation of x-ray diffraction intensity

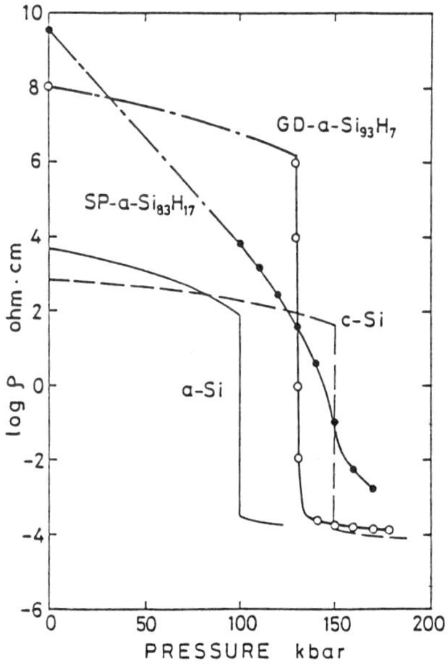

Fig. 12. Variation of resistivity with pressure for GD and SP a-Si:H, a-Si, and c-Si.

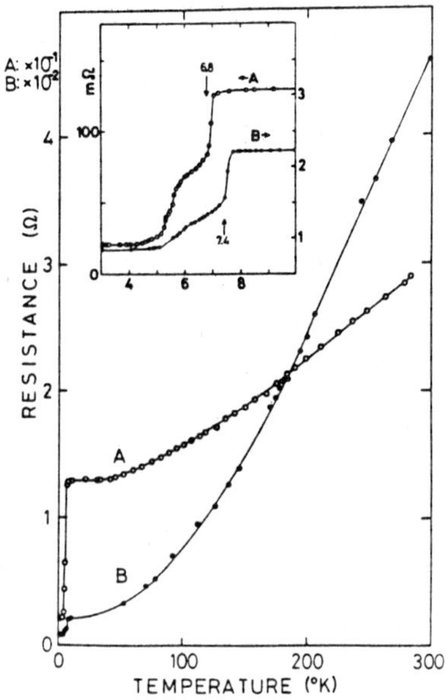

Fig. 13. Variation of resistance with temperature for the high-pressure phases of a-Si:H and c-Si at 170 kbar.

profiles with pressure for evaporated a-Si is shown in Fig. 14. Above 100 kbar the diffraction profiles reveal the growth and development of new peaks over the amorphous background. The first and second peaks at 120 kbar correspond to the (101) and (211) reflections of β-Sn structure, and are reversible to the amorphous state after releasing pressure. The high pressure metallic states are characteristic of a heterogeneous system which contains the tetrahedrally bonded and highly coordinated metastable phases.

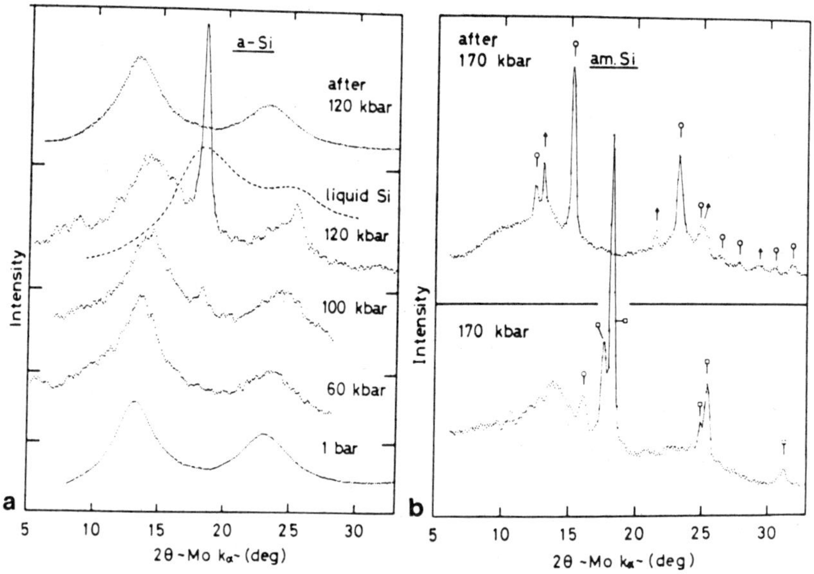

Fig. 14. Variation of x-ray diffraction patterns with pressure for a-Si, (a) 120 kbar and below; (b) 170 kbar and above. □, β-Sn; O, BC-8; ▲, diamond.

Evaporated or sputtered a-InSb under pressure shows the transition to metallic state at about 13 kbar. The high-pressure x-ray studies have shown successive structural transitions to a rocksalt phase at about 13 kbar and a β-Sn phase at about 29 kbar. After releasing pressure, the rocksalt phase can be retained at atmospheric pressure. On the other hand, c-InSb shows the transition to a β-Sn structure at 29 kbar which is reversible at room temperature.

It is surprising that the rocksalt phase of InSb can be deposited onto glass substrate by tetrode sputtering with anode voltage of 60 V in argon pressure of 10^{-4} Torr. It transforms to the zinc-blende phase at 488 K. The valence-band x-ray photoemission spectra observed for the rocksalt and zinc blende phases of InSb are shown in Fig. 15. These spectra are well understood on the basis of the density of states calculated by pseudopotential method (Mele and Joannopoulos, 1981; Kobayashi et al., 1983). The valence band spectrum of the rocksalt phase is interpreted as indicating a mixture of covalent and ionic as well as metallic character. The density of states at Fermi energy arises predominantly from the s-like antibonding state.

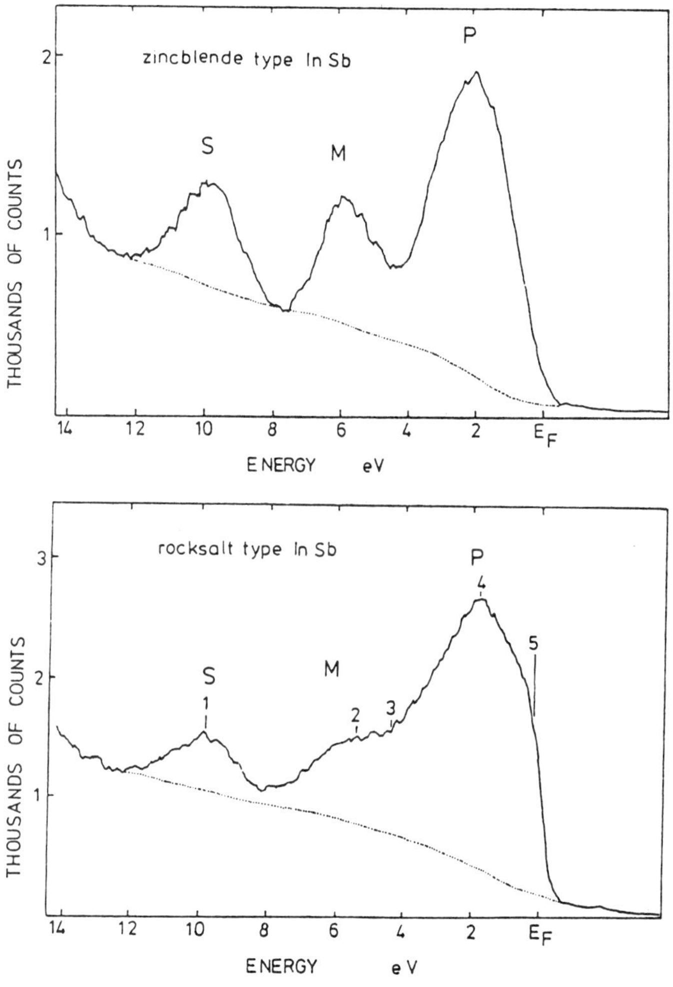

Fig. 15. Valence-band x-ray photoemission spectra for the zinc-blende and rocksalt phases of InSb.

CONCLUSION REMARKS

It has been demonstrated that molecular and covalent solids at high pressure become a metal with the overlap between the conduction and valence bands. In particular, it has been proved that diatomic molecular iodine shows the metallization with a continuous decrease in energy gap to zero at 180 kbar, and the dissociation into a monatomic phase at 210 kbar. The iodine studies have served as a prototype for theoretical analysis of the insulator-metal transitions in molecular solids, in particular hydrogen. In trigonal Se and Te at high pressure, the insulator-metal transition is accompanied by a change of structure to puckered layers. The earliest studies on the insulator-metal transitions in diamond and zinc-blende crystals have shown that the high-pressure phases are identified as a β-Sn or rocksalt strucutre. Recently the transition pressures have been predicted reasonably by pseudopotential calculations for the structural energies of various phases as a function of atomic volume. Amorphous solids are known to be metastable phases which contain considerable strain energies in terms of the variation of local and medium range order. As a consequence, the insulator-metal transition, accompanied by structural changes to metastable phases, occurs at lower pressure than those of the crystals.

REFERENCES

Aoki, K., Shimomura, O., and Minomura, S., 1980, Crystal structure of high-pressure phase of tellurium, J. Phys. Soc. Japan, 48:551.

Aoki, K., Shimomura, O., Minomura, S., Koshizuka, N., and Tsushima, T., 1980, Raman scattering of trigonal Se and Te at high pressure, J. Phys. Soc. Japan, 48:906.

Asaumi, K., Shimomura, O., and Minomura, S., 1976, Pressure-induced structural transformations of amorphous InSb, J. Phys. Soc. Japan, 41:1630.

Balchan, A. S., and Drickamer, H. G., 1961, Effect of pressure on the resistance of iodine and selenium, J. Chem. Phys., 34:1948.

Bridgman, P. W., 1952, The resistance of 72 elements, alloys and compounds to 100,000 KG/CM^2, Proc. Am. Acad. Atrs Sci., 81:165.

Chakravarty, S., Rose, J. H., Wood, D., and Ashcroft, N. W., 1981, Theory of dense hydrogen, Phys. Rev. B, 24:1624.

Doerre, G., and Joannopoulos, J. D., 1979, Electronic states of Te above the high-pressure phase transition, Phys. Rev. Lett., 43:1040.

Drickamer, H. G., 1965, The effects of high pressure on the electronic structure of solids, in:"Solid State Physics," F. Seitz and D. Turnbull, eds., Academic Press, New York, p.1.

Edwards, A. L., and Drickamer, H. G., 1961, Effect of pressure on the absorption edges of some III-V, II-VI, and I-VII compounds, Phys. Rev., 122:1149.

Edwards, A. L., Slykhouse, T. E., and Drickamer, H. G., 1959, The effect of pressure on zinc blende and wurtzite structures, J. Phys, Chem. Solids, 11:140.

Friedli, C., and Ashcroft, N. W., 1977, Combined representation method for use in band-structure calculations: Application to highly compressed hydrogen, Phys. Rev. B, 16:662.

Jamieson, J. C., and McWhan, D. R., 1965, Crystal structure of tellurium at high pressure, J. Chem, Phys., 43:1149.

Jayaraman, A., Dernier, P. D., and Longinotti, L. D., 1975, Study of the valence transition in SmS induced by alloying, temperature, and pressure, Phys. Rev. B, 11:2783.

Jayaraman, A., Mcwhan, D. B., Remeika, J. P., and Dernier, D. D., 1970, Critical behavior of the Mott transition in Cr-doped V_2O_3, Phys. Rev. B, 2:3751.

Joannopoulos, J. D., Schlüter, M., and Cohen, M. L., 1975, Electronic structure of trigonal and amorphous Se and Te, Phys. Rev. B, 11:2186.
Kobayashi, T., Shindo, K., and Nara, H., 1982, Charge density of metallic InSb with NaCl-type structure, Proceeding of Sagamore Conference on Charge-Spin and Momentum Density, p.21.
Lynch, R. W., and Drickamer, H. G., 1966, Effect of pressure on the lattice parameters of iodine, stannic iodine, and ρ-Di-iodobenzene, J. Chem. Phys., 45:1020.
McCann, D. R., and Cartz, L., 1972, High-pressure phase transformations in hexagonal and amorphous selenium, J. Chem. Phys., 56:2552.
McMahan, A. K., Hord, B. L., and Ross, M., 1977, Experimental and theoretical study of metallic iodine, Phys. Rev. B, 15:726.
McWhan, D. B., and Remeika, J. P., 1970, Metal-insulator transition in $(V_{1-x}Cr_x)_2O_3$, Phys. Rev. B, 2:3734.
Mele, E. J., and Joannopoulos, J. D., 1981, Electronic states of the zinc-blende and rocksalt phase of InSb, Phys. Rev. B, 24:3145.
Minomura, S., 1978, Pressure-induced convalent-metallic transitions, in: "High Pressure and Low Temperature Physics," C. W. Chu and J. A. Woolam, eds., Plenum Publishing, New York, p.483.
Minomura, S., 1981, Pressure-induced transitions in amorphous silicon and germanium, J. de Physiq., 42:C4-181.
Minomura, S., 1984, Pressure effects on the local atomic structure, in: "Semiconductor and Semimetals," Academic Press, New York, 21A:273.
Minomura, S., 1984, Pressure-induced phase transitions in tetrahedrally bonded semiconductors, Mat. Res. Soc. Sym. Proc., 22:277.
Minomura, S., Aoki, K., Shimomura, O., and Tanaka, K., 1976, Electronic processes in amorphous Se and some lone pair semiconductors at high pressure, in:" Electronic Phenomena in Non-Crystalline Semiconductors," B. T. Kolomiets, ed., Academy of Sciences of USSR, Leningrad, p.289.
Minomura, S., and Drickamer, H. G., 1962, Pressure-induced phase transitions in silicon, germanium and some III-V compounds, J. Phys. Chem. Solids, 23:451.
Minomura, S., Shimomura, O., Asaumi, K., Oyanagi, H., and Takemura, K., 1977, High-pressure modifications of amorphous Si, Ge and some III-V compounds, in:"Amorphous and Liquid Semiconductors," W. E. Spear, ed., University of Edinburgh, Edinburgh, p.53.
Minomura, S., Tsuji, K., Oyanagi, H., and Fujii, Y., 1980, Effect of hydrogen on the structure and pressure-induced transition of amorphous silicon-hydrogen alloys, J. Non-Cryst. Solids, 35&36:513.
Morita, A., Soma, T., and Takeda, T., 1972, Perturbation theory of covalent crystals. I. Calculation of cohesive energy and compressibility, J. Phys. Soc. Japan, 32:29.
Mott. N. F., 1949, The basis of the electron theory of metals, with special references to the transition metals, Proc. Phys. Soc., 62:416.
Mott. N. F., 1968, Metal-insulator transition, Rev. Mod. Phys., 40:677.
Paul, W., and Warschauer, D. M., 1963, The role of pressure in semiconductor research, in:"Solid under Pressure," W. Paul and D. M. Warschauer, eds., McGraw-Hill, New York, p.179.
Riggleman, B. M., and Drickamer, H. G., 1962, Temperature coefficient of resistance of iodine and selenium at high pressure, J. Chem. Phys., 37:446.
Riggleman, B. M., and Drickamer, H. G., 1963, Approach to the metallic state as obtained from optical and electrical measurements, J. Chem. Phys., 38:2721.
Ross, M., and McMahan, A. K., 1982, Systematics of the s→d and p→d electronic transition at high pressure for the elements I through La, Phys. Rev. B, 26:4088.
Samara, G. A., and Drickamer, H. G., 1962, Pressure induced phase transitions in some II-VI compounds, J. Phys. Chem. Solids, 23:457.

Shevchik, N. J., Cardona, M., and Tejeda, J., 1973, X-ray and far-uv photoemission from amorphous and crystalline films of Se and Te, Phys. Rev. B, 8:2833.

Shimomura, O., Asaumi, K., Sakai, N., and Minomura, S., 1976, Pressure-induced semiconductor-metal transitions in amorphous InSb, Phil. Mag., 34:839.

Shimomura, O., Minomura, S., Sakai, N., Asaumi, K., Tamura, K., Fukushima, J., and Endo, H., 1974, Pressure-induced semiconductor-metal transitions in amorphous Si and Ge, Phil. Mag., 29:547.

Shimomura, O., Takemura, K., Fujii, Y., Minomura, S., Mori, M., Noda, Y., and Yamada, Y., 1978, Structure analysis of high-pressure metallic state of iodine, Phys. Rev. B, 18:715.

Slykhouse, T. E., and Drickamer, H. G., 1958, The effect of pressure on the optical absorption edge of germanium and silicon, J. Phys. Chem. Solids, 7:210.

Suchan, H. L., Widerhorn, S., and Drickamer, H. G., 1959, Effect of pressure on the absorption edges of certain elements, J. Chem. Phys., 31:355.

Takemura, K., Fujii, Y., Minomura, S., and Shimomura, O., 1979, Pressure-induced structural phase transition of iodine, Solid State Commun., 30:137.

Takemura, K., Minomura, S., Shimomura, O., Fujii, Y., 1980, Observation of molecular dissociation of iodine at high pressure by x-ray diffraction, Phys. Rev. Lett., 45:1881.

Takemura, K., Minomura, S., Shimomura, O., Fujii, Y., and Axe, J. D., 1982, Structural aspects of solid iodine associated with metallization and molecular dissociation under high pressure, Phys. Rev. B, 26:998.

Van Straaten, J., Wijngaarden, R. J., and Silvera, I. F., 1982, Low-temperature equation of state of molecular hydrogen and deuterium to 0.37 Mbar: Implications for metallic hydrogen, Phys. Rev. Lett., 48:97.

Van Vechten, J. A., 1973, Quantum dielectric theory of electronegativity in covalent systems. III. Pressure-temperature phase diagrams, heat of mixing, and distribution coefficients, Phys. Rev. B, 7:1479.

Vereshchagin, L. F., Yakovlev, E. N., and Timofeev, Yu. A., 1975, Possibility of transition of hydrogen into the metallic state, Soviet Physics - JETP Lett., 21:85.

Wilson, A. H., 1931, The theory of electronic semi-conductors, Proc. Roy. Soc., 133:458.

Yin, M. T., and Cohen, M. L., 1980, Microscopic theory of the phase transformation and lattice dynamics of Si, Phys. Rev. Lett., 45:1004.

THE METAL-INSULATOR TRANSITION AND SUPERCONDUCTIVITY IN AMORPHOUS MOLYBDENUM-GERMANIUM ALLOYS

Shozo Yoshizumi

Dept. of Materials Science and Engineering
Stanford University
Stanford, CA 94305

David Mael and Theodore H. Geballe†

Department of Applied Physics
Stanford University
Stanford, CA 94305

Richard L. Greene

IBM Research Laboratory
San Jose, CA 95193

ABSTRACT

The metal-insulator (M-I) transition in amorphous molybdenum-germanium ($a\text{-}Mo_xGe_{1-x}$) alloys is found at $x \sim 0.10$. The electrical conductivity measurement has shown exponential behavior due to the variable-range hopping in the insulating and $\sqrt{T}$-dependence in the metallic regime at low temperature. Differential tunneling conductance has shown an anomalous behavior in the tunneling density of states near the Fermi level as a function of energy and composition. Low-temperature specific heat has been measured and indicates there is a finite density of states through the M-I transition. Localization, electron interactions and superconductivity near the M-I transition are discussed.

INTRODUCTION

The metal-insulator (M-I) transition has long been one of the most challenging problems in condensed matter science. Mott[1] first called attention to this problem in trying to understand the insulating nature of NiO in 1949. In 1958 Anderson[2] presented a new idea that the single-electron wave function becomes localized due to disorder and the system becomes insulating at T=0°K as the disorder is increased. Our interest

†also at Bell Communications Research, Murray Hill, New Jersey 07974

in studying strong localization and the M-I transition stems from the opportunity of employing a-Mo_xGe_{1-x} as a model system. A-Mo_xGe_{1-x}has been actively studied at Stanford for the past several years from a number of aspects such as superconductivity,[3-5] structure,[6] and weak localization in 2-dimensions.[7] Partial radial distribution functions show that Mo atoms are not incorporated substitutionally into the random tetrahedral network of Ge atoms in the Ge-rich phase.[6] The Ge random tetrahedral network and Mo-modified structure coexist near the M-I transition.[6]

We have undertaken a study of a-Mo_xGe_{1-x} as a function of concentration through the M-I transition. Using recently developed calorimetric capability we compare transport, tunneling and specific heat measurements made on the same films. Our objective is to evaluate important physical parameters from the experimental data and to compare them with recent theories of localization and electron interactions. Further, we can compare our results with other amorphous metal-semiconductor systems, a-Au-Si,[8] a-Au-Ge,[9,10] and a-Nb-Si,[11] which have been recently studied in the same regime.

EXPERIMENTAL

A-Mo_xGe_{1-x} films (~2500 Å) were deposited onto single crystal sapphire substrates at room temperature by magnetron cosputtering from pure Mo and pure Ge targets; their amorphous structure was confirmed by X-ray diffraction and TEM studies. The substrates were mounted on a rotating table (300 RPM) to achieve compositional homogeneity in the films. The concentration of the films was determined by electron microprobe analysis. The films were ion-etched to produce a bridge pattern for four-point measurement using standard photolithography. Resistivity of the films was measured by either DC or AC (100 hz) down to 30 mK using a dilution refrigerator.

Tunneling junctions were fabricated on the a-Mo_xGe_{1-x} films by depositing a 40 Å Al-barrier, subsequently oxidizing it in air and depositing a Pb counter-electrode. Differential tunneling conductance of the tunneling junctions was measured at 1.5°K using the four-probe AC bridge method in 1.7 kOe magnetic field in order to suppress the Pb superconductivity.

To measure the specific heat, thick a-Mo_xGe_{1-x} films (1.5-2.0 μm) were deposited in the rotating mode onto the backside of silicon-on-sapphire substrates, on the front side of which a bolometer was fabricated in the silicon film. The specific heat of the films was measured by the relaxation time constant method. The total absolute accuracy, as confirmed by runs on known standards, is 1% in the temperature region of our interest. The contribution of the sapphire, aluminum contact pads, and electrical lead wires to the toal heat capacity ranges from 75% of the total (at ~10°K) to 84% of the total (at ~2°K). Thus, we are limited to sample specific heat accuracy of 4-7%. The renormalized density of states $N^*(0)$ and the Debye temperature θ_D were calculated from the experimental data.

RESULTS AND DISCUSSION

Electrical Conductivity

The resistivity (log ρ) is plotted as a function of temperature (1/T) for samples beteeen 9.1 and 23.1 at% Mo in Fig. 1. The resistivities ρ

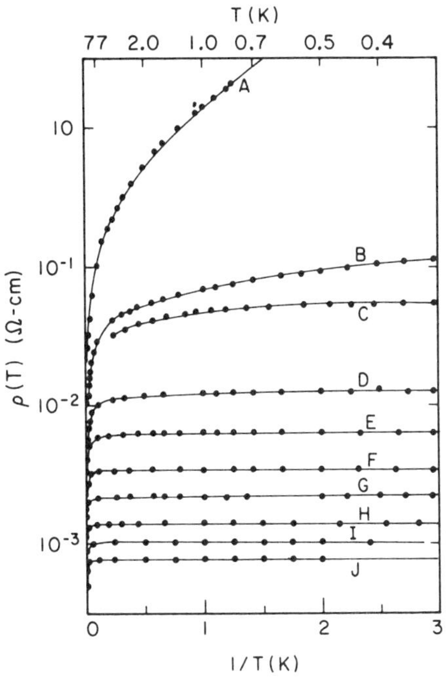

Fig. 1. Resistivity ρ vs. temperature 1/T for a-Mo_xGe_{1-x}. For curve A,x=0.091;B=0.103; C,x=0.106;P,x=0.116;E,x=0.123; F,x=0.135;G,x=0.150;H,x=0.168; I,x=0.191;J,x=0.231.

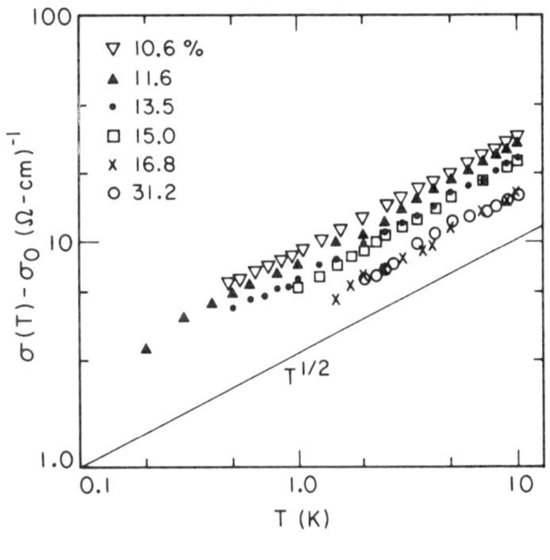

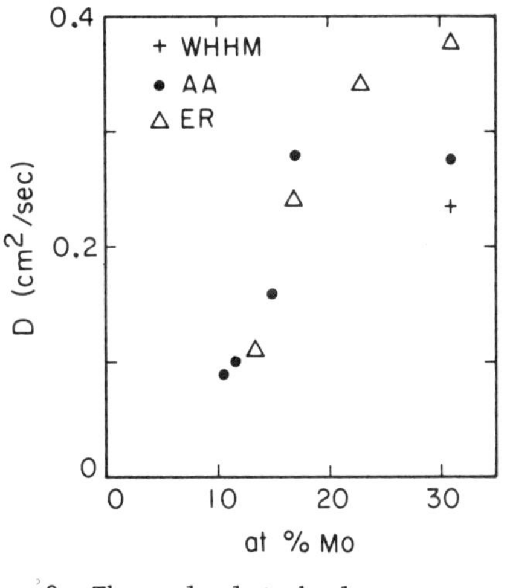

Fig. 2. The extrapolated conductivity at 0°K σ_0 is subtracted from the data σ(T) and plotted vs. temperature T below 10°K. The exponent of the power-low dependence is 0.50 ± 0.05.

Fig. 3. The calculated electron diffusion constant D vs. at% Mo. D is calculated from $(dH_{c2}/dT)_{Tc}$ and WHHM theory, the plot in Fig. 2 and the electron interaction theory by Altshuler and Aronov (AA), and Einstein relation (ER).

of the metallic samples saturate; on the other hand, ρ's of insulating samples increase exponentially at low temperatures.

The sample 9.1 at% Mo is insulating; its conductivity follows the variable-range hopping model of Mott[12] below 20°K

$$\sigma(T) = A \exp(-B/T^{1/4}), \quad (1)$$

where A and B are constants. The localization length for 9.1 at% Mo is 20 Å which is calculated using B and measured density of states of $\sim 1.0 \times 10^{22} [\text{eV-cm}^3]^{-1}$ from the specific heat experiment.

In the metallic samples, the temperature-dependent part of the conductivity $\Delta\sigma(T)$ in zero magnetic field can be expressed in the following form:

$$\Delta\sigma(T) = \sigma(T) - \sigma_0 = AT^n, \quad (2)$$

where σ_0 (zero temperature conductivity) is determined by a least-square extrapolation of $\sigma(T)$ vs. $\sqrt{T}$ down to T = 0°K. Above 30°K $\Delta\sigma(T)$ is nearly proportional to temperature ($n \simeq 1.0$), for some reason which is not entirely clear, but the possibility of electron-phonon interactions has been suggested.[9] Below 10°K, $\Delta\sigma(T)$ is proportional to $\sqrt{T}$ as shown in Fig. 2. However, the proportionality constant A increases somewhat as the M-I transition is approached in contrast with the finding that A is constant in a-Au-Ge[9] and a-Nb-Si.[11] Our accuracy in determining A is limited by the ability to measure the film thickness which introduces an estimated error of ± 5%, which is clearly smaller than the difference between 10.6 and 16.8 at% Mo.

There are three different experiments which, under differing assumptions, can be related to the electron diffusion constant D. According to the theory of Altshuler and Aronov[13] for interacting electrons with strong spin-orbit scattering,

$$\Delta\sigma(T) = 2.73 \sqrt{T/D} \ (\text{ohm-cm})^{-1}, \quad (3)$$

where T is in Kelvin and D is in cm^2/sec. D can be estimated using the experimental values of A for different concentrations. The increase in $\Delta\sigma(T)$ near the M-I transition requires the dependence of D shown in Fig. 3. D decreases almost a factor of three between 19 and 10.6 at% Mo and then is nearly constant between 19 and 31 at% Mo. Secondly, at 31 at% Mo D can be estimated from the measured superconducting critical field slope $(dH_{c2}/dT)_{T_c}$ using WHHM theory,[14,15]

$$D = 11.0/(dH_{c2}/dT)_{T_c}, \quad (4)$$

where $(dH_{c2}/dT)_{T_c}$ is in kOe/K. As can be seen in Fig. 3 there is fair agreement for the 31 at% Mo sample. Thirdly, using the Einstein relation,

$$\sigma_0 = e^2(dn/d\mu)D, \quad (5)$$

where in the most recent theories $dn/d\mu$ is the thermodynamic density of states,[16] D's are obtained from experimental values of σ_0 (from the transport) and $dn/d\mu$ (from the linear temperature coefficient of the specific heat) (Fig. 3). The values of D's decrease as the sample is closer to the M-I transition, which is in qualitative agreement with the result obtained from Equation (3). Since, as we discuss below, specific heat data indicate that $dn/d\mu$ is finite at the M-I transition, it follows that σ_0 goes to zero due to a vanishing D.

The 10.3 at% Mo sample is particularly interesting because it is very close to the critical concentration x_c, at which at T=0°K the M-I transition occurs. $\Delta\sigma(T)$ of the sample is proportional to $T^{1/3}$ (instead of $\sqrt{T}$) between 1 and 10°K, and below 1°K it follows Mott's variable-range hopping model,[12] $\sigma(T)=A\exp(-B/T^{1/4})$. This cross-over is consistent with the scaling fit which will be discussed below.

The scaling theory of the localization of noninteracting electrons[17] predicts that σ_0 near the M-I transition behaves as

$$\sigma_0 \alpha\ (x/x_c-1)^{\nu}, \tag{6}$$

where x is Mo concentration. The critical exponent ν is found to be ~1 with x_c=0.104 for a-Mo_xGe_{1-x} (Fig. 4). The value of unity for ν has been found in other amorphous systems.[8,11]

Altshuler and Aronov[13] examined the electron interactions with strong spin-orbit scattering where a term from the localization effects was included and derived $\nu = 1$. Our experimental result is consistent with their model since strong spin-orbit scattering is expected in the presence of Mo atoms.

Tunneling

The normal state differential tunneling conductance, dI/dV, which is proportional to single-particle tunneling density of states $N_1(E)$, of the tunneling junctions was measured as a function of DC bias V_b at 1.5°K under 1.7 kOe magnetic field (Fig. 5). The single-particle density of states has a minimum at E_F and can be expressed, following McMillan[18] as

$$N_1(E) = N_1(0)[1 + \sqrt{E/\Delta}] \quad (E<\Delta), \tag{7}$$

where E is measured from E_F and Δ is a correlation energy. Equation (7) is in agreement with the electron interaction model of Altshuler and Aronov.[19] It would be expected that the correlation gap begins to open near x_c. The results shown in Fig. 5 are in agreement with those in refs. 18 and 19. Similar electron interaction effects have already been observed by others in amorphous and disordered systems.[8,10,11,20] The correlation energies Δ are calculated from the plots shown in Fig. 2 and Fig. 5. As found in other amorphous and disordered metals,[11,20,21] Δ is roughly inversely proportional to ρ_0^2 (Fig. 6). $\Delta\ \alpha\ \rho_0^2$ implies that $\Delta\sigma(T)$ = const. $\sqrt{T}$, independent of the sample conductivity σ_0. This independence is not consistent with the results shown in Fig. 2, and hence is not consistent with the view that D varies according to Equation (3). There may be problems[16] in the scaling theory of McMillan[18] due to the distinction between the tunneling density of states and the thermodynamic density of states.

Specific Heat

The most striking result of the specific heat measurements is that the value for the coefficient of the electronic specific heat γ^* appears to increase monotonically with Mo concentration over the range of 0 to 80 at% (Fig. 7). The unrenormalized (bare) density of states can be written as

$$N_b(0) = \frac{N^*(0)}{1+\lambda}\ \alpha\ \frac{\gamma^*}{1+\lambda}, \tag{8}$$

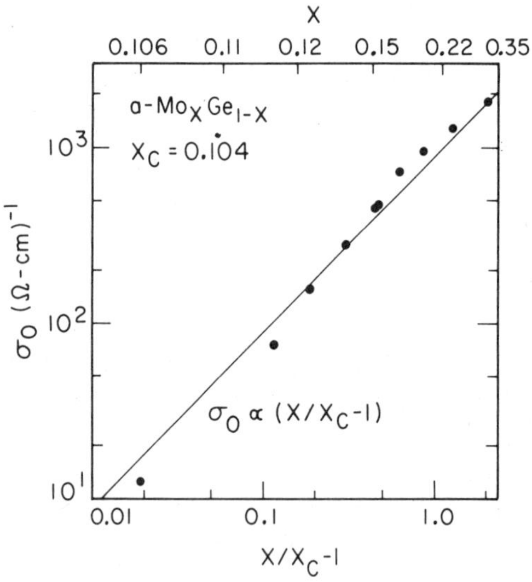

Fig. 4. The extrapolated conductivity at 0°K σ_0 vs (x/x_{c-1}), where x is the Mo concentration and the critical concentration x_c=0.104. The critical exponent is 1.0 ± 0.1.

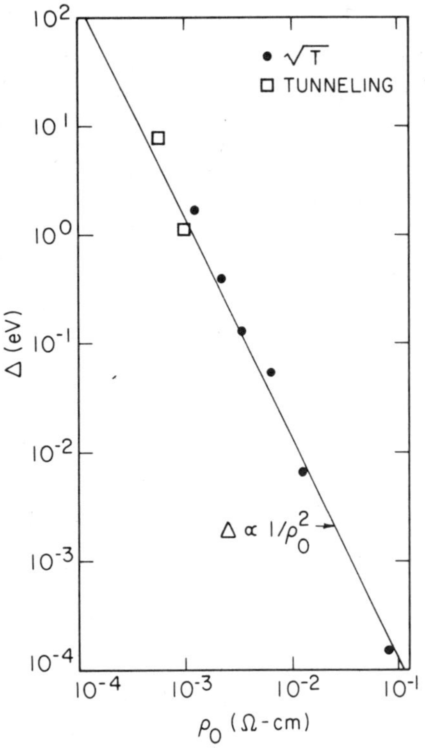

Fig. 6. The correlation energy Δ vs. the extrapolated resistivity at 0°K ρ_0; Δ from the slope of $\Delta\sigma(T)$ vs. $\sqrt{T}$ (dots) and Δ from tunneling (squares). Δ is roughly inversely proportional to ρ_0^2 .

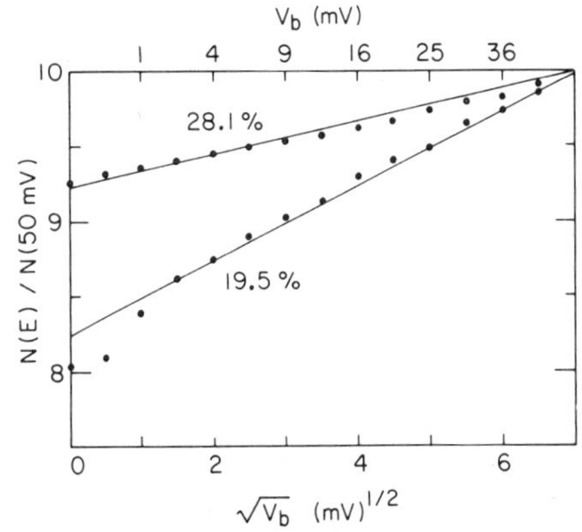

Fig. 5. The normalized tunneling density of states N(E)/N(50mV) vs. square root of the bias voltage $\sqrt{V_b}$ for 28.1 and 19.5 at% Mo metallic samples.

where $N^*(0)$ is the dressed (thermodynamic) density of states and λ is a renormalization constant (electron-phonon coupling constant in the simple case). Unless λ becomes infinite at the M-I transition, $N_b(0)$ is continuous and finite right through the M-I transition, in contrast to results derived from tunneling.[8,10,11,20] In fact, Fig. 7 implies that γ^*, and hence the thermodynamic density of states shows no critical regions, or discontinuities, as a function of Mo concentration. Unlike the tunneling density of states, the thermodynamic (specific heat) density of states measures the number of states available for thermal excitation at E_F. Within our experimental limits, this is not affected by the M-I transition.

Superconductivity

One of the important issues is how the superconducting transition temperature T_c in disordered systems changes near the M-I transition where strong localization ($k_F l \sim 1$) is expected. In the high Mo concentration, which is in the weakly localized regime, T_c decreases linearly with decreasing Mo concentration from 7.5°K (78 at% Mo) at a rate of ~0.18°K/at% Mo.[3] In this region our experiment[5] shows the ratio of electron-phonon coupling constant λ to the bare density of states $N_b(0)$ is constant, which is consistent with the Varma-Dynes tight-binding model.[22] An extrapolation of the linear behavior of T_c in this regime yields the disappearance of T_c near 35 at% Mo. However, our measurements show that T_c exists down to 13.5 at% Mo. Further, there appears to be a metallic phase between the superconducting phase and the insulating phase. T_c's and resistivity at 0°K, ρ_0, are plotted in Fig. 8(a). A rapid decrease in T_c is found with decreasing Mo concentration near 15 at% Mo, where a rapid increase in ρ_0 also occurs.

Fukuyama et al.[23] examined the critical temperature, T_c, and the upper critical field, H_{c2}, for three-dimensional superconductors in the weakly localized regime (WLR) ($k_F \ell \gg 1$). Their results indicate an appreciable suppression of T_c even in the WLR which is not affected by strong spin-orbit scattering. Our experimental data for high Mo concentration samples to which their theory might be relevant are not in agreement with their results unless an unphysically large effective mass m^* is assumed. Apparently localization alone cannot explain the degradation of T_c in a-Mo_xGe_{1-x} even in the WLR.

Kapitulnik and Kotliar[24] have considered the importance of fluctuations in reducing T_c in the strongly localized regime (SLR). They find a breakdown of the mean field theory and a large increase of the critical region (defined by the Ginzburg criterion). Since we have only been able to measure the specific heat in the SLR above T_c, the best analysis we can make is to assume the applicability of McMillan's equation[25] and to analyze for the electron-phonon coupling constant λ and the repulsive Coulomb interaction μ^* for three limiting cases. Using McMillan's equation in the form

$$T_c = \frac{\theta_D}{1.45} \exp\left[- \frac{1.04(1+\lambda)}{\lambda-\mu^*(1+0.62\lambda)} \right], \tag{9}$$

we take the measured T_c and use $N^*(0)$ and θ_D from the heat capacity measurements and the following assumptions:

<u>$\mu^* = 0.1$ = constant</u>. This approximation should be good in the high Mo concentration (weakly localized region). μ^*=0.1 is the value found by tunneling spectroscopy with nitrogen stabilized amorphous Mo.[26]

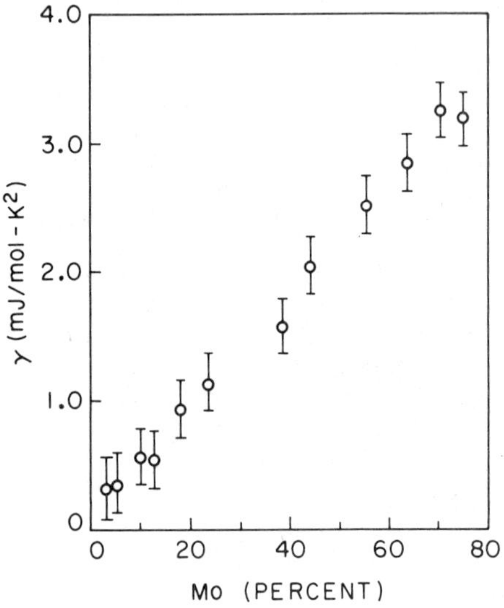

Fig. 7. The coefficient of the electronic specific heat γ^* is simple monotonic functions of electron density or disorder and finite at the M-I transition, which occurs at ~10% Mo.

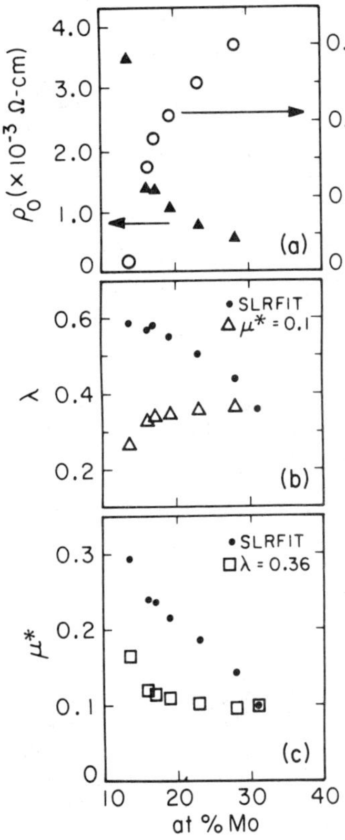

Fig. 8. (a) The extrapolated resistivity at 0°K ρ_o (triangle) and the superconducting transition temperature T_c (circle) vs. at% Mo.
(b) The electron-phonon coupling constant λ vs. at% Mo for the strongly localized regime fit (SLRFIT) by Anderson et al. (dot) and for the constant $\mu^*=0.1$ (triangle).
(c) The repulsive Coulomb interaction μ^* vs. at% Mo for the SLRFIT by Anderson et al. (dot) and for constant $\lambda=0.36$ (square).

Then λ decreases almost linearly with Mo concentration >31 at% where $\lambda=0.36$. Deviation from linearity occurs below 31 at% Mo, then λ decreases very slowly until a sharp drop occurs near 13.5 at% Mo (Fig. 8(b)).

$\lambda = 0.36 =$ constant. In the SLR it is no longer valid to assume μ^* is constant. In fact, it is probably better to consider λ to be constant and to assign the observed variation in T_c to μ^*. Below 31 at% Mo, μ^* then increases gradually. An increase of only 65% in μ^* is needed to suppress superconductivity at 13.5 at% Mo (Fig. 8(c)).

Enhanced μ^* Model. Undoubtedly both λ and μ^* increase in the SRL. Anderson et al.[27] proposed a theory of degradation of T_c in high T_c superconductors due to the increase in μ^* caused by localization effects. In their model, one of the important parameters is a critical resistivity ρ_c above which localization effects increase. We choose ρ_c is 5.50×10^{-4} ohm-cm at 31 at% Mo where the marked deviation from the Varma-Dynes model occurs. Using Equation (4) in ref. 27, μ^* is calculated and plotted in Fig. 8(c). λ is then estimated from the McMillan's equation (Fig. 8(b)).

SUMMARY

One of the advantages of using amorphous systems to study the M-I transition relies upon the fact that very homogeneous films can be produced over a wide range of compositions.[5]

We have studied the behavior of amorphous Mo_xGe_{1-x} alloys from the insulating to the weakly localized metallic state. In the insulating state, where determination of the partial radial distribution functions shows that the Mo atoms are not incorporated substitutionally into the random tetrahedral network of Ge atoms,[6] the transport is thermally activated in conformance with Mott's variable-range hopping model.[12] The M-I transition in a-Mo_xGe_{1-x} occurs at $x \sim 0.10$ where the Ge random tetrahedral network and the Mo-modified structure coexist.[6] The M-I transition is continuous and has the form of

$$\sigma_0 \propto (x/x_c - 1) \qquad (10)$$

with the critical exponent of unity, which is consistent with the theory of Altshuler and Aronov[13] for electron interactions with strong spin-orbit scattering. The thermodynamic density of states as measured by specific heat is continuous and finite through the M-I transition regime. From the Einstein relation the diffusion constant D is found to decrease rapidly and of course to vanish at the M-I transition. The superconductivity disappears for Mo concentrations 13% leaving a ~3% metallic non-superconducting region. The similar behavior was found in a-Au-Si[8] and a-Nb-Si,[11] where there is a range of composition which is metallic and nonsuperconducting. Some limits on the values of the Coulomb pseudopotential, μ^*, and the electron-phonon coupling constant, λ, have been set.

The detailed roles of both localization and electron interactions near the M-I transition need to be investigated further. It is hoped that our studies on the M-I transition in a-Mo_xGe_{1-x} will be generalizable to other systems and will stimulate new theoretical and experimental investigation.

ACKNOWLEDGEMENTS

The work at Stanford University was mainly supported by the Center for Materials Research under NSF-MRL Grant DMR 83-16982. The work on the specific heat was supported by the Air Force Office of Scientific Research Grant F49620-83-C-0014. We have benefited from interaction with H. Fukuyama (University of Tokyo), M. R. Beasley (Stanford), and R. C. Dynes (AT&T Bell Laboratories).

REFERENCES

1. N. F. Mott, Proc. Phys. Soc. A, 62:416 (1949).
2. P. W. Anderson, Phys. Rev., 109:1492 (1958).
3. W. L. Carter, S. J. Poon, G. W. Hull, Jr., and T. H. Geballe, Sol. St. Comm., 30:41 (1981); W. L. Carter, Ph.D. dissertation, unpublished, Stanford Univ. (1983).
4. S. Yoshizumi, W. L. Carter, and T. H. Geballe, J. Non-Crystal. Solids, 61&62:589 (1984).
5. D. Mael, W. L. Carter, S. Yoshizumi, and T. H. Geballe, submitted to Phys. Rev. B.
6. J. Kortright and A. Bienenstock, J. Non-Crystal. Solids, 61&62:273 (1984); J. Kortright, Ph.D. dissertation, unpublished, Stanford Univ. (1984).
7. J. M. Graybeal and M. R. Beasley, Phys. Rev. B, 29:4167 (1984); J. M. Graybeal, Ph.D. dissertation, unpublished, Stanford Univ. (1985).
8. N. Nishida, T. Furubayashi, M. Yamaguchi, K. Morigaki, and H. Ishimoto, Sol. St. Electronics, in press.
9. B. W. Dodson, W. L. McMillan, J. M. Mochel, and R. C. Dynes, Phys. Rev. Lett., 46:46 (1981).
10. W. L. McMillan and J. M. Mochel, Phys. Rev. Lett., 46:556 (1981).
11. G. Hertel, D. J. Bishop, E. G. Spencer, J. M. Mochel, and R. C. Dynes, Phys. Rev. Lett., 50:743 (1983).
12. N. F. Mott, Phil. Mag., 19:835 (1969).
13. B. L. Altshuler and A. G. Aronov, Sol. St. Comm., 46:429 (1983).
14. N. R. Werthamer, E. Helfand, and P. C. Hohenberg, Phys. Rev., 147:295 (1966).
15. K. Maki, Phys. Rev., 148:362 (1966).
16. P. A. Lee, Phys. Rev. B, 26:5882 (1982).
17. E. Abraham, P. W. Anderson, D. C. Liccicardello, and T. V. Ramakrishnan, Phys. Rev. Lett., 42:673 (1979).
18. W. L. McMillan, Phys. Rev. B, 24:2739 (1981).

19. B. L. Altshuler and A. G. Aronov, Sov. Phys. JETP, 50:968 (1979).

20. R. C. Dynes and J. P. Garno, Phys. Rev. Lett., 46:137 (1981).

21. R. W. Cochrane and J. O. Strom-Olsen, Phys. Rev. B, 29:1088 (1984).

22. C. M. Varma and R. C. Dynes, p. 507 in: "Superconductivity in d- and f-Band Metals," D. H. Douglass, ed., Plenum Press, New York (1968).

23. H. Fukuyama, H. Ebisawa, and S. Maekawa, J. Phys. Soc. Jpn., 53:3560 (1984).

24. A. Kapitulnik and G. Kotliar, Phys. Rev. Lett., 54:473 (1985).

25. W. L. McMillan, Phys. Rev., 167:331 (1968).

26. D. Kimhi and T. H. Geballe, Phys. Rev. Lett., 45:1039 (1980); D. Kimhi, Ph.D. dissertation, unpublished, Stanford Univ. (1980).

27. P. W. Anderson, K. A. Muttalib, and T. V. Ramakrishnan, Phys. Rev. B, 28:117 (1983).

ON THE NATURE OF THE METAL-INSULATOR TRANSITION IN METAL-RARE-GAS MIXTURE FILMS

Hans Micklitz

Inst.f.Experimentalphysik IV
Ruhr-Universität
4630 Bochum, W-Germany

The nature of the metal-insulator (MI) transition in frozen metal-rare-gas mixtures has been the subject of extensive research in recent years.[1-9] Two different types of MI transition can occur in such mixtures: (1) the classical percolation transition, typical for granular systems and (2) the Anderson or Anderson-Mott transition due to localization in disordered systems. One can distinguish between these two types of MI transition by measuring the electrical dc conductivity σ as a function of metal atomic concentration x. For both transitions the conductivity at T=0 K starts at the critical metal atomic concentration x_c and develops continuosly according to a power-law $\sigma \propto (x-x_c)^{\nu}$. The characteristic exponent ν as predicted by theory, however, is different for these two types of MI transition. For the classical percolation transition $\nu = 1.7 - 2.0$,[10,11] while a value between 1.0 and 0.5 is predicted for an Anderson or Anderson-Mott transition.[12-15]

Almost all metal-rare-gas mixtures studied so far show a MI transition of type (1), i.e. have a conductivity exponent ν as predicted by classical percolation theory for granular systems. A typical example is the system Sn-Ar.[7] Fig.1 shows the σ-data measured at $T \simeq 5$ K as a function of Sn atomic concentration x. The data points below $x_c = 0.32$ show variable range hopping ($\ln\sigma \propto T^{-1/4}$).[16] The conductivity exponent ν is 1.6 ± 0.1. The critical concentration x_c=0.32 corresponds to a Sn volume fraction of 0.26, slightly higher than that predicted by percolation theory (x_p=0.18).[17] In addition to

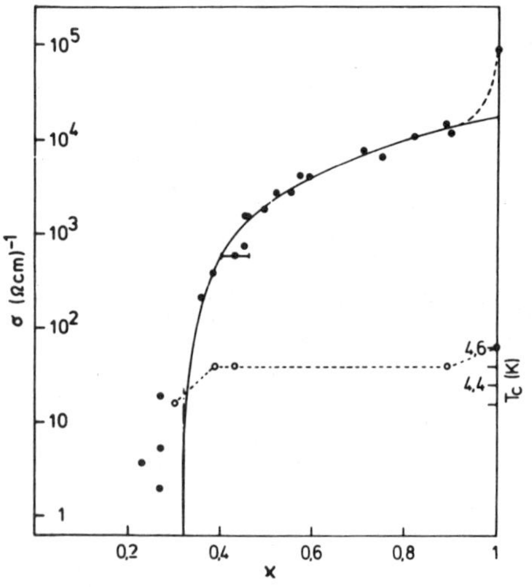

Fig.1 Conductivity σ (solid symbols) and transition temperature T_c (open symbols) in Sn-Ar mixtures as a function of Sn atomic concentration x. The solid line is a fit $\sigma \propto (x-x_c)^{\nu}$ with $x_c=0.32$ and $\nu=1.6$. The other lines are only guides to the eye.

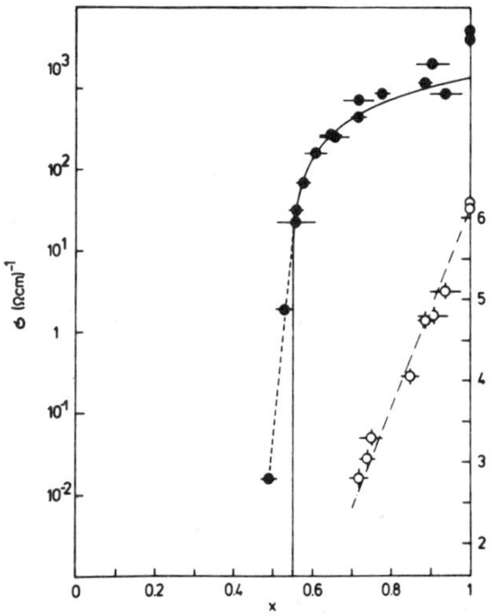

Fig.2 Conductivity σ (solid symbols) and transition temperature T_c (open symbols) in Bi-Kr mixtures as a function of Bi atomic concentration x. The solid line is a fit $\sigma \propto (x-x_c)^{\nu}$ with $x_c=0.55$ and $\nu=1.07$. The other lines are only guides to the eye.

the superconducting transition temperature T_c was measured in this system as a function of x (see open symbols in fig.1). T_c stays almost constant in the entire metallic region and just drops close to x_c. Such a behavior is typical for a percolating system since percolation as a classical phenomenon does not by itself affect T_c.

Very recently a metal-rare-gas mixture has been found which for the first time shows unambiguosly that the MI transition in this system belongs to type (2), i.e. is caused by localization. This system is Bi-Kr.[18] Fig.2 shows σ and the superconducting transition temperature T_c as a function of Bi atomic concentration x. The solid line through the data points is a least-squares fit $\sigma \propto (x-x_c)^{\nu}$ with $x_c=0.55$ and $\nu=1.07 \pm 0.1$. Both the characteristic exponent ν and the critical concentration x_c (corresponding to a Bi critical volume fraction $v_c=0.42$) speak against a percolation transition. The exponent ν, however, is in good agreement with the critical exponent obtained from a pure localization theory,[12] a scaling approach by McMillan for an Anderson-Mott transition,[13] or the value given by Oppermann for his time reversal invariant model.[14] Similar exponents have been found in amorphous metal-semiconductor mixtures.[19-21] The superconducting transition temperature T_c (open symbols in fig.2) shows a strong decrease with decreasing x. This is in contrast to the $T_c(x)$-behavior of the Sn-Ar system as discussed above (see also fig.1). It is an additional indication that this system is not a percolating one but a system where localization is important. The observed T_c decrease can be interpreted as a decrease in the density of states $N(E_F)$ at the Fermi energy. The detailed analysis yields $N(E_F) \propto (x-x_c)^s$, with $s=0.65 \pm 0.2$ or $\eta=3-s/\nu$ $=2.4 \pm 0.3$.[22,23] A similar value for η was found from tunneling experiments in amorphous Nb-Si mixtures.[21]

The question arises now if one can understand this completely different behavior of Bi-Kr compared to that of Sn-Ar. We think that the answer lies in the different length scale for disorder in these two systems. Quench-condensed Bi films are known to be amorphous,[24] the same will be true for Bi-Kr mixtures. Thus the characteristic length scale for disorder in this system is the interatomic distance. Sn-Ar mixtures, on the other hand, probably consist of a random network of Sn clusters embedded in an insulating Ar matrix. The characteristic length scale for this system is that of the characteristic Sn cluster diameter and will be much larger than that in Bi-Kr mixtures. In the following we want to give a short sketch how the length scale for disorder enters in the description of the MI transition and in the crossover from classical percolation to localization. More details about this approach can be found in Ref.25. The basic concept follows the idea of Khmelnitskii[26] who described localization in a percolating structure by introducing a dimensionless parameter $y=\rho_0 e^2/L\hbar$. L is the characteristic minimum length scale over which a resistivity ρ_0 can be defined, i.e. L is the characteristic length of the building blocks constituting the metal-insulator mixture. Fig.3 shows the schematic phase

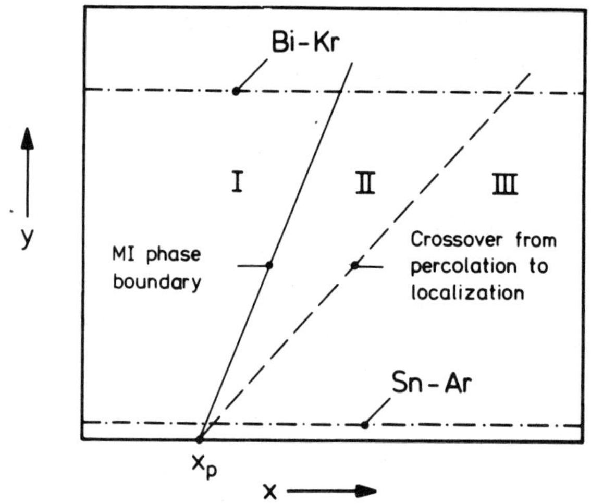

Fig.3 Schematic phase diagram of the metal-insulator transition in a system where there is a percolation-localization crossover. x_p is the critical concentration for the percolation threshold. The metal-insulator phase boundary is given by the slanted solid line. The slanted dashed line is a measure of the percolation-localization crossover in the metallic phase. The horizontal dashed-dotted lines indicate the y values for Sn-Ar and Bi-Kr mixtures respectively. Phase I: insulating phase. Phase II: metallic phase in which localization is dominant. Phase III: metallic phase in which classical percolation theory can be applied.

diagram of the MI transition in metal-insulator mixtures as a function of this parameter y and the metal atomic concentration x. The critical atomic concentration for the classical percolation transition is given by x_p. The MI transition occurs at x_p only if $y \rightarrow 0$, i.e. only in granular systems with an essentially infinite cluster seize L. For systems with a finite length scale for disorder, i.e. systems with $y > 0$, localization always sets in before the classical percolation threshold x_p is reached. The slanted solid line in fig.3, separating phases II and I, is the MI phase boundary. The MI transition is always due to localization which means that it is always an Anderson or Anderson-Mott transition. It takes place when the resistance of a cube of seize ξ_p is of the order $\hbar/e^2$. ξ_p is the so-called percolation correlation length. The slanted dashed line in fig.3, separating phases III and II, indicates crossover from percolation to localization in the metallic phase. We made the following "Ansatz" for the criterion of this crossover: it occurs when the localization correlation length becomes larger than the percolation correlation length at the localization transition. For systems with large y, i.e. large ρ_o/L, this crossover takes place for x values far above x_p. The system Bi-Kr, for example, has $y \approx 1$ (indicated by the upper horizontal dashed-dotted line in fig.3), i.e. the crossover occurs already at $x_c \simeq 1$. In other words, localization will be dominant for all metal atomic concentrations on the metallic side of the MI transition in this system. Granular systems with a cluster seize of the order $L \simeq 100$ Å and $\rho_o \simeq 10^{-5}$ $(\Omega \cdot cm)$, on the other hand, have an y value of the order 10^{-3} (indicated by the lower horizontal dashed-dotted line in fig.3). Crossover from percolation to localization thus occurs that close to x_p that from the experimental point of view one observes only percolation behavior. Sn-Ar is an example for such a system. For metal-rare-gas mixtures with a somewhat smaller metal cluster seize the crossover from percolation to localization should be observable. This seems to be the case in the Hg-Xe system.[27]

The above described approach to the problem of percolation-localization crossover provides an unified view of apparently disparate results on different types of metal-rare-gas mixtures. A more quantitative test of this approach may be possible by studies of granular systems of well defined metal clusters embedded in an insulating matrix as a function of cluster seize. Such studies are under way.

It is a great honor for us to dedicate this paper to Professor Sir Nevil Mott on the occasion of his 80th birthday. His continuos interest in the metal-insulator transition of metal-rare-gas mixtures has always been a strong motivation for our work. All the experiments reported here have been performed by R.Ludwig as part of his doctoral thesis. Many stimulating discussions with friends and colleagues are greatfully acknowledged.

REFERENCES

1. Z.Shanfield, P.A.Montano, and P.H.Barrett, Phys.Rev. Lett.35:1789(1975)
2. D.J.Phelps, and C.P.Flynn, Phys.Rev.B14:5279(1976)
O.3.Cheshnovski,U.Even, and J.Jortner, Solid State Commun.22:745(1977)
4. R.Ryberg, and O.Hunderi, J.Phys.C10:3559(1977)
5. K.Epstein,E.D.Dahlberg, and A.M.Goldman, Phys.Rev. Lett.43:1889(1979)
6. R.Römer,F.Siebers, and H.Micklitz, Solid State Commun.36:881(1980)
7. R.Ludwig,F.S.Razavi, and H.Micklitz, Solid State Commun.39:363(1981)
8. R.Ludwig, T.Paul, and H.Micklitz, Z.Phys.B47:31(1982)
9. A.I.Eatah, and A.A.Ghani Awad, J.Chem.Phys.79:1552 (1983); A.I.Eatah,N.E.Cusack, and J.G.Wright,Phys. Lett.51a: 1889(1979)
10. D.Stauffer, Physica (Utrecht)106A:177(1981)
11. "Percolation Structure and Processes",Annal of the Israel Physical Society Vol.5, G.Deutscher, R.Fallen, and J.Adler, ed., Adam Hilger, Bristol and the Israel Physical Society, Jerusalem (1983)
12. E.Abrahams,P.W.Anderson,D.C.Licciardello, and T.V. Ramakrishnan, Phys.Rev.Lett.42:673(1979); Y.Imry, Phys.Rev.Lett.44:469(1980); F.Wegner, Z.Phys.B25: 327(1976); D.Vollhardt, and P.Wölfle,

Phys.Rev.Lett.45:842(1980); A.McKinnon, and B.Kramer, Phys.Rev.Lett.47:1546(1981); D.Belitz, A.Gold, and W.Götze, Z.Phys.B44:273(1981)
13. W.L.McMillan, Phys.Rev.B24:2739(1981) 14. G.S.Grest, and P.A.Lee, Phys.Rev.Lett.50:693(1983)
15. R.Oppermann, Z.Phys.B49:273(1983)
16. N.F.Mott, "Metal-Insulator Transition", Taylor and Francis, London (1974)
17. S.Kirkpatrick, Rev.Mod.Phys.45:574(1973)
18. R.Ludwig, and H.Micklitz, Solid State Commun.50: 861(1984)
19. B.W.Dodson, W.L.McMillan, and J.M.Mochel, Phys. Rev.Lett.46:46(1981); W.L.McMillan, and J.Mochel, Phys.Rev.Lett.46:556(1981)
20. N.Nishida, M.Yamaguchi, T.Furubayashi, K.Morigaki, H.Ishimoto, and K.Ono, Solid State Commun.44:305 (1982)
21. G.Hertel,D.J.Bishop,E.G.Spencer,J.M.Rowell, and R.C.Dynes, Phys.Rev.Lett.50:743(1983)
22. R.Ludwig, Doctoral Thesis, Ruhr-Universität Bochum, 1984
23. R.Ludwig, and H.Micklitz, in "Proceedings of the Internatinal Conference on Localization, Interaction, and Transport Phenomena in Impure Metals", L.Schweitzer and B.Kramer, ed., PTB Braunschweig (1984) p.283
24. W.Buckel, Z.Phys.138:136(1954)
25. G.Deutscher,A.M.Goldman, and H.Micklitz, Phys.Rev.B (in print)
26. D.E.Khmelnitskii, Pis'ma Zh.Teor.Fiz 32:248(1980) (JETP Letters 32:229(1980))
27. K.Epstein,A.M.Goldman, and A.M.Kadin, Phys.Rev.B27: 6685(1983)

ELECTRICAL CONDUCTIVITY OF DISCONTINUOUS METAL FILMS

C.J. Adkins

Cavendish Laboratory
Madingley Road
Cambridge CB3 OHE, U.K.

Abstract

We describe investigations into the process of transport at low temperatures in discontinuous metal films. A theoretical analysis using effective medium theory shows that observed temperature dependencies cannot be explained by the conventional model with realistic distributions of relevant parameters. Nor can they be explained by inhomogeneity as is shown by a series of experiments in which the scale of inhomogeneity is measured and shown to be generally much too small. Field-effect measurements show a very small symmetric effect at low temperatures that can only be explained if metal grains are subject to large random potentials. The presence of these potentials is shown directly in another experiment designed to measure the energetics of grain charging. We argue that new models for transport have to be developed which incorporate random potentials as an essential feature.

INTRODUCTION

Discontinuous metals consist of inhomogeneous mixtures of metal and non-metal. Three-dimensional composites are generally produced by co-evaporation or co-sputtering of a metal and an insulator. They are known as *cermets*. Two-dimensional *discontinuous metal films* are produced during early stages of film growth by evaporation or sputtering: the deposited metal first forms isolated islands which only later join up to form a continuous film. In this case, the intervening insulator consists partly of the substrate and partly of the space between islands above the substrate which may be left free or filled with an insulating overlayer.

The electrical properties of such systems vary continuously as the composition is changed. When the concentration of metal is small, the metal forms small isolated islands embedded in an insulating matrix, and the electrical conductivity is small and highly activated. As the proportion of metal is increased, the islands grow and coalesce, the activation energy falls, and eventually continuous metallic paths extending through the material are established. At this stage the system undergoes a *metal - non-metal transition* to the metallic state. For higher metal concentrations the structure becomes that of isolated insulating inclusions in a metallic matrix, the conductivity remains metallic but continues to improve as the proportion of insulator is reduced to zero.

The basic physics of conduction in the regime of activated conductivity has been believed to be that of the model of Neugebauer and Webb (1962) in which there are two essential components: firstly, charge transfer between islands is by tunnelling; secondly, activation arises from the need to supply non-negligible electrostatic energy to place a single electronic charge on an island. (Here we are not concerned with *very* small-grained systems in which the spacing of quantum levels within the grains becomes important.) Both these aspects of the conduction process are associated with features of the real physical systems that may in principle be observed directly by electron microscopy. Tunnelling rates will depend on the distance between grains, and the energy required to produce a 'carrier' by charging a grain will depend on the capacitance of the grain to its surroundings which in turn depends on the grain size and its environment. Discontinuous metals therefore appear, *prima facie*, ideal as model systems for exploring the electronic structure of disordered systems and for testing the ideas developed by Sir Nevill Mott and others.

In this paper, I review some of the work of the Cambridge group. A particular feature of their experimental programme, which has concentrated on discontinuous metal films, has been exploitation of electrostatic charging so as to obtain independent control of carrier concentration. The technique has been most successful in providing new information about the physical nature of these systems.

THE TRADITIONAL MODEL APPLIED TO A SYSTEM WITH DISTRIBUTED PARAMETERS

The traditional model of Neugebauer and Webb applied to an idealized system consisting of a regular array of identical islands predicts a conductivity showing simple activation. At low temperature, the proportion of charged islands will be given by a Boltzmann factor and the mobility of the carriers will be proportional to the tunnelling rate, so that the conductivity σ will be of the form

$$\sigma \;\propto\; \exp[-(2\alpha R + E_C/kT)] \qquad (1)$$

where R is the grain separation, α the tunnelling exponent, E_C the charging energy, k Boltzmann's constant and T temperature. Real physical systems do not generally show simple activation. Instead, over moderate ranges of temperature and large ranges of conductivity, experimental results may be fitted to the form $\ln\sigma \propto 1/T^x$. For cermets, it is usually found that $x \approx 0.5$. For discontinuous films, x varies between 1 (simple activation) and 1/3, simple activation tending to be associated with higher temperatures and with films having small islands of relatively uniform size.

Fractional temperature dependence has been attributed to the distributed nature of R and E_C in real systems. However, there are difficulties with such explanations. The treatments of Sheng *et al.* (1973) and of Heinrichs *et al.* (1976) require a correlation between grain size and grain separation which may sometimes exist in cermets but is certainly not present in discontinuous films. Another class of theory does not require such correlation but invokes unphysical distributions of E_C (e.g. Hill and Coutts 1977). A theoretical investigation was therefore made of the consequences of applying the simple physics of Neugebauer and Webb to systems with realistic distributions of R and E_C (Adkins 1982).

The approach was to represent the system by a network of resistors, each representing the conductance between a pair of neighbouring islands. Using the Einstein relation for mobility, these elementary conductances were expressed as a sum of contributions from the various charge-state configurations that could be initial states for a charge transfer process. These contributions were written

$$s_i \;=\; p_i \nu_i e^2 R^2/2kT \qquad (2)$$

where p_i is the probability of occurrence of the initial charge configuration, ν_i the associated tunnelling rate, and e the electronic charge. Treating contributions as independent implies low temperature: $kT < E_C$. The p_i are obtained by elementary statistical mechanics. To obtain the ν_i, normal-metal normal-metal tunnelling theory is used, but evaluation is non-trivial since, in contrast to common tunnelling situations the potentials of the metals (the islands) do not remain constant in the tunnelling process but change by the relevant charging potentials.

To deal realistically with the distributions of E_C and R, both were taken to be described by log-normal distributions (Grandquist and Buhrman 1975). The two charging energies involved in a given elementary conductance can be convolved to give a single effective activation energy E, again log-normally distributed, so that each conductance becomes a function of only two distributed parameters E and R.

The bulk conductivity was then obtained by using symmetric effective-medium theory (Landauer 1978) generalized to a continuous distribution of conductances, an approach which in this case is closely similar to a critical percolation path analysis. The criterion for the bulk conductivity is that the bulk conductivity must divide the

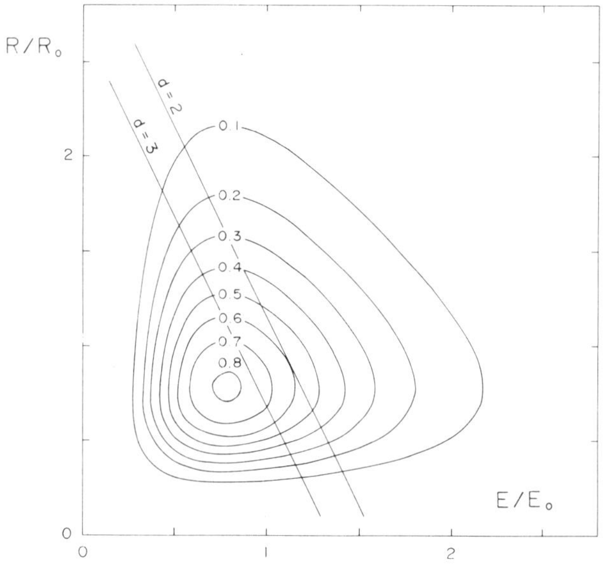

Figure 1. Realistic distributions of tunnelling distance and activation energy with superimposed lines corresponding to the bulk conductivity for two and three dimensions.

distribution of elementary conductances in a certain proportion. Figure 1 illustrates such a division for a particular temperature, the temperature determines the gradient of the dividing lines according to equation (1). The solutions were found numerically.

The result of this analysis was that for realistic values of parameters, the low-temperature conductivity would always be very close to simple activation, and it was concluded that the distributed nature of the structure of these systems cannot generally explain observed temperature dependencies of conductivity.

MACROSCOPIC INHOMOGENEITY

An alternative explanation for fractional temperature dependence might be found in the presence of resistive inhomogeneity on a scale comparable to specimen dimensions so that, in effect, measured resistances would correspond to differently characterized paths competing in parallel across the sample. If the metal - non-metal transition is viewed as a typical critical phenomenon, a diverging scale of inhomogeneity would be expected as the transition is approached from either side. A series of experiments was devised to reveal the scale of resistive inhomogeneity in discontinuous metal films (Benjamin *et al.* 1984).

Thin-film Capacitor Measurements

In these experiments a discontinuous metal film formed one electrode of a parallel-plate capacitor, the other electrode being a continuous film. The idea was to monitor the flow of charge onto the capacitor after application of a potential step. The charging current was interpreted by comparison with the charging behaviour of an ideal system in which resistance and capacitance are uniformly distributed over the whole area of a capacitor, a system for which the diffusion equation can be solved exactly.

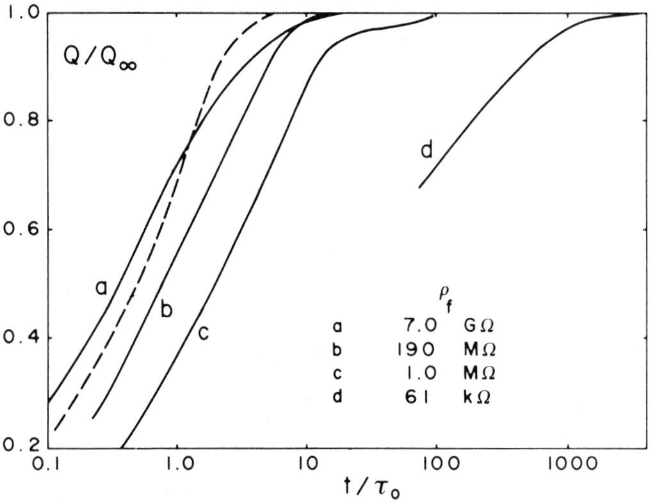

Figure 2. Inhomogeneity revealed by the capacitor-charging experiments. The broken curve shows charge as a function of time if there were no inhomogeneity.

The effect of inhomogeneity on charging is complex. Firstly, an uncharged area will only show up if it contributes a significant reduction in capacitance. Analysis of the electrostatics shows that the reduction in capacitance falls off as a/d where a is its linear dimension and d the separation of the capacitor electrodes. Since we wished to probe small-scale inhomogeneity, this result required small values of d which were obtained by using thin layers of anodic tantalum oxide as dielectric between a tantalum sheet as the continuous electrode and the discontinuous film deposited on top of the oxide. Secondly, the smaller the size of an abnormally resistive region, the larger the potential gradient driving charge onto it, and this partially compensates the increased resistivity by increasing the rate of charging. Abnormally slow charging of abnormally resistive regions will therefore only be observed if the resistivity is very much greater than the bulk value.

Results of measurements with films of different resistivity on 100 nm oxide layers are shown in figure 2. The charging of highly resistive films ($\rho \gg 1$ GΩ) was close to that of the ideal uniform system, although there were some excess currents at long times. However, as resistivity was reduced towards the metal - non-metal transition there was a progressive and eventually dramatic increase in delayed charging. In sample d the charge on the film was still 27% below its final value at times $t/\tau_0 \approx 100$ where τ_0 is the characteristic charging time of a uniform film of the same bulk resistivity. In another series of measurements, a dramatic increase in delayed charging was found in similar films as oxide thickness was reduced, thus demonstrating the sensitivity to oxide thickness and indicating the size of inhomogeneity in those samples.

Quantitatively, the results implied that inhomogeneity in highly resistive films was on a scale of a few tens of nanometres, i.e. on the scale of small groups of islands, whereas, as the resistivity was reduced towards the transition, the scale of inhomogeneity increased to greater than some hundreds of nanometres.

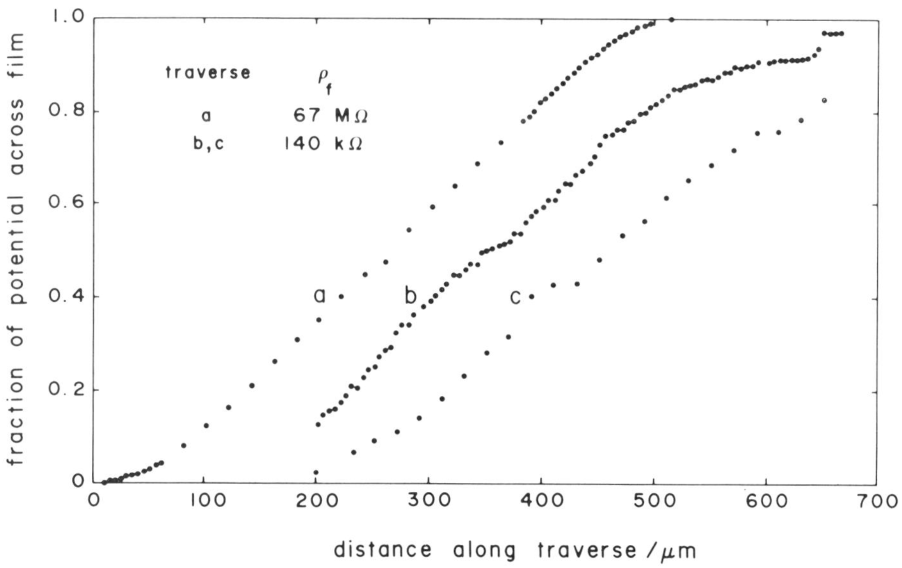

Figure 3. Inhomogeneity revealed by measurement of potential profiles across current-carrying films.

Direct Measurement of Inhomogeneity

The results of the previous section suggested that it might be possible to observe inhomogeneity by direct electrical measurements if sufficient spatial discrimination could be achieved.

The first set of measurements used micromanipulation of electrolytically sharpened tungsten points to *probe* the potential distribution in discontinuous films across which a potential difference was maintained. Figure 3 shows one traverse of a highly activated film and two traverses of a film near the metal - non-metal transition. The former shows a smooth profile and, at the ends of the traverse, one may note the increased conductivity that results from spreading of material at contact electrode edges. The other film shows clearly inhomogeneity with structure up to some 100 μm in scale.

Resistive inhomogeneity will result in tortuous current paths and development of potential differences in the direction normal to current flow. A second set of experiments used measurement of *transverse potentials* developed across current-carrying discontinuous films 1 mm wide. Both highly-activated and continuous films showed very small transverse potentials that were consistent with geometrical asymmetries. In contrast, films near the metal - non-metal transition could show large transverse potentials, implying inhomogeneity on a scale certainly not less than the size of the transverse contacts (150 μm). In one film that was just discontinuous, the ratio of transverse to longitudinal potential gradients was measured at 4.2 K as a function of field along the film. The reduction of inhomogeneity with increasing field was consistent with critical resistive links in the percolation path being spaced about 40 μm apart along the length of the film, again indicating inhomogeneity of similar scale.

Finally, a series of measurements was made using different *electrode configurations* with the expectation that closely-spaced, wide electrodes would give results corresponding to apparently reduced resistivity due to bridging of the narrow gap by regions of film of anomalously low

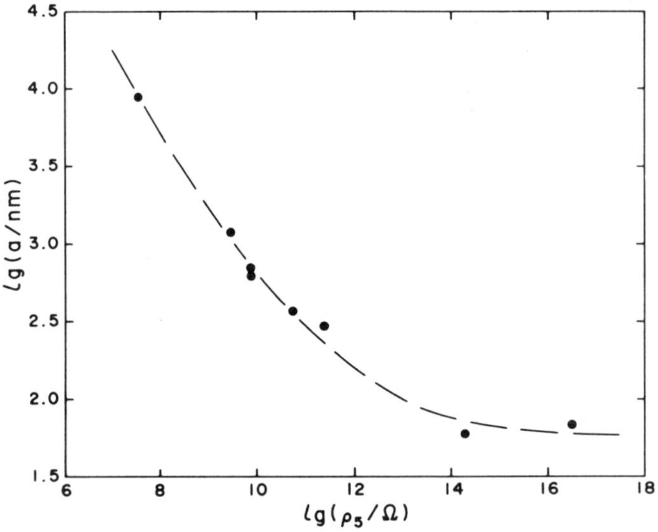

Figure 4. Distance a between critical resistive links as a function of bulk resistivity ρ_5 as estimated from onset of non-ohmic conduction.

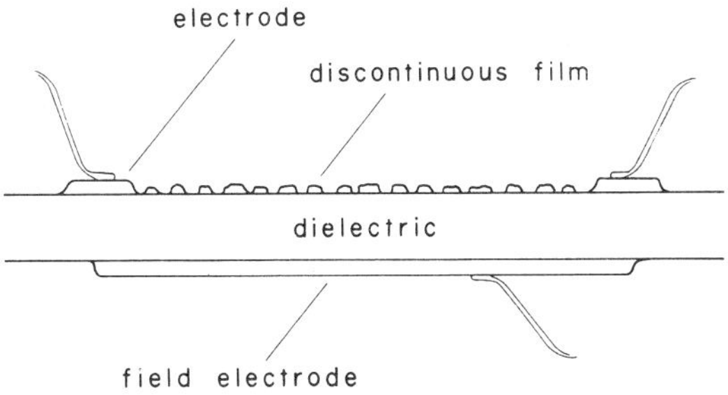

Figure 5. Experimental configuration for field effect experiments.

resistivity. For these experiments, electrodes were formed by lithographic lift-off techniques to eliminate spreading of metal at electrode edges, and spacings down to 50 μm were used. Reduced apparent resistivities were observed, reductions increasing with narrower geometries and with reduction of film resistivity towards the metal - non-metal transition. The scale of inhomogeneity implied by the results was consistent with the previously-described experiments. Again, measurements of non-ohmic conduction were used to estimate the distance between critical links giving the results in figure 4 which show the scale of inhomogeneity increases as the bulk resistivity is reduced towards the transition.

All these results show that inhomogeneity is present in these systems at a scale that corresponds to small groups of islands in highly-activated films and that the scale increases to macroscopic dimension as the metal - non-metal transition is approached. However, fractional temperature-dependence cannot be explained in general by inhomogeneity because it is present with well-activated films for which these experiments show the range of inhomogeneity to be negligible compared with typical sample dimensions.

FIELD-EFFECT MEASUREMENTS AND POTENTIAL DISORDER

Another way of probing the electronic structure in granular metals is to examine the effect on the conductivity of discontinuous metal films of electrostatic charging by application of an electric field normal to the plane of the films. This induces carriers independently of thermal

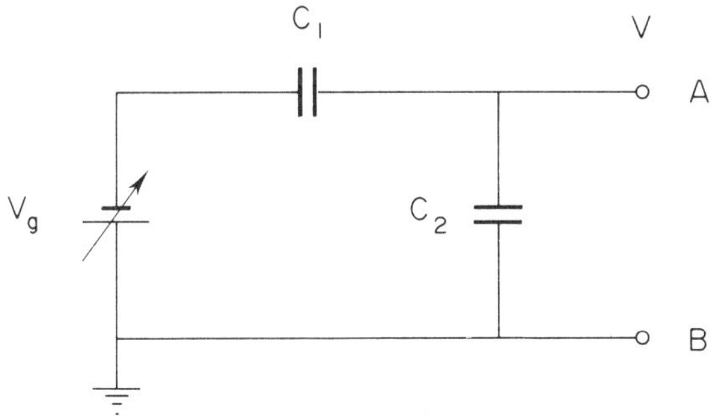

Figure 6. Circuit analogue for grain charging in the independent-carrier approximation.

activation and the resulting modification of conductivity is known as the *field effect* (Adkins *et al.* 1984).

The experimental configuration is shown in figure 5. The discontinuous film is kept near ground potential and carriers are induced by applying a potential V_g to the field or gate electrode on the other side of the substrate. The energetics of grain charging may be analysed in the independent-carrier approximation by reference to the electrical analogue of figure 6 in which the two capacitor plates connected to A represent a metal island, C_1 the capacitance between the island and the gate electrode and C_2 its capacitance to the surrounding film which will, on average, be at zero potential. Since the island is electrically connected to ground through the rest of the film it would wish to remain at zero potential as V_g is varied. However, charge localization means that the charge on it must be an integral number of elementary charges. It can be shown that the energy of the system U varies periodically as shown in figure 7(a) each parabola representing the energy of a different charge state n as a function of V_g. The energy minima occur when V_g corresponds precisely to the various integer values of n. At $T = 0$ the charge state changes discontinuously as parabolae intersect (b), and in each charge state the island potential varies linearly with V_g between transitions (c). At finite temperatures thermally-activated transitions between neighbouring states remove discontinuities in physical properties which, according to this model, will vary periodically with V_g. It is a

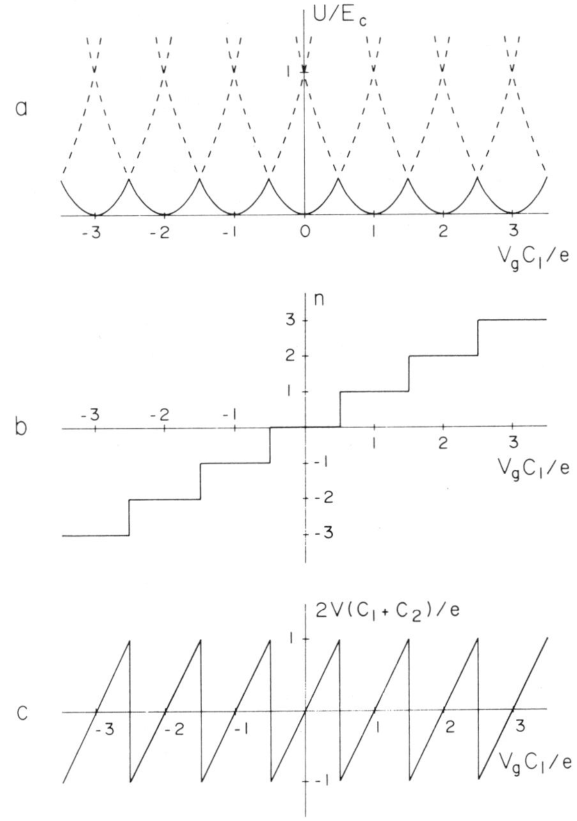

Figure 7. The physics of grain charging in field-effect experiments in the independent carrier approximation.

straightforward calculation to use statistical mechanics to find the probabilities that an island is in its various charge states and hence to deduce the effect of charging on conductivity in a uniform system; but without reference to the detailed calculations it is obvious that at low temperatures, $kT \ll E_C$, the relative change in conductance should be enormous as the concentration of induced carriers can be very much greater than the thermal background. (The maximum $\delta\sigma/\sigma$ should be of order $\exp(E_C/kT)$.) In practice, a maximum change of a few per cent is typical. Figure 8 shows an unusually large response in a film at 4.2 K with $E_C/kT \approx 8$. It was also found that $\delta\sigma/\sigma$ varied *linearly* with $1/T$ suggesting that the effect of the applied field was only to reduce very slightly the activation energy required for transport.

An obvious element left out of the simple analysis is that the real systems have metal islands with a distribution of charging energies. This would cause de-phasing of the responses of the islands as V_g is increased thus wiping out structure in the conductivity at large V_g. However, regardless of island size, all responses should be in phase at $V_g = 0$ and a large response of the whole system should remain about zero bias.

This lead Adkins *et al.* to postulate the presence of large *random potentials* in the system. Since an island can only change potential by a maximum of E_C/e before making a transition to the next charge state, the energies of neutral states must form a band of maximum width of order $2E_C/e$. The random potentials are attributed to impurity states, surface states and charge exchange between metal and substrate.

If the potential disorder completely randomized grain energies there would, of course, be no field effect at all (because island responses would be dephased at all values of V_g). The fact that a small field effect is present lead Benjamin to suggest that it resulted from *electrostatic relaxation* by which the system relaxed in relation to the mean charge states of the islands, this relaxation becoming frozen-in as the system is cooled. This would lower slightly the configurational energy of the system for the state in which is cooled, inhibiting thermally-activated formation of carriers. Application of the gate field then forces the system from this relaxed state, reducing slightly the energy required for subsequent carrier mobility.

This idea was tested by cooling samples with a constant non-zero gate voltage applied. It was then found that the conductivity minimum was

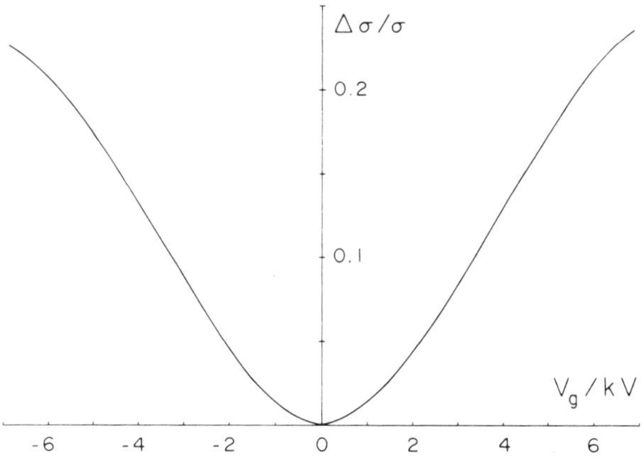

Figure 8. A large field effect measured at 4.2 K in a film with $E_C/kT \approx 8$.

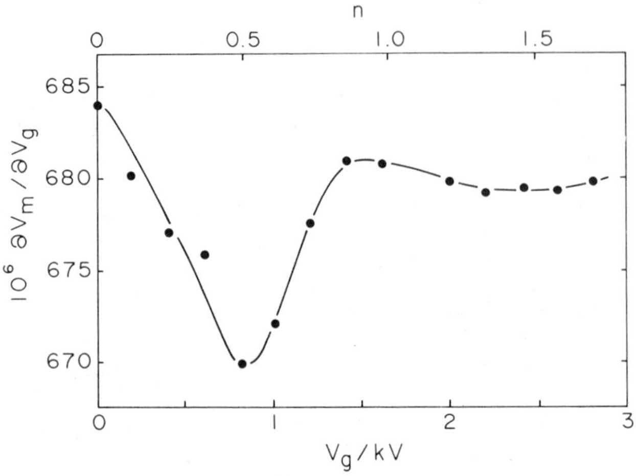

Figure 9. Variations of mean potential in a discontinuous film during initial stages of charging.

shifted to the value of V_g at which the sample had been cooled. Provided the sample was kept at low temperatures, this state persisted; but annealing to higher temperatures with $V_g = 0$ caused gradual loss of the displaced minimum and reappearance of that at zero bias. The authors concluded that potential disorder plays a vital role in discontinuous metals and emphasized the need for a complete restructuring of the theory of transport.

ENERGETICS OF CHARGING OF ISLANDS

The energetics of charging of islands has been investigated directly by monitoring potential during the charging process (Gardner and Adkins 1985). These experiments used a configuration like that of figure 5 except that another plane metal electrode is added on the side of the

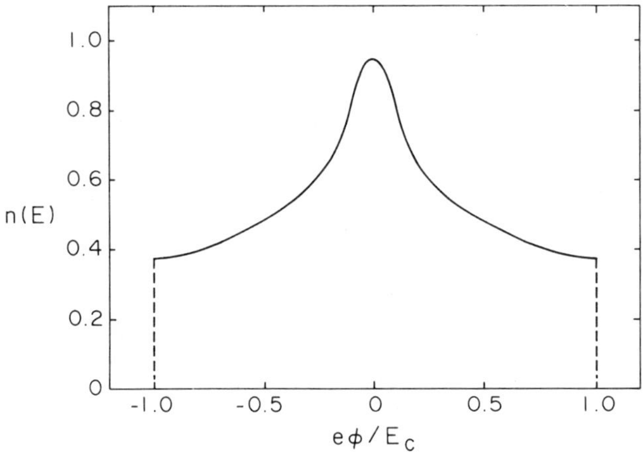

Figure 10. Distribution of potentials ϕ for charging of islands resulting from potential disorder. The sharp cut-offs are a result of assuming all islands to have the same capacitative charging energy.

discontinuous metal film remote from the gate. This 'monitor' electrode is connected to an amplifier of very high input impedance so that, by capacitative coupling, its potential V_m then follows changes in the *mean* potential in the plane of the discontinuous film. In the measurements the discontinuous film is grounded. There are then two contributions to changes in V_m as V_g is changed:
(a) There is a small direct capacitative coupling because the experimental film by being discontinuous cannot provide complete screening of the monitor from the gate. The size of this contribution is determined by geometrical factors and gives a strictly constant contribution to $\partial V_m/\partial V_g$.
(b) There is a contribution from the fact that the potentials of the islands do not remain zero (because of charge quantization) but fluctuate about zero as V_g is changed in the way illustrated in figure 7. Provided that the response of the islands is not entirely washed out by disorder, there should therefore be structure in $\partial V_m/\partial V_g$ as a function of V_g. Such structure is observed (figure 9).

These results may again be analysed by statistical treatment of the energetics illustrated in figure 7. As with the field effect, the structure is much weaker than predicted by the model, again implying the presence of potential disorder. Alternatively, if potential disorder is *assumed* to be present, results like those of figure 9 can be deconvolved to obtain a crude representation of the density of states. An example is shown in figure 10. The discontinuous drops at $E/E_C = \pm 1$ result from assuming all grains to have a single charging energy in this simple analysis; they are, of course, unphysical. What is clear, however, is that the grain energies do form a band with width of order $2E_C$ and that there is a maximum in the density of states at the centre of the band. We believe these to be the first measurements to give direct information about densities of states in granular metal systems.

DISCUSSION

The investigations described above demonstrate the inadequacy of the conventional models for transport in discontinuous metals in the regime where electrostatic charging energies limit conduction. The analysis of the conventional model by effective-medium theory with realistic distributions of relevant parameters shows that conductivity should always be close to simple activation. Yet fractional temperature dependence is usually observed. Models like those of Sheng *et al.* (1973) which depend on a correlation between particle size and particle separation cannot provide a general explanation for the reasons explained earlier, nor can the behaviour be attributed to gross electrical inhomogeneity because our experiments have measured the scale of inhomogeneity and shown that, except very near to the metal - non-metal transition, its scale is far too small. Models such as that of Celasco *et al.* (1978), which invoke a temperature-dependent tunnelling barrier resulting from charge exchange with surface and bulk donor levels, may apply at high temperatures where asymmetric field effects have been observed (Hill 1964) but they cannot apply at low temperatures where electrostatic relaxation becomes frozen out and the field effect becomes small and symmetric. The reason for fractional temperature dependence at low temperatures therefore remains an open question.

However, some of our investigations, notably the field-effect and charging-potential measurements, show clearly that the metal islands in discontinuous metals are subject to random potentials that spread their energies into a band of width of order $2E_C$. Cavicchi and Silsbee (1984) have also recently had to invoke random potentials to explain their measurements on small metal particles incorporated into tunnelling barriers. We believe that progress in understanding low-temperature transport in these systems now requires the development of new models in which the presence of random potentials is an essential component.

REFERENCES

Adkins C J 1982 *J. Phys. C: Solid State Phys.* **15** 7143

Adkins C J, Benjamin J D, Thomas, J M D, Gardner J W and McGeown A J 1984 *J. Phys. C: Solid State Phys.* **17** 4633

Benjamin J D, Adkins C J and Van Cleve J E 1984 *J. Phys. C: Solid State Phys.* **17** 559

Cavicchi R E and Silsbee R H 1984 *Phys. Rev. Lett.* **52** 1453

Celasco M, Masoero A, Mazzetti P and Stepanescu A 1978 *Phys. Rev.* **B17** 2553

Gardner J W and Adkins C J 1985 to be published

Grandquist C G and Buhrman R A 1975 *J. Appl. Phys.* **47** 5

Heinrichs J, Kumar A A and Kumar N 1976 *J. Phys. C: Solid State Phys.* **9** 3249

Hill R M 1964 *Nature* **204** 35

Landauer R 1978 *Proc. Conf. Electrical Transport and Optical Properties of Inhomogeneous Media* (New York: AIP) pp. 2–45

Neugebauer C A and Webb M B 1962 *J. Appl. Phys.* **33** 74

Sheng P, Abeles B and Arie Y 1973 *Phys. Rev. Lett.* **31** 44

METAL-NONMETAL TRANSITION AND THE CRITICAL POINT PHASE TRANSITION IN FLUID CESIUM

F. Hensel, S. Jüngst, F. Noll and R. Winter

Institute of Physical Chemistry
Philipps-University Marburg
D-3550 Marburg, W.-Germany

INTRODUCTION

The study of the electronic and thermophysical properties of fluid alkali metals near the gas-liquid critical point has attracted interest for decades for many technical and scientific reasons. It is the subject of numerous theoretical and experimental papers and various review articles (see e.g. Mott, 1974; Cusack, 1978; Hensel, 1980, 1982, 1984; Freyland, 1981; Yonezawa and Ogawa, 1982; Alekseev and Iakubov, 1983; Freyland and Hensel, 1985). However, considerable controversy still surrounds the exact nature of the gas-liquid critical point phase transition, in particular with regard to the role of the metal-nonmetal transition which occurs at reduced densities of the monovalent metals. The critical data are not known with sufficient certainty. A reliable theoretical prediction of such data is not possible.

Part of the difficulty arises from the fact that the high cohesive energies of metals place the critical region at temperatures and pressures too high for easy experimental investigation. The resulting problems of temperature and pressure measurement and control together with the highly reactive nature of fluid alkali metals have chiefly limited the accuracy with which properties have been measured in the past.

The second serious difficulty is that the well established rules and methods for the determination of critical point data of normal, insulating fluids can not be applied for fluid metals just as they are. The fundamental difference of dealing with the critical point phase behaviour of metals is the existence of competing interactions. Regardless of the way in which the effective interactions of atoms in a metal are described the description must change with density. It is obvious that at low temperatures, e.g. near the triple point, two transitions occur at once when the metallic liquid condenses from the insulating vapour. One of these transitions is the liquid-vapour and the other is the metal-insulator. This implies that the interatomic cohesion must change from metallic cohesion in the liquid to van der Waals' interaction between neutral atoms and molecules in the dilute vapour. However, this coincidence does not necessarily occur up to the gas-liquid critical point. At such high temperatures and densities thermal excitation effects and many particle interaction must become significant in the alkali metal vapour phase. Consequently, in addition to the van der Waals' interaction between neutral atoms and molecular clusters, the charge neutral interaction and the Coulomb interaction between charges play an important role. By contrast, for most insulating molecular substances the intermolecular interactions can be described by reference to a single, density independent pair potential.

This contrast has been discussed in a number of theoretical papers (Landau and Zeldovitch, 1943; Mott, 1961, 1978; Krumhansl, 1965; Ebeling et al., 1976; Nara et al., 1977; Yonezawa and Ogawa, 1982) which studied the interrelation between the liquid-vapour phase transition and the metal-nonmetal transition. However, the existing theoretical attempts to model the statistical mechanics of the metal-nonmetal transition in fluids are still insufficient to provide a clear-cut answer from theory. At present no theory exists which incorporates both the fluid aspects and the variation in the electronic structure from extended to localized states.

Several attempts have been made to explore this problem experimentally, but the subject has remained elusive. As the experimental difficulties in the critical region are rather severe this is not surprising. The main problem was that the analysis of electrical, magnetic and thermophysical data close to the critical point of the alkali metals was hampered by the presence of spurious effects due to temperature gradients.

However, this situation is now changing. Improvements and extensions of experimental technique make it possible to measure the equation of state (Jüngst, 1985) and the electrical conductivity (Noll, 1985) of cesium relatively close to the critical point with quite high precision and comparatively accurate temperature control and optimal elimination of temperature gradients. These data seem to be accurate enough to permit the first determination of the asymptotic behaviour of the thermophysical properties of an alkali metal near the gas-liquid critical point.

THE CRITICAL POINT PHASE TRANSITION OF CESIUM

Many attempts have been made to determine the critical point of fluid cesium (Dillon et al., 1965; Renkert et al., 1969; Oster and Bonilla, 1970; Bonilla and Silver, 1970; Korshunov et al., 1975; Franz et al., 1980). However, these works did not approach the immediate critical region, but used rules or laws obeyed for insulating liquids in the extrapolation of experimental results and the estimate of critical data. Whether such rules are valid for liquid metals is, however, one of the points at issue.

Therefore we have undertaken new work in order to obtain a clearer understanding of the behaviour of fluid cesium in the critical region. Specifically, our study has the following aims. The first is to determine precisely the critical parameters temperature T_c, pressure p_c and density d_c. The second aim is to explore carefully the densities of the coexisting liquid and vapour phases as a function of temperature near the gas-liquid critical point. These data should yield the first information about the exactness of the "law of rectilinear diameter" applied to liquid alkali metals. This law appears to hold for most insulating liquids. It has also been used by many workers in the extrapolation of experimental results and the prediction of critical data of metals. The third aim is to study the question how the gross changes in the electronic structure at the metal-nonmetal transition which are presumably relevant in the vicinity of the critical point, influence the gas-liquid critical point phenomena of cesium.

In pursuit of these aims new and very accurate measurements of the equation of state (Jüngst, 1985) and electrical conductivity (Noll, 1985) were performed for fluid cesium close to its gas - liquid critical point. For the equation of state measurements two methods have been employed. In the first experiment we measured the pressure (p) - temperature (T) coordinates for a very large number of isochores in the liquid and vapour phase close to the critical point. Whether the average density of an isochore was greater (liquid L) or smaller (vapour V) than critical was easily found from the isochore slopes $\left(\frac{\partial p}{\partial T}\right)_L$ or $\left(\frac{\partial p}{\partial T}\right)_V$ respectively. $\left(\frac{\partial p}{\partial T}\right)_L$ or $\left(\frac{\partial p}{\partial T}\right)_V$ are greater or smaller than the slope of the vapour pressure curve, $\left(\frac{\partial p}{\partial T}\right)_{sat.}$, at the intersection of the two curves. From these intersections points of the coexistence curve for orthobaric liquid and vapour densities have been obtained. As the critical point was approached with increasing temperature, the slopes $\left(\frac{\partial p}{\partial T}\right)_L$, $\left(\frac{\partial p}{\partial T}\right)_V$, and $\left(\frac{\partial p}{\partial T}\right)_{sat.}$ became indistinguishable. From this observation the critical constants T_c=1652°C, p_c=92,5 bar, d_c=0,36 g/cm³ have been determined.

Similar values have been obtained from the second experiment which measured density (d) - pressure (p) coordinates for a series of isotherms. Approximate values of the critical data have been deduced from an examination of the shapes of the isotherms. The new critical data, especially the critical temperature T_c, deviate strongly from the old estimates. The hitherto existing critical temperatures T_c from five separate determinations were given between 1740°C and 1785°C.

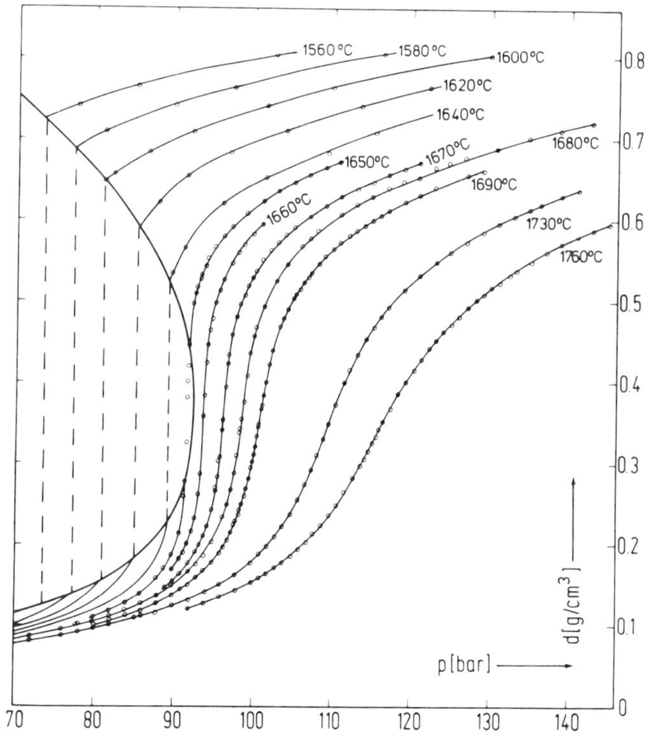

Fig. 1. Equation of state data of fluid cesium near the critical point.

A selection of equation of state data from both experiments is given in fig.1 in form of density isotherms plotted versus pressure. These data approach the critical point close enough to yield a first impression of the asymptotic behaviour of the thermophysical properties of metals near the critical point.

Fig.2 shows as an example a plot of the coexisting vapour (d_V) and liquid (d_L) densities of cesium together with the curve of average densities $\bar{d} = \frac{1}{2}(d_L+d_V)$ versus temperature. The form of the coexistence curve is clearly asymmetric compared to those of simple nonconducting fluids. The asymmetry, however, is very similar to that observed for the metal-ammonia (Chieux et al, 1980) and electron-hole liquid phase diagram (see e.g.Thomas, 1984). The most striking feature of fig.2 is the breakdown of the law of rectilinear diameter. The fact that this law does not hold indicates already an unusual behaviour of the critical exponents β_L and β_V which are used in the power law analysis of coexistence curves in the asymptotic expression

$$\left|\frac{d_{L,V} - d_c}{d_c}\right| \propto \left|\frac{T_c - T}{T_c}\right|^{\beta_{L,V}}$$

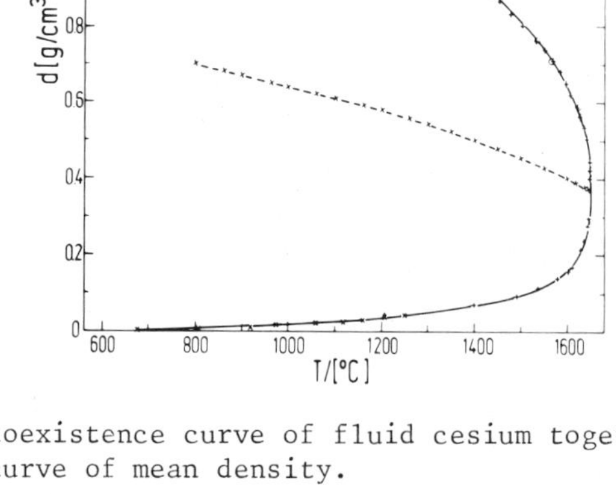

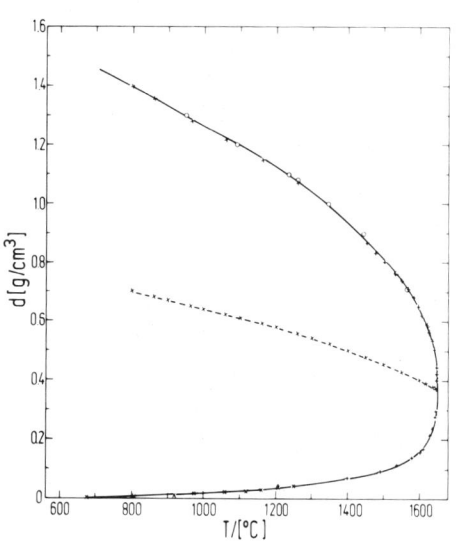

Fig. 2. The coexistence curve of fluid cesium together with the curve of mean density.

Here $ß_L$ is the exponent of the liquid (d_L)-, $ß_V$ of the vapour (d_V)-branch of the coexistence curve. Experimentally it is found (fig.3) that the liquid branch can be fitted with $ß_L \approx 0.46$ over a relatively large temperature range, whereas the value $ß_V$ for the vapour branch depends strongly on the relative temperature distance from the critical point. Close to T_c $ß_V$ approaches a value close to that observed for $ß_L$. These values are comparable with ß=0.5 obtained from classical mean field theory. A similar behaviour is shown by the isothermal compressibility χ_T of cesium (Jüngst, 1985). The power law analysis for the divergence of χ_T along the critical isochore of cesium according to the asymptotic expression

$$\chi_T \propto \left| \frac{T_c - T}{T_c} \right|^{-\gamma}$$

yields a critical exponent γ very close to the mean field value 1.

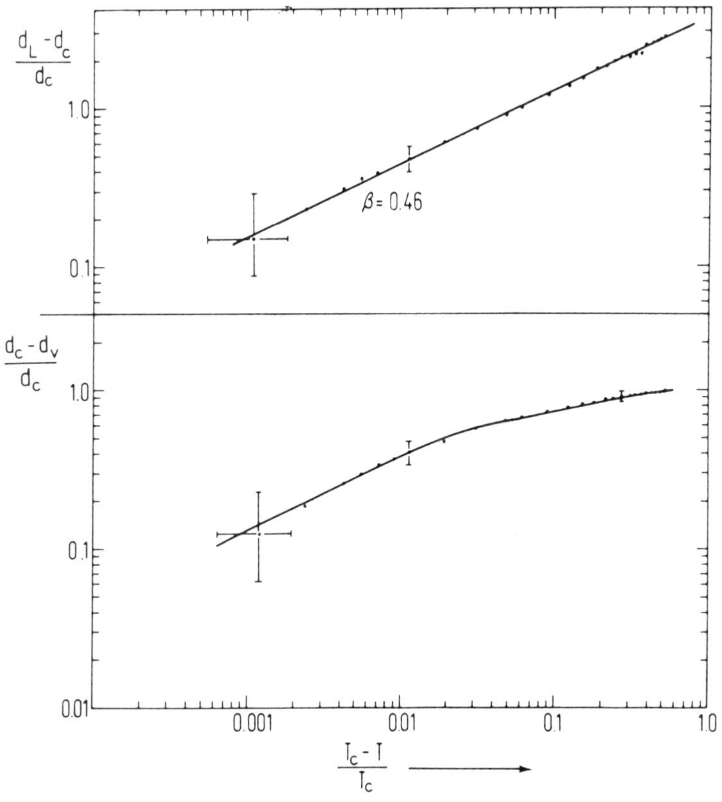

Fig. 3. Power law analysis for the asymptotic behaviour of the coexisting liquid and vapour densities.

Similar observations have been made for mercury (Götzlaff, 1983; Hensel, 1984; Schmutzler et al., 1985) for which close to the critical point the experimentally deduced exponents are also consistent with the classical mean field theory. To our knowledge this has never been observed before for other one-component liquids.

It seems that long range interatomic interactions or the presence of competing interactions close to the critical point of fluid cesium reduces the upper critical dimensionality from 4 to 3 resulting in critical exponents very close to mean field values. It must be pointed out, however, that at present, because of the severe experimental problems connected with the high critical temperatures of metals, reliable experimental data can only be obtained for the region $|(T_c-T)/T_c| \gtrsim 10^{-3}$, so that it may be argued that the indices could change at lower values of $|(T_c-T)/T_c|$ to the normally for insulating fluids observed non-classical values.

THE METAL-NONMETAL TRANSITION

The new conductivity experiment (Noll, 1985) was undertaken in order to extend the earlier experimental investigation (Franz et al., 1980) to the

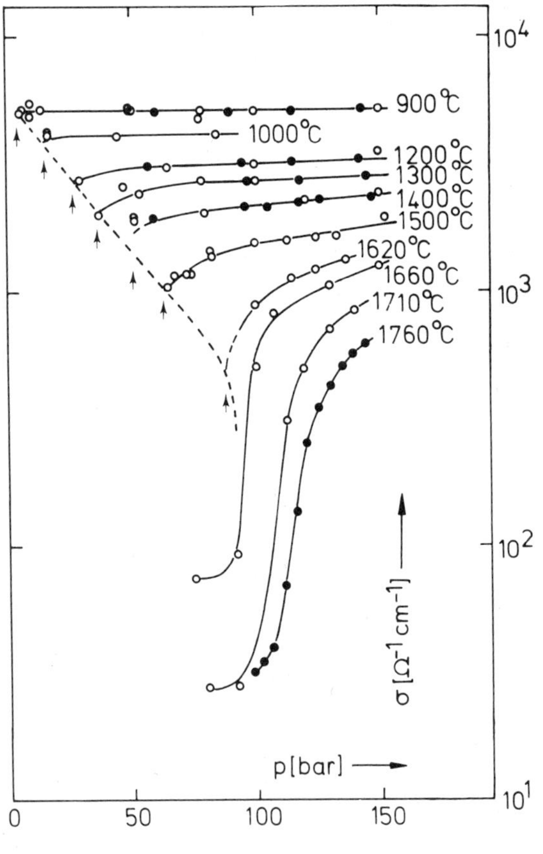

Fig. 4. Conductivity isotherms of fluid cesium.

newly determined critical region and in particular to explore the interrelation between the electronic transition from a metal to a nonmetal and the liquid-vapour phase transition. Fig.4 gives a selection of data in form of conductivity isotherms plotted versus pressure. The conductivities of the coexisting liquid phase are given by the dashed line; the arrows indicate the abrupt transition to the vapour phase. Apart from the liquid-vapour phase transition no discontinuous changes are indicated in σ. The steep fall in σ near the critical point suggests that the locus of the metal - nonmetal transition is close to the critical region.

A central question for the understanding of the metal-nonmetal transition in monovalent metals is the role of electronic correlation effects. As is well known two effects have to be considered. The first involves the effects of long range screening between charges whereas the second considers the effect of the intraatomic electron-electron repulsion of two electrons at the same site, the Hubbard energy. Brinkman and Rice (1970) studied the effect of the Hubbard energy in a metal and showed that near the transition the electron gas should be highly correlated, i.e. a small fraction of the atomic sites should be doubly occupied, the spin on the other sites resonating between the two possible positions. They predicted for this correlated metal enhanced values for the paramagnetic susceptibility.

The first experimental evidence for large electronic correlation effects in expanded cesium was provided by the magnetic susceptibility measurements

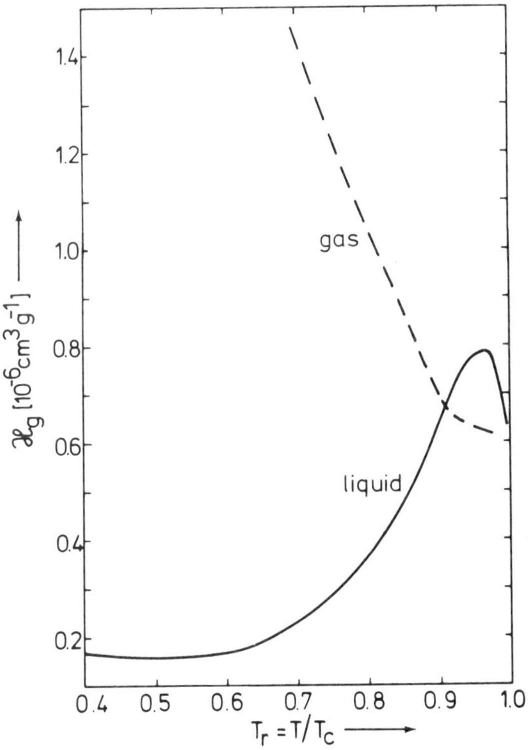

Fig. 5. Total mass susceptibility of liquid (solid curve) and gaseous (dashed curve) cesium along the coexistence curve as a function of the reduced temperature $T_r=T/T_c$.

of Freyland (1979, 1980). His results are replotted in fig.5. It is a revised plot of the measured total mass susceptibility along the liquid (solid curve)-gas (dashed curve) saturation line as a function of the reduced temperature $T_r=T/T_c$, i.e. the density is an implicit variable. The revision is stimulated by the availability of the new and more accurate data for the equation of state close to the critical point (fig.1 and fig.2). A strong enhancement of the total mass susceptibility with increasing T_r, i.e. decreasing density, is observed for liquid cesium until a susceptibility peak is observed at $T_r=0.97$ (the corresponding density of the coexisting liquid is about twice the critical density). Recent nuclear magnetic resonance measurements of the knight shift in cesium (El-Hanany et al., 1983) also showed the low density enhancement, confirming that the effect arises from the electron spin contribution.

On the low density side of the transition below d_c the temperature - or density dependence of the vapour susceptibility (dashed curve in fig.5) shows a strong diamagnetic deviation from the Curie limit for localized s=1/2 electrons. Consequently, Freyland (1979) concluded that spin pairing processes which lead to aggregated species like dimer molecules or higher neutral or charged molecular clusters are an important feature approaching the critical point transition from the nonmetallic vapour side.

What is more interesting in connection with the possible interrelation between the metal-nonmetal transition and the gas-liquid critical point transition in cesium is the susceptibility peak observed for the expanded coexisting liquid around $2d_c$ ($T_r \approx 0.97$) which probably defines the onset of the transition from a paramagnetic metal to antiferromagnetic states. We believe that these results indicate that nonmetallic states play already an important role in the liquid below $2d_c$. In the liquid atomic centers are mobile and so an analogue to the antiferromagnetic ground state can be the state of molecular clusters.

The only theoretical works bearing on the problem of cluster formation in the transition region are those by Alekseev and Iakubov (1983), Hernandez et al. (1984), Hernandez (1984, 1985a, 1985b) and Redmer and Röpke (1985). Hernandez studied the role played by thermally generated positive molecular ions in the density dependence of the electrical and magnetic properties of Hg, Cs and Rb in the nonmetallic and semimetallic regime. Redmer and Röpke (1985) performed a theoretical calculation of the equation of state and the critical parameters of cesium. The basis for their treatment is an extension and improvement of former works by Ebeling et al.(1973, 1979). The intention was to investigate the influence of bound states (atoms and molecules). The essential result of all calculations is that a high concentration of molecular clusters can exist on both branches of the coexistence curve near the critical point.

REFERENCES

Alekseev, V. A., and Iakubov, I. T., 1983, Physics Reports, 96:1.

Brinkman, W. F., and Rice, T. M., 1970, Physical Review, B2:4302, ibid B2: 1324.

Chieux, P., Damay, P., Dupuy, J., and Jal, J. F., 1980, J.Phys. Chem., 84:1211

Cusack, N. E., 1978, in: "Metal Non-Metal Transitions in Disordered Systems," L.R.Friedman and D.P.Tunstall, ed., Edinburgh.

Dillon, I. G., Nelson, P. A., and Swanson, B.S., 1966, J.Chem.Phys., 44(11): 4229.

Ebeling, W., and Sändig, R., 1973, Ann.Phys.(Leipzig) 28:289.

Ebeling, W., Kraeft, W. D., and Kremp, D., 1976, "Theory of Bound States in Plasmas and Solids," Akademie-Verlag, Berlin.

Ebeling, W., Meister, C.V., Sändig, R., and Kraeft, W. D., 1979, Ann.Phys. (Leipzig), 36:321.
El-Hanany,W., Brennert, G. F., and Warren, W. W., Jr., 1983, Phys.Rev.Lett., 50:540.
Franz, G., Freyland, W., and Hensel, F., 1980, J.Phys. 41, Colloque C-8: 70.
Freyland, W., 1979, Phys.Rev.B., 20:5104.
Freyland, W., 1980, J.Phys., 41, Colloque C-8: 74.
Freyland, W., and Hensel, F., 1985 in: "The Metallic and the Non-Metallic States of Matter: An Important Facet of the Chemistry and Physics of Condensed Matter," P.P. Edwards and C.N.R. Rao, eds., Taylor and Francis, London.
Götzlaff, W., 1983, Diplom-Thesis, Universität Marburg.
Hensel, F., 1980, Angew.Chem., 92:598, Angew.Chem.Int.Ed.Engl., 19:593.
Hensel, F., 1982, Proc. 8th Symposium on Thermophysical Properties J. v. Sengers, ed., (ASME, New York) p.151.
Hensel, F., 1984 in: "Nato ASI Series C," 30:401, J.V.Acrivos, N. F. Mott and A.D. Yoffe, eds., Reidel, Dordrecht.
Hernandez, J. P., Schönherr, G., Götzlaff, W., and Hensel, F., 1984, J.Phys.C., 17:442.
Hernandez, J. P., 1984, Phys.Rev.Lett., 53:2320.
Hernandez, J. P., 1985a, Phys.Rev.B. to be publ.
Hernandez, J. P., 1985b, private communication.
Jüngst, S., 1985, Doctoral-Thesis, Universität Marburg
Korshunov, V. S., Vetchinin S. P., Senchenkov A. P. and Asinovskii E. I., 1975, High Temp., 13:477.
Krumhansl, J. A. 1965 in: "Physics of Solids at High Pressures," C.T. Tomizuka and R.M. Emrick, eds., Academic Press, New York.
Landau, L., and Zeldovitch, G., 1943, Acta Phys.Chim.USSR, 18:194.
Mott, N. F., 1961, Philos.Mag., 6:287.
Mott, N. F., 1974, "Metal-Insulator Transitions", Taylor and Francis, London.
Mott, N. F., 1978, Philos.Mag., 37:377.
Nara, S., Ogawa, T., and Matsubara, T., 1977, Prog.Theo.Phys., 57:1474.
Noll, F., 1985, Diplom-Thesis, Universität Marburg.
Oster, G.F., Bonilla C. F., 1970, Proc.5th Symp. Thermophys. Prop., 486.
Redmer, R., and Röpke, G., 1985, Physica A, to be publ.
Renkert, H., Hensel, F., and Franck, E.U., 1971, Ber.Bunsenges. Phys.Chem., 75:507.
Schmutzler, R.W., Seyer, P., and Hensel, F., 1985, to be publ.
Silver, I. L., Bonilla C. F., 1970, Proc.5th.Sym.Thermophys.Prop., 461.
Thomas, G. A., 1984, J.Phys.Chem., 88:3749.
Yonezawa, F., and Ogawa, T., 1982, Prog.Theo.Phys.(Japan) Suppl., 72:1.

THE SEMICONDUCTOR-TO-METAL TRANSITION IN LIQUID SE-TE ALLOYS

M. Cutler and H. Rasolondramanitra

Department of Physics
Oregon State University
Corvallis, OR 97331

ABSTRACT

A logical distinction is made between two aspects of the transition: electronic and thermodynamic. Published experimental data for the density is used to provide a quantitative expression for the conversion of the liquid from a low temperature to a high temperature form. This description of the thermodynamic transition indicates a change in the molecular structure, for which we propose a specific mechanism: the replacement of the normal two-fold (2F) bonded neutral chalcogen atoms in the molecular chains by pairs of ions (D_p centers) consisting of a 3-fold D^+ positive ion bonded to a one-fold D^- ion. The required concentrations of D_p centers is consistent with existing information concerning the concentrations of dissociated ion pairs and their expected tendency to form associated pairs. Electrostatic interactions are expected to cause clustering of D_p centers within a chain, and to cause an electrostatic coordination which can increase the average coordination number Z to a value > 2. This can explain the fact that Z > 2 in molten Te. We discuss the implications of the model for the effect of the transition on the electronic structure.

1. INTRODUCTION

There has been much discussion of the transition from semiconductor to metallic behavior of the liquid alloys Se_XTe_{100-X} following the work of Perron,[1] which clearly defined their electronic behavior.[2,3,4] The transition is inferred from the dependence of the electrical conductivity σ on the temperature T. At low T and large X, σ has an activation energy characteristic of semiconductor behavior, and at the opposite extreme σ increases slowly with T, which is characteristic of poor metals. The boundary between the semiconductor regime II and the metallic regime III is commonly taken to be $\sigma \simeq 300$ ohm^{-1}cm^{-1}. (Another semiconductor domain called region I will be referred to later, in which the activation energy is smaller than in region II.) This value of σ agrees with Mott's minimum metallic conductivity σ_M,[5] and when σ becomes as large as σ_M, one expects the Fermi energy E_F to enter the band in which transport occurs. At this point, changes in other electronic properties occur which also indicate a transition to metallic behavior. The thermopower S decreases from values greater than (k/e) to values less than (k/e),[1] and the behavior of the paramagnetic susceptibility χ_p changes from one described by the Curie law to one described by the Pauli law.[6]

Changes in other physical properties are found or inferred in the same range of T and X, which have also been referred to as a semiconductor-to-metal transition. This reflects the tacit assumption that they are caused by the same mechanism. But these changes are logically distinct from the electronic changes discussed in the preceding paragraph, and they may not necessarily be the result of the same phenomenon. To avoid confusion, the latter will be referred to as the electronic transition. The other changes will be categorized as the thermodynamic or the structural transition. We define the electronic transition temperature T_E as the temperature at which $\sigma = 300$ ohm^{-1}cm^{-1}. The dependence of T_E on X is shown in Fig. 1.

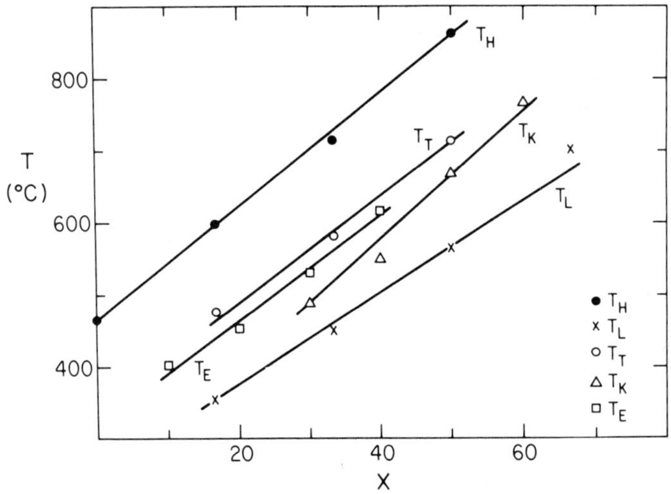

Fig. 1. Dependence of characteristic temperatures (defined in the text) on composition in Se_XTe_{100-X}.

The thermodynamic transition refers to an anomalous behavior of the equation-of-state parameters, such as the atomic volume V_A and the adiabatic compressibility K_S, as a function of T. It is most clearly delineated by the measurements of the density by Thurn and Ruska.[7] Plots of the atomic volume V_A vs. T, based on their data, are shown in Fig. 2. It is seen that there are ranges of T, which depend on X, in which dV_A/dT is negative. The lower and upper boundaries of the temperature range can be defined by the temperature T_L, at which V_A is a maximum, and T_H where V_A is a minimum. We define the thermodynamic transition temperature T_T as the average of T_L and T_H. These three temperatures are also plotted in Fig. 1, and it is seen that T_T is rather close to T_E. The thermal expansion of liquids is closely connected with an increase in the compressibility. Measurements by Takimoto and Endo[8] show that the adiabatic compressibility K_S has a maximum at a temperature T_K, whose dependence on X is also plotted in Fig. 1. T_K is somewhat higher than T_L.

There is much more uncertainty about the existence and character of a structural transition, which refers to experimental information concerning the molecular structure such as the radial distribution function. Its existence has been inferred from the fact that the coordination number Z = 2 in liquid selenium,[9] and Z = 3 in liquid tellurium near the melting point.[10] The structure factor of the alloys at 500°C shows a continuous change with X,[11] largely in the vicinity of X = 40. But clearcut informa-

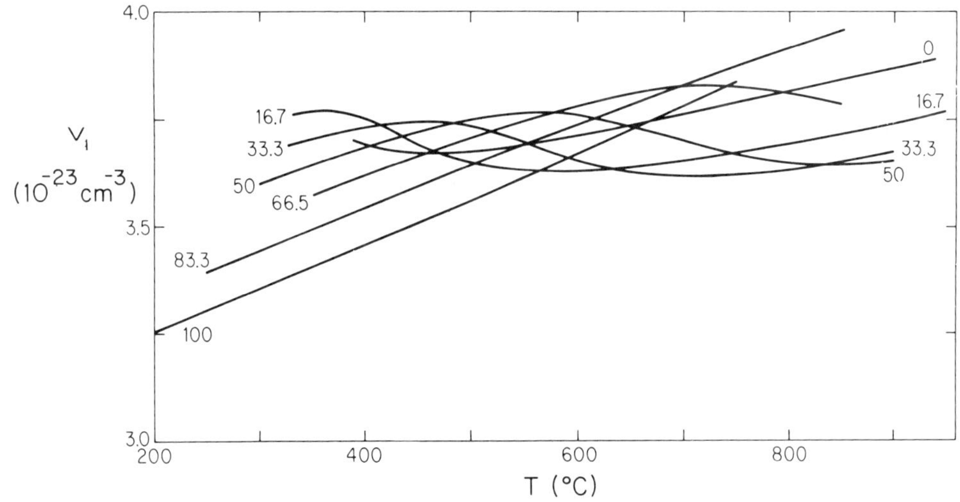

Fig. 2. Dependence of atomic volume V_A on temperature at various compositions X, from data of Thurn and Ruska (Ref. 7).

tion about the behavior of Z in an alloy requires resolution of the three pair correlation functions, and that information is not yet available. It has been commonly assumed that the decrease in volume between T_L and T_H is accompanied by an increase in Z from 2 to 3.[2,7] Recently, Raman scattering measurements have been reported and interpreted to indicate that Z changes from 2 to 3 when X is changed from 30 to 20.[12]

Cabane and Friedel[13] have offered an explanation of Z = 3 in Te in terms of a structure similar to arsenic, in which each atom has three equivalent bonds, and this interpretation has been extended by other authors to the Se-Te alloys. There is a fundamental difficulty in this model because three equivalent bonds on each chalcogen atom would lead to an electronic structure in which the Fermi energy E_F is below the middle of a conduction band formed from the antibonding orbitals. This would make it very difficult to explain the observed behavior of the electrical conductivity σ and the thermopower S. S is positive, and the systematic decrease in S as σ increases, when $X \rightarrow 0$ and T increases, seems to require an explanation in terms of a movement of E_F downward into the valence band. That implies that the bond number (the average number of bonds per atom) $B \leq 2$.[14] There is also a conflict between the three-bond model and analysis of the structure factor, which indicates disorder within the first neighbor shell.[15]

Thurn and Ruska[7] analyzed the $V_A(T)$ curves in terms of a transformation of the alloys from a low temperature (L) form to a high temperature (H) form, each of which has an atomic volume V_L or V_H which is linear in T. Using this model, they inferred the fraction of atoms C_v which has been converted from the L form to the H form as a function of T, and compared this behavior with the law of mass action (LMA) for the equilibrium between L and H atoms. The determination of $C_v(T)$ depends sensitively on the specification of $V_L(T)$ and $V_H(T)$, and their paper describes their procedure for doing this only in a general way. In Sec. 2 we determine the behavior of $C_v(T)$, with the specification of V_L and V_H made explicit. We find that the LMA expression is obeyed within the uncertainties in the analysis. However, there are other reasons to doubt the validity of the LMA equation, and we use the LMA formulation primarily to provide an empirical description of $C_v(T)$ which can be interpolated or extrapolated with respect to T and X.

In a recent paper[16] (referred to in the following text as I), we were able to deduce, from the electrical behavior in the semiconductor regions

II and I, quantitative information about the concentrations of bond defect atoms. This was done in ranges of X and T for which C_v is as large as 0.5. This provides a basis for examining whether the thermodynamic transition is caused by the generation of bond defects. The electronically active bond defect atoms, whose concentrations were determined in I, include one-fold bonded (1F) dangling bond atoms D^*, 1F negative ions D^-, and three-fold bonded (3F) positive ions D^+. The concentrations are found to be too low to account for the behavior of C_v. But there is reason to believe that there will be a considerably larger concentration of associated ion pairs, which we will refer to as dipole pairs D_p. Their concentrations are likely to be large enough to account for the behavior of C_v. The main purpose of this paper is to present a molecular model for the L-to-H transformation based on the replacement of the normal two-fold bonded (2F) atoms (C_2) in chains by dipole pairs, which are also effectively 2F units. In Sec. 3, we describe the basis of the D_p model, and examine the effects of the electrostatic interactions between neighboring D_p centers on a chain. These interactions are found to be strong enough to cause appreciable clustering of D_p units within a chain. They also give rise to an "electrostatic coordination" effect which can increase Z from 2 to 3, while B = 2. This provides an attractive resolution of the discrepancy, noted earlier, between the expected values for these two parameters. In the final section we discuss some of the remaining difficulties and uncertainties in the proposed model.

2. THE THERMODYNAMIC TRANSITION

The curves in Fig. 2 have shapes which correspond to temperature-limited sections of the model curve shown in Fig. 3. This model curve has two linear ranges, in which V_A is equal to the low temperature V_L (T) or the high temperature V_H (T):

$$V_L = V_{L0} + V_{L1}\,T \quad , \tag{1}$$

or

$$V_H = V_{H0} + V_{H1}\,T \quad , \tag{2}$$

where V_{L0}, V_{L1}, V_{H0}, and V_{H1} are constants. Since a nearly linear increase with T is the normal behavior of the atomic volume of a liquid, it is reasonable to assume that in the intermediate region B in Fig. 3, V_A is linearly related to a transformation of the atoms from a low temperature L form to a high temperature H form. The degree of conversion at a given T is determined by the relative values of V_L, V_A, and

V_H. Thus we identify the fraction of atoms in the H form with the contraction C_v, which is defined by

$$C_v = (V_L - V_A)/(V_L - V_H) \quad . \tag{3}$$

Assuming that there is equilibrium corresponding to a homogeneous monatomic transformation L → H, the law of mass action yields

$$\frac{C_v}{1-C_v} = K\ (T) = \exp\ [-\frac{E}{kT} + \frac{S}{k}] \ , \tag{4}$$

where the equilibrium constant K is expressed on the right in terms of the enthalpy E and entropy S of formation of an H atom from an L atom.

In order to determine C_v, it is necessary to evaluate the constants in Eqs. 1 and 2 for V_L and V_A. Unfortunately, the experimental data clearly define the linear sections only at large X for V_L, and only at small X for V_H. These constants are plotted vs. X in Fig. 4. It is seen that the temperature coefficients are nearly constant, with V_{H1} considerably smaller than V_{L1}. The T = 0 limits, V_{LO} and V_{HO}, decrease with X, as one would expect from Vegard's law. We have therefore assumed constant values V_{L1} = .001 and V_{H1} = .000525 in units $10^{-23}cm^{-3}/K$, corresponding to the average values, and used a linear extrapolation to obtain V_{LO} = 3.290 - .00507X and V_{HO} = 3.244 - .00636X in units $10^{-23}cm^{-3}$. These parameters were used in calculating $C_v(T)$ for compositions X = 0 to 66.7, for which the contraction is appreciable in

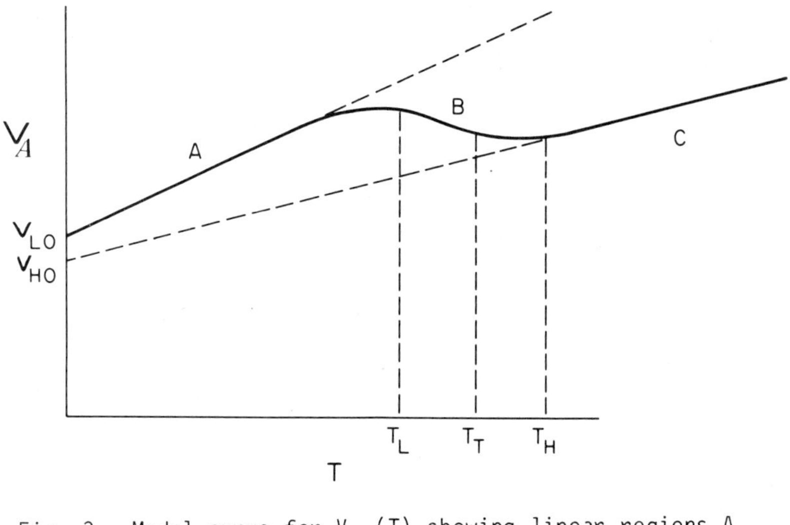

Fig. 3. Model curve for V_A (T) showing linear regions A, where $V_A = V_L$, and C, where $V_A = V_H$ and the transition region B where both L and A forms are present.

the range of measurement, and the corresponding values of ln K are plotted vs. T^{-1} in Fig. 5. It is seen that most of the points tend to fall on a straight line for each composition in accordance with Eq. 4, but there are systematic deviations at low C_v for $X \leq 33.3$ and at large C_v for $X \geq 33.3$. Since these deviations occur when $C_v \ll 1$ or $1 - C_v \ll 1$, it is possible to attribute them to uncertainties in the coefficients of V_L and V_H caused by the extrapolation with respect to composition. In Fig. 6, an alternative set of points, obtained by using $V_{LO} = 3.020$ instead of 3.039 for X = 50, and by using the experimentally determined $V_{H1} = .000517$ instead of the average value .000525 for X = 16.7, are shown to fall better on straight lines.

Actually, a straight line is not necessarily to be expected, since the monatomic reaction L → H may not be the correct description of the transformation. In Sec. 3, we describe the behavior of C_v implied by our ion pair model, for which a different equation of equilibrium is appropriate. Thurn and Ruska[4] also made an analysis in terms of Eq. 4. They did not provide quantitative information about their determination of V_L and V_H. But they presented curves for log K vs. T^{-1} for X = 33.3 and 50 which are similar to ours, with a positive curvature in the region of low C_v. They attributed this to clustering of H and L atoms to form semiconducting and metallic regions in the liquid. We shall discuss the possible role of clustering in Sec. 4.

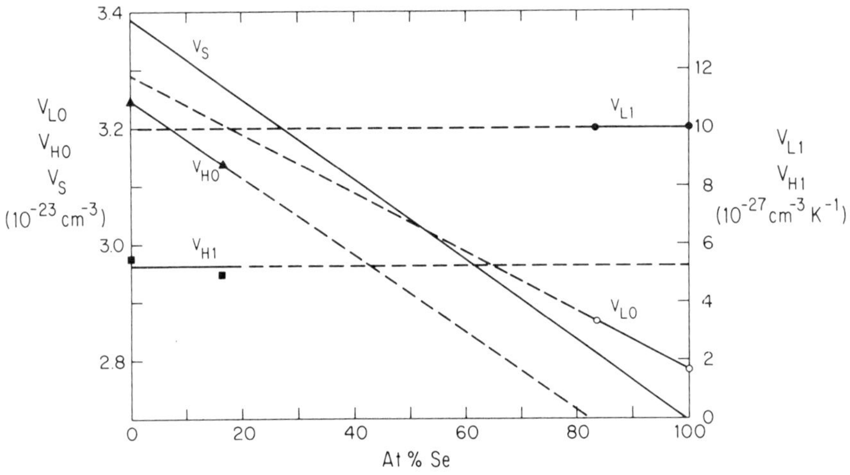

Fig. 4. Plots of the coefficients V_{LO}, V_{L1}, V_{HO}, and V_{H1} versus composition. The dashed lines are extrapolations of the experimentally derived solid lines. V_S is the atomic volume of the crystalline solid.

The parameters E and S, listed in table 1, were obtained by fitting the straight line plots in Fig. 5 to Eq. 4. The values seem too large to be physically reasonable when applied to a monatomic reaction, since $E \simeq 0.8$ eV per atom is comparable to the enthalpy of vaporization, and values of S/k much larger than 1 seem impossible to justify physically for a transformation in a condensed phase. Therefore we regard Eq. 4 primarily as an empirical description of the behavior of C_V. The values of E in table 1 lie close to an average value of 0.815 eV, and we find that the values of S, when plotted vs. X, fall on a line with the equation $S/k = 15.0 - 0.104\,X$. These results, together with Eq. 4, provide a means of interpolating or extrapolating the behavior of C_V with respect to X

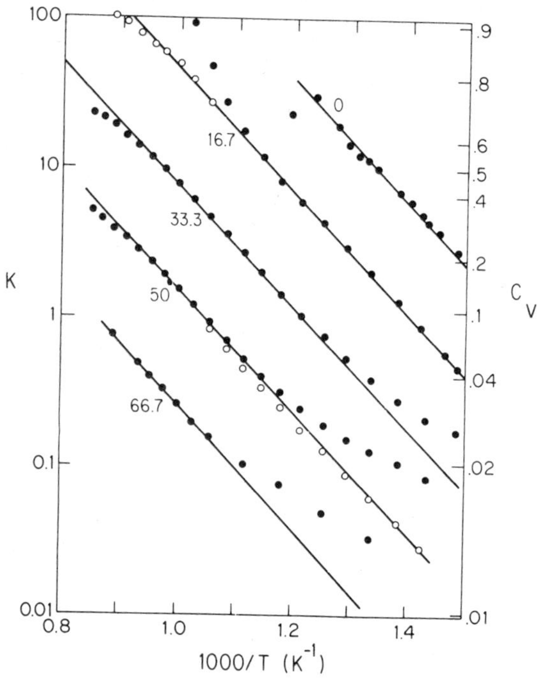

Fig. 5. The temperature dependence of the L-H equilibrium constant K (Eq. 4) for various compositions X. The corresponding values of the contraction C_V are indicated on the right.
The open circles show straight linear plots obtained with small modifications in the values of the coefficients describing V_L and V_H.

Table 1. Parameters describing $C_V(T)$ according to Eq. 4

X	0	16.7	33.3	50	66.7
E(ev)	.817	.811	.805	.81	.83
S/k	15.0	13.2	11.4	9.78	8.25

and T. In this framework, E/S gives the value of T_T as a function of X.

The contraction which occurs between T_L and T_H has generally been assumed to be caused by an increase in the packing density due to an increase in Z. However, the postulate that it reflects conversion between L and H forms with atomic volumes given by Eqs. 1 and 2 leads to a different conclusion. That is because $V_L - V_H$ extrapolates at T = 0 to a value which is small compared to the value in the vicinity of T_T. In Figure 4, we have plotted V_{L0} and V_{H0} vs. X, together with the atomic volume V_S of the crystalline solid, which was interpolated between X = 0 and 100, using Vegard's law. It is seen that the differences between the three curves are small, and they are much smaller than $(V_{L1} - V_{H1})T_T$. Thus the contraction is caused primarily by the decrease in the coefficient of thermal expansion β. β is related to the isothermal compressibility K_T by the thermodynamic equation

$$\beta = K_T(\partial S/\partial V)_T \quad . \qquad (5)$$

There is no apparent reason to expect that the volume derivative of the entropy $(\partial S/\partial V)_T$ to change much with the molecular structure. So it seems reasonable to ascribe the decrease in β, and hence the contraction, to an increase in the intermolecular forces which causes the compressibility to decrease. Such a change can be expected in the model described in the next section, since the replacement of neutral atoms in the chains by ions will cause the interchain forces of the van der Waal type to be superseded by the stronger forces which occur between ions and the induced dipoles on their neighbors.

3. THE DIPOLE PAIR MODEL

Liquid Se-Te alloys are expected to contain high concentrations of bond defect atoms at high T, and the question naturally arises whether the L-to-H transformation is caused by the generation of bond defects. Several previous authors have offered explanations of the metal transition in which threefold bond defect atoms play a role. In a recent study (I), we have obtained quantitative information about the concentrations of electronically active bond defect atoms in liquid Se_XTe_{100-X}, in the range 50 to 70 at% Se, from an analysis of the transport behavior. The results for X = 50 are of particular interest since they extend up to a temperature $\simeq T_T$, where half of the atoms are expected to be in the H form. We show in Fig. 6 the behavior of the concentrations d^+ of D^+ centers, d^- of D^- centers, and d^* of D^* centers. (Concentrations are normalized to the concentration of atoms $N_a = 2.7 \times 10^{22} cm^{-3}$). The figu e also has plots of the coefficient of thermal expansion β and the adiabatic compressibility K_S. It is seen that at T_T, $d^+ = d^- =$ .04 and $d^* = .03$, which adds up to 0.11. This is considerably less than $C_v = 0.5$. However, there is theoretical and experimental evidence that concentrations of dissociated (electronically active) ions pairs $\simeq$ .04 may be accompanied by relatively large concentrations of associated (electronically inactive) ion pairs.

Ion pairs in chalcogen alloys have been frequently discussed in the context of amorphous solids, where the interest is focussed on their pos-

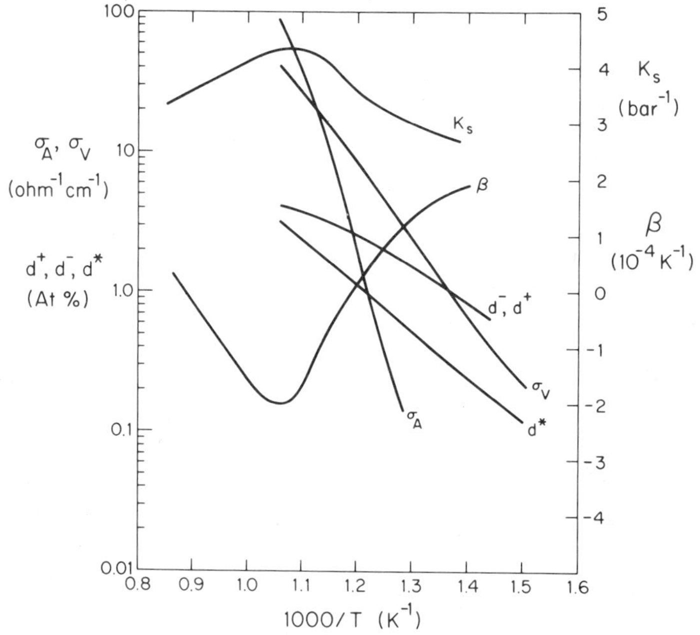

Fig. 6. Temperature dependence of various parameters (described in text) for $Se_{50}Te_{50}$.

sible formation directly from pairs of neutral bond defect atoms as the result of a greater stability.[17,18] In liquids, this issue does not arise because each center can be formed directly from a normal C_2 atom. The nomenclature valence alternancy pairs (VAP) for dissociated pairs, and intimate valance alternancy pairs (IVAP) has been used in the discussions of amorphous solids,[18] but we prefer to use a simpler terminology--dissociated and associated ion pairs. In liquids, another distinction is also important: the associated pair can be in the same molecule, i.e., directly bonded to each other, or they can be on neighboring chains. Because of entropy differences arising from polymer statistics, the intramolecular associated pairs, which are effectively 2F centers, are expected to have much higher concentrations than the intermolecular associated pairs, which are effectively 4-fold centers. We shall therefore ignore the latter. The directly bonded dipole pairs are labeled D_p with a concentration d_p.

The theory of ion pair association is discussed in a paper by one of us in relation to the liquid semiconductor alloy $(Se_{.5}Te_{.5})_{1-y}M_y$, where M is one of several monovalent metals (Na, Tl, Cu, Ag).[19] (That paper will be referred to as II). The negative ions in these alloys differ from the D^- ions of interest here, but the concentrations, temperature range, and dielectric screening are similar. It was shown that one can expect a simple LMA relation for the equilibrium $D^+ + D^- = D_p$ to be obeyed in a good approximation:

$$d^+d^-/d_p = \exp\left[-E_d/kT + S_d/k\right] . \qquad (6)$$

The entropy S_d and the enthalpy E_d have theoretical values which arise from the use of a modified Bjerrum model for the interaction of the D^+ and D^-_d ions. $S/k \lesssim 1$, and E_d is given by

$$E_d = (e^2/K)(r_N^{-1} - r_I^{-1}) \quad , \qquad (7)$$

where r_N is the distance of closest approach of the ions, r_I is the radius of the sphere with a volume equal to the average volume per dissociated ion pair, and K is the dielectric constant. The model cannot be accurate when the ions are in close proximity as in a D_p center. Therefore r_N and S_d should be regarded as adjustable parameters reflecting the particular structure of the D_p center.

Although the values of r_N and S_d are not known for the D_p centers, the values obtained in II for the ion species formed by doping $Se_{50}Te_{50}$ with Cu, Tl, or Ag are representative of what can be expected. These values (E_d - 0.49eV and S_d = 0.39k) lead to the result

$d_p = 0.46$ at T_T. This is larger than the value $C_v/2 = 0.25$, required by the hypothesis that the L-to-H transition is due to generation of ion pairs. If the data in Fig. 6 for d^+ and d^- are used in Eq. 6, together with $C_v/2$ for d_p, we find that the LMA expression is satisfied with $S_d = 2.92k$ and $E_d = 0.24eV$. The smaller value of E_d can be partly ascribed to a larger value of the screening term e^2/Kr_I in Eq. 7, and partly to a weakening of the bonds in the D_p center.[20] The larger value of S_d can likewise be attributed to smaller force constants in the bonds.

Thus our hypothesis is that as T increases, an increasing fraction of C_2 units in the polymer chains are replaced by diatomic D_p units. The D^* and (dissociated) D^- centers create chain ends and the (dissociated) D^+ centers create chain branches in the usual way. How strong are the electrostatic interactions between the D_p centers? Taking into account the fact that the bond angles are ~ 90°, it is apparent that there is an attractive interaction between two D_p centers which yields a minimum Coulomb energy when they are next nearest neighbors on a chain with a configuration shown in Fig. 7a. In this configuration, the D^- atom of $(D_p)_a$ is at approximately one bond length away from the D^+ atom of $(D_p)_b$. Assuming, for simplicity, 90° bond angles and equal bond lengths r_b, the energy of the two dipole pairs is reduced by $.24e^2/Kr_b$ as compared to an infinite separation distance. Using estimates for the dielectric constant K and the bond length r_b based on linear interpolation between values for Se and Te, the ratio of e^2/Kr_b to kT_T is found to be nearly constant with changing X, and equal to ~ 8.5. Therefore the energy of attraction $\simeq 2\ kT_T$, so that there will be an appreciable tendency to cluster within a chain, particularly at $T < T_T$. When $C_v \geq 2/3$, more than half of the chain units are D_p centers, and most of them will be in next-nearest-neighbor positions. They will tend to be arranged as in Fig. 7a, with an electrostatic link between the D^- atom of $(D_p)_a$ and the D^+ atom of $(D_p)_b$. The barrier for rotation of $(D_p)_a$ about the D_a^+-C bond (as in Fig. 7b) is $0.41e^2/Kr_b$, or about 3.4 kT_T. Therefore our model predicts that at large values of C_v, a large fraction of the dipole pairs will have an electrostatic coordination shown in Fig. 7a or 7c. This has the effect of increasing the average coordination number for the two D_p centers to 3, as compared to the average number of bonds per atom, which remains equal to 2. Thus, the ion pair hypothesis offers a resolution to the apparent conflict between the structural and electronic behavior of liquid Te, discussed in Sec. 1.

What is the expected effect of ion-pair formation on the electronic structure? To answer this, we start by reviewing the conclusions about the electronic structure just below T_E, derived in paper I. It was found that the larger activation energy of σ ($\simeq$ 1.3 eV) found in region II, as compared with a lower temperature range (region I) where the activation energy $\simeq$ 0.8eV, is caused by the appearance of acceptor band transport with conductivity σ_A, in addition to the valence band transport with conductivity σ_V. These two contributions to σ for X = 50 are shown in Fig. 6. The acceptor band derives from D^- and D^* centers, and the

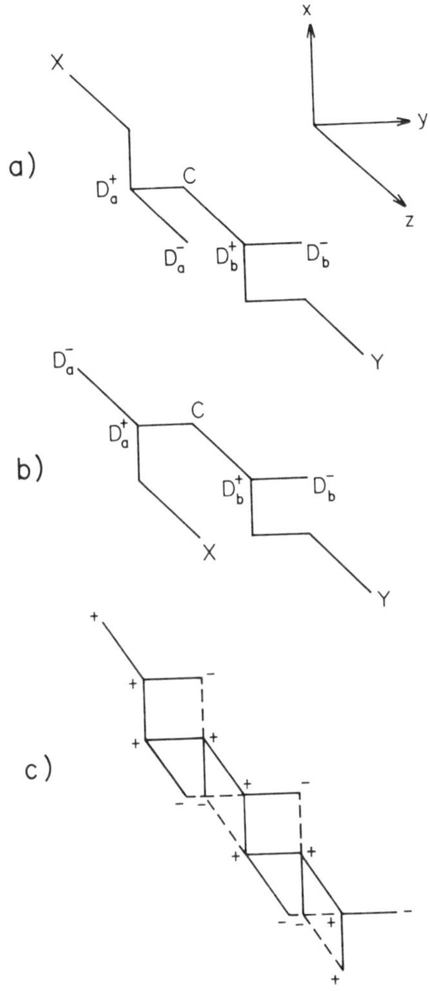

Fig. 7. a) Attractive interaction between D_p centers on chain. b) Configuration at the top of the barrier for rotation about the axis $D^+{}_a$-C. c) Ionic coordination (dashed lines) at large concentrations of D_p centers.

rapid increase in σ_A with T is due to the increase in d^* and d^-, causing an increase in the mobility in addition to the increased density of acceptor band states. The analysis was based on a narrow band model (compared to kT) for the acceptor states. Actually, the distribution of electrostatic energies as described by the Bjerrum model leads to a two-peaked density of states as shown in Fig. 8a. As described in II, the peak at the higher energy, which corresponds to dissociated ion pairs (D^+ and D^-) is below E_F. The lower peak corresponds to the associated ion pairs. These states are electronically inactive because of their greater distance from E_F, and this justifies the narrow band model in the experimental range of paper I. The separation in energy between the two peaks corresponds to the value of E_d in Eq. 7.

According to our hypothesis, the acceptor band will contain the non-bonding p electrons of the D^- ions in the ion pairs. There are four states for each such ion, and they correspond to two lone-pair electrons of each of the two C_2 atoms from which the pair of ions was created. Thus the L-to-H transformation involves a transfer of electrons from the lone-pair band, labelled π in Fig. 8, into the acceptor band, most of them in the lower peak region corresponding to the associated pairs. The diagram

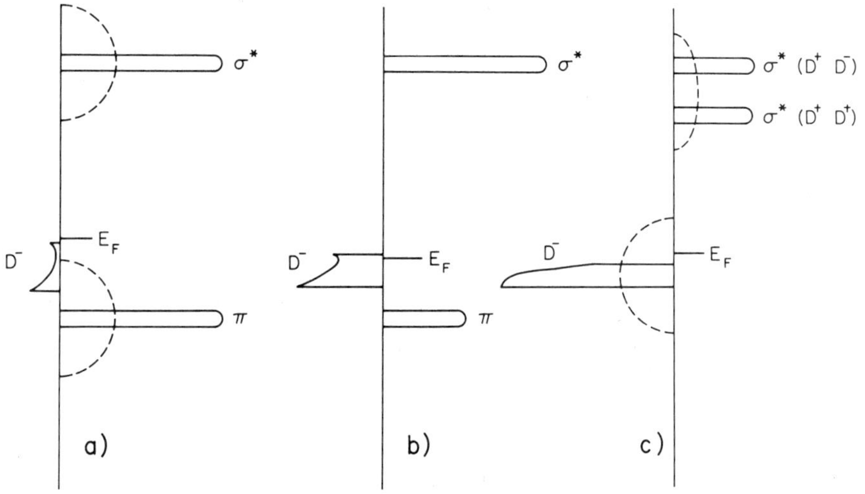

Fig. 8. Schematic behavior of electronic bands at a) low, b) intermediate, and c) large concentrations of D_p centers. Solid (dashed) lines indicate the density of states with transfer integral off (on). The density of states is plotted on the left for the acceptor band, and on the right for the valence (π) and conduction (σ^*) bands.

in Fig. 8b depicts roughly the situation expected for X = 50 at T_T, where $C_v \simeq 0.5$, $d_p \simeq 0.25$, and half of the original lone-pair valence band states are shifted into the acceptor band. In this diagram, we've indicated only the expected effect of the electrostatic energy on $N_A(E)$, and ignored broadening due to the electronic transfer integral. As the dissociated ion concentration increases, the separation E_d between the peaks decreases and ultimately the upper peak disappears when $E_d \lesssim 4$ kT.[19] This is expected to occur when $d^- \simeq .08$ and $r_I \simeq 5$ Å. The Bjerrum model is expected to become inaccurate in this range for many reasons, one of which is that the macroscopic dielectric constant cannot be used when r_I is comparable to the bond length. In addition, when d_p approaches its maximum value 0.5, the C_2 atoms disappear, and a D^- ion must attach itself to a D_p center. It therefore cannot be distinguished from the D^- component of that D_p center, and the distinction between dissociated and associated ion pairs has nearly completely disappeared.

Our ion pair model for the L to H transformation does not lead to any direct connection between the electronic and thermodynamic transition. The electronic transition occurs when E_F moves into the band in which conduction occurs, i.e., the acceptor band. This is governed by the values of d^* and d^-, which are not directly related in our model to the behavior of d_p.

The transfer of states from the C_2 lone-pair band to the acceptor band ensures that the width of the acceptor band will be large (compared to kT) when d_p approaches its maximum value 0.5. It is interesting to note another change in the electronic structure predicted by the model. When the conversion from C_2 to D_p is complete, half of the bonds will be between pairs of D^+ ions, and the other half will be between D^+D^- pairs. This will shift downward the energies of the bonding and antibonding states of the former, as indicated schematically in Fig. 8c. This shift may be large enough to destroy the gap between the acceptor band (now the valence band) and the conduction band due to the antibond states of the D^+D^+ bonds.

4. DISCUSSION

As noted earlier, the values of S and E in Table 1 are too large to be interpreted as themodynamic parameters for monatomic equilibrium between L and H atoms. Several authors have suggested that the equilibrium involves clusters of atoms which are large enough to constitute heterogeneous regions, the L clusters being semiconductor and the H clusters metallic.[21,22] They have analyzed experimental data for physical parameters such as the Knight shift, electrical conductivity etc., in terms of microscopically heterogeneous models. In order for this to be valid, the minimum number of atoms in a cluster must be ~ 30. If equilibrium between L and H forms involves n-atom clusters, the values of S and E in Table 1 should be divided by n to give the entropy and enthalpy change per atom. On the other hand, if the L to H transition corresponds simply to a reaction converting two C_2 atoms to D_p units, with $C_v = 2d_p$, the equation of equilibrium can be shown to be

$$\frac{C_v(1-C_v)}{2\,(1-C_v)^2} = \exp\left(\frac{-g_p}{kT}\right) \quad . \tag{8}$$

In this equation the presence of dissociated ion pairs and dangling bond atoms is ignored, and g_p is the free energy of the reaction $2C_2 \rightarrow D_p$. Eq. 8 is consistent with Eq. 4 if C_v is restricted to values $\lesssim 0.5$, and the result reduces the enthalpy and entropy per atom from the values in Table 1 by a factor 1/2. But Eq. 8 indicates that the slope in Fig. 5 should decrease by a factor 1/2 at $C_v \gtrsim .5$, and this is not observed.

Whatever the reaction is, the transformation between L and H forms is complete within a temperature interval $(T_H - T_L) \lesssim 300$ K, so that one should find an increase in the heat capacity ΔC_p between T_L and T_H whose integrated value is equal to the enthalpy per atom. While this paper was being written we have learned of C_p measurements by Takeda, Okazaki and Tamaki for compositions $0 \leq X \leq 50$ which show evidence of this excess ΔC_p.[23] The values of the atomic enthalpy of transformation from their measurements were ~ 0.08eV. This indicates that the L-to-H transformation involves groups of ~ 10 atoms. Clusters of this size are too small to justify heterogeneous models for interpreting the electronic parameters. It seems possible that the attractive electrostatic interactions between D_p units in the chain can lead to an equilibrium in which 3-4 D_p units are formed at a time. As indicated in Sec. 3, these would be separated by C_2 atoms at small concentrations. The theory for equilibrium required to describe this process is a good deal more complicated than the LMA, and it is currently being investigated.

Although the electrostatic coordination discussed in Sec. 3 can account for $Z > 2$ as required by diffraction data for Te, it is difficult to reconcile our model, as it stands, with the conclusions of Magana and Lannin.[12] They infer from their Raman spectra that there is a structural transformation corresponding to a decrease in the number of bonds per atom when X is increased from 20 to 30, but they find no evidence of a change in bonding as the temperature is varied at either of these compositions. Although the electrostatic coordination predicted by the D_p model is weakened at high T, some dependence of Z on T is expected.

In summary, a specific model has been proposed for the molecular basis of the thermodynamic transition in liquid Se-Te alloys. This involves replacement of neutral 2F C_2 atoms in the chain structure by 2F diatomic D_p ion pair units. This model is in accord with the concentrations of dissociated ion pairs below the transition temperature deduced in I. It also provides a resolution of the problem posed by the observation of a coordination number >2 in molten tellurium in the face of electronic evidence that the bond number <2. The comparison of thermodynamic information from the density measurements with heat capacity measurements indicates that the transformation procedes with clusters of ~ 10 atoms on the average. Such clusters are a natural consequence of the electrostatic interactions between the D_p units in the chain, but further work is necessary to determine whether detailed agreement can be obtained between the behavior of C_v and ΔC_p. Another way in which the model may be tested more directly would be to look for evidence of change transfer in the form of electrostatic shifts of X-ray core levels of the Se and Te atoms.

ACKNOWLEDGEMENTS

Part of this work was done while one of us (MC) was a visitor at the Cavendish Laboratory of Cambridge University. This work has been supported by National Science Foundation grant DMR 8023682 and DMR 8320547.

REFERENCES

1. J. C. Perron, Adv. Phys., 16:657 (1967).
2. F. Hensel, "Amorphous and Liquid Semiconductors", W. E. Spear, Ed., Univ. of Edinburgh, Edinburgh (1977), p. 815.
3. H. Endo, J. Non-Cryst. Solids 61-62:1 (1984).
4. R. Fainchtein and J. C. Thompson, Phys. Rev. 27:5967 (1983).
5. N. F. Mott and E. A. Davis, "Electronic Processes in Non-crystalline Materials", 2nd Ed., Clarendon Press, Oxford (1979), p. 30.
6. J. A. Gardner and M. Cutler, Phys. Rev. B14:4488 (1976).
7. H. Thurn and J. Ruska, J. Non Cryst. Solids 22:331 (1976).
8. K. Takimoto and H. Endo, Phys. Chem. Liq. 12:141 (1982).
9. G. Tourand, J. de Physique 34:937 (1973).
10. G. Tourand, Phys. Letters 54A:209 (1975).
11. R. Bellisent, Nuclear Instr. and Methods 199:289 (1982).
12. J. R. Magaña and J. S. Lannin, Phys. Rev. B29:5663 (1984).
13. B. Cabane and J. Friedel, J. Phys. 32:73 (1971).
14. M. Cutler, Solid State Commun. 13:1293 (1973).

15. J. E. Enderby and M. Gay, J. Non-Cryst. Solids 35-36:1269 (1980).
16. M. Cutler and H. Rasolondramanitra. J. Non-Cryst. Solids 61-62:1097 (1984).
17. R. A. Street and N. F. Mott, Phys. Rev. Letters 35:1293 (1975).
18. M. Kastner, D. Adler, and H. Fritzsche, Phys. Rev. Letters 37:1504 (1976).
19. M. Cutler, Phil. Mag., B49:83 (1984).
20. D. Vanderbilt and J. D. Joannopoulos, Phys. Rev B22:2927 (1980).
21. M. H. Cohen and J. Jortner, J. de Physique 35:C4-345 (1974).
22. Y. Tsuchiya and E. F. W. Seymour, J. Phys. C15:L687 (1982).
23. S. Takeda, H. Okazaki, and S. Tamaki, to be published in J. Phys. Soc. Japan and Phys. Rev. B.

LOCALIZATION AND THE METAL-NONMETAL TRANSITION IN LIQUIDS

W. W. Warren, Jr.

AT&T Bell Laboratories

Murray Hill, N.J. 07974

INTRODUCTION

In 1966 Mott introduced the idea of a "pseudogap" or minimum in the electronic density of states of a disordered system.[1] When the pseudogap becomes sufficiently deep, he argued, a metal-nonmetal transition should occur as the states at the Fermi level become localized by disorder. Although these ideas were eventually extended to a variety of disordered materials,[2] their initial application was an attempt to explain the unusual electrical properties of a *liquid*, elemental mercury. In subsequent papers[2-7] Mott considered a whole class of electronically conducting liquids which exhibit nonmetallic properties under certain conditions. Examples are liquid alloys which resemble semiconductors at special stoichiometric compositions, expanded elemental metals, elemental chalcogens, metal-molten salt solutions, and metal-ammonia solutions. In each case the conduction electron density can be controlled with an appropriate thermodynamic variable (composition, temperature, or pressure) to bring the system from a metallic to a nonmetallic state. Mott's ideas and continuing interest had a stimulating effect on the whole field and by 1976 nearly half the papers presented at the Third International Conference on Liquid Metals dealt with the development of nonmetallic properties in electronic liquids.[8]

It became apparent quite early in this period that nuclear magnetic resonance (NMR) offers a particularly powerful probe for microscopic study of electron localization effects at the metal-nonmetal transition in liquids.[9-13] The magnetic hyperfine interaction is sufficiently strong that NMR parameters are often dominated by couplings between nuclei and unpaired electrons. The experiments then provide information related to the electron density or the density of states at the Fermi level and, most important, they reveal the time scale of the electron-nuclear interaction. The latter is a direct measure of the localization effect. NMR investigation of liquids has the further advantage of motional averaging on the NMR time scale. The averaging removes broadening effects inherent in the study of solid amorphous systems such as glasses or doped semiconductors. The main drawback of liquid-state studies is, of course, the need for high sample temperatures.

Interest in the metal-nonmetal transition in liquids has continued and deepened up to the present time. The initial NMR studies of metal-tellurium alloys[11-13] have now been extended to several groups of electronic liquid including the liquid chalcogens and their alloys,[14-16] expanded metals,[17,18] metal-ammonia solutions,[5,19,20] metal-molten salt solutions,[21-24] and the so-called "ionic alloys".[25-27] Some of these systems have also been studied using related hyperfine techniques such as perturbed angular correlations of γ-rays (PAC) or the β-decay of polarized nuclei.[28-30] While some features of Mott's original model have been confirmed, particularly in tellurium alloys, the detailed behavior of liquid systems is richly varied. Electron localization appears to be universal at sufficiently low electron concentrations, but the forms of the localized states and, probably, the mechanism of the metal-nonmetal transition are quite different in different systems.

The preparation of this *Festschrift* presents a timely opportunity to summarize the current state of this subject in which Mott's ideas and stimulation have played such a central role. It is clearly inappropriate as well as impractical in this space to review thoroughly the field. Rather, I will focus on the NMR experiments and will compare selected experimental results for various types of liquid system in the light of the basic theoretical concepts.

THEORY: NUCLEAR MAGNETIC RESONANCE IN ELECTRONIC LIQUIDS

In liquids with sufficiently high concentrations of unpaired electrons the NMR properties are usually dominated by the magnetic hyperfine interaction. The strongest interactions are with those electrons whose wave functions have at least partial atomic s-character at the resonant nuclei. Then the contact magnetic hyperfine interaction between the nuclear spin $\vec{I}^j$ and the total electronic spin $\vec{S}(\vec{R}_j)$ at the nuclear sites $\vec{R}_j$ is

$$H_c = (8\pi/3)\gamma_n\gamma_e\hbar^2\sum_{j=1}^{N}\vec{I}\cdot\vec{S}(\vec{R}_j) \qquad (1)$$

where γ_n and γ_e are, respectively, the nuclear and electronic gyromagnetic ratios. In an applied field H_0, the nuclear magnetic resonance at $\omega_0 = \gamma_n H_0$ is shifted by an amount $\gamma_n \Delta H$ where ΔH is a local field produced by the polarization of the electron spins. Spin fluctuations and translational motion of the electrons introduce time dependence to $\vec{S}(\vec{R}_j)$ and lead to longitudinal and transverse relaxation of the nuclear spins. These effects are manifest as spin-lattice relaxation and NMR linewidth, respectively.

The nuclear electric quadrupole interaction may also contribute to nuclear relaxation for resonant nuclei $I > 1/2$. The frequency spectra of the fluctuating local electric field gradients responsible for this effect normally reflect the nuclear motions in the liquid. The strength of the interaction (magnitude of the electric field gradients) depends on details of the electronic structure. As a rough rule it may be said that electric quadrupole relaxation is primarily useful as a probe of the dynamic liquid structure while the magnetic hyperfine effects relate exclusively to the electron system. In a few favorable cases both kinds of information can be obtained from the same NMR experiment.[31]

Resonance Shifts

The shift of the Zeeman energy of a nuclear spin is given by the average value

$$\Delta E_j = \langle (8\pi/3)\gamma_n\gamma_e\hbar^2 I_z^j S_z(\vec{R}_j)\rangle \qquad (2)$$

where the applied field is taken along the z-direction. For a liquid it is assumed that an individual nucleus samples all local environments possible for that chemical element on the relevant time scale. The necessary condition is that the nuclear exchange rate ω_{ex} exceed the spread in frequencies associated with different values of ΔE_j, i.e. $\omega_{ex} >> \delta(\Delta E_j)/h$. There is then a single, average shift

$$\overline{\Delta E} = N^{-1}\sum_{j=1}^{N} \Delta E_j . \tag{3}$$

The spin operator expressed in second-quantized notation is

$$S_z(\vec{R}_j) = \frac{1}{2}\sum_{\alpha} |\phi_\alpha(\vec{R}_j)|^2 [a^+_{\alpha\uparrow}a_{\alpha\uparrow} - a^+_{\alpha\downarrow}a_{\alpha\downarrow}] \tag{4}$$

where the $\phi_\alpha(\vec{r})$ form a complete set of orbital basis states and the operators $a^+_{\alpha\sigma}$ and $a_{\alpha\sigma}$, respectively, create and destroy an electron in orbital state α with spin σ. The resonance shift expressed as a fraction of the applied field H then becomes

$$\Delta H/H = (8\pi/3)N^{-1}\sum_{j=1}^{N}\sum_{\alpha} |\phi_\alpha(\vec{R}_j)|^2 V\chi'_z(\alpha,\alpha;0) \tag{5}$$

where V is the normalization volume and, for spin S = 1/2,

$$\chi'_z(\alpha,\alpha;0) = (\gamma_e\hbar/2H)\langle a^+_{\alpha\uparrow}a_{\alpha\uparrow} - a^+_{\alpha\downarrow}a_{\alpha\downarrow}\rangle \tag{6}$$

is the contribution of the state α to the real part of the volume magnetic susceptibility at frequency $\omega = 0$. The usual electronic paramagnetic susceptibility is just

$$\chi'(0) = \sum_{\alpha}\chi'_z(\alpha,\alpha\ ;0) \tag{7}$$

Evaluation of Eq. (7) depends on the nature of the electronic states and electron statistics. If the electrons are degenerate, then $\chi'_z(\alpha,\alpha;0) \neq 0$ only for states whose energies E_α lie near the Fermi energy E_F. Further, if the states are extended, motional averaging gives

$$|\phi_\alpha(\vec{R}_j)|^2 \cong \langle|\phi_\alpha(0)|^2\rangle \text{ for all } j \tag{8}$$

and we get the usual result for the Knight shift[32]

$$K \equiv \Delta H/H = (8\pi/3)\langle|\phi_F(0)|^2\rangle V\chi'(0) \tag{9}$$

where $\langle|\phi_F(0)|^2\rangle$ is the average probability amplitude at the resonant nuclei of states whose energies lie near E_F. Note that Eq. (9) has been obtained without the usual assumption that the $\phi_\alpha(\vec{r})$ are Bloch states $(\alpha \equiv \vec{k})$.

A result similar to Eq. (9) is obtained in the limit of strong (atomic) localization. In this case the state indices α, are correlated with the indices j:

$$|\phi_\alpha(\vec{R}_j)|^2 = |\phi(0)|^2\delta_{\alpha j} \tag{10}$$

Then, if the susceptibility is independent of the site, Eq. (5) yields

$$\Delta H/H = (8\pi/3)\langle|\phi(0)|^2\rangle V\chi'(0). \tag{11}$$

The susceptibility in this limit is of the Curie or Curie-Weiss form.

Nuclear Relaxation

The longitudinal nuclear relaxation rate, $1/T_1$ is determined by low frequency fluctuations of the transverse electronic magnetization.[33]

$$1/T_1 = (16/9)\pi^2\gamma_e^2\gamma_n^2\hbar^2[J^1(\omega_0) + J^{-1}(\omega_0)] \tag{12}$$

where

$$J^{\pm 1}(\omega) = \int_{-\infty}^{\infty} dt\, e^{i\omega t} N^{-1}\sum_{j=1}^{N}\langle S_\pm(\vec{R}_j, t)S_\mp(\vec{R}_j, 0)\rangle \tag{13}$$

and

$$S_+(\vec{R}_j) = \sum_{\alpha,\beta} \phi_\alpha^*(\vec{R}_j)\phi_\beta(\vec{R}_j)a^+_{\alpha\uparrow}a_{\beta\downarrow} \tag{14a}$$

$$S_-(\vec{R}_j) = \sum_{\alpha,\beta} \phi_\alpha^*(\vec{R}_j)\phi_\beta(\vec{R}_j)a^+_{\alpha\downarrow}a_{\beta\uparrow}. \tag{14b}$$

Evaluation of Eq. (13) in the low frequency limit yields the general result

$$J^1(\omega) = \frac{2\hbar V}{(\gamma_e\hbar)^2}\frac{1}{N}\sum_{j=1}^{N}\sum_{\alpha,\beta}|\phi_\alpha(\vec{R}_j)|^2|\phi_\beta(\vec{R}_j)|^2\frac{kT}{\hbar\omega}\chi''_\pm(\alpha,\beta;\omega) \tag{15}$$

where

$$\chi''_\pm(\alpha,\beta;\omega) = \frac{\pi(\gamma_e\hbar)^2}{V}\frac{\hbar\omega}{kT}f_{\alpha\uparrow}(1-f_{\beta\downarrow})\delta(\hbar\omega+E_{\alpha\uparrow}-E_{\beta\downarrow}) \tag{16}$$

and $f_{\alpha\sigma}$ is the occupation probability for an electron in a state of energy $E_{\alpha\sigma}$. A similar expression applies for $J^{-1}(\omega)$. A typical nuclear spin relaxation event thus involves a mutual nuclear-electron spin flip and an orbital transition $\alpha \to \beta$. The spectral functions $J^{\pm 1}(\omega)$ are usually evaluated for Bloch states ($\alpha \equiv \vec{k}$, $\beta \equiv \vec{k} + \vec{q}$).[34] However, this is unnecessarily restrictive and is obviously inappropriate for liquids where $\vec{k}$ is not a good quantum number. We need to assume only that the states are extended, so that Eq. (8) applies, and that the electrons are degenerate. In this case, from Eq. (15)

$$J^1(\omega) = \frac{2\hbar V}{(\gamma_e\hbar)^2}\langle|\phi_F(0)|^2\rangle^2\frac{kT}{\hbar\omega}\sum_{\alpha,\beta}\chi''_\pm(\alpha,\beta;\omega) \tag{17}$$

The corresponding result for Bloch states was given by Moriya.[35]

The integrated susceptibility for noninteracting nearly free electrons is

$$\sum_{k,q} \chi_{\pm}''(k,q;\omega) = \pi V N(E_F)^2(\gamma_e \hbar)^2 \hbar\omega \tag{18a}$$

$$= 2\pi V \chi'(0) N(E_F)\hbar\omega \tag{18b}$$

since $\chi'(0) = (\gamma_e \hbar)^2 N(E_F)/2$. Substitution into Eqs. (12) and (17) yields the familiar Korringa result for the relaxation rate in a metal[36]

$$\left(\frac{1}{T_1}\right)_{\text{Korr.}} = (64/9)\pi^3 \hbar^3 \gamma_e^2 \gamma_n^2 kT \langle|\phi_F(0)|^2\rangle^2 V^2 [N(E_F)]^2 \tag{19}$$

or, using Eq. (9), the "Korringa Relation"

$$\left(\frac{1}{T_1}\right)_{\text{Korr.}} = (4\pi k/\hbar)(\gamma_n/\gamma_e)^2 K^2. \tag{20}$$

To discuss liquids, we must be more careful as Holcomb[37] has emphasized. Equations (9) and (17) remain correct for extended states, but the dynamic susceptibility must be approximated. A simple model[11] containing the essential physics is based on the assumption that the spin correlations decay exponentially with a characteristic time τ

$$\langle S_{\pm}(\vec{R}_j, t) S_{\mp}(\vec{R}_j, 0)\rangle \propto \exp(-t/\tau) \tag{21}$$

and in the low frequency limit

$$\sum_{\alpha,\beta} \chi_{\pm}''(\alpha,\beta;\omega) \cong 2\chi'(0) \frac{\omega\tau}{1+\omega^2\tau^2} \to 2\chi'(0)\omega\tau. \tag{22}$$

Also, for $\omega\tau \ll 1$ motional narrowing gives $1/T_1 = 1/T_2$.[33] Comparison of Eq. (22) with Eq. (18b) shows that to within a factor π

$$\frac{1}{T_1} \cong \left(\frac{1}{T_1}\right)_{\text{Korr.}} \frac{\tau}{\hbar V N(E_F)}. \tag{23}$$

Now, in a metal in which the mean free path λ exceeds the average separation between atoms a, the characteristic time is roughly the time for a Fermi velocity electron to move a distance a, i.e.

$$\tau \cong a/v_F \cong \hbar V N(E_F). \tag{24}$$

Thus Korringa relaxation is expected, even in disordered systems, as long as $\lambda > a$. If this is not true and $\tau > a/v_F$, a relaxation enhancement is expected:

$$\eta \equiv \frac{(1/T_1)}{(1/T_1)_{\text{Korr.}}} \cong \frac{\tau}{\hbar V N(E_F)} > 1. \tag{25}$$

In the highly localized limit represented by Eq. (10), evaluation of Eqs. (12), (15), and (22) using a Curie susceptibility $\chi'(0) = n_e(\gamma_c \hbar)^2/4\,kT$

leads to the following expression for the relaxation rate:

$$1/T_1 = (c_s/2\hbar^2)\langle A^2\rangle\tau \tag{26}$$

where c_s is the mole fraction of localized spins and A is the hyperfine coupling, $(8\pi/3)\gamma_e\gamma_n\hbar^2|\phi(0)|^2$.

Electron-Electron Interactions

In the foregoing discussion we have neglected the effects of the many-body interactions among the electrons. In the metallic limit, interaction effects have a noticeable effect, mainly through exchange enhancement of the susceptibility. Because $\chi'(0)$ and the integrated q-dependent dynamic susceptibility, Eq. (18a), are enhanced by different amounts, the Korringa relation is not precisely obeyed.[35] Values of η in the range 0.5-0.7 are typical of simple metals, both solid and liquid. Closer to the metal-nonmetal transition interactions have still stronger effects. In a system containing one electron per atom, in fact, electrons can be localized at the Mott transition by the effects of coulomb interactions.[38] Near the transition, in the so-called "highly correlated metal", the susceptibility and Knight shift are enhanced by a factor $1/2\xi$ where ξ is the fraction of doubly - occupied sites.[39] The q-dependent susceptibility, and Korringa enhancement can be expected to change in this regime, especially if the system shows a tendency toward antiferromagnetic exchange.

THE IOFFE-REGEL LIMIT AND DIFFUSIVE TRANSPORT

When the conduction electron density is sufficiently high, ionic potentials are screened so strongly that scattering is relatively weak. This is the case in most molten elemental metals and many liquid alloys. For these "nearly free electron" metals, the electron mean free path exceeds the near-neighbor separation and the conductivity is adequately predicted by the Ziman and Faber-Ziman theories.[40] From the arguments in the preceding section, we would not expect Korringa enhancement in the NFE metals. The main reasons for the small deviations from the Korringa relation in metals are the effects of electron-electron interactions and, in transition metals, modifications to Eq. (20) necessitated by the dominance of d-electrons.

It was pointed out by Ioffe and Regel[41] and emphasized by Mott that there exists a fundamental limit to scattering theory when the scattering becomes so strong that $\lambda \sim a$. The conductivity at this point is, approximately,[42]

$$\sigma_{IR} \cong (1/3)(e^2/\hbar a) \tag{27}$$

In typical liquids, $a \sim 2.5$-3.0Å, and $\sigma_{IR} \cong 3000\ (\Omega\ \text{cm})^{-1}$. Below the Ioffe-Regel limit transport becomes diffusive. The author has shown[11] that if the electron spin and charge diffuse with the same characteristic time, there is a simple approximate relation between the Korringa enhancement and the conductivity

$$\sigma \cong \sigma_{IR}/\eta. \tag{28}$$

This result is obtained by using the "hydrodynamic" form for the dynamic susceptibility, Eq. (22), with the correlation time determined by the electron diffusion coefficient $\tau = a^2/6D$.

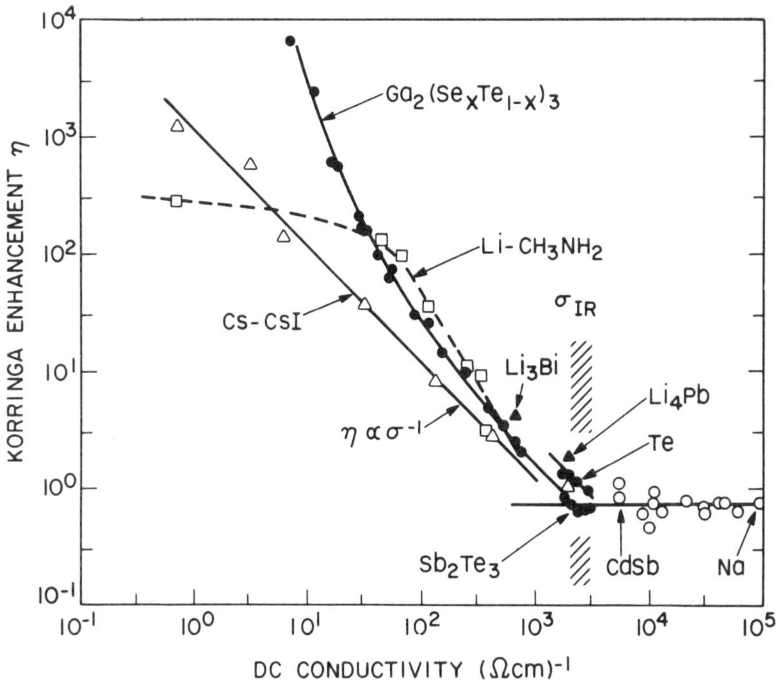

Fig. 1. Log-log plot of Korringa enhancement versus DC electrical conductivity, corrected where appropriate for ionic conductivity. Shaded region denotes approximate location of the Ioffe-Regel limit. Data sources: Cs-CsI, Ref. 24; Li-CH_3NH_2, Refs. 20, 66; $Ga_2(Se_xTe_{1-x})_3$, Ref. 43; Li_3Bi and Li_4Pb, Ref. 67; Te, Ref. 14; Sb_2Te_3, Ref. 11; CdSb, Ref. 68; other metallic liquids (o), Ref. 69.

Equation (28) has two important implications. First, it shows explicitly that the onset of Korringa enhancement coincides with the Ioffe-Regel limit, i.e. $\eta > 1$ if $\sigma < \sigma_{IR}$. Second, when the conductivity is less than σ_{IR}, σ should vary roughly as $1/\eta$ in the diffusion range. These features were originally confirmed in studies of liquid semiconducting metal-chalcogen alloys[11] and have since been verified in a number of liquid systems of quite different chemical character. A few examples are shown in the log-log plot of η versus given in Fig. 1. In general, Eq. (28) holds for a range of conductivity about one order of magnitude below σ_{IR}. At lower conductivities, the correlation may break down for a number of reasons, including incipient localization. For example, in liquid $Ga_2(Se_xTe_{1-x})_3$, the conductivity begins to be dominated by higher mobility electrons further from E_F than those responsible for nuclear relaxation.[43] In Cs-CsI, however, Eq. (28) holds over three decades indicating that the "magnetic" and "conducting" electrons are the same.

Götze and Ketterle[44] have recently described a mode-coupling theory for nuclear relaxation in disordered conductors. Their theory yields Eq. (28) in the strong scattering limit and predicts the detailed behavior in the region of enhancement onset. Various higher order terms provide for deviations from Eq. (28).

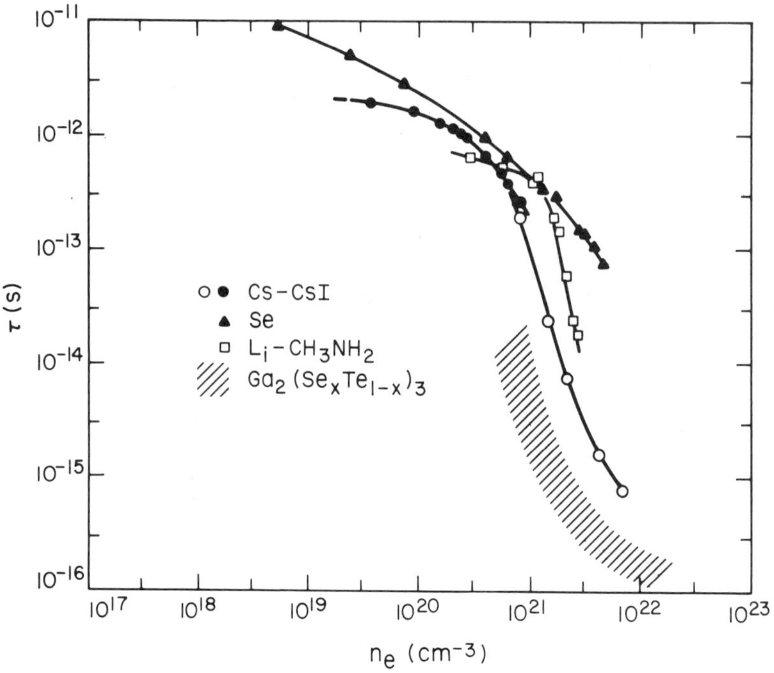

Fig. 2. Log-log plot of nuclear hyperfine correlation time τ versus electron density n_e for various liquid systems. Solid points, strong localization model, Eq. (26); open points, itinerant - weak localization model, Eq. (23). Shaded region indicates uncertainty in n_e values derived from resonance shift for $Ga_2(Se_xTe_{1-x})_3$. Data sources are given in Fig. 1 except Se, Ref. 16.

LOCALIZATION

Localization in Liquids

Discussions of electron localization have usually defined localization as the absence of electronic diffusion[45] or vanishing DC electrical conductivity at 0 K.[3] It is important to state, therefore, what is meant by localization in liquids at relatively high temperatures. We need also to consider the structures of some of the localized states that can form in the liquid state.

The characteristic which distinguishes liquids from other forms of disordered matter is the time scale associated with changes in the local atomic arrangement. For *non-associated* liquids including most elemental liquid metals and simple ionic melts like the alkali halides, single-particle diffusive motions lead to rearrangement of the local structure on the picosecond time scale (10^{-12}s). *Associated* or molecular liquids have somewhat longer characteristic times for intermolecular correlations depending on the size of the associated units.

In metallic liquids, electrons move through a disordered structure that looks static on the electronic time scale (10^{-15}-10^{-16}s). The "slow" structural rearrangements simply lead to an ensemble average of properties such as NMR which are measured over long times (10^{-8}s). With progressively stronger localization, electrons slow down with respect to ionic motions until their *residence time* near a given ion or group of ions is limited by the lifetime of the ionic configuration. This is "localization" in a liquid. Mott[4] recognized that electrons localized in this sense can move without thermally activated hopping. True "localization, that is, complete absence of long-range diffusion in the laboratory frame, is impossible in liquids. Nevertheless, the three to four orders of magnitude ratio of electronic times in a metal to those for ionic diffusion is sufficient to give practical meaning to liquid-state localization.

Hyperfine correlation times for several liquid systems are plotted against electron density n_e in Fig. 2. These diverse materials all show rapid increases of τ as n_e decreases; localization times $\gtrsim$ 1 ps are found for $n_e \sim 10^{20}$ cm^{-3}. The results for Cs-CsI and Li-methylamine show a saturation effect with constant τ-values in the ps range at low electron density. Localized electrons in these systems are solvated by groups of ions or molecules and it is clear that the localization lifetime is limited by rearrangement of the host liquid structure. For Se, on the other hand, localized states are believed to be dangling bonds on broken polymeric chains. The polymers move relatively slowly and carry the localized electron with them. Thus longer correlation times are possible and values up to 10 ps were observed.

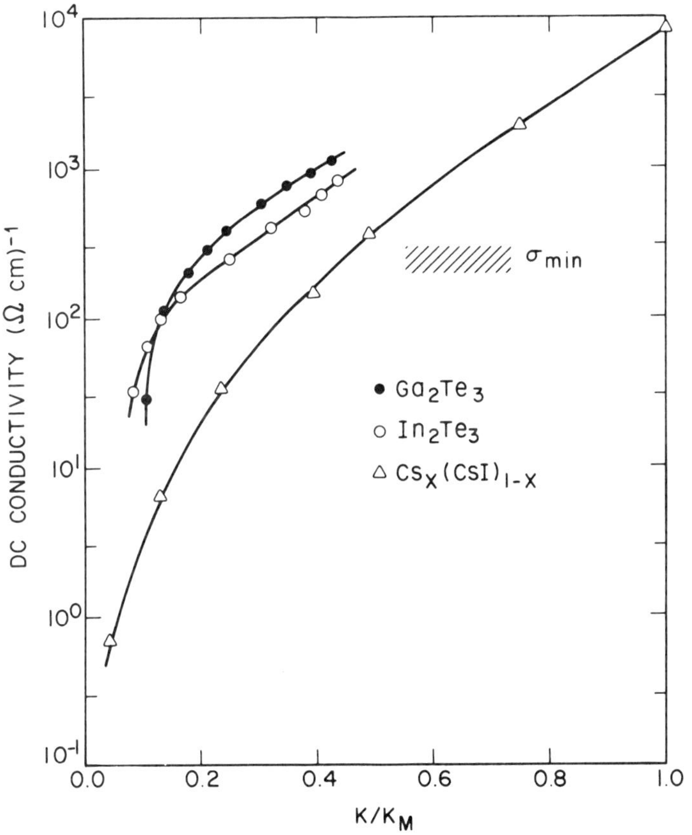

Fig. 3. Semi-log plot of DC electrical conductivity versus resonance shift normalized to the metallic limit K/K_M. Data sources: Ga_2Te_3 and In_2Te_3, Ref. 11; $Cs_x(CsI)_{1-x}$, Refs. 24 and 49.

The foregoing results hint at the rich variety of specific forms localization can take. Mott[46] proposed that Anderson localization should lead to multi-site states in which the probability amplitude on individual sites decays with the distance from a central site. The strong localization limit in this case is essentially an atomic state. The model provides a simple means of describing the continuous evolution of localized states from extended states in terms of gradually decreasing correlation lengths. However, in liquids showing the strongest localization effects, there is little evidence of a dominant role for structural disorder. The states encountered (broken bonds, F-centers) are closely related to defect states found in crystals in which translational symmetry is broken at only a single point. It is plausible that weakly localized multisite states form in the range of incipient localization ($\tau \sim 10^{-14}$s) only to be replaced by solvation or defect states in the dilute limit, but this has not been shown experimentally. Except for expanded liquid alkali metals, atomic states are not found in the extreme nonmetallic limit.

Onset of Localization - The Minimum Metallic Conductivity

Because the metal-nonmetal transitions observed in liquids are continuous, it is difficult to establish a boundary between diffusive transport in extended states and weak localization. Mott[3,42,47] has long argued that this transition is marked by a minimum conductivity σ_{min} for transport in extended states. The value of σ_{min} is derived from the Anderson criterion[45] for localization by disorder. It is important to keep in mind that σ_{min} is distinct from the Ioffe-Regel limit. In fact,[47]

$$\sigma_{min} \cong (1/10)\sigma_{IR} \qquad (29)$$

so that for liquids, $\sigma_{min} \cong 300\ (\Omega\ \text{cm})^{-1}$. The concept of the minimum metallic conductivity implies a discontinuous drop of the DC conductivity to zero at 0 K. This is in contrast to the scaling theory assumption[48] of a continuous decrease to zero near the metal-nonmetal transition. At high temperature, the transition would necessarily be broadened by thermal fluctuations and it is impossible to distinguish between these possibilities although one might expect to observe residual features in the electronic properties when $\sigma \sim \sigma_{min}$.

The best evidence for σ_{min} in liquids is found in experiments on the liquid III-VI semiconductors Ga_2Te_3 and In_2Te_3.[11] When the conductivity is plotted against the Knight shift (Fig. 3); a clear shoulder is seen for $\sigma \cong 100\ (\Omega\ \text{cm}^{-1})$. This value of the conductivity and the estimated density of states at the same point, $N(E_F)/N(E_F)_{\text{free elec.}} \sim 0.2$, are in reasonably good agreement with Mott's estimates for the localization onset.

A conductivity anomaly at $\sigma \sim \sigma_{min}$ is not a general feature of the metal-nonmetal transition in liquids. For example, conductivity data for $Cs_x(CsI)_{1-x}$,[49] shown in Fig. 3, and other alkali metal - alkali halide solutions show only a smooth decrease with no hint of any "special" value of σ. The onset of a divergence in the dielectric constant (polarization catastrophe) in $K_x(KCl)_{1-x}$ does provide a criterion for a transition at $x \cong 0.1$,[50] but at this concentration, $\sigma \sim 0.1\ \sigma_{min}$. Thus the importance of disorder in localizing electrons may be quite different in the metal-tellurium alloys and metal-salt solutions.

Mott, in fact, predicted that such differences might occur as a consequence of differing orbital symmetry of the electronic states.[51] In particular, he argued that the degree of disorder normally found in liquids should be insufficient to localize states having mainly s-character. In contrast, the highly directional character of p-states requires both the correct radial and angular correlations between neighbors in order to preserve extended states. Chemical bonding in the III-VI compounds is of the covalent s-p hybrid form in the crystal and it is reasonable to expect a high degree of p-character in the liquid as well. That there is evidence of an effect near σ_{min} in Ga_2Te_3 and In_2Te_3 but not in the alkali metal-halide solutions is consistent with Mott's suggestion. The excess electrons in the halide systems are expected to have mainly alkali metal s-character.

Inhomogeneous Models

Our discussion up to this point is based on the implicit assumption that the liquid system is microscopically homogeneous. This means that there is some continuous mono-modal distribution of parameters which characterize the sites and the local electronic structure. This is the viewpoint taken by Mott. Others, including Hodgkinson[52] and Cohen and Jortner[53] have invoked inhomogeneous models. Cohen and Jortner argued, in fact, that separation into distinct metallic and nonmetallic regions is a general property of systems undergoing metal-nonmetal transitions. Electron localization is then interpreted as the isolation of small metallic regions and the transition is essentially a percolation transition that occurs when the metallic fraction increases sufficiently. With suitable parameters, the inhomogeneous model can be used to fit much of the experimental data, but there is little structural evidence that the required fluctuations actually exist in the systems considered. The main exception to this is the situation close to a critical point where critical fluctuations may span the metal-nonmetal transition.

Tsuchiya et al.[54] have recently applied an inhomogeneous model to liquid In_2Te_3 and Ga_2Te_3. They argue that partial preservation of covalent bonds across the melting transition leads to regions of distinct metallic or nonmetallic character. As the temperature is raised, the nonmetallic regions are said to dissociate leading eventually to a percolation transition to the metallic state. The NMR shift K is used to determine the metallic fraction p at each temperature according to

$$p = K/K_M \tag{30}$$

where K_M is the shift in the fully metallic limit. The values of p(T) obtained in this way lead to a reasonably consistent interpretation of the electrical transport properties. But, as we now show, the inhomogeneous model is in poor agreement with the observed nuclear relaxation rates.

Assuming rapid motional averaging between the metallic and nonmetallic regions, as is implicit in Eq. (30), the relaxation rate in an inhomogeneous system is

$$1/T_1 = p(1/T_1)_{Korr.} + (1-p)(1/T_1)_{NM} \tag{31}$$

where $(1/T_1)_{Korr.}$ is the Korringa rate corresponding to K_M and $(1/T_1)_{NM}$ is the rate in the nonmetal. It is straightforward to show that the Korringa enhancement is

$$\eta = \frac{1}{p} + \frac{(1-p)}{p^2}\,\frac{(1/T_1)_{NM}}{(1/T_1)_{Korr.}} \tag{32}$$

If electric quadrupolar contributions are eliminated, as is possible for $^{69,71}Ga$ or ^{133}Cs, the nonmetallic rate should be much less than the Korringa rate so that the second term in Eq. (32) is negligible except for very small values of p.

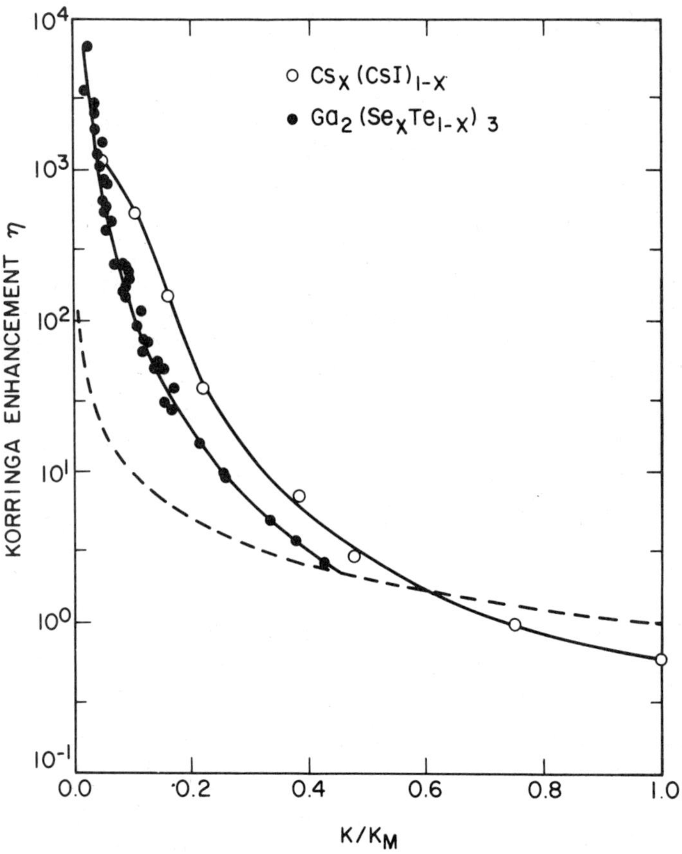

Fig. 4. Semi-log plot of Korringa enhancement versus normalized resonance shift K/K_M. Broken line denotes behavior expected for inhomogeneous system, Eq. (32), as described in text.

Experimental data for η in $Ga_2(Se_xTe_{1-x})_3$ and $Cs_x(CsI)_{1-x}$ are plotted against K/K_M in Fig. 4 and compared with $\eta = 1/p$. It is clear that the inhomogeneous model seriously underestimates the degree of Korringa enhancement. Tsuchiya et al.[54] noticed a similar discrepancy with the relaxation data in In_2Te_3 although the quadrupole effect cannot be excluded in that case. In order that Eq. (32) agree with the observed relaxation enhancement for $p \cong 0.1$, it would be necessary that $(1/T\ )_{NM} \gtrsim (1/T_1)_{Korr}$. There is no reason to expect such rapid *magnetic* relaxation in the nonmetal. The strong observed nuclear relaxation enhancement thus presents a challenge to the inhomogeneous model which can apparently not be met by any reasonable choice of model parameters.

ELECTRON-ELECTRON INTERACTIONS

Among liquid systems, the expanded alkali metals stand out as liquids in which interactions are likely to play an important role because there is precisely one electron per atom. Indeed a number of magnetic effects have been observed and attributed to the coulomb repulsion between electrons. Freyland[55] observed a strong enhancement of the magnetic susceptibility of cesium at low density. The effect is consistent with the expected enhancement of the density of states in a highly correlated metal.[39,56] The Knight shift is also enhanced,[18] as expected from Eq. (9), but the effect is weakened by a surprising decrease in $<|\phi_F(0)|^2>$ in the low density metal.[57]

Korringa enhancement measurements in cesium[18] show an increase in η at low density, similar to the effect at the Ioffe-Regel limit, but which appears to have a different origin. The onset of enhancement in cesium occurs before the Ioffe-Regel limit in a range where the Hall coefficient remains at the free electron value.[58] A similar effect was observed recently in sodium at densities for which the conductivity exceeds σ_{IR} by more than an order of magnitude.[59] The change in the Korringa enhancement in the alkali metals has been interpreted[18,59] as a change in the enhancement of the q-dependent dynamic susceptibility associated with a trend from ferromagnetic to antiferromagnetic exchange at low density. This development is consistent with the expectation of an antiferromagnetic state in a hypothetical crystal[60] and with the known occurrence of singlet Cs_2 dimers in the dense vapor.[55]

Simple considerations suggest that interaction effects are less important in the alkali metal-alkali halide solutions such as $Cs_x(CsI)_{1-x}$. At low metal concentrations, the ratio 1/x of available sites to the number of electrons becomes large. Thus, just as in a compensated semiconductor, conduction can take place without double occupation of the sites. In systems such as the chalcogen alloys or expanded mercury, the potential for interaction effects is higher and their roles in these systems is by no means clear.

Finally we consider briefly the effective coulomb attraction or "negative U" effect.[61] Alternating pairs of doubly-occupied and vacant sites form the low temperature states of such systems as liquid Se and mixed-valence salts like $In^+(InCl_4)^-$. The alternating valence state is stabilized by structural relaxation and is nonmagnetic. At higher temperatures, singly-occupied localized states can be generated by thermal excitation and these have a strong effect on NMR properties.[16,62,63] It is possible that mixed-valence effects also play a role in the chemical bonding responsible for semiconducting behavior in liquids like In_2Te_3. Among the solutions with solvated electrons, there is good evidence for diamagnetic species in Na-NaBr[64] and the ion Na^- has been reported in NMR studies of Na-hexamethylphosphoramide.[65]

CONCLUSIONS

As a result of NMR studies of a large number of liquid systems, two features emerge as general properties of liquids undergoing metal-nonmetal transitions. One is the onset of diffusive transport at the Ioffe-Regel limit. This point is marked by the appearance of Korringa enhancement that increases rapidly as the conductivity decreases. The other universal feature is localization at low electron densities. Localization times comparable with the characteristic times for rearrangement of the liquid structure have been measured in several systems. There are, however, a variety of specific structures for localized states including solvated electrons, dangling bonds and free atoms or ions.

The most difficult aspect of the subject remains the metal-nonmetal transition itself. The pseudogap model of Mott seems most appropriate for semiconducting alloys where progressively stronger chemical bonding continuously decreases the density of states at the Fermi level. In such systems the gap is a consequence of partial preservation of solid-like local order with an appreciable covalent contribution to the bonding. The minimum metallic conductivity σ_{min} is a useful concept for such systems but the experiments in liquids cannot distinguish between a "hard" (discontinuous) and a "soft" mobility edge.

With regard to future research, a major unsolved problem concerns the role of interactions. Except for the expanded alkali metals, the effects of interactions near the metal-nonmetal transition in liquids are poorly understood. Similarly, the relationship between the metal-nonmetal transition and the phase transitions in liquids (liquid-liquid, liquid-gas) is still only vaguely discernible. It is an intriguing and as yet unexplained fact that the conductivity close to the critical point in a number of liquid systems lies close to Mott's value of σ_{min}.

REFERENCES

1. N. F. Mott, Phil. Mag. 13:989 (1966).
2. See, for example, N. F. Mott and E. A. Davis, "Electronic Processes in Non-Crystalline Materials" (Clarendon, Oxford, 1979), second edition.
3. N. F. Mott, Adv. Physics 16:49 (1967).
4. N. F. Mott, Phil. Mag. 24:1 (1971).
5. J. V. Acrivos and N. F. Mott, Phil. Mag. 24:19 (1971).
6. N. F. Mott, Phil. Mag. 26:505 (1972).
7. N. F. Mott, Phil. Mag. 29:613 (1974).
8. R. Evans and D. A. Greenwood, ed., "Liquid Metals 1976" (Institute of Physics, Bristol and London, 1977) Conference Ser. No. 30.
9. W. W. Warren, Jr., J. Non-Crystalline Solids 4:168 (1970).
10. W. W. Warren, Jr., Sol. State Comm. 8:1269 (1970).
11. W. W. Warren, Jr., Phys. Rev. B 3:3708 (1971).
12. D. Brown, D. S. Moore, and E. F. W. Seymour, Phil. Mag. 23:1249 (1971).
13. W. W. Warren, Jr., J. Non-Crystalline Solids 8-10:241 (1972).
14. W. W. Warren, Jr., Phys. Rev. B 6:2522 (1972).
15. E. F. W. Seymour and D. Brown in "Properties of Liquid Metals", ed. S. Takeuchi (Taylor and Francis, London, 1973) p.399.
16. W. W. Warren, Jr. and R. Dupree, Phys. Rev. B 22:2257 (1980).
17. U. El-Hanany and W. W. Warren, Jr., Phys. Rev. Lett. 34:1276 (1975).
18. U. El-Hanany, G. F. Brennert, and W. W. Warren, Jr., Phys. Rev. Lett. 50:540 (1983).
19. J. P. Lelieur in "Electrons in Fluids", ed. J. Jortner and N. R. Kestner (Springer-Verlag, Berlin, 1973) p.305.
20. P. P. Edwards, J. Phys. Chem. 84:1215 (1980).
21. R. Dupree and W. W. Warren, Jr., in "Liquid Metals 1976", ed. R. Evans and D. A. Greenwood (Institute of Physics, Bristol and London, 1977) Conference Ser. No. 30, p.454.
22. R. Dupree and J. A. Gardner, J. Phys. (Paris), Colloque C8-41:C8-20 (1980).
23. S. Sotier and W. W. Warren, Jr., J. Phys. (Paris), Colloque C8-41:C8-40 (1980).
24. W. W. Warren, Jr., S. Sotier, and G. F. Brennert, Phys. Rev. B 30:65 (1984).
25. R. Dupree, D. Kirby, W. Freyland, and W. W. Warren, Jr., Phys. Rev. Lett. 45:130 (1980).
26. C. van der Marel, W. Geertsma and W. van der Lugt, J. Phys. F 10:2305 (1080).
27. R. Dupree, D. J. Kirby and W. Freyland, Phil. Mag. B 46:595 (1982).
28. M. von Hartrott, J. Hohne, D. Quitmann, J. Rossbach, E. Weihreter, and F. Willeke, Phys. Rev. B 19:3449 (1979).
29. R. L. Rasera and J. Gardner, Phys. Rev. B 18:6856 (1979).
30. C. van der Marel, P. Heitjans, H. Ackerman, B. Bader, P. Freiländer, G. Kiese, and H.-J. Stockmann, J. Non-Crystalline Solids 61-62:213 (1984).

31. See, for example, Ref. 11.
32. C. H. Townes, C. Herring, and W. D. Knight, Phys. Rev. 77:852 (1950).
33. A. Abragam, "The Principles of Nuclear Magnetism" (Clarendon, Oxford, 1961) Chap. VIII.
34. See, for example, J. Winter, "Magnetic Resonance in Metals" (Clarendon, Oxford, 1971), p.41.
35. T. Moriya, J. Phys. Soc. Japan 18:516 (1963).
36. J. Korringa, Physica (Utrecht) 16:601 (1950).
37. D. F. Holcomb, in "The Metal-Nonmetal Transition in Disordered Systems", ed. L. R. Friedman and D. P. Tunstall (Scottish Universities Summer School in Physics, Edinburg, 1978), p.251.
38. N. F. Mott, Proc. Phys. Soc., (London) Sect. A62:416 (1949).
39. W. F. Brinkman and T. M. Rice, Phys. Rev. B 2:4302 (1970).
40. J. M. Ziman, Phil. Mag. 6:1013; T. E. Faber and J. M. Ziman, Phil. Mag. 11:153 (1965).
41. A. F. Ioffe and A. R. Regel, Progr. in Semiconductors 4:237 (1960).
42. N. F. Mott, Phil. Mag. 26:1015 (1972).
43. W. W. Warren, Jr. and G. F. Brennert, in "Amorphous and Liquid Semiconductors", ed. J. Stuke and W. Brenig (Taylor and Francis, London, 1974), p.1047.
44. W. Götze and W. Ketterle, Z. Phys. B 54:49 (1983).
45. P. W. Anderson, Phys. Rev. 109, 1492 (1958).
46. N. F. Mott, Phil. Mag. 22:7 (1970).
47. N. F. Mott, Phil. Mag. 44:265 (1981).
48. E. Abrahams, P. W. Anderson, D. C. Licciardello, and T. V. Ramakrishan, Phys. Rev. Lett. 42:673 (1979).
49. S. Sotier, H. Ehm, and F. Maidl, J. Non-Crystalline Solids 61-62:95 (1984).
50. W. Freyland, K. Garbade, and E. Pfeiffer, J. Phys. Chem. 88:3745 (1984).
51. N. F. Mott, Phil. Mag. 19:835 (1969).
52. R. J. Hodgkinson, Phil. Mag. 23:673 (1971).
53. M. H. Cohen and J. Jortner, Phys. Rev. Lett. 30:699 (1973); Phys. Rev. A 10:978 (1974).
54. Y. Tsuchiya, S. Takeda, S. Tamaki, Y. Waseda, and E. F. W. Seymour, J. Phys. C 15:2561 (1982); Y. Tsuchiya and E. F. W. Seymour, J. Phys. C 16:5815 (1953).
55. W. Freyland, Phys. Rev. B 20:5104 (1979).
56. W. W. Warren, Jr., Phys. Rev. B 29:7012 (1984).
57. W. W. Warren, Jr., U. El-Hanany, and G. F. Brennert, J. Non-Crystalline Solids, 61-62:23 (1984).
58. U. Even and W. Freyland, J. Phys. F 5:L104 (1975).
59. L. Bottyan, R. Dupree, and W. Freyland, J. Phys. F 13:L173 (1983).
60. L. M. Sander, H. B. Shore, and J. H. Rose, Phys. Rev. B 24:4879 (1981).
61. P. W. Anderson, Phys. Rev. Lett. 34:953 (1975).
62. K. Ichikawa and W. W. Warren, Jr., Phys. Rev. B 20:900 (1979).
63. W. W. Warren, Jr., G. Schönherr, and F. Hensel, Chem. Phys. Lett. 96:505 (1985).
64. W. W. Warren, Jr., S. Sotier, and G. F. Brennert, Phys. Rev. Lett. 50:1505 (1983).
65. P. P. Edwards, S. C. Guy, D. M. Holton, and W. McFarlane, Phil. Mag. B 47:367 (1983).
66. Y. Nakamura, T. Toma, and M. Shimoji, Phys. Lett. A 60:373 (1977).
67. G. Kiese, P. Heitjans, H. Ackerman, B. Bader, W. Butler, P. Freiländer, C. van der Marel, H. Ruppersberg, and H.-J. Stöckman, in "Ionic Liquids, Molten Salts, and Polyelectrolytes", ed. K.-H. Bennemann, F. Brouers, and D. Quitmann (Springer-Verlag, Berlin, 1982) p. 117.
68. A. Kornblit and W. W. Warren, Jr. (unpublished).
69. G. C. Carter, L. H. Bennett, and D. J. Kahan, "Metallic Shifts in NMR" (Pergamon, Oxford, 1977), Part I.

DIFFUSION AND CONDUCTION NEAR THE PERCOLATION TRANSITION IN A FLUCTUATING MEDIUM

D. Beaglehole and M.T. Clarkson

Physics Department
Victoria University of Wellington
Wellington, New Zealand

INTRODUCTION

It is a pleasure to contribute this paper to the honour of Professor Sir Nevill Mott. Professor Mott gave encouragement to the first author during his student days at Cambridge, and this encouragement was deeply appreciated.

The paper describes the molecular diffusion, the conductivity and the dielectric constant of a microemulsion near a percolation transition. While the microemulsion does not become metallic, it shows features of the metal-insulator transition. The conductivity σ changes by six orders of magnitude over a narrow range of conducting component composition, and is accompanied with a divergence in the dielectric constant. The molecular diffusion D of the conducting component however changes more gradually, by only two orders of magnitude. In a uniform medium D and σ are related by the charge mobility. The difference in behaviour in the present system we associate with fluctuations in the microemulsion, which allow D to increase but do not contribute to the conductivity until the percolation threshold is reached.

The microemulsion system studied here, one for which many of the physical properties have been characterised by Langevin and colleagues[1,2], is formed from water and oil (toluene), and a surfactant sodium dodecyl-sulphate and cosurfactant butanol. Salt ($NaC\ell$) is added to the water, and as a function of salinity S the mixture changes from a two phase system at low S, to three phases at mid salinities, to two phases at high salinities. The two phases at low salinities are essentially pure oil in contact with a microemulsion consisting of oil drops surrounded by surfactant in brine, while at high salinities the two phases are essentially pure brine in contact with a microemulsion of water drops surrounded by surfactant in toluene. The two-to-three phase transitions, at values of S = 5.4 and 7.4 (wt % of NaCℓ in water) show features of critical transitions - the microemulsion in particular showing a peak in light scattering at these transition salinities - while the microemulsion in the three-phase region is thought to be "bicontinuous"[3], made up of continuous paths of both oil and brine components.

MOLECULAR DIFFUSION

The self-diffusion coefficients for the water, oil and surfactant molecules in the microemulsion phase have been measured by the Pulsed Field Gradient NMR method[4], which essentially determines D from the distance the molecule moves in a magnetic field gradient in an observation time Δ. In the experiment Δ can be varied from 1 to 100 msec, to check that the diffusion is unrestricted, i.e. $r^2 \propto \Delta$, and this was found to be the case for all samples. The data[5] are shown in Figure 1. The open circles show the diffusion coefficient of the toluene molecules, the closed circles for the water molecules. Similar but less complete data have been obtained by Chatenay et al[6].

Consider the toluene molecules. At low salinities when the oil is in the form of isolated droplets in the continuous water medium, D falls until $S \sim 4$, then rises smoothly through the three-phase region until D reaches values corresponding to the diffusion of toluene molecules in the continuous toluene phase. No discontinuities or divergences occur at the phase boundaries. Similarly the diffusion coefficient for water molecules falls smoothly from values for water in a continuous water phase at low salinities to a minimum at around $S \sim 9\frac{1}{2}$.

The crosses shown in Figure 1 are values of D deduced from light scattering measurements[7], which are carried out on a time scale of microseconds. These therefore measure the Brownian motion diffusion of the droplets. Near the extremes of the two phase regions the circles and crosses approach each other (although the circles always show larger diffusion coefficients, particularly for the oil rich microemulsion at high salinity). The Brownian droplet diffusion coefficient is given by $D_d = kT/_{6\pi\eta R^2}$ for a droplet of radius R in a medium of viscosity η. The molecular diffusion coefficient near the extremes are in inverse proportion to the emulsion viscosities, and suggest $R \sim 10$ nm, an increasing radius causing the weak minima in D.

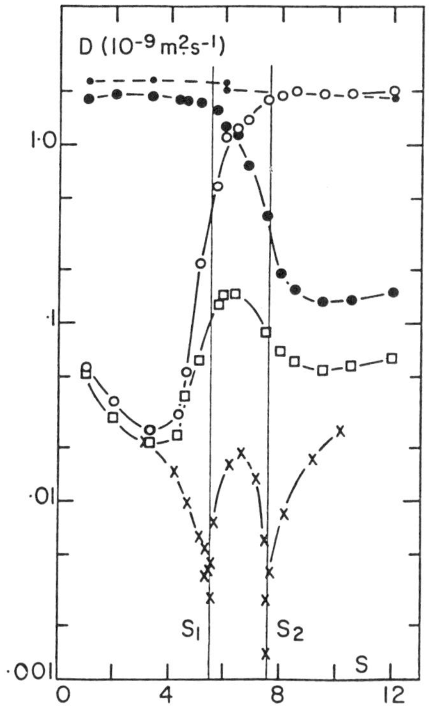

Fig. 1. Molecular diffusion coefficient versus brine salinity for oil (open circles), water (filled circles) and surfactant (squares) in the microemulsion. Also shown are values (dots + dashes) for oil in the pure oil phase, and water in the brine phase. The crosses are light scattering measurements[7].

Consider further the oil molecular diffusion data. The subsequent smooth increase in D past the minimum at S = 4 is evidence that the oil molecules are moving considerably greater distances than the oil droplets. This can only occur if the droplets come into contact, providing the oil molecules with the freedom to diffuse an extra distance ~ R with each coalescence. D is then $D_d + \frac{1}{6} n_c R^2$, where n_c is the coalescence rate. The smooth increase in D then indicates dynamic fluctuations in the microemulsion. The same comments apply equally to the diffusion of water molecules near the phase boundary at high salinities.

The collision rate of droplets undergoing Brownian motion was given by Smoluchowski[8], namely $n_c = 8\pi D_d n R$, where n is the number of droplets per unit volume. This leads to $D = D_d(1+f)$, with f the droplet volume fraction, which near S_1 and S_2 has a value ~ 0.1. Thus Brownian motion coalescene can explain roughly a 10% increase in D, but not the larger smooth increase to the three-phase values. We have suggested that fluctuations in droplet shape[9] will contribute to the coalescence rate, and to the distance molecules can move during molecular redistribution.

At the phase boundary S_1 $D_{oil} \sim \frac{1}{3} D_{free}$, while at the midpoint of the three-phase region $D_{oil} \sim \frac{2}{3} D_{free}$, with corresponding values for the water molecular diffusion coefficient at the midpoint and S_2. D_{free} here is the diffusion coefficient for the oil (water) molecule in the continuous oil (water) phase. This is good evidence that the molecular motion is one dimensional (and unrestricted in spatial direction) at S_1 and S_2, and two dimensional at the midpoint.

CONDUCTIVITY AND DIELECTRIC CONSTANT

The electrical properties of the microemulsions have been studied over the frequency range 5 Hz to 10 MHz, taking particular care in the low frequency region to eliminate electrode capacitance effects. This new data will be described in detail in a subsequent publication, but here we present the low frequency conductivity σ and dielectric constant ε variations with salinity. Pouchelon's[7] values have been used to relate salinity to volume-fraction composition.

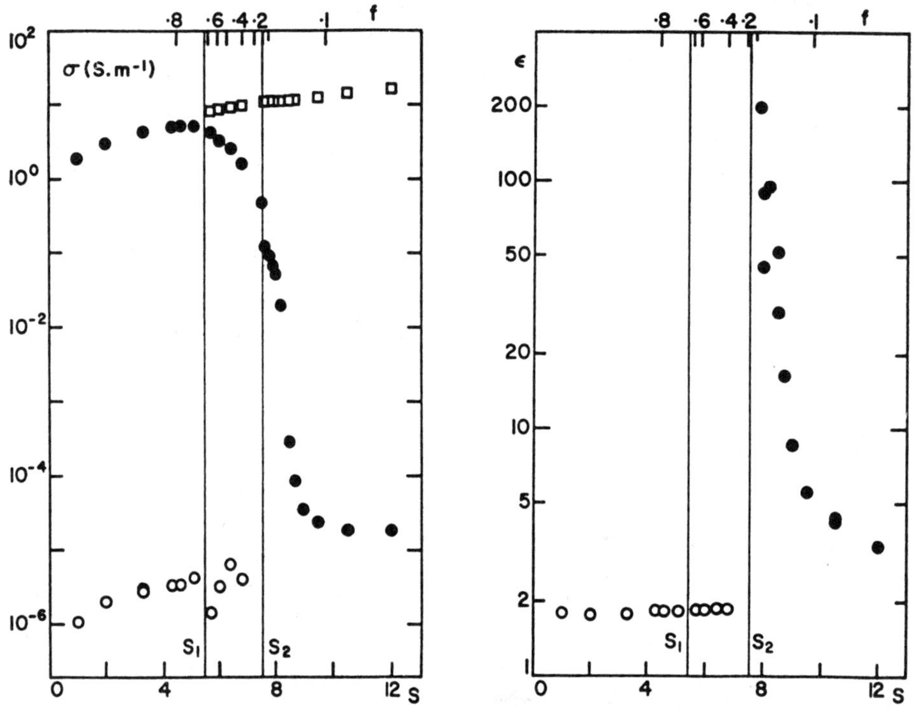

Fig. 2. σ and ε versus salinity and brine volume fraction. Filled circles, microemulsion; open circles, oil phase; squares, brine phase.

The electrical properties are shown in Figure 2, where both salinity and brine volume fractions are shown on the horizontal axis. It can be seen that the conductivity falls abruptly near S_2, to reach values typical of the insulating toluene phase by S = 9. The dielectric constant could be determined at higher salinities, and shows a rapid increase on the insulating side of the conductivity transition. The large change in conductivity and the divergence of the dielectric constant are characteristic of a percolation transition. The conductivity variation is similar to that reported by Cazabat et al[1] and Pouchelon[7] on the same system, and by Laguës and Sauterey[10] on another microemulsion system (which does not have the phase behaviour of the present system).

We look first at the volume fraction dependence of the conductivity and dielectric constant near the percolation transition. Since the conductivity of the water component depends upon salinity, we use the function $\sigma_r = \sigma_\mu/\sigma_b$ where σ_b is the brine conductivity at the same salinity. Calculations with the Bruggerman symmetric model[11] for the conductivity of an inhomogeneous mixture show that the percolation transition occurs halfway up the conductivity curve on a log scale, and the dielectric constant rises to a peak at the same point. These two measures of f_c, the percolation threshold volume fraction, almost coincide in our data; from σ $f_c = 0.135 \pm .001$, from ε $f_c = 0.132 \pm .003$ (salinity respectively 8.30 ± 0.05 and 8.45 ± 0.1).

σ and ε near the threshold are expected to vary as[12,13,14]

$$\sigma \propto (f-f_c)^t \quad f > f_c$$
$$\sigma \propto (f_c-f)^{-s} \quad f < f_c$$
$$\varepsilon \propto (f_c-f)^{-s^1} \quad f < f_c$$

where it has been argued[12] that s and s^1 exponents should be equal. Numerical calculations have suggested that $s \sim 0.7$ and $t \sim 1.7$. Our data, Figure 3, gives the values

$s = 1.8 \pm 0.2, \quad t = 1.3 \pm 0.1 \qquad s^1 = 0.8 \pm 0.1$

The conductivity variation is similar to that of Laguës and Sauterey, as we have mentioned, who have found s in the range 1.1-1.3, and t in the range 1.6-1.8. The value of s^1 characterising the dielectric constant divergence is close to that reported by Grannan et al[15] for Ag particles in a KCℓ matrix, $s^1 = 0.73 \pm .07$. Thus it appears that ε below the percolation transition and σ_r above the transition are following the expected theoretical variations, while the exponent for σ_r below f_c does not.

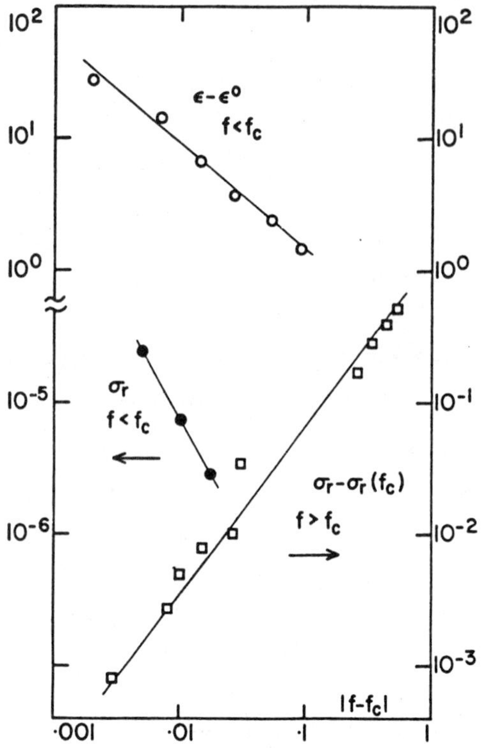

Fig. 3. Logarithmic plot of relative conductivity σ_r and dielectric constant $\varepsilon-\varepsilon^0$ versus $|f-f_c|$, where ε^0 is the dielectric constant of the oil phase.

Laguës and Sauterey[10] have suggested that the anomalous s exponent occurs as a result of "stirred" percolation, as against the usual "fixed" percolation, but such an argument should apply equally to the ε exponent. The following remarks based upon Cheshnovsky, Even and Jortner[16] elucidate the difference between s and s^1 exponents. ε is related to the frequency dependent conductivity $\sigma(\omega)$ through the Kramers Kronig relationship

$$\varepsilon - 1 = \sigma(o) \int_o^\infty \frac{\sigma(\omega)/\sigma(o)-1}{\omega^2} \, d\omega$$

The dependence of ε upon f will be the same as that of σ only if there is no f dependence of the frequency integral. Experimentally we observe a frequency dependence of σ_r which does depend upon volume fraction, and this can explain the difference between the s and s^1 exponents.

Interestingly, f_c occurs below the two-to-three phase transition, and the latter therefore must occur not at the onset of percolation but rather when most of the microemulsion consists of continuous paths.

COMPARISON BETWEEN DIFFUSION AND CONDUCTION

The diffusive motion of the charge-carrying ion will be similar to that of the water molecule, and thus the data for the self diffusion of water in the microemulsion, closed circles in Figure 1, should be a measure of the ion mobility. Since σ is also proportional to the ion mobility at first sight we might expect σ and D to show a similar variation. Indeed Mitescu et al[17] comment that this has been rigorously proven. It is clear from Figures 1 and 2 that this is not the case, but this is shown more explicitly in Figure 4, where $D(\mu)/D_b$ and σ_r/f are shown versus f. D_b is the free diffusion value for water in brine, while we have divided σ by f to allow for the varying amount of conducting medium. Above the percolation threshold the curves are indeed very similar, but below the threshold the conductivity curve shows a much greater variation. Isolated

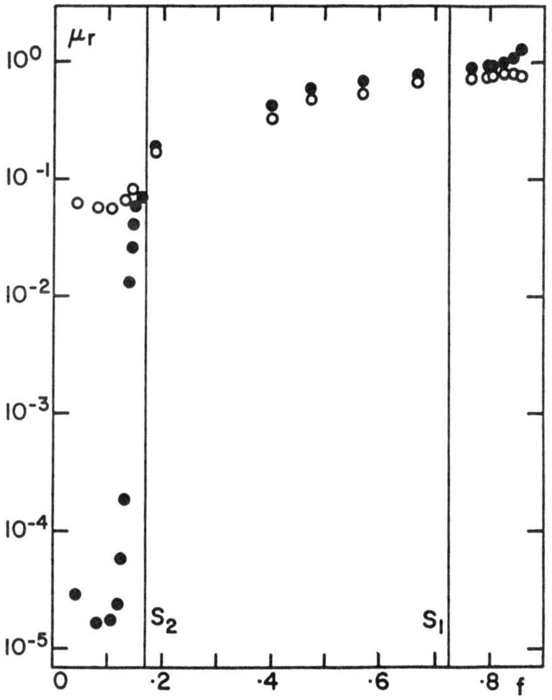

Fig. 4. Relative mobilities in the microemulsion versus brine volume fraction. Open circles, relative water mobility from diffusion coefficient $\mu_r = D(\mu)/D_b$; filled circles, relative ion mobility from conductivity $\mu_r = \sigma_r/f$.

water droplets at high salinities are probably neutral in their net charge, and thus will not contribute to the conductivity. However, even when the Brownian droplet coefficients D_d (crosses in Figure 1) are subtracted from the molecular diffusion coefficients, the difference between σ and D remains.

The transient coalescence and breakup of the water droplets which provided extra paths for diffusion will not provide additional conducting paths for the low frequency conductivity. A similarity between D and σ (essentially the Nernst-Einstein relationship) is therfore not to be expected in a fluctuating medium.

REFERENCES

1. A.M. Cazabat, D. Chatenay, D. Langevin and J. Meunier, Faraday Discuss. Chem. Soc. 76 291 (1982).
2. A. Pouchelon, J. Meunier, D. Langevin, D. Chatenay and A.M. Cazabat, Chem. Phys. Letters 76 277 (1980).
3. L.E. Scriven, Nature 263 123 (1976).
4. P.T. Callaghan, Aust.J.Phys. 37 359 (1984), P.T. Callaghan and O. Soderman, J. Phys. Chem. 87 1737 (1983).
5. M.T. Clarkson, P.T. Callaghan and D. Beaglehole submitted to Phys. Rev. Letters.
6. D. Chatenay, P. Guering, W. Urbach, A.M. Cazabat, D. Langevin, J. Meunier, L. Léger and B. Lindeman. Proceedings Bordeaux International Meeting on Surfactant in Solutions, K.L.Mittal (Ed) (Plenum), to be published.
7. A. Pouchelon, Thesis, Universite P & M. Curie, Paris (1983).
8. M. von Smoluchowski, see J.Th. G. Overbeek in "Colloid Science" ed. H.R. Kruyt (Elsevier 1952) Vol.1 p280.
9. S.A. Safran, J. Chem. Phys. 78 2073 (1983).
10. M. Laguës and C. Sauterey, J. Phys. Chem. 84 3503 (1980)
11. R. Landauer, in "Electrical Transport and Optical Properties of Inhomogeneous Mixtures", Eds. J.C. Garland and D.B. Tanner, A.I.P. Conference Proceedings 40 (New York 1978).
12. D.J. Bergman and Y. Imry, Phys. Rev. Letters 39 1222 (1977).
13. J.P. Straley, Phys. Rev. B15 5733 (1977).
14. M.H. Cohen, J. Jortner and I. Webman, in "Electrical Transport and Optical Properties of Inhomogeneous Mixtures", Eds. J.C. Garland and D.B. Tanner, A.I.P. Conference Proceedings 40 (New York 1978).
15. D.M. Grannan, J.C. Garland and D.B. Tanner, Phys. Rev. Letters 46 375 (1981).
16. O. Cheshnovsky, U. Even and J. Jortner, Phil. Mag. 44 1 (1981).
17. C.D. Mitescu, H. Ottari and J. Roussenq in "Electrical Transport and Optical Properties of Inhomogeneous Mixtures", Eds. J.C. Garland and D.B. Tanner, A.I.P. Conference Proceedings 40 (New York 1978).

COUNTER-CATION ROLES IN Ru(IV) OXIDES WITH PEROVSKITE OR PYROCHLORE STRUCTURES

J.B. Goodenough, A. Hamnett, and D. Telles

Inorganic Chemistry Laboratory
South Parks Road
Oxford OX1 3QR, UK

ABSTRACT

The Ru(IV) oxides with perovskite or pyrochlore structures contain a subarray of Ru(IV) octahedra sharing common corners, and Ru-O-Ru interactions introduce narrow π^* bands of 4d parentage. At most counter cations, only acceptor orbitals are energetically accessible; these compete with the Ru-t_2 orbitals for interaction with the O-p orbitals and thus modulate the widths of the narrow π^* bands. On Bi(III) and Pb(II) ions, $6s^2$-core states interact directly with Ru-t_2 states. This paper discusses the transition from ferromagnetism in metallic $SrRuO_3$ to Curie-Weiss paramagnetism down to 4 K in metallic $CaRuO_3$. It presents structural and transport data on the pyrochlore system $Bi_{2-x}Gd_xRu_2O_7$, which exhibits a first-order transition from the phase of metallic, Pauli paramagnetic $Bi_2Ru_2O_7$ to the phase of semiconducting, magnetic $Gd_2Ru_2O_7$. A change in the effective radius of the Bi(III) ion occurs at the transition; cooperativity and ordering among Bi:6s-Ru:t_2 interactions is discussed.

INTRODUCTION

Oxides of ruthenium are technologically important as catalysts, electrocatalysts, and resistors; they are of theoretical interest because narrow Ru(IV)-4d bands of isostructural compounds may be strongly or weakly correlated depending upon the competitive bonding at any counter cation that is present. This paper discusses the role of the counter cation in cubic perovskites and in pyrochlores.

Among the pyrochlores, the rare-earth ruthenium oxides $Ln_2Ru_2O_7$ (Ln = Pr-Lu) and $Y_2Ru_2O_7$ are all semiconductors with a spontaneous ruthenium atomic moment [1] whereas $Bi_2Ru_2O_7$ is a metallic Pauli paramagnet with a nearly temperature-independent resistivity in the temperature interval $150 < T < 500$ K [2] and $Tl_2Ru_2O_7$ exhibits a first-order semiconductor-metal transition near 125 K [3].

The perovskites $CaRuO_3$ and $SrRuO_3$ are metallic, but with a spontaneous ruthenium atomic moment. Their magnetic properties, like their transport properties, suggest they have itinerant 4d electrons even though each exhibits a Curie-Weiss paramagnetism [4,5]: $CaRuO_3$ has a $\mu_{eff} = 2.97\ \mu_B$ with a Weiss constant $\theta_p = 161$ K. The spin-only μ_{eff} for S = 1 is 2.83 μ_B; and a low-spin configuration $t_2^4e^0$ on an octahedral-site Ru(IV) ion has S = 1. However, a localized-electron configuration should exhibit a cooperative Jahn-Teller distortion at low temperatures and a giant magneto-crystalline anisotropy associated with an orbital angular momentum L = 1. Neutron-diffraction data to 4.2 K [5] showed no Jahn-Teller distortion. Moreover, $SrRuO_3$ is ferromagnetic ($T_c = 160\ K \simeq \theta_p$) [4], but its magnetization appears to be reduced from a localized electron value and to increase with externally applied magnetic field H_a. High-field (to $H_a = 12.5$ T) studies [5] on a polycrystalline sample give a normal coercivity for the hysteresis loop (signalling a modest crystalline anisotropy) and a magnetization at 4 K corresponding to an increase in the ruthenium atomic moment from $\mu_{Ru} \simeq 0.9\ \mu_B$ at $H_a = 2.7$ T to 1.55 μ_B at $H_a = 12.5$ T, where $d\mu_{Ru}/dH_a$ remained positive and relatively large. On the other hand, $CaRuO_3$ remains paramagnetic to 4.2 K [6].

Solid solutions in the system $Ca_{1-x}Sr_xRuO_3$ have been prepared [5]. Although θ_p varies continuously with x in two straight-line steps, changing to a smaller slope ($d\theta_p/dx$) at $x \simeq 0.37$ where θ_p has just become positive, nevertheless the system behaves as if it were magnetically inhomogeneous. In the range $0.4 < x < 1$ a $\theta_p < T_c$ is found and the magnetization falls rapidly as x decreases from x = 1; for small x, where θ_p is negative, there appears to be no long-range magnetic order.

Attempts to prepare a $PbRuO_3$ perovskite yielded a defect pyrochlore [7], and subsequent neutron-diffraction studies have established the composition $Pb_2Ru_2O_{6.5}$ with ordering of the oxide-ion vacancies on half of the O' sites of the $A_2B_2O_6O'$ pyrochlore structure [8]. Metallic $Pb_2Ru_2O_{6.5}$ has no spontaneous atomic moment and an apparent mixed V/IV valence on the ruthenium.

Attempts to prepare a $BaRuO_3$ perovskite were frustrated by the formation of hexagonal polytypes and a lack of sufficient pressure to convert the 9R structure all the way to the cubic phase [9]. The hexagonal polytypes are metallic without spontaneous atomic moments.

RuO_2 itself, which has the rutile structure, is also metallic without any spontaneous atomic moment [10].

From these experimental results, it is clear that counter cations interstitial to an RuO_3 array of corner-shared Ru(IV) octahedral sites can modulate the character of the Ru-4d electrons in a physically significant manner, changing them from weakly correlated itinerant electrons to strongly correlated electrons exhibiting spontaneous atomic moments and a correlation splitting between donor and acceptor Ru(IV) configurations that may exceed the bandwidth. This paper reviews evidence from photoelectron spectroscopy (PES) for the evolution from weak to strong correlations among the Ru-4d electrons in several ruthenium oxides with the pyrochlore or perovskite structure, and it presents new data on the pyrochlore system $Bi_{2-x}Gd_xRu_2O_7$. These data and those for $T\ell_2Ru_2O_7$ are interpreted in terms of the modulating role of the counter cation.

PHOTOELECTRON SPECTROSCOPY

Valence-region He(I) photoelectron spectra of several ruthenium oxides have been presented and discussed elsewhere [11]. Fig. 1 shows the spectra for $Bi_2Ru_2O_7$ and $Y_2Ru_2O_7$. The primary photoemission is dominated by the filled $O{:}2p^6$ valence band, which has a maximum intensity about 5 eV below the Fermi energy E_F. This dominant photoemission merges with a strong background of secondary-electron emission beyond 10 eV below E_F. Structure close to the Fermi energy is due to the Ru-4d states.

Expanded scans in the neighborhood of E_F, taken as zero energy on the binding-energy coordinate, are shown in Fig. 2 for $Pb_2Ru_2O_{7-y}$, $Bi_2Ru_2O_7$, $Y_2Ru_2O_7$, $SrRuO_3$ and $CaRuO_3$. Spectra of the metallic compounds terminate with a sharp discontinuity at the Fermi energy; the edge width is determined by thermal and instrumental broadening. The one semiconductor, $Y_2Ru_2O_7$, has a very small density of states at E_F.

The 4d spectrum of $Pb_2Ru_2O_{7-y}$ and, to a lesser extent, that of $Bi_2Ru_2O_7$ exhibit a shape characteristic of free electrons with a one-electron bandwidth W sufficiently larger than the correlation energy U for conventional band theory to be appropriate. This observation is consistent with the lack of any spontaneous atomic moment in these metallic compounds.

On the other hand, the 4d spectrum of the magnetic semiconductor $Y_2Ru_2O_7$ can be interpreted in terms of the multiplet states accessible upon ionization of a localized $t_2^4(^3T_{1g})$ configuration [12], the weak 4d component close to E_F being associated with the $^4A_{2g}$ final state and the structure at higher binding energy with the $^2T_{1g}$ and ^2E_g states. Thus a

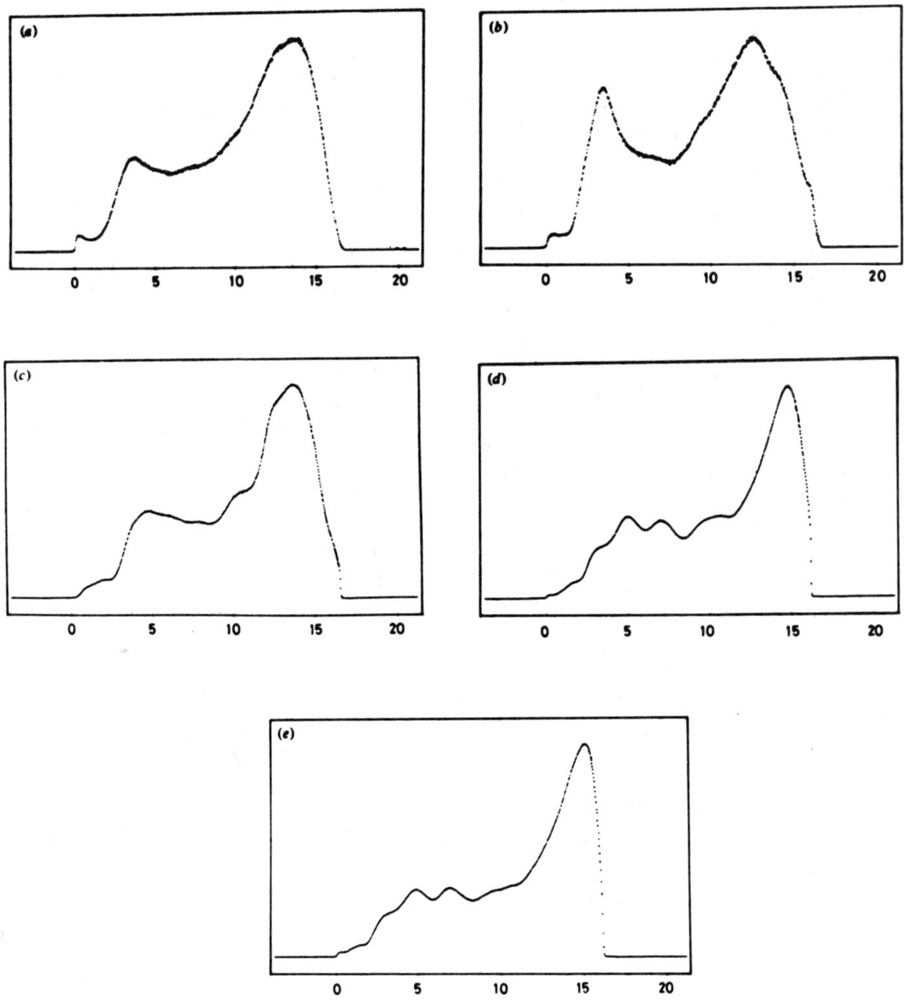

Fig. 1. Valence region He(I) PES of (a) $Pb_2Ru_2O_{7-y}$, (b) $Bi_2Ru_2O_7$, (c) $Y_2Ru_2O_7$, (d) $CaRuO_3$, and (e) $SrRuO_3$. Binding energies relative to the Fermi energy. After [11].

correlation splitting of the 4d bands into upper and lower Hubbard bands by the electron-electron intraatomic interactions appears to be the appropriate model for $Y_2Ru_2O_7$.

The spectra of the two perovskites $CaRuO_3$ and $SrRuO_3$ are more complex, as might be expected for metals with spontaneous ruthenium magnetic moments. These spectra show evidence of a multiplet structure like that of $Y_2Ru_2O_7$ superposed on a free-electron spectrum like that of $Pb_2Ru_2O_{7-y}$.

The Ru-3d core spectra of $Pb_2Ru_2O_{7-y}$ and $Bi_2Ru_2O_7$ consist of two overlapping sets of spin-orbit doublets, the one at lower binding energy being the more intense. In $CaRuO_3$ the doublet at higher binding energy is more intense, and in $Y_2Ru_2O_7$ only the component at higher binding energy is present. Interpretation of these two doublets [11] is based on a model of Kotani and Toyazawa [13] in which the higher-energy doublet corresponds to an 'unscreened' final state, the doublet at lower energy to a 'screened' final state. The 'screened' state represents trapping of an electron from the conduction band by a Coulomb attraction to the core-state hole, and the relative probability of reaching a screened final state increases with W/U [14]. Thus the progressive increase in the intensity of 'screened' final-state components in the series $Y_2Ru_2O_7$, $CaRuO_3$, $Bi_2Ru_2O_7$ is consistent with the conduction-band u.v. photoelectron spectra; it provides convincing evidence that the changes in the character of the 4d electrons are not due to changes in stoichiometry, but to the intrinsic properties of the Ru(IV)-4d electrons.

Inspection of the expanded spectra of Fig. 2 for $CaRuO_3$ and $SrRuO_3$ reveals that the free-electron component is weaker in $CaRuO_3$ than in $SrRuO_3$ and hence that the Ru-4d bands are narrower in $CaRuO_3$. This observation is consistent with a stronger Ca-O competition for the $O{:}p_\pi$ orbitals responsible for formation of the narrow, partially filled π^* band deriving from $Ru{:}t_2$-$O{:}p_\pi$-$Ru{:}t_2$ interactions [15]. This chemical argument has been shown to be applicable to perovskites containing first-row transition elements and counter alkaline-earth or rare-earth counter cations [16]. It is amplified in the next section. Here attention is called to the curious fact that evolution from spontaneous ferromagnetism in $SrRuO_3$ to a negative Weiss constant in the paramagnetic susceptibility of $CaRuO_3$ - but without apparent magnetic order to 4.2 K - is associated with a narrowing of the Ru(IV)-π^* band.

Finally, the A-cation core-level spectra provide additional information. The Pb-4f and Bi-4f peaks show a pronounced asymmetry whereas the Y-3d doublet does not. For metallic compounds, the asymmetry index for the core line of a given element is proportional to the square of the

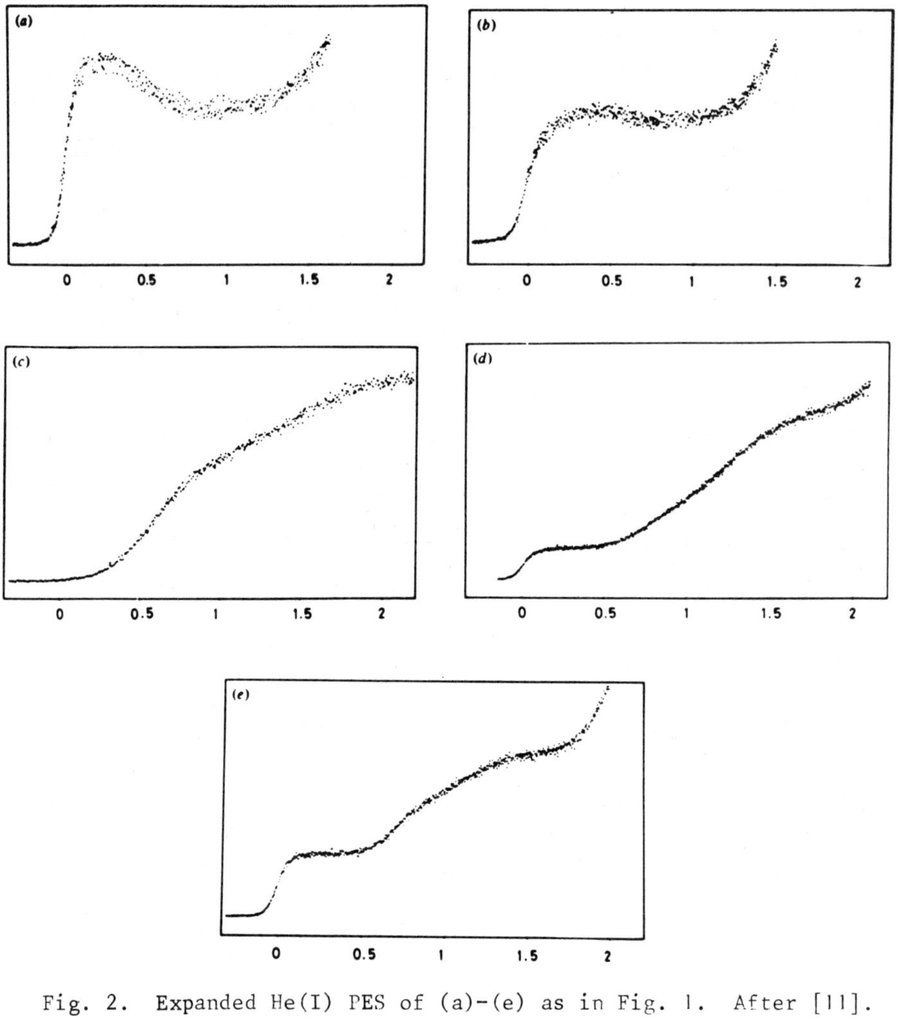

Fig. 2. Expanded He(I) PES of (a)-(e) as in Fig. 1. After [11].

partial density of states at the Fermi energy that is provided by the valence orbitals of that element [17]. Thus the line asymmetries for the 4f signals of Bi and Pb in the pyrochlore ruthenates provide direct evidence for a significant 6s partial density of states at the Fermi energy. Moreover, the observation that the asymmetry is more pronounced for $Pb_2Ru_2O_{7-y}$ than for $Bi_2Ru_2O_7$ shows that the Pb-6s partial density of states at the Fermi energy of $Pb_2Ru_2O_{7-y}$ is greater than the Bi-6s partial density of states at E_F in $Bi_2Ru_2O_7$. These results establish that broadening of the narrow, strongly correlated Ru(IV)-π^* band in $Y_2Ru_2O_7$ to a broader, weakly correlated Ru(IV)-π^* band in isostructural $Bi_2Ru_2O_7$ is associated with interactions between the Bi^{3+}:$6s^2$ core electrons and the Ru(IV)-π^* band [11].

ENERGY DIAGRAMS

A. Perovskites

In the ideal cubic-perovskite structure of composition ABO_3, Fig. 3, the B cations occupy an array of corner-shared octahedra having B-O-B angles of 180°. In an $A^{2+}B^{4+}O_3$ perovskite, the Madelung energy stabilizes

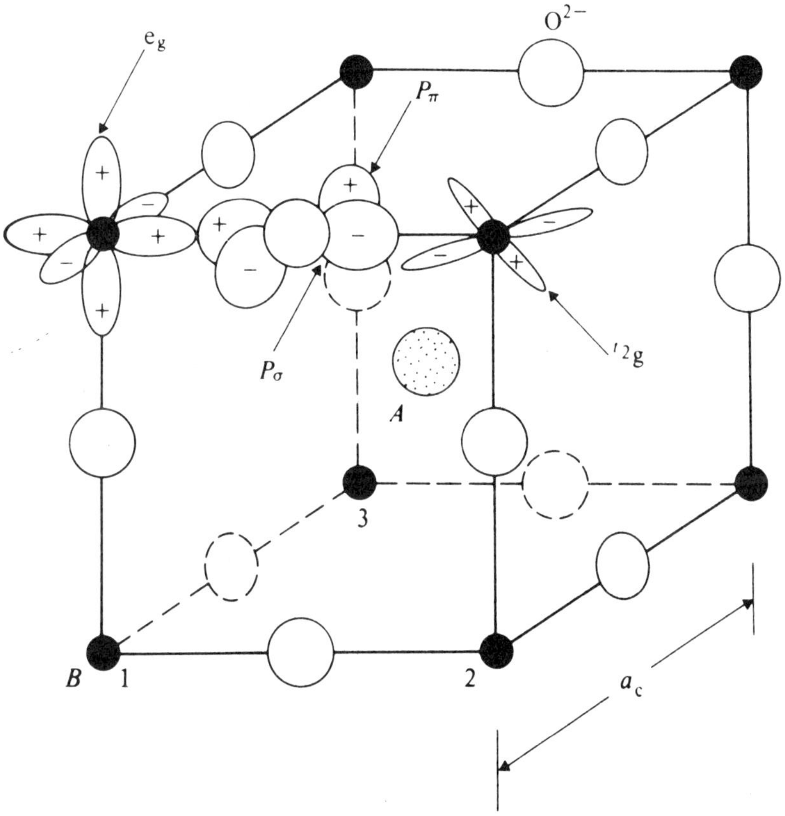

Fig. 3. The cubic-perovskite structure.

the oxide-ion 2p band, labelled $O:2p^6$, some 7.5 eV below vacuum, as has been demonstrated for $SrTiO_3$ [18]. The bottom of the s band of a transition-metal B cation lies some 6 eV above the top of the $O:2p^6$ valence band, and the Fermi energy lies in the energy gap between these two bands. The acceptor s bands of an alkaline-earth A cation like Ca^{2+} or Sr^{2+} are at an even higher energy. Therefore the number of "4d" electrons in $CaRuO_3$ and $SrRuO_3$ are unambiguously four per Ru(IV) ion.

The octahedral-site crystalline fields split the d-state manifold of a transition-metal B cation into less stable, antibonding (with respect to Ru-O interactions) e orbitals that σ-bond and antibonding t_2 orbitals that π-bond with nearest-neighbor anions. For second-row and third-row transition-metal cations, the crystal-field splitting of the e and t_2 orbitals is larger than the energy of any intraatomic electron-electron interaction Δ_{ex} that would stabilize the highest spin configuration, so an octahedral-site Ru(IV) ion is in a low-spin state: $t_2^4e^0$.

With a 180° B-O-B angle, the e orbitals of Ru atoms on opposite sides of an oxide ion each overlap the same p_σ orbital and are orthogonal to the remaining two p_π orbitals; the t_2 orbitals of these two Ru atoms each overlap the same p_π orbitals and are orthogonal to the p_σ orbital. Therefore where the $B:e\text{-}O:p_\sigma\text{-}B:e$ and $B:t_2\text{-}O:p_\pi\text{-}B:t_2$ interactions are strong enough to delocalize the d electrons, as appears to be the case in $CaRuO_3$ and $SrRuO_3$, the localized t_2 orbitals are transformed into π^* band orbitals of t_2 parentage and the localized e orbitals into σ^* band orbitals of e parentage [19]. The σ^* bands remain above the π^* bands because of the crystal-field splitting, so the Ru(IV) localized-electron low-spin configuration $t_2^4e^0$ becomes transformed to Ru(IV):$\pi^{*4}\sigma^{*0}$.

The orbitally threefold-degenerate π^* band is narrow, so the electrons within it may be correlated by electron-electron interactions. As argued by Hubbard [20], the most important correlations are due to intraatomic electron-electron interactions of magnitude U. If the width of the π^* band is $W_\pi < U$, then the Ru(IV) donor and acceptor bands may be split by a finite energy gap, and the compound is semiconducting. In this limit, the electron correlations induce a spontaneous atomic moment on the Ru(IV) ions; the maximum spin contribution to this moment is found in the localized-electron limit, where it is $\mu_{Ru} = 2\mu_B$ for a low-spin $t_2^4e^0$ configuration.

On the other hand, a $W_\pi \simeq U$ may give rise to a situation where the donor and acceptor bands overlap, but the intraatomic correlations are nevertheless strong enough to induce a spontaneous atomic moment on the

Ru(IV) ions. In this case, the compound is intrinsically metallic and magnetic. It is this intermediate case that appears to apply to the π^* bands of $CaRuO_3$ and $SrRuO_3$.

To this point we have neglected the role of the A cation. The empty s orbitals of the alkaline-earth ions Ca^{2+} and Sr^{2+} overlap the oxide-ion p_π orbitals (which σ-bond with the A cations and π bond with the B cations). The A:s orbitals also overlap the B:t_2 orbitals. The A:s-O:p_π interactions appear to dominate the A:s-B:t_2 interactions in most cases; and they compete with the B:t_2-O:p_π-B:t_2 interactions. The stronger this competition, the narrower the π^* bands. Since the strength of the competition varies inversely with the energy separation of the O:p_π donor orbitals and the A:s acceptor orbitals (second-order perturbation), it follows that, in the cubic-perovskite structure, the π^* bandwidth W_π should decrease with A-cation sequence Ba^{2+}, Sr^{2+}, Ca^{2+}. The PES observation of a narrower π^* band in $CaRuO_3$ than in $SrRuO_3$ is consistent with this argument. Moreover, the fact that the π^* bands are two-thirds filled and strongly correlated in these perovskites makes comparison of their magnetic properties particularly significant.

Fig. 4 presents semiempirical energy diagrams for the perovskites $CaRuO_3$ and $SrRuO_3$. Placement of the Fermi energy E_F relative to the top of the $O^{2-}:2p^6$ valence band is estimated from the PES spectra of Fig. 1. Since the separation of the π^* and σ^* bands in low-spin $\pi^{*6}\sigma^{*0}$ systems is generally 2.0 to 2.5 eV, the bottom of the σ^* bands is placed about 2.5 eV above E_F.

Paramagnetic $CaRuO_3$ is shown in Fig. 4 with a minimum at E_F in the π^*-band density of states N(E) because strong correlations produce a Curie-Weiss law with a Curie constant corresponding to a Ru(IV) spin $S \simeq 1$. The empty π^* states are the band of acceptor states associated with the Ru(IV) cation array; the filled π^* states are the band of donor states associated with this array. Since $CaRuO_3$ is metallic, these two bands must overlap.

Ferromagnetic $SrRuO_3$, on the other hand, is represented for a temperature $T < T_c$. As the measured ferromagnetic moment at 4 K is reduced from a localized-electron value to an observed $\mu_{Ru} \simeq 1\ \mu_B$, we assume that the interatomic interactions are strong enough to stabilize minority-spin bonding states relative to majority-spin antibonding states. This assumption permits a quantitative prediction of the spin-only component of the magnetization in itinerant-electron ferromagnets [21]. For localized electrons, the predicted moments are

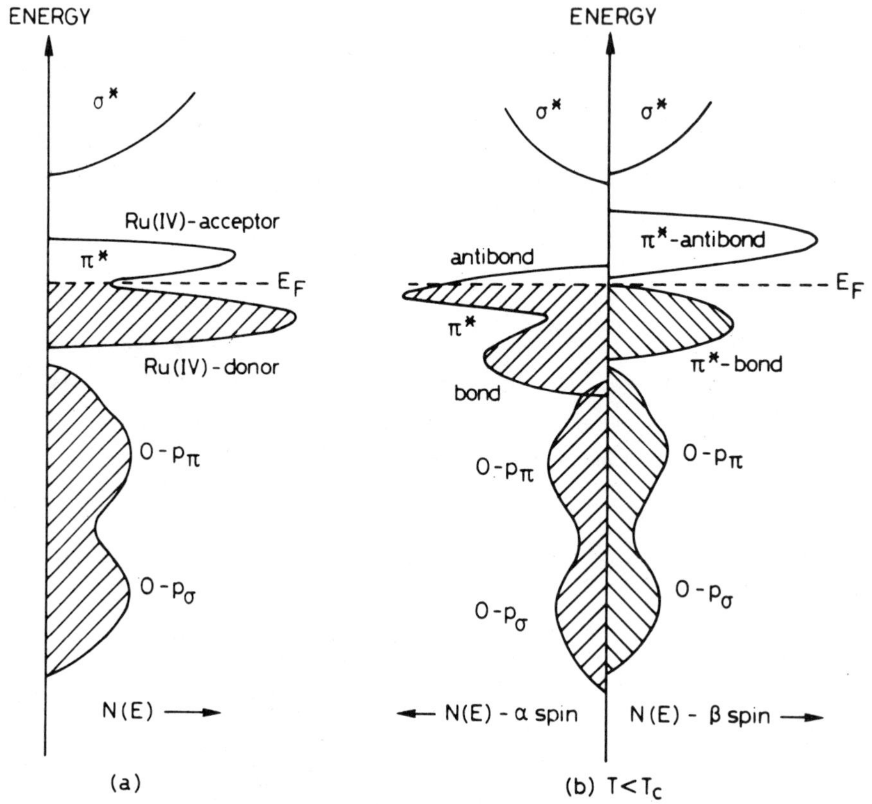

Fig. 4. Semiempirical energy diagrams for (a) $CaRuO_3$, (b) $SrRuO_3$.

$$\mu_B^F = \nu n_\ell \mu_B \text{ for } 0 \leqslant n_\ell \leqslant 1$$
$$\mu_B^F = \nu(2-n_\ell)\mu_B \text{ for } 1 \leqslant n_\ell \leqslant 2 \quad (1)$$

where ν is the orbital degeneracy of the band and n_ℓ is the band-occupancy number, which includes the twofold spin degeneracy ($0 \leqslant n_\ell \leqslant 2$). Since only bonding states are filled if $n_\ell < \frac{1}{2}$ and only antibonding states are empty if $n_\ell > 3/2$ in the ferromagnetic state, reduced ferromagnetic moments only occur in the interval $\frac{1}{2} < n_\ell < 3/2$:

$$\mu_B^F = \nu(1-n_\ell)\mu_B \text{ for } \tfrac{1}{2} < n_\ell \leqslant 1$$
$$\mu_B^F = \nu(n_\ell - 1)\mu_B \text{ for } 1 \leqslant n_\ell < 3/2 \quad (2)$$

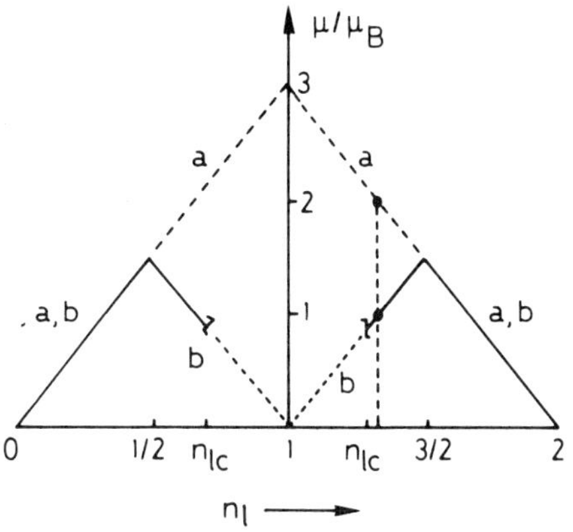

Fig. 5. Spin-only atomic moment vs band occupancy for a π^* band: (a) full and (b) reduced.

For a threefold-degenerate π^* band, the full and reduced moments vary with band occupancy as shown in Fig. 5. The Ru(IV):π^{*4} band is two-thirds filled, $n_\ell = 4/3$ and $\nu = 3$, so the reduced-spin contribution to the ferromagnetic moment is predicted to be $\mu_{Ru} = 1\ \mu_B$. In fact the π^* bands of $SrRuO_3$ may not be quite broad enough to give the full reduction in moment, so E_F is shown just overlapping the minority-spin bonding (with respect to the Ru-Ru interactions) band. Such an overlap could account for the observed increase in μ_{Ru} with applied field H_a.

To "explain" the evolution of magnetic properties on going from $SrRuO_3$ to $CaRuO_3$, it is necessary to recognize that, in fact, a ferromagnetic configuration cannot be stable relative to an antiferromagnetic configuration in a half-filled band [21]. In a collinear-spin antiferromagnetic configuration, the magnitudes of the atomic moments are

$$\mu_B^{AF} = \nu(n_\ell - \delta)\mu_B \quad \text{for } \tfrac{1}{2} < n_\ell \le 1$$
$$\mu_B^{AF} = \nu(2 - n_\ell - \delta)\mu_B \quad \text{for } 1 \le n_\ell < 3/2 \tag{3}$$

where δ is a fraction that increases with W/U. It arises from the fact that the interatomic electron-transfer reactions are only Pauli allowed if the transferred spin is antiparallel to the net atomic spin on the acceptor ion. Within the band-occupation intervals $\frac{1}{2} < n_\ell \le 1$ and $1 \le n_\ell < 3/2$ there must be a cross-over between ferromagnetic and antiferromagnetic ground-state configurations.

The simplest hypothesis is that the ground state having the higher atomic moment would be the more stable, which would define the cross-over occupancy as that where $\mu_B^F = \mu_B^{AF}$ or, from equations (2) and (3), at

$$n_{\ell c} = (1+\delta)/2 \quad \text{or} \quad (3-\delta)/2 \tag{4}$$

Values of $\delta = 0.5 \pm 0.2$ are commonly found in metals and alloys [22]. For $\delta = 0.5$, an $n_{\ell c} = 1.25$ is less than the $n_\ell = 1.33$ for the Ru(IV):π^{*4} band, so the retention of a ferromagnetic ground state in $SrRuO_3$ is not unreasonable. On the other hand, the value of W_π/U was shown to be smaller in $CaRuO_3$ than in $SrRuO_3$, so a smaller δ in $CaRuO_3$ should increase $n_{\ell c}$ toward, if not beyond, the $n_\ell = 1.33$. In fact, a $\delta = 0.3$ would make $n_{\ell c} = 1.35$ and thereby stabilize an antiferromagnetic ground state.

The Weiss constant θ_p obtained from the paramagnetic Curie-Weiss law is, in a Weiss molecular-field theory, $\theta_p = CW_m$, where C is the Curie constant and W_m is the Weiss interatomic exchange constant; W_m is positive for ferromagnetic interactions and negative for antiferromagnetic interactions. The observed decrease in θ_p with x in the system $Sr_{1-x}Ca_xRuO_3$ [5] from a positive value in $SrRuO_3$ to a negative value in $CaRuO_3$ is thus consistent with a decreasing W_π/U.

What remains surprising is the lack of any long-range magnetic order down to 4.2 K in $CaRuO_3$. In metal alloys, where M-M rather than M-O-M interactions are operative, a large exchange striction associated with magnetic ordering would lift the π^*-band degeneracy so as to stabilize half-filled orbitals split into bonding and antibonding bands by the antiferromagnetic order; the remaining π^* orbital would be filled and nonbonding with respect to Ru-Ru interactions [22]. In the perovskite structure, such a lifting of the orbital degeneracy is apparently inhibited by the antibonding character of the Ru-O interactions; the π^*-band degeneracy can only be lifted by a disproportionation reaction of the type Ru(IV) + Ru(IV) $\rightarrow$ Ru(V) + Ru(III), which is inhibited by the electron correlations. Thus the lack of any long-range magnetic order seems to reflect an itinerant-electron "frustration" that is present where $n_\ell \simeq n_{\ell c}$ and there is no energetically feasible way to remove this frustration.

B. Pyrochlores

Like the cubic-perovskite structure, the pyrochlores $A_2B_2O_6O'$ have a BO_3 array of corner-shared octahedra, but the B-O-B angles are reduced from 180° to about 130°, Fig. 6. Moreover, the interstitial space is

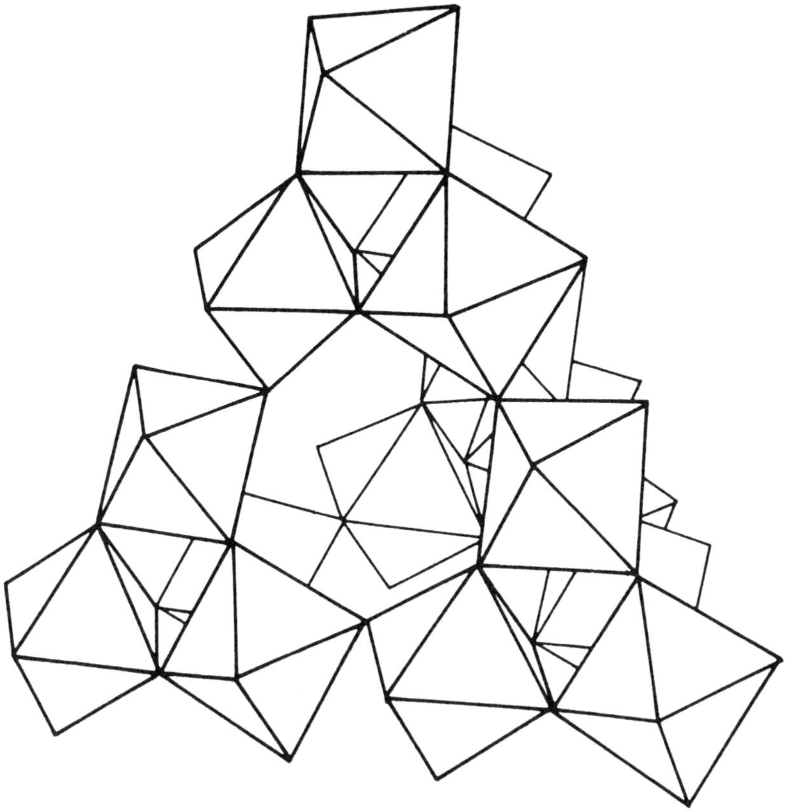

Fig. 6. View of B_2O_6-framework linkage in cubic pyrochlore.

occupied by the oxide ions O' as well as the A cations. Reducing the B-O-B angle from 180° destroys the separation of the $O{:}2p^6$ band into more stable p_σ and less stable p_π bands, and the O' ions add additional O'-p bands. However, the PES spectra of Fig. 1 indicate that the $O{:}2p^6$ bands lie in about the same position relative to E_F in the pyrochlore structure as they do in the perovskite structure. Moreover, the octahedral-site crystalline fields produce a low-spin Ru(IV):$t_2^4e^0$ configuration that is transformed by Ru-O-Ru interactions into orbitally threefold-degenerate π^* bands of t_2 parentage and twofold-degenerate σ^* bands of e parentage that are analogous to the π^* and σ^* bands of the perovskite structure. However, reduction of the B-O-B angle reduces the Ru-O-Ru overlap integrals, so W_π is smaller. Moreover, the lowest acceptor states of the A-site cations Y^{3+} or a Ln^{3+} ion are d states that are of lower energy than the s bands of the B cation, so the A:d-O:p interactions from these ions are more competitive with Ru:t_2-O-Ru:t_2 interactions than the A:s-O:p_π interactions in $CaRuO_3$ and $SrRuO_3$. It follows that in the $Ln_2Ru_2O_7$ and $Y_2Ru_2O_7$ pyrochlores, the ratio W_π/U should be even smaller than in $CaRuO_3$. Therefore we must

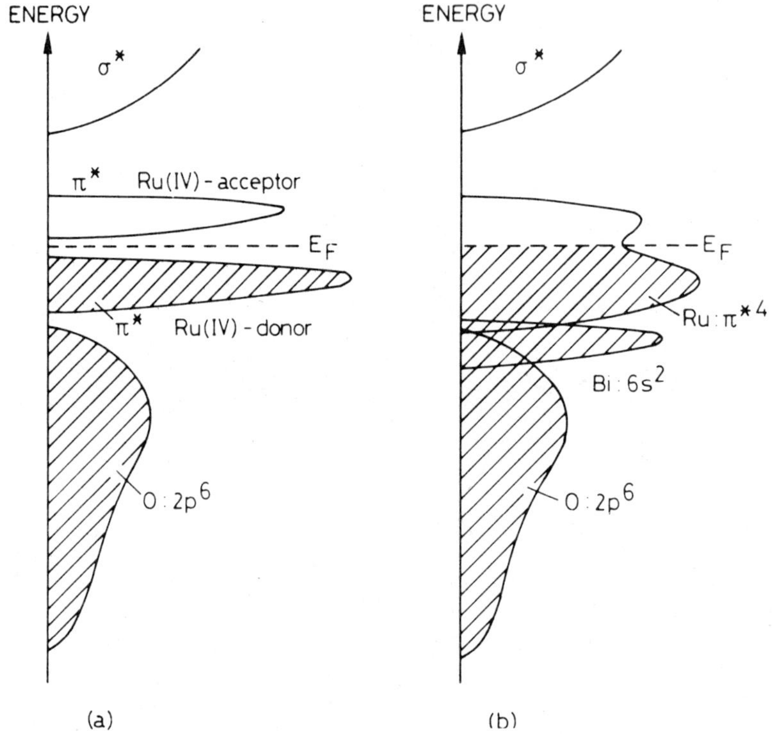

Fig. 7. Semiempirical energy diagram for two pyrochlores: (a) $Y_2Ru_2O_7$ and (b) $Bi_2Ru_2O_7$.

expect a stronger correlation splitting, one that is strong enough to separate the Ru(IV) acceptor states from the Ru(IV) donor states by a finite energy gap E_g.

The semiconducting and magnetic properties of these pyrochlores are quite consistent with the model we have developed. Fig. 7(a) presents a semiempirical energy diagram for the pyrochlore $Y_2Ru_2O_7$. For the rare-earth pyrochlores it is only necessary to add the localized $4f^n$ configurations. In all cases, the $Ln^{3+}:4f^n$ donor state lies well below E_F and the $Ln^{3+}:4f^n$ acceptor level $4f^n/4f^{n-1}$ lies above E_F; these levels are separated by a large correlation energy $U_{4f} > 8$ eV.

The A cations Pb^{2+} and Bi^{3+}, on the other hand, have filled $6s^2$ cores having band energies that, according to the PES of Fig. 1, overlap the $O^{2-}:2p^6$ band and have a band edge within about 2 eV of E_F. We must, therefore, enquire into the influence of the $6s^2$ core electrons on the π^* bands;

we have already pointed out that the core-level Bi:4f and Pb:4f PES gave direct evidence of a 6s partial density of states at E_F, which signals an important A:6s-Ru:t_2 covalent mixing in these compounds.

Each A cation occupies an eightfold-coordinated site formed by squashing an O-atom octahedron along a trigonal axis and capping the resulting six-membered, puckered ring on either side by O' ions on the trigonal axis. In $Bi_2Ru_2O_7$, there are no anion vacancies, and the $6s^2$ electrons are not stabilized by vacancy trapping. Nevertheless, there is no evidence from Fig. 1 for an overlapping of Bi:6s and Ru:π^* bands as proposed by Sleight et al [23]. However, location of the Bi(III) ion in the center of the six-membered, puckered ring of O atoms allows a direct overlap of the Bi:6s and Ru:t_2 orbitals, as in the perovskite structure. The proximity of the energies of the donor Bi:$6s^2$ and the Ru(IV);π^* acceptor bands means that there must be an important Bi:6s-Ru:t_2 covalent mixing. Hence the 6s partial density of states at E_F in $Bi_2Ru_2O_7$.

The Bi:6s-Ru:t_2 interactions extend the Ru:t_2 orbitals out over the Bi atoms and enhance the Ru:t_2-O:p_π overlap while reducing U. Therefore the ratio W_π/U is increased relative to its value in $Y_2Ru_2O_7$; in fact, the interatomic interactions are apparently enhanced sufficiently that strong electron correlations are suppressed at room temperature and above. However, the π^* conduction band remains sufficiently narrow that the conducting electrons appear to be strongly scattered by the Bi(III) vibrations, so the conductivity is nearly temperature-independent down to 150 K, below which the Bi(III) thermal parameter settles down to a normal value [24].

Given the contrasting features of isostructural $Y_2Ru_2O_7$ and $Bi_2Ru_2O_7$, which are summarized in Fig. 7, we thought it would be interesting to explore the metal-semiconductor transition in the pyrochlore system $Bi_{2-x}Ln_xRu_2O_7$. For this purpose we chose Ln = Gd. The Gd^{3+}:$4f^7$ configuration lies below the O^{2-}:$2p^6$ band, and the Gd^{2+}:$4f^8$ configuration lies above the Gd:5d band, so they contribute little to W_π; therefore we can use Fig. 7(a) in our discussion of the system.

EXPERIMENTAL

A. Synthesis

The starting materials were RuO_2 from B.D.H., Gd_2O_3 and Bi_2O_3 from Koch-Light, Ltd, all of supplier purity 99.999%. Stoichiometric amounts of these oxides, which were weighed at 150°C (Bi_2O_3) or 450°C, were ground in a porcelain mortar and pestle for 10 m. In order to obtain a more homogeneous mix, the mixture was pelletized at 2 tonnes cm^{-2} and then reground.

$Bi_2Ru_2O_7$ was successfully prepared by heating the oxides at 600°C for 8 h, at 700°C for 12 h, and finally at 965°C for 24 h with regrindings between successive firings, the product being quenched after the final firing. This procedure minimizes the loss of volatile Bi_2O_3 [2]. $Gd_2Ru_2O_7$ was fired in open air in a carbolite furnace at 1100°C for 36 h with regrindings every 8 h. Reacting the two end members stoichiometrically in alumina boats at temperatures increasing from 965°C for $x = 0$ to 1100°C for $x = 2$ for 24 h gave solid solutions with the lattice-parameter variation shown in Fig. 8. As the compositional range of the two-phase region was approached, it was increasingly difficult to obtain homogeneous samples.

The lattice parameters were all refined with a matrix-least-squares program that required 5 cycles to reach self-consistent results with reliability factors $1.8 \times 10^{-7} < R < 5 \times 10^{-7}$. The variation with composition is in agreement with Végard's law, but with a different coefficient for two distinct compositional domains.

Bulk chemical analysis of the Bi/Ru, Gd/Ru, and Bi/Gd ratios were obtained by electron microprobe analysis, and Bi contents were measured by atomic absorption spectroscopy. All samples, with the exception of $Bi_{0.45}Gd_{1.55}Ru_2O_7$, were fairly homogeneous. Measured Bi/Gd ratios varied with x well within experimental error. Microprobe analysis of $Bi_{0.45}Gd_{1.55}Ru_2O_7$, which is located in the shaded region of Fig. 8, showed

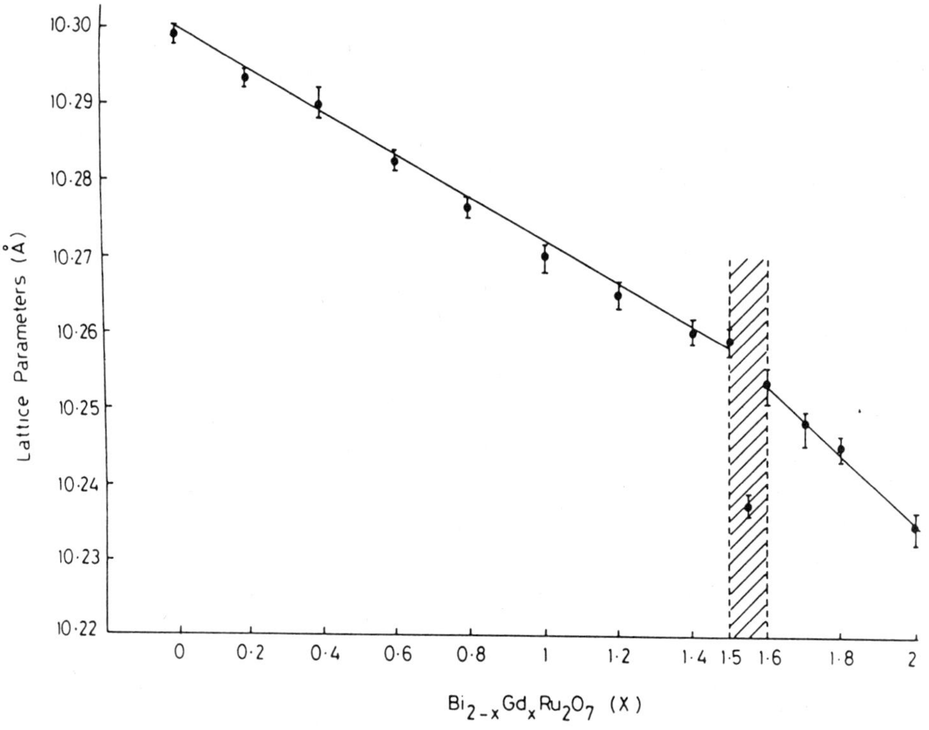

Fig. 8. Lattice parameter vs composition for the system $Bi_{2-x}Gd_xRu_2O_7$.

evidence of segregation into Bi-rich and Gd-rich regions.

B. Transport Data

Four-probe measurements of d.c. conductivity σ were made on a 13-mm disc 1 mm thick with Pt-foil electrodes. Ohmic contact between electrodes and sample were obtained with a Ga/In eutectic mixture. Temperature was controlled to $\pm$ 0.1 deg over the range $25 < T < 500^{o}C$, and measurements could be made in a controlled atmosphere. The σ vs T^{-1} curves obtained are shown in Fig. 9. A change from metallic to semiconducting behavior is observed on passing across the two-phase region of Fig. 8, from $x = 1.5$ to $x = 1.6$.

Seebeck data were taken on rectangular ($2.5 \times 0.5 \times 0.1$-0.2 cm) polycrystalline samples pressed under 10 tons and sintered for 24 h. The temperature difference across the sample varied from 1 to $2^{o}C$. The Seebeck coefficients obtained are shown in Fig. 10.

DISCUSSION

From Fig. 7, we may interpret the evolution from metallic conductivity in $Bi_2Ru_2O_7$ to semiconducting behavior in $Gd_2Ru_2O_7$ to be an illustration of correlation (Mott-Hubbard) splitting of a narrow Ru(IV):π^{*4} band. However, in the mixed compositions the Bi(III) and Gd(III) ions perturb differently the t_2 orbitals on neighboring Ru(IV) ions, and it is probably significant that the critical composition $x_c = 1.55$ is close to the theoretical percolation limit $x_p = 1.5$ [25] for continuous Ru-O-Ru pathways through the crystal via Ru(IV) ions that neighbor Bi(III) ions. On the other hand, the abrupt change in the slope of the lattice-parameter change with x (Végard's law) at x_c in Fig. 8 and the small two-phase region separating the metallic and semiconducting domains suggests a first-order transition between weakly and strongly correlated π^* electrons throughout the crystal. It would thus appear that the transition from strongly to weakly correlated π^* electrons is a cooperative, long-range phenomenon.

According to the model of Fig. 7, the enlargement of the ratio W_π/U in $Bi_2Ru_2O_7$ is due to a covalent admixture of Bi:6s character into the Ru:t_2 orbitals; since the $6s^2$ shell is full, the 6s orbitals do not compete for the donor O-p electrons. This model does not require a significant change in the Ru-O bond distance or the Ru-O-Ru bond angle. Structural refinements on $Bi_2Ru_2O_7$ and two $Ln_2Ru_2O_7$ pyrochlores have shown that the Ru-O bond distances are 2.0 ± 0.1 Å in all cases and that the Ru-O-Ru bond angles are all about 130° [23]. These data also support the idea that the Bi:$6s^2$ core electrons have a strong influence on the Ru-π^* bands.

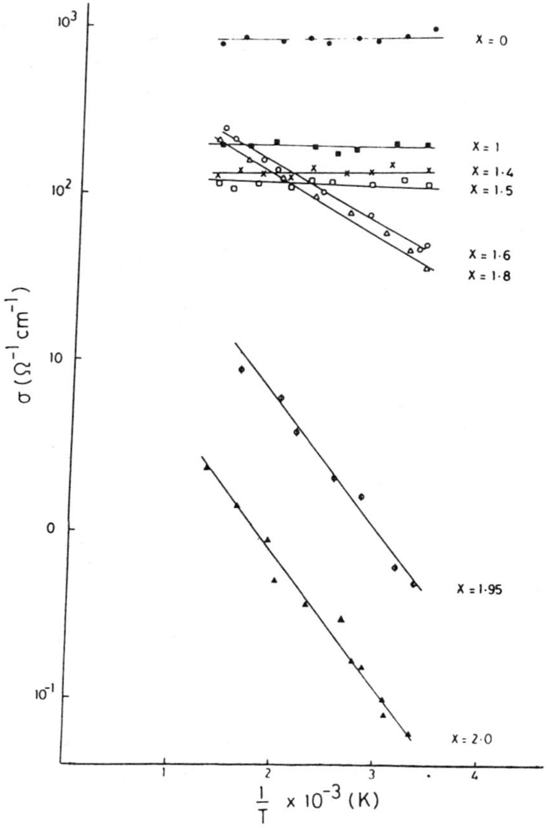

Fig. 9. D.C. conductivity σ versus reciprocal temperature for the system $Bi_{2-x}Gd_xRu_2O$.

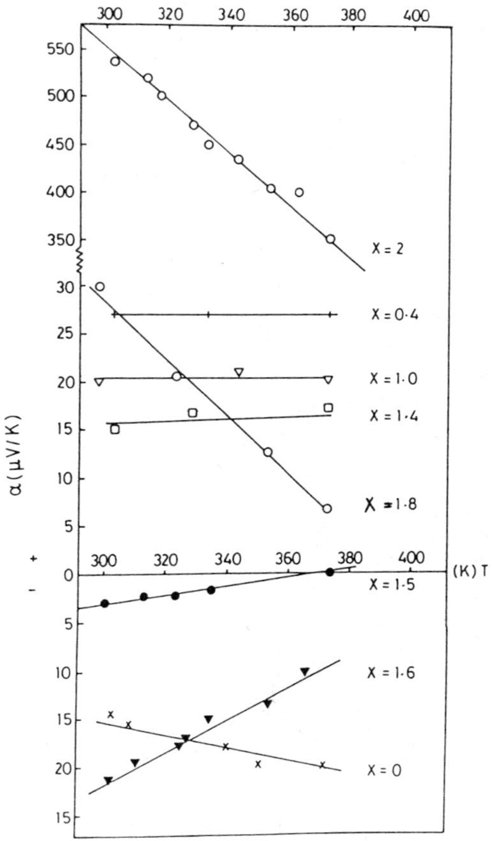

Fig. 10. Seebeck coefficient α versus temperature for members of the system $Bi_{2-x}Gd_xRu_2O$.

If this model is correct, then action on the Ru-π^* bands by the Bi:$6s^2$ core electrons must be counterbalanced by a reaction on the Bi:$6s^2$ core by the Ru-t_2 orbitals. This form of reaction would manifest itself as a reduction in the effective radius of the Bi(III) ion, particularly if the $6s$-π^* admixture were stronger with the empty vs filled π^* states. Admixture with empty π^* states results in a net transfer of charge from the Bi:$6s^2$ core out into the Ru-t_2 orbitals.

The significance of Fig. 8 is the evidence it provides for two different A-cation effective radii. From the slopes of the lattice-parameter variation with x, it is clear that the differences in the effective radii of Bi(III) and Gd(III) are much greater in the semiconducting phase than in the metallic phase. To show that this deduction may be a more general phenomenon, we compare in Fig. 11 the lattice-parameter variations with x of two other Bi-Gd pyrochlores: metallic $Bi_{2-x}Gd_xIr_2O_7$ and semiconducting $Bi_{2-x}Gd_xTi_2O_7$. The slope of Végard's law for the metallic system parallels that for the metallic phase of $Bi_{2-x}Gd_xRu_2O_7$; the slope for $Bi_{2-x}Gd_xTi_2O_7$ exceeds that of the semiconducting phase of the system $Bi_{2-x}Gd_xRu_2O_7$. The plots demonstrate clearly that the ionic radius of a Bi(III) ion may be significantly altered by the energetic availability of acceptor orbitals at the counter cations.

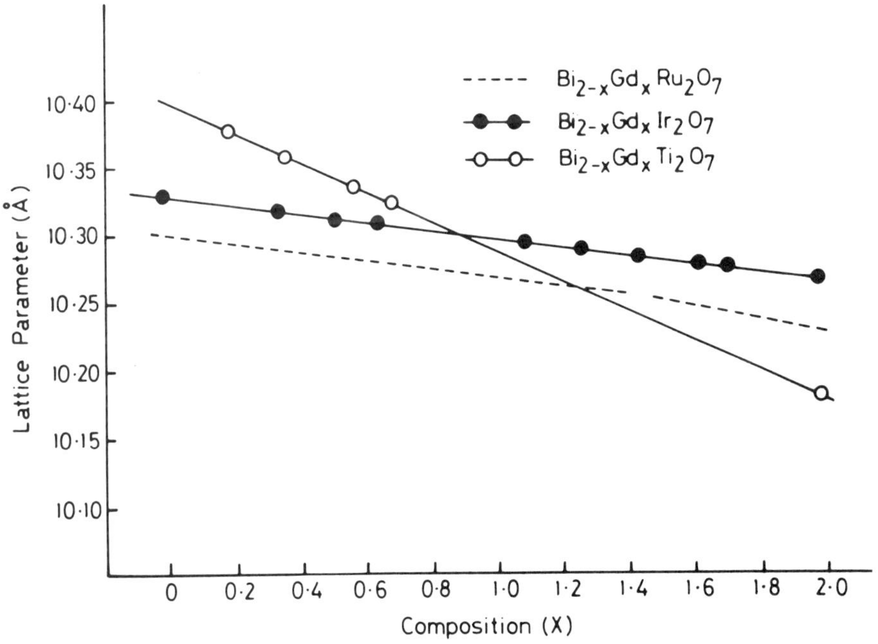

Fig. 11. Lattice parameter a_o vs composition x for the systems $Bi_{2-x}Gd_xB_2O_7$, B = Ru, Ir, Ti.

The first order character of the metallic-semiconductor phase change implied by Fig. 8 also suggests that the Bi:$6s^2$-Ru:π^* interactions are cooperative and that this type of interaction can give rise to cooperative phase changes as a function of temperature. For example, the transition near 150 K in $Bi_2Ru_2O_7$ itself could well reflect an ordering of the 6s-t_2 interactions that removes the π^*-orbital degeneracy.

Finally, these observations appear to be applicable to $T\ell_2Ru_2O_7$, but with an important modification. The $T\ell^{3+}$:6s core states are empty, but the bottom of the Tℓ:6s conduction band must lie close to E_F. Indeed, the peculiar properties of the system $T\ell_2Ru_2O_{7-\varepsilon}$ indicate trapping of $T\ell^+$:6s electrons at O'-atom vacancies [3]. An empty 6s band represents a band of acceptor states that compete for the O-p as well as the Ru-π^* states for covalent admixture. Therefore the first-order semiconductor-metal transition observed near 125 K [3] may reflect either an ordering that stabilizes Tℓ:6s-O:p interactions or one that stabilizes interactions of the empty Tℓ:6s orbitals with occupied Ru:π^{*4} orbitals. If the former ordering prevails, the semiconducting phase must be antiferromagnetic; if the latter ordering is operative, the low-temperature phase would be diamagnetic. A neutron-diffraction study could resolve this issue.

REFERENCES

1. M.A. Subramanian, G. Aravamudan, and G.V. Subba Rao, "Oxide Pyrochlores - A Review", Progr. Solid State Chem. 15, 55 (1983).
2. R.J. Bouchard and J.L. Gillson, "A New Family of Bismuth-Precious Metal Pyrochlores", Mater. Res. Bull. 6, 669 (1971); J.F. Weiher and W. Bindloss, "Magnetic Susceptibility of Precious-Metal Pyrochlores", Bull. Am. Phys. Soc. 17, 316 (1972).
3. H.S. Jarrett, A.W. Sleight, J.F. Weiher, J.L. Gillson, C.G. Frederick, G.A. Jones, R.S. Swingle, D. Swartzfager, J.E. Gulley, and P.C. Hoell, "Metal-Semiconductor Transition in $T\ell_2Ru_2O_7$" in Valence Instabilities and Related Narrow-Band Phenomena, R.D. Parks, ed., Plenum Press, N.Y. (1977) p.545.
4. A. Callaghan, C.W. Moeller, and R. Ward, "Magnetic Interactions in Ternary Ruthenium Oxides", Inorg. Chem. 5, 1572 (1966).
5. J.M. Longo, P.M. Raccah, and J.B. Goodenough, "Magnetic Properties of $SrRuO_3$ and $CaRuO_3$", J. Appl. Phys. 39, 1327 (1968).
6. J.M. Longo and G. Shirane (private communication, 1969); T.C. Gibb, R. Greatrex, N.N. Greenwood, and P. Kaspi, "Ruthenium-99 Mössbauer Studies of the Magnetic Properties of Ternary and Quaternary Ruthenium(IV) Oxides", J. Chem. Soc. Dalton, 1253 (1973).
7. J.M. Longo, P.M. Raccah, and J.B. Goodenough, "Preparation and Properties of Oxygen-Deficient Pyrochlores", Mater. Res. Bull. 4, 191 (1969).
8. R.A. Beyerlein, H.S. Horowitz, J.M. Longo, M.E. Leonowicz, J.D. Jorgensen, and F.J. Rotella, "Neutron-Diffraction Investigation of Ordered Oxygen Vacancies in the Defect Pyrochlores $Pb_2Ru_2O_{6.5}$ and $PbT\ell Nb_2O_{6.5}$", J. Solid State Chem. 51, 253 (1984).
9. J.M. Longo and J.A. Kafalas, "Pressure-Induced Structural Changes in the System $Ba_{1-x}Sr_xRuO_3$", Mater. Res. Bull. 3, 687 (1968);

J.B. Goodenough, J.A. Kafalas, and J.M. Longo, "High-Pressure Synthesis" in Preparative Methods in Solid State Chemistry, P. Hagenmuller, ed., Academic Press (1972) Chap. 1.

10. D.B. Rogers, R.D. Shannon, A.W. Sleight, and J.L. Gillson, "Crystal Chemistry of Metal Dioxides with Rutile-Related Structures", Inorg. Chem. 8, 841 (1969).
11. R.G. Egdell, J.B. Goodenough, A. Hamnett, and C.C. Naish, "Electrochemistry of Ruthenates I - Oxygen Reduction on Pyrochlore Ruthenates", J. Chem. Soc. Faraday Trans. 1, 79, 893 (1983); P.A. Cox, R.G. Egdell, J.B. Goodenough, A. Hamnett, and C.C. Naish, "The Metal to Semiconductor Transition in Ternary Ruthenium(IV) Oxides: A Study by Electron Spectroscopy", J. Phys. C16, 6221 (1983).
12. P.A. Cox, "Fractional Parentage Methods for Ionisation of Open Shells of d and f Electrons", Struct. Bonding (Berlin) 24, 59 (1975).
13. A. Kotani and Y. Toyazawa, "Photoelectron Spectra of Core Electrons in Metals with an Incomplete Shell", J. Phys. Soc. Japan 37, 912 (1974).
14. P.A. Cox, "Many-Electron Effects in the Core-Level Photoelectron Spectra of Narrow-Band Solids: Some Model Calculations" in Inner-Shell and X-Ray Physics of Atoms and Solids, D.J. Fabian, H. Kleinpoppen, and L.M. Watson, eds., Plenum Press, N.Y. (1981) p. 549.
15. J.B. Goodenough, "Localized vs Collective d Electrons and Néel Temperatures in Perovskite and Perovskite-Related Structures", Phys. Rev. 164, 785 (1967).
16. J.B. Goodenough, "Valence Band Approach to Magnetic Semiconductors" in New Developments in Semiconductors, P.R. Wallace, R. Harris, and J.J. Zuckerman, eds., Noordhoff Int. Publ., Leyden, (1973) p. 107.
17. J.C.W. Folmer and D.K.G. De Boer, "XPS Core-Level Line Shapes in Metallic Compounds: A Probe for the Nature of the Electrons at the Fermi Level", Solid State Commun. 38, 1135 (1981).
18. J.B. Goodenough, A. Hamnett, M.P. Dare-Edwards, G. Campet, and R.D. Wright, "Inorganic Materials for Photoelectrolysis", Surface Science 101, 531 (1980).
19. J.B. Goodenough, "Metallic Oxides", Progr. Solid State Chem. 5, 145 (1972).
20. J. Hubbard, "Electron Correlations in Narrow Energy Bands", Proc. Roy. Soc. (London) A276, 238 (1963).
21. J.B. Goodenough, "Conceptual Phase Diagrams and Their Applications to Itinerant-Electron Magnetism" in Magnetism and Metallic Compounds, J.T. Lopuszanski, A. Pekalski, and J. Przystawa, eds., Plenum, N.Y. (1976) p.35.
22. J.B. Goodenough, Magnetism and the Chemical Bond, Interscience and John Wiley, N.Y. (1963).
23. A.W. Sleight and R.J. Bouchard, "Precious-Metal Pyrochlores" in NBS Special Publication 364, Proc. 5th Materials Research Symp., Solid State Chem., R.S. Roth and S.J. Schneider, Jr., eds., U.S. Dept. of Commerce (1972) p. 227.
24. W. Bindloss and A.W. Sleight, "Structural Transition and Electrical Resistivity of $Bi_2Ru_2O_7$", Bull. Am. Phys. Soc. [11] 17, 356 (1972).
25. S. Kirkpatrick, "Percolation and Conduction", Revs. Mod. Phys. 45, 574 (1973).

THE MOTT MOBILITY EDGE AND THE MAGNETIC POLARON†

S. von Molnár and T. Penney

IBM T. J. Watson Research Center
P. O. Box 218
Yorktown Heights, New York 10598

Abstract

This paper reviews the metal-insulator transition in a magnetic semiconductor with random potential fluctuations. The transition is explained in terms of Mott's mobility edge in disordered systems and the magnetic polaron.

Introduction

In this paper, we will describe how Mott's concept of a mobility edge (1),(2),(3) leads to the formation of magnetic polarons and localization in a magnetic system with coulombic disorder, $Gd_{3-x}v_xS_4$. In this material, the Fermi energy can be tuned through the mobility edge by magnetic field, temperature, or composition. Transport of charge carriers in many magnetic semiconductors is dominated by potential fluctuations of both coulombic and magnetic origin. In contrast to the coulombic case, the binding energies of states localized by spatial fluctuations in magnetic order may be both temperature and magnetic field dependent.(4),(5),(6),(7)

The first part of this paper describes the physical properties of materials in which the electrons are localized by coulombic disorder and extends these ideas to magnetic materials. In particular, the properties of paramagnetic $Ce_{3-x}v_xS_4$ studied by Cutler and Mott(8) are applied to a magnetic semiconductor, $Gd_{3-x}v_xS_4$ (9),(10). Such a comparison is helpful in describing the anomalous properties of the magnetic solid in terms of the well established model of the band tail and mobility edge.(1),(2),(3) The second part of the paper deals with the recent studies of magnetotransport in

$Gd_{3-x}v_xS_4$ [11],[12],[13]. It will be shown that the magnetic semiconductor exhibits a metal-insulator (MI) phase transition at a characteristic magnetic field, H_c.[11] On the insulator side, localization occurs with the formation of magnetic polarons: small regions of ferromagnetic Gd spin alignment in an antiferromagnetic host, in which the conduction electron spin is trapped. On the metal side, the electrons undergo anomalous diffusive motion.

Structure, Band Tails and the Mobility Edge

$Ce_{3-x}v_xS_4$crystallizes in the Th_3P_4 structure, which was first described by Meisel.[14] In 1949 Zachariasen[15], on the basis of x-ray powder diffraction and density data, showed that the defect Ce_2S_3 compound crystallized with the Th_3P_4 structure, in the space group $I\bar{4}3d - T_d^6$ with four molecules per unit cell. In this space group, the 12-fold cationic sites are fixed while the 16-fold anionic sites are determined by a parameter, μ. Both Zachariasen's and Meisel's analyses fixed the value of μ at 1/12 or 0.083. The deficit Ce_2S_3 was described by Zachariasen as having 10 2/3 Ce atoms statistically distributed over the twelve cation sites, whereas the sixteen fold anion sites were filled. On the basis of these results, Zachariasen predicted that the 4/3 vacant cation sites per unit cell could be filled and therefore the structure should exist over a single phase region extending from S:Ce = 1.5 to S:Ce = 1.33. The ex-

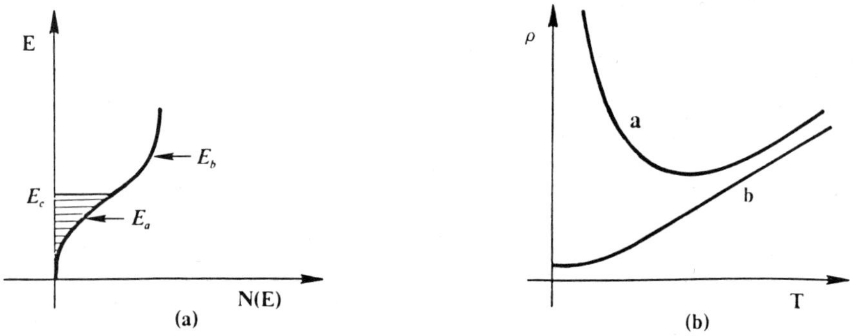

Fig. 1. a) Schematic density of states N(E). The shaded region represents localized states, E_c is the mobility edge, and E_a and E_b give the positions of the low temperature Fermi energy, E_F, for small and large electron concentration, respectively. b) Schematic plots of resistivity versus temperature for $E_F = E_a$ and $E_F = E_b$, respectively (after Cutler and Mott, ref. 8).

istence of randomly placed vacancies in these crystalline materials produces the random potential fluctuations and concomitant energy band tails so familiar in amorphous and highly doped semiconductors. Cutler and Mott[8] have demonstrated the applicability of the concepts of Anderson localization[16] in a band tail and the mobility edge to this class of crystalline materials.

The existence of a rigid band tail in Th_3P_4 structures is clearly a good assumption as long as the donor concentration is small compared to the total number of vacancies ($\sim 2.1 \times 10^{21} cm^{-3}$ in $Ce_{3-x}v_xS_4$). Cutler and Mott[8] have argued that when $E_F(0)$, the low temperature Fermi energy, lies below E_c, (position E_a in Fig. 1a) conduction is thermally activated and approaches 0 as $T \rightarrow 0$. When $E_F(0)$ is above E_c, conduction occurs in extended states and remains finite as T approaches 0 (see Fig. 1). $E_F(0)$ is strictly a function of the number of vacancies replaced by the rare earth atoms Ce or Gd. In the case of $Ce_{3-x}v_xS_4$ Cutler and Mott found experimentally that $E_f(0) = E_c$ for a critical concentration of $n_c \approx 8 \times 10^{19} cm^{-3}$.[8] In fact, a major conclusion of the Cutler and Mott work was the existence of n_c, which has formed the basis for all subsequent analyses of the metal-insulator transition in three dimensions.

Magnetic Properties and the Polaron

In $Gd_{3-x}v_xS_4$, it is supposed that $|E_c - E_F|$ is not only a function of concentration (and temperature) but also a function of the magnetization. Gd_2S_3 is an antiferromagnetic insulator of unknown structure whereas with increased n, i.e., with increased Gd, the compound approaches a ferromagnetic metal.[9],[10],[17] These magnetic data and their variation with carrier concentration indicate the presence of localized magnetic polarons, i.e., donor electrons, surrounded by clusters of Gd ions having a net moment, imbedded in an antiferromagnetic background.[9],[10] An alternative interpretation by Kamijo, et. al.[18] involving Wigner localization has been suggested. For the present purposes, the important features of the magnetic data are: a) the high temperature susceptibility, χ, follows a Curie-Weiss law (resulting in the extrapolated θ values, Table 1); b) the general trend is from negative to positive θ's, as the carrier concentration increases, indicating an increase in ferromagnetic coupling; c) in the ordered state, the magnetization has a pronounced field dependence which shows the coexistence of both ferro and antiferromagnetic order (see Fig. 2). The 4.2 K magnetic data for samples with composition close to Gd_2S_3 show, at low

Table 1

No.	x	N_p**	n(cm^{-3})*	M_0/M_s+**	θ(K)**
1	0.321	24	$(2.5 \pm 0.2) \times 10^{20}$	0.33	+22
2	0.325	27	$(1.6 \pm 0.5) \times 10^{20}$	0.23	+16
3	0.329	34	$(8.7 \pm 0.8) \times 10^{19}$	0.16	+10

+ M_s = 190 emu/gm is the saturation magnetization.

*From ref. 9.

**From ref. 10.

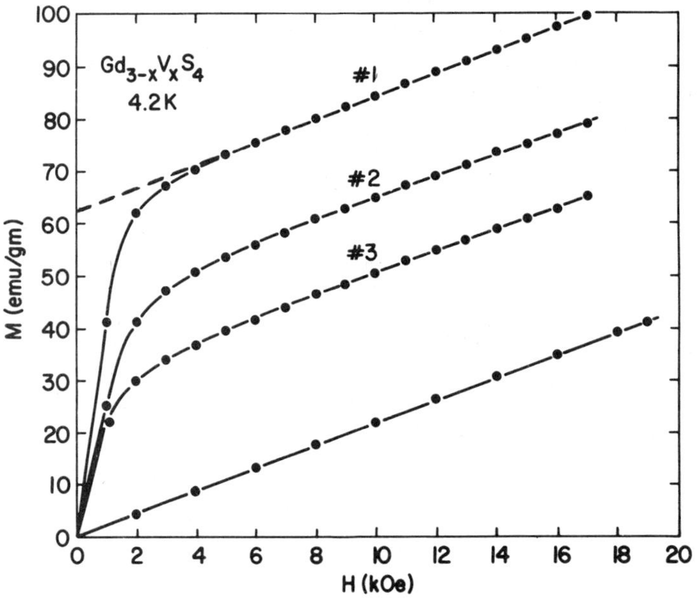

Fig. 2. Magnetization, M, vs applied field, H, for samples 1, 2, and 3 together with the pure antiferromagnetic insulator. The dashed line indicates the extrapolation to M_0 (after ref. 10).

field, a rapid rise similar to domain rotation in a ferromagnet. At higher fields the data are described by $M = M_0 = \chi_L H$, where χ_L is nearly the same for all samples, and is, therefore, characteristic of the antiferromagnetic background. The M_0 value (the 4.2K intercept) is the magnetization of the ferromagnetic polarons and increases with increasing carrier concentration, n (Table I).

It should be noted that these magnetic polarons are large on the scale of a lattice constant. If all polarons were the same size and all the Gd spins in a polaron were colinear, then the number of Gd ions per polaron, N_p, can be calculated from the ratio of M_o to the electron density n. That is, $N_p = (M_o/M_s)/(n/N_{Gd})$, where M_s is the saturation magnetization and N_{Gd} is the Gd concentration. Values for N_p are listed in Table I. The polarons have a profound effect on the transport properties, as can be seen in Fig. 3. Two observations are to be noted: 1) There is a giant negative magnetoresistance which, for sample 3 at 4.2 K, reaches 7 orders of magnitude at 20 kOe. No M-I transition is achieved. 2) In the more highly doped sample 2 an insulator-metal transition appears to be achieved at 32 kOe. This follows, since the activation energy $\Delta E \equiv E_c - E_F$, defined by the low temperature slope in Fig. 3 goes continuously to 0.

There is a direct relationship between this activation energy and the magnetic state of the systems.[10] This can be demonstrated most clearly by plotting the change in activation energy $\delta E = |\Delta E(0) - \Delta E(H)| \equiv T \ln\{\rho(H = 0)/\rho(H)\}$ vs M/M_s as in Fig. 4. The figure shows that the variation of ΔE is linear with M and therefore with H for fields above ~ 4 kOe (see Fig. 2). Furthermore, the intercept at $\delta E = 0$ occurs at $M/M_s \simeq 0.22 = 42$ emu which is in excellent agreement with the extrapolated value of M_o and is consistent with the idea that the binding energy is due to magnetic interactions. Specifically, the magnetic part of the activation energy depends on the difference between the magnetization in the ferromagnetic polaron and the antiferromagnetic background. This difference decreases upon application of an H field because the background magnetization increases as $\chi_L H$. Since χ_L is nearly temperature-independent at low temperature, then δE depends linearly upon H. This is clearly demonstrated by the parallel slopes in Fig. 4.

The concepts of the mobility edge and the magnetic polaron together explain the transport in Fig. 3 including the metal-insulator transitions. Sample 3 is non-metallic because the Fermi energy lies below the mobility edge at $T \rightarrow 0$ (see Fig. 1). At low temperatures, the slopes of ln ρ vs $1/T$ are constant because χ_L and, therefore, the magnetic contribution to the binding energy are nearly temperature independent. With increasing temperature, however, χ_L and the binding energy decrease, and the resistivity drops more dramatically. The change in sign of the slope at the highest measuring temperatures reflects the fact that $\Delta E = |E_F(T) - E_c| \simeq kT$. Sample No.

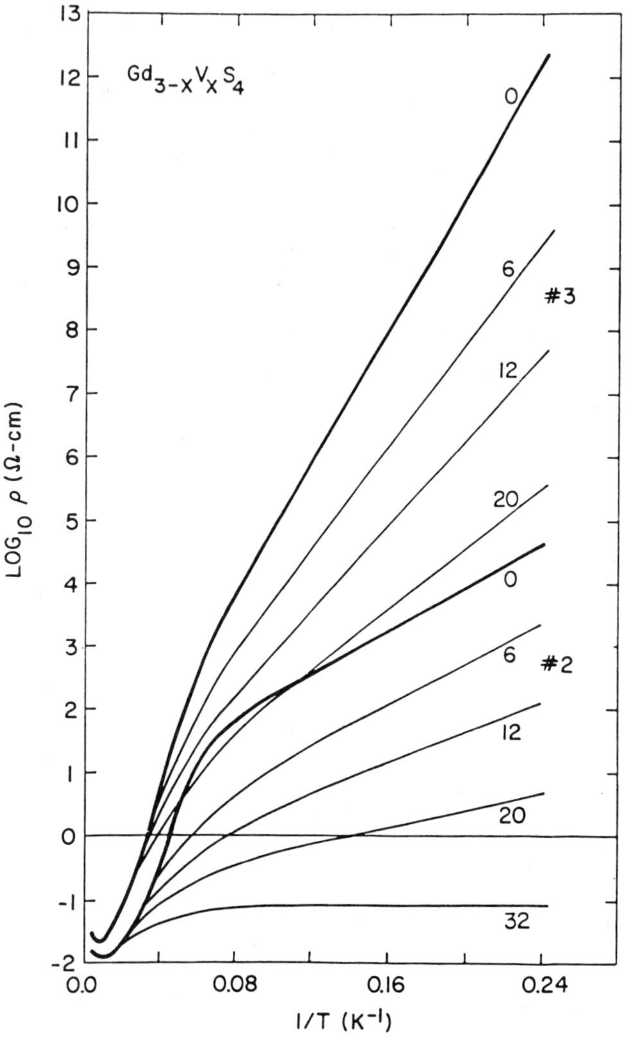

Fig. 3. Log resistivity vs reciprocal temperature for samples 2 and 3 in applied fields H=0, 6, 12, 20, and 32 kOe (from ref. 10).

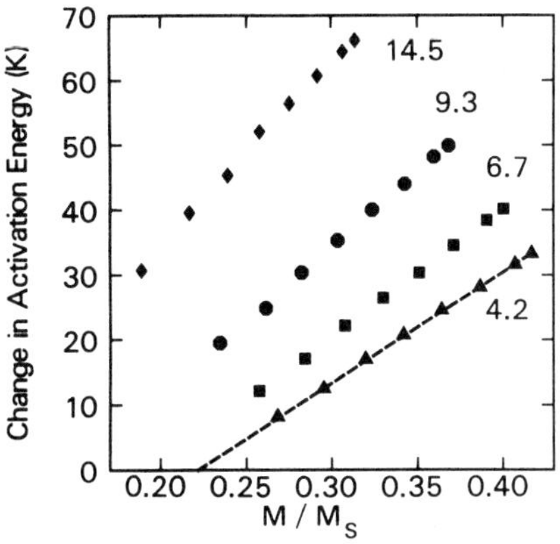

Fig. 4. Decrease in activation energy for resistivity vs M/M_s at various temperatures (after ref. 10). $M_s = 190$ emu.

2 has similar features, but, most importantly, it is seen to have undergone the M-I transition by 32 kOe. In zero field, the electron concentration of this sample is slightly below the critical concentration, n_c, for which $E_F(0) \equiv E_c$. The transition in field may be understood with the help of Fig. 5. Schematically the random potential due to the vacancies would produce the band tail depicted in Fig. 5a in the absence of magnetic polarons. E_F lies above E_c and the material is metallic. The effect of the many-body magnetic interaction is to lower the energies of the occupied states with respect to the unoccupied. The result of this shift is that E_F is now below E_c and the material is an insulator. Since the states in the immediate vicinity of E_c are unfilled and not shifted, E_c remains constant (fig. 5b). This picture is supported by the fact that n_c for this system is about twice that for paramagnetic $Ce_{3-x}v_xS_4$ [19]. With a magnetic field, the energies of the empty states split and one subband shifts downwards as indicated in Fig. 5c, bringing E_c with it. The result is that E_F now is positioned above E_c, i.e., a metal. In this diagram (Figs. 5a, b, and c) only one subband is shown for the sake of clarity.

It has been shown by an extension of Mott's mobility edge to a magnetic system how the magnetic field can vary the quantity $E_c - E_F$ and that the metal-insulator transition occurs at $E_c = E_F$. For a study of the region very close to the transition, it is necessary to go to much lower temperatures. The remainder of the paper will review these experiments and their interpretation.

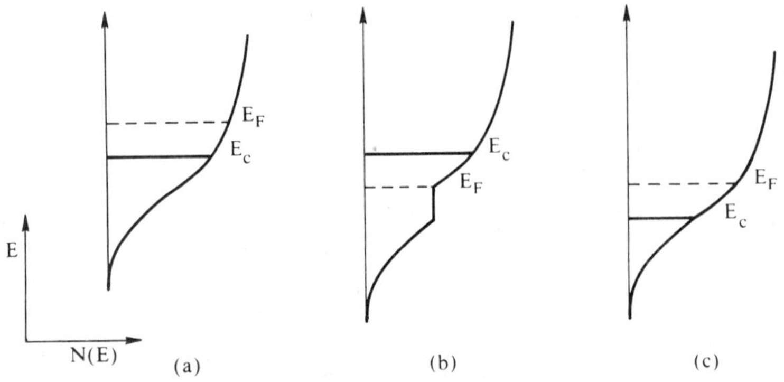

Fig. 5. Schematic density of states, N(E) vs E for: a) no polaron, H=0; b) polaron, H=0; c) polaron, $H >> H_c$. Note that the energy position of E_c and E_F is inverted in going from b to c.

Transport Studies (1983-84)

Recent work on the metal-insulator transition in magnetic semiconductors is contained in three magnetotransport studies on $Gd_{3-x}v_xS_4$ [11],[12],[13] and which focus on several theoretical developments.

Mott predicted over a decade ago that in a metal at zero temperature, the conductivity cannot be arbitrarily small. Instead, when the Yoffe-Regel criterion ($k_F\ell \gtrsim 1$) is violated, the conductivity will drop abruptly from a finite value

$$\sigma_{min} = C\frac{e^2}{\hbar a_E} \tag{1}$$

to zero.[20] In Eq. 1, C is a dimensionless constant (between 0.025 and 0.050), and a_E is approximately the distance between electrons at the critical concentration for the metal insulator (MI) transition, $a_E \simeq n_c^{-1/3}$. (In circumstances where another length limits the diffusion of the carriers, then it replaces a_E). However, Abrahams, et al.[21] predicted in 1979 that (at T=0) the transition from metal to insulator should be smooth;

$$\begin{aligned} &\sigma \sim (E_F - E_c)^\nu, \quad E_F > E_c \\ &\sigma = 0, \qquad\qquad\quad\;\; E_F < E_c. \end{aligned} \tag{2}$$

That is, the conductivity should scale continuously to zero with an exponent $\nu \simeq 1$. Both of these theories describe the behavior of non-interacting electrons moving in random potentials. The steadily advancing theory of interactions among the carriers in a random potential[22] also arrives at a continuous MI transition through more complex scaling arguments. However, recent arguments by Mott and Kaveh[23] suggest that a discontinuous transition can be masked in systems with large conduction band degeneracy.

Recent magneto-resistivity measurements in $Gd_{3-x}v_xS_4$ (where v = vacancy) established that a continuous transition from metal to insulator occurred with increasing applied magnetic field, H.[11],[12] Figs. 6a and 6b summarize the most significant aspects of the experiments. In particular, the conductivity, σ, is linear in H above a critical field, H_c, and no discontinuity is observed, even at the lowest experimentally attainable temperature, 6mK (Fig. 6b).

As is apparent from Fig. 6, there exist three regions in H for which a study of $\sigma(T)$ is of interest. For $H < H_c$, the material is an insulator. Activated transport in this regime has been studied[12] in sample 2. The results indicate that the temperature

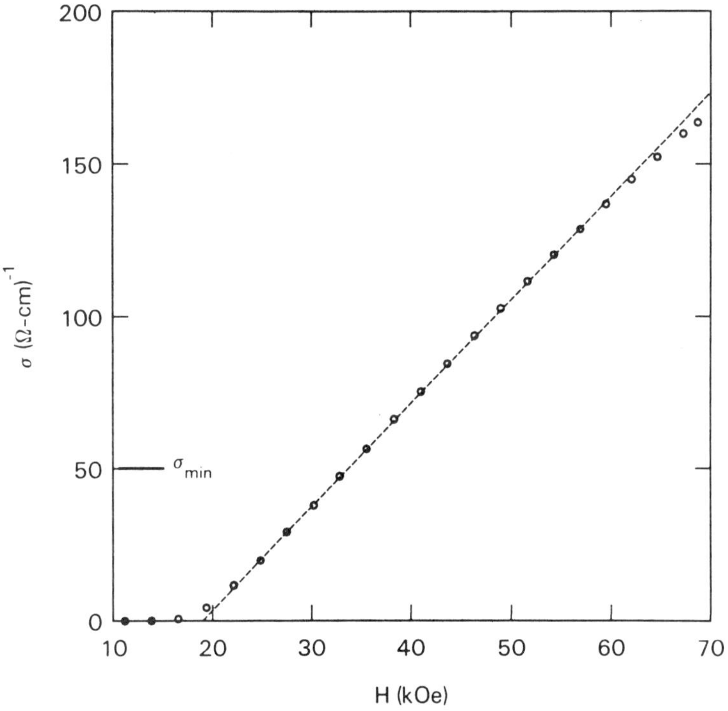

Fig. 6a. Magnetic field dependence of conductivity of $Gd_{3-x}v_xS_4$ sample No. 1 at T=300 mK. The fitted line obeys the equation $\sigma = [3.4(\Omega - cm - kOe)^{-1}][H - 19.0 \pm 0.1 kOe]$. An estimate of σ_{min} according to Eq. 1 of the text is also indicated (from ref. 11).

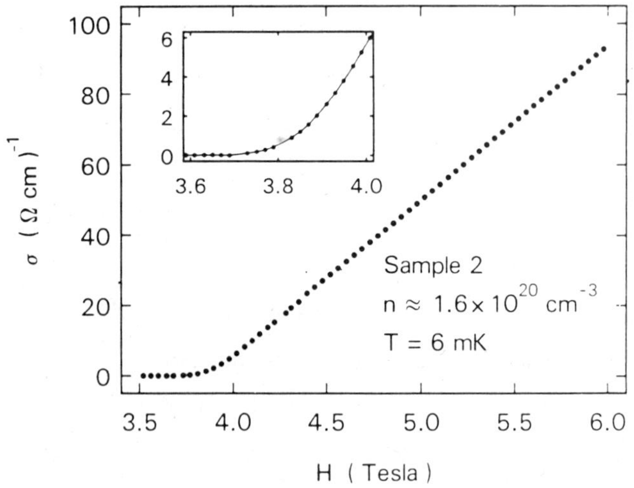

Fig. 6b. The conductivity of sample 2 as a function of the magnetic field at T = 6 mK. The critical field defined by a linear extrapolation of the data to $\sigma = 0$ is $H_c = 3.9$ Tesla $\equiv$ 39kOe. The insert is an expanded view of the transition region showing the rounding of the transition (from ref. 12).

dependence for this system does not follow Mott's variable range hopping law, i.e. $\log \sigma(T) \propto T^{1/4}$, expected for non-interacting electrons.[(24)] The experimental dependence is approximately $T^{1/2}$ over the limited range in T where such fits could be made (see Fig. 7), which is an indication of variable range hopping including electron-electron interactions.[(25)]

The temperature dependence of the conductivity at the metal-insulator transition, $H = H_c$, and on the metallic side, $H > H_c$ is complex. Both the theories of pure localization[(21),(26)] and those which include interactions in the presence of a large magnetic field ($g\mu_B H >> kT$) [(22)]give essentially similar dependencies. In analogy to the original treatment for three dimensional systems, substitution of $(H - H_c)$ for $(E_F - E_c)$ in Eq. 2 yields:

$$\sigma(0) \sim (H - H_c)^{\nu}, \qquad \nu \sim 1 . \tag{3}$$

The substitution is justified on the basis that the behavior of $|E_F - E_c| \propto H$ is equally valid on both sides of the M-I transition.[(11)] As has already been pointed out, Eq. 3 is verified by the data presented in Figs. 6a and 6b. Furthermore, since the temperature dependence appears to rule out pure localization,[12] subsequent discussion will focus on the predictions of interaction theory[22], i.e.

$$\Delta \sim (H - H_c)^r , \qquad r \sim 2 , \tag{4}$$

$$\sigma(T) = \sigma(0) + A\sqrt{T} , \qquad kT < \Delta , \tag{5}$$

$$\sigma(T) \sim BT^{1/3}, \qquad kT > \Delta . \tag{6}$$

Here, A and B are constants, and Δ is identified as a correlation gap, i.e., a depression of the single particle density of states at the Fermi energy. A schematic description of the expected behavior together with measurements at $H = H_c$ is given in

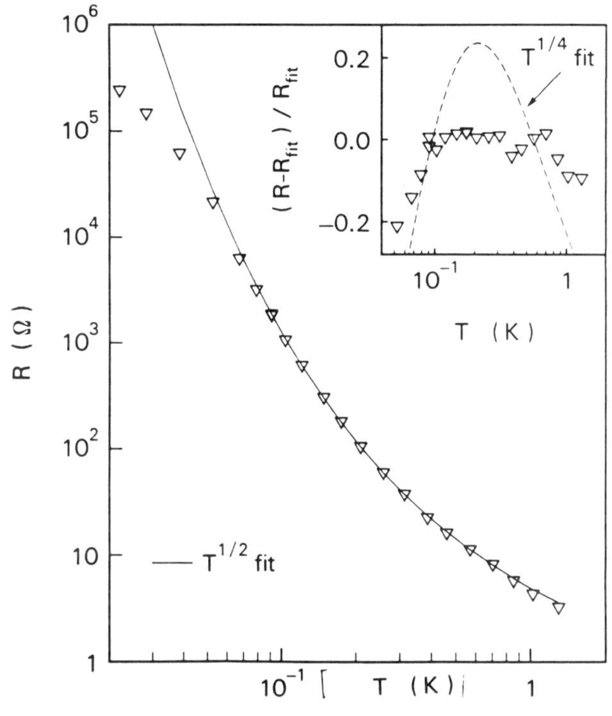

Fig. 7. The resistance data of sample 2 for H=3.5T illustrating the change in activation law at very low temperatures. The solid line is the fit to resistance predicted by the theory of variable range hopping with interactions ($R = R_0 \exp[(T_0/T)^{1/2}]$ with $R_0 = 0.38\Omega$ and $T_0 = 6.5K$). The insert is a "deviation graph" which compares the fit quality for the exponents 1/2 and 1/4. The triangles are the deviation of the data from the best fit to a $T^{-1/2}$ law, and the dashed line represents the best fit of a $T^{-1/4}$ law. (from ref. 12).

Fig. 8. In particular, the predicted low temperature variation at the transition, $H = H_c$ or $E_c = E_F$, is not $\sqrt{T}$ but follows Eq. 6, since $\Delta \rightarrow 0$ and $kT > \Delta$ is always satisfied.

Although the predicted crossover behavior has been observed in several cases [27],[28],[29], other temperature studies on the metallic side of the transition focus on the $\sqrt{T}$ dependence, Eq. 5. These fall into two classes, one in which the parameter A, Eq. 5, is strongly varying with $E_c - E_F$. [28],[30],[31] The other class includes amorphous and disordered metals such as Nb_xSi_{1-x}, [29],[32] Au_xSi_{1-x} in a magnetic field[33] and alloys containing transition elements.[34] Here, the value of A can be regarded as roughly constant for a very wide range of materials,varying between ~ 4 and $8(\Omega - cm)^{-1} K^{-1/2}$. $Gd_{3-x}v_xS_4$ appears to fall into this last category for $H >> H_c$. As an example, Fig. 9 displays the data for a sample from the No. 1 boule in which $H_c = 1.45T$. This value is smaller than for the sample depicted in Fig. 6a, indicating smaller $|E_c - E_F|$ and, by inference, a slightly larger carrier concentration. $\sigma(T)$ data were obtained at $H = H_c = 1.45T$ and 1.76, 4.0 and 6.0 T and are displayed as σ vs $\sqrt{T}$. There are obvious deviations from the $\sqrt{T}$ law for the 1.76 and 4T curves.

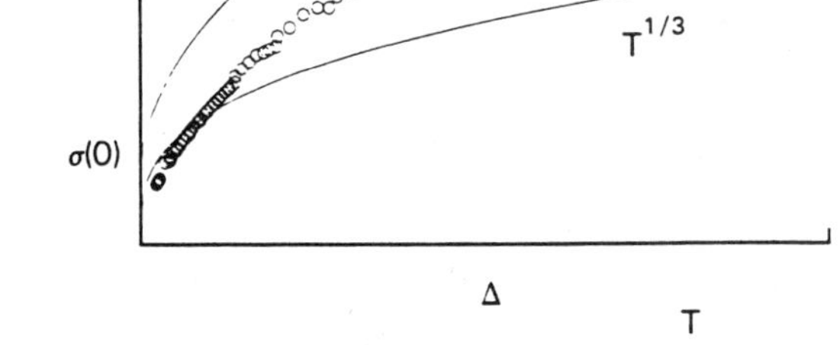

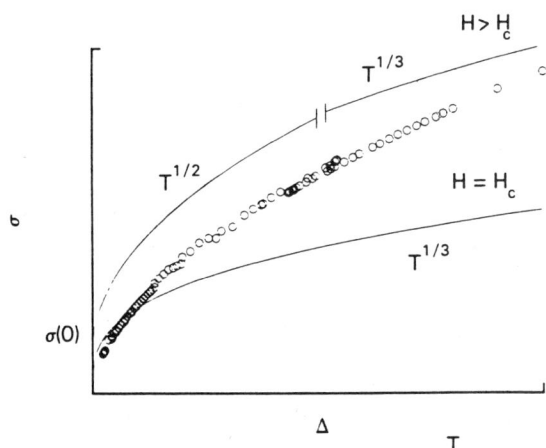

Fig. 8. Schematic representation of the expected variation of σ with T according to equations (5) and (6). The superimposed data are of sample 1 at $H = H_c$. They follow a $T^{1/2}$ law more closely than a $T^{1/3}$ law (from ref. 13).

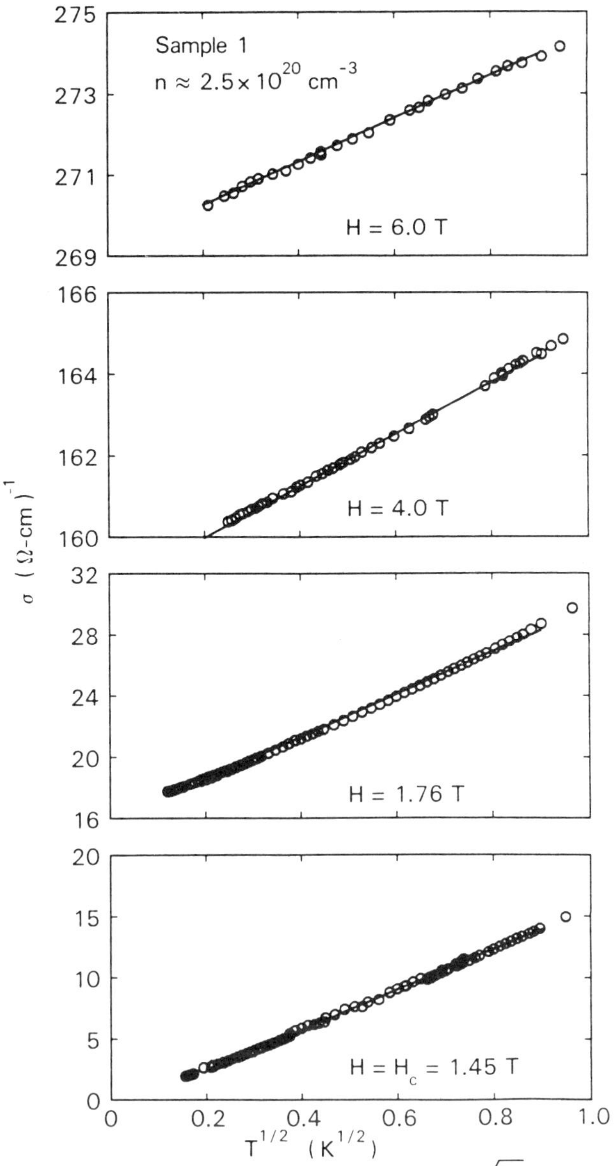

Fig. 9. The conductivity of sample 1 as a function of $\sqrt{T}$ for various fields. H_c = 1.45 Tesla (from ref. 13).

The results of a Newtonian fit to the data, in which the power of T is allowed to vary, i.e., $\sigma(T) = \sigma(0) + A'T^m$, are summarized in Table 2, together with the parameters derived from the forced fit (to a $\sqrt{T}$ law). It is obvious that the parameters $\sigma(0)$ and A are similar, regardless of the fitting procedure used. Plots of $\sigma(0)$ vs H yield a straight line intercepting the H axis at ~ 1.45T as expected. The parameters A and A′, however, are monitonically decreasing functions of H/H_c, reaching

Table 2

Sample 1*		$\sigma = \sigma(0) + A\sqrt{T}$		$\sigma = \sigma(0) + A'T^m$		
H(T)	H/H_c	$\sigma(0)(\Omega - cm)^{-1}$	$A[(\Omega - cm)^{-1}K^{-1/2}]$	$\sigma(0)(\Omega - cm)^{-1}$	$A'[(\Omega - cm)^{-1}K^{-1/2}]$	m
1.45	1.0	− 0.8 ± 0.1	16.3 ± 0.1	− 0.3 ± 0.1	16.0 ± 0.1	0.54 ± 0.01
1.76	1.21	15.8	13.9	16.7 ± 0.1	13.5 ± 0.1	0.60 ± 0.01
4.0	2.76	158.7	6.4	159.2 ± 0.2	6.0 ± 0.1	0.59 ± 0.03
6.0	4.14	269.2	5.3	269.1 ± 0.2	5.4 ± 0.2	0.49 ± 0.06

*From ref. 13

$\sim 6(\Omega - cm)^{-1}\ K^{-1/2}$ [34] far away from H_c. It is also significant that m > 0.5 near H_c, approaching the limiting $\sqrt{T}$ law only far from the transition. Evidently, no discernible crossover is observed.

Conclusion

This paper has reviewed published magnetic and magnetotransport data on the magnetic semiconductor $Gd_{3-x}v_xS_4$. The bound magnetic polaron model gives a microscopic explanation for the large effects observed. The metal-insulator transition has been studied at temperatures down to T = 6 mK. The existence of this transition is understood in terms of Mott's idea of a critical energy, E_c, separating localized from delocalized states. In the region very close to the transition, the conductivity varies smoothly from insulator to metal. Mott and Kaveh have suggested[23] that a discontinuous transition can be masked in materials with large conduction band degeneracy. However, the results for this system, $\sigma(0) \propto (H - H_c)^1$, are consistent with the scaling theories which arise from consideration of either non-interacting electrons (localization) or interacting electrons in a random potential. The observed $\ln \rho \propto (T_0/T)^{1/2}$ behavior for $H < H_c$, however, supports the view that the MI transition is driven by interactions. The temperature dependence for $H > H_c$ is of the form $\sigma(T) = \sigma(0) + A'T^m$, with $m \gtrsim 0.5$. Agreement with theory [22] (m = 0.5) is achieved only for $H/H_c >> 1$. The parameter A' is a monotonically decreasing function of H/H_c, saturating near the "universal" value, ref. 34. One surmises, therefore, that the behavior is simple only for $H/H_c >> 1$, i.e., $kT << (E_F - E_c)$, and that as the critical field H_c is approached, a more complex description of transport is required. The "crossover" description, Eqs. 5 and 6 and Fig. 8, is ruled out, however. At $H = H_c$ the temperature dependence is not the theoretically predicted[22] $\sigma(T) = BT^{1/3}$.

Acknowledgements

The studies summarized here were performed in collaboration with many colleagues, including A. Briggs, J. Flouquet, F. Holtzberg, T. Kasuya, G. Remenyi, L. J. Tao, S. Washburn, and R. A. Webb. Early measurements were performed with the technical assistance of H. Lilienthal, P. Lockwood, J. Rigotty and A. Torressen. We are grateful to all of them for their important contributions.

References

†Portions of this brief review are to be published in *Acta Physica Polonica, Suppl.*

1. N. F. Mott, Adv. Phys. **16**, 49 (1967).

2. N. F. Mott and E. A. Davis, "Electronic Processes in Non-Crystalline Materials", Clarendon Press, Oxford, 1971.

3. N. F. Mott. Phil. Mag. **29**, 613 (1974).

4. T. Kasuya and A. Yanase, Rev. Mod. Phys. **40**, 684 (1968).

5. S. von Molnár and T. Kasuya, in "Proc. 10th Int. Conf. on the Physics of Semiconductors, Cambridge, Mass., 1970", S. P. Keller, J. C. Hensel and F. Stern, eds., CONF-700801 (U.S. AEC Div. of Tech. Info., Springfield, VA 1970), p. 233; T. Kasuya, ibid, p. 243.

6. J. B. Torrance, M. W. Shafer, and T. R. McGuire, Phys. Rev. Lett. **29**, 1168 (1972).

7. Descriptions of magnetic trapping effects in the currently popular Mn doped II-VI semiconductors are given in T. Dietl and J. Spałek, Phys. Rev. Lett. **48**, 355 (1982) and P. A. Wolff and J. Warnock, J. Appl. Phys. **55**, (6), 2300 (1984).

8. M. Cutler and N. F. Mott, Phys. Rev. **181**, 1336 (1969).

9. S. von Molnár and F. Holtzberg, in *Magnetism and Magnetic Materials - 1972*, edited by C. D. Graham, Jr., and J. J. Rhyne, AIP Conference Proceedings No. 10 (American Institute of Physics, New York, 1973), p. 1259.

10. T. Penney, F. Holtzberg, L. J. Tao and S. von Molnár, in *Magnetism and Magnetic Materials - 1973*, edited by C. D. Graham, Jr., and J. J. Rhyne, AIP Conference Proceedings No. 18 (American Institute of Physics, New York, 1974), p. 908.

11. S. von Molnár, A. Briggs, J. Flouquet and G. Remenyi, Phys. Rev. Lett. **51**, 706 (1983).

12. S. Washburn, R. A. Webb, S. von Molnár, F. Holtzberg, J. Flouquet, and G. Remenyi, Phys. Rev. **B30**, 6224 (1984).

13. S. von Molnár, J. Flouquet, F. Holtzberg and G. Remenyi, accepted for publication in Solid State Electronics.

14. K. Meisel, Z. Anorg. Chem. **240**, 300 (1939).

15. W. H. Zachariasen, Acta Cryst. **2**, 57 (1949).

16. P. W. Anderson, Phys. Rev. **109**, 1492 (1958). 318 (1966).

17. F. Holtzberg, T. R. McGuire, S. Methfessel, and J. C. Suits, J. Appl. Phys. **35**, 1033 (1964).

18. A. Kamijo, A. Takase, Y. Isikawa, S. Kunii, T. Suzuki and T. Kasuya, J. de Physique **41**, C5-189 (1980).

19. M. Cutler and J. F. Leavy, Phys. Rev. **133**, A1153 (1964).

20. N. F. Mott, Philos. Mag. **26**, 1015 (1972); K. F. Berggren, J. Phys. C. **15**, L45 (1982); M. Kaveh and N. F. Mott, ibid., **15**, L697 (1982).

21. E. Abrahams, P. W. Anderson, D. C. Licciardello, and T. V. Ramakrishnan, Phys. Rev. Lett. **42**, 673 (1979).

22. C. Castellani, C. Di Castro, P. A. Lee, M. Ma, Phys. Rev. B **30**, 527 (1984).

23. N. F. Mott and M. Kaveh, Philos. Mag. **47B**, L17 (1983).

24. N. F. Mott, J. Non-Cryst. Solids **1**, 1 (1969).

25. A. L. Efros and B. I. Shklovskii, J. Phys. **C8**, L49 (1975).

26. Y. Imry, J. Appl. Phys. **52**, 1817 (1981).

27. Y. Imry and Z. Ovadyahu, J. Phys. **C15**, L327 (1982).

28. G. A. Thomas, M. Paalanen, and T. F. Rosenbaum, Phys. Rev. **B27**, 3897 (1983).

29. D. J. Bishop, E. G. Spencer, R. C. Dynes, to be published in Solid State Electronics.

30. T. F. Rosenbaum, R. F. Milligan, M. A. Paalanen, G. A. Thomas, R. N. Bhatt, and W. Lin, Phys. Rev. **B27**, 7509 (1983).

31. G. A. Thomas, A. Kawabata, Y. Ootuka, S. Katsumoto, S. Kobayashi, and W. Sasaki, Phys. Rev. **B24**, 4886 (1981).

32. G. Hertel, D. J. Bishop, E. G. Spencer, J. M. Rowell, and R. C. Dynes, Phys. Rev. Lett. **50**, 743 (1983).

33. N. Nishida, T. Furubayashi, M. Yamaguchi, K. Morigaki, and H. Ishimoto, to be published in Solid State Electronics.

34. R. W. Cochrane and J. O. Strom-Olsen, Phys. Rev. **B29**, 1088 (1984).

METAL-NONMETAL TRANSITIONS AND THERMODYNAMIC PROPERTIES

Fumiko Yonezawa and Tohru Ogawa

Department of Physics
Keio University
Yokohama 223, Japan

Institute of Applied Physics
University of Tsukuba
Ibaraki 305, Japan

ABSTRACT

In this article, we discuss the relationship between the electronic and thermodynamic properties of a system (crystalline or disordered) in connection with the metal-nonmetal (M-NM) transitions. We show that a first-order thermodynamic transition is expected at a M-NM transition point. When our analysis is applied to expanded metallic fluids in the vicinity of the liquid-gas critical point, it predicts a possible existence of two different kinds of thermodynamic phase transitions —— first the atomic gas-to-liquid condensation for which electrons play no explicit role, and secondly the electron-induced first-order phase transition accompanying a M-NM transition. We assert that the latter could be detected in those thermodynamic properties which are defined by higher-order derivatives of the free energy.

INTRODUCTION

Concerning the metal-nonmetal (M-NM) transitions, one of the striking features, the importance of which has been somehow overlooked, is the fact that a naive consideration predicts a possible occurrence of a first-order phase transition accompanying the M-NM transition. The free energy F per atom as a function of atomic volume v changes from a metallic to nonmetallic behaviour at the volume v_t where the M-NM transition takes place, whatever the mechanism of the M-NM transition may be. This is the case both for crystalline and disordered systems. In other words, a transition from metal to nonmetal implies that the nature of the cohesive force changes there and the free energy must reflect the change in the cohesive force. This in turn is described as follows. The transition point v_t is defined as the volume at which the free energy for a metallic system intersects with that for a nonmetallic system and accordingly the favourable branch shows a kink at v_t (see Fig.1), thus making it possible to draw a common tangent to the free energy-volume curve with two points, v_1 and v_2, of contact. As a result, there appears the two-phase region between v_1 and v_2, typical of the first-order phase transition.

In this paper, we give some detailed explanation for the occurrence of an electron-induced first-order phase transition and discuss its bearings on the thermodynamic properties of expanded metallic fluids.

METAL-NONMETAL TRANSITIONS AND FREE ENERGY

In this section, we show how a M-NM transition would affect the free energy of the system under consideration.

Interplay between Two Terms of Free Energy

The discontinuous change in volume, or equivalently the appearance of the first-order phase transition, which accompanies the M-NM transition, was previously predicted[1~6] and a few theoretical analyses were proposed[7~10] for a case of the M-NM transition due to electron correlation found in some crystalline transition metal oxides.

Following the idea of ref.10, let us show how the first-order phase transition with a discontinuous change of volume is expected to occur. We assume that the total free energy F_{tot} of a system is given by the sum of the electron part F_e and the lattice part F_L. A necessary condition for the system to undergo a first-order phase transition with a discontinuous volume change is that there exists a region in which the second derivative of F_{tot} per atom with respect to atomic volume v=V/N becomes negative, where N is the number of atoms and V is the volume of the system. This corresponds

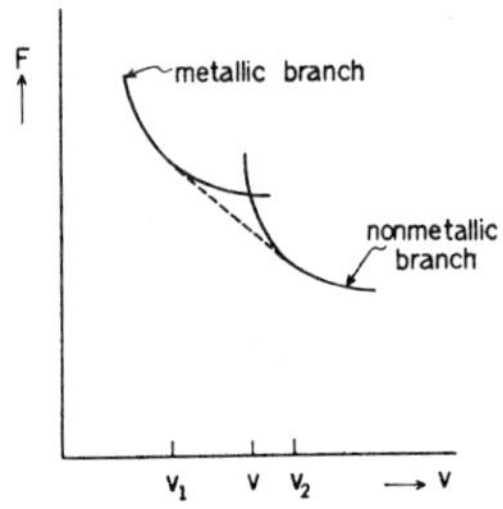

Fig.1. Free energy vs volume.

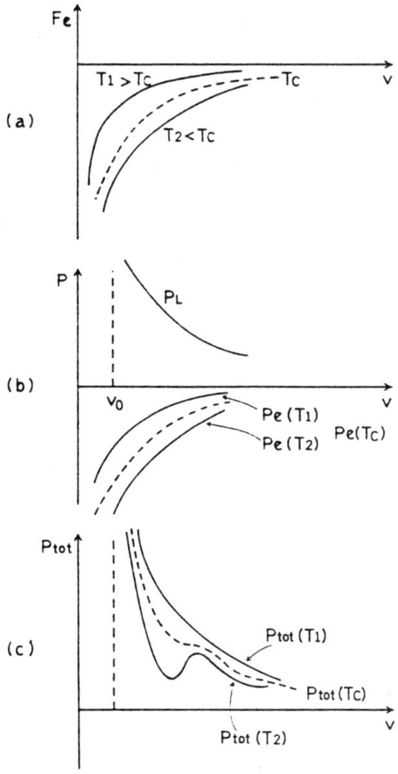

Fig.2. (a) F_e vs v.
(b) P_L and P_e vs v.
(c) P_{tot} vs v.
$T_1 > T_c > T_2$.
T_c = the critical temperature

to a violation of the thermodynamical stability condition for the total compressibility, since the violation is expressed as

$$\left(\frac{\partial^2 F_{tot}}{\partial v^2}\right)_T = \left(\frac{\partial^2 Fe}{\partial v^2}\right)_T + \left(\frac{\partial^2 F_L}{\partial v^2}\right)_T = -\left(\frac{\partial P}{\partial v}\right)_T < 0 \qquad (1)$$

The violation sometimes referred to as the Lifshitz instability. The second term of (1) gives the main contribution to the compressibility and is considered to be positive. On the other hand, a simple model as given below suggests the possibility for the first term of (1) to become negative.

Let us write the electron Hamiltonian H_e as a sum of a kinetic term H_K and an interaction term H_I, and assume that only $H_K$is dependent on volume. When t(v) denotes the transfer energy defining the band width, we can write

$$H_e = H_K + H_I = t(v)H_{K'} + H_I \,. \qquad (2)$$

We assume that $t(v)>0$ while $H_K<0$ and $H_{K'}<0$. The electron free energy F_e due to electrons is then defined by

$$F_e = -T\, ln\, \{\mathrm{Tr}\, \exp(-H_e/T)\}\,, \qquad (3)$$

which yields

$$\frac{\partial^2 F_e}{\partial t^2} = -\frac{1}{T}\frac{1}{t}<(H_K - <H_K>)^2> < 0\,, \quad \frac{\partial Fe}{\partial t} = \frac{1}{t}<H_K> < 0, \qquad (4)$$

where the angular brackets indicate the temperature average as defined by $<A>\equiv \mathrm{Tr}\{A\exp(-H_e/T)\}/\mathrm{Tr}\{\exp(-H_e/T)\}$.

The first term of (1) is now written as

$$\frac{\partial^2 F_e}{\partial v^2} = \frac{\partial^2 Fe}{\partial t^2}\left(\frac{dt}{dv}\right)^2 + \frac{\partial Fe}{\partial t}\frac{d^2t}{dv^2}\,, \qquad (5)$$

where the first term is negative (see Eq.(4)). As for the second term, it is negative if $d^2t/dv^2 > 0$ since Eq.(4) guarantees that $\partial Fe/\partial t<0$, or it is zero if $d^2t/dv^2=0$. In most cases, it is highly expected that $d^2t/dv^2>0$, or small enough to be ignored safely. This leads to the conclustion that $\partial^2F_e/\partial v^2<0$ and therefore the free energy due to electrons can work to introduce instability.

When $|\partial^2F_e/\partial v^2|>\partial^2F_L/\partial v^2$, the system is in an unstable state and the phase separation takes place. The situation is illustrated in Fig.2. The free energy F_e decreases as the volume v decreases since it is easier for electrons to move around freely for smaller v and the transfer energy t(v) increases monotonically as the decrease of v. As can be seen from Eq.(3), the dependence of F_e on volume is more marked for low temperatures than for high temperatures (Fig.2(a)). Since the pressure P is defined by the negative of the first derivative of F_{tot} with respect to v, the volume dependence of P_e and P_L takes the form as shown in Fig.2(b). While P_L vs v curve is independent of temperature, the dependence of P_e on v varies from temperature to temperature, the dependence being more marked for lower temperatures. Accordingly, the total pressure P_{tot} is a monotonically decreasing function of v for high enough temperatures (T_1). On the other hand, P_{tot} has two extrema for low enough temperatures (T_2), thus giving the unstable region $(\partial P/\partial v)_T = -(\partial^2F/\partial v^2)_T$ is positive. The temperature at which the two extrema coincides is identified as the critical temperature and the corresponding pressure as the critical pressure.

This kind of situation is expected to occur whenever an interplay exists between the repulsive and attractive parts of the total free energy, and it is not necessary to invoke for the effects of electron correlation. The simplest example which occurs to our mind may be the van der Waals equation of state as shown in Fig.3. In the figure, v_0 is the van der Waals volume, a is a constant and T_c is the critical temperature.

It must be noted here that, in the above-mentioned two examples, the first-order phase transitions are not the outcome of singularites in the total free energy $F_{tot}(v)$ as a function of v. Instead, $F_{tot}(v)$ itself is regular throughout the v region of interest, but the convexity violates the thermodynamic stability condition, which eventually yields the two-phase region, the condensation and the critical point.

A few attempts[7~10] have been made to study the thermodynamic properties of the Hubbard model. The motivation of these attempts was to explain the phase diagram of chromium doped V_2O_3. Although these attempts give some elaborate analysis, the essential point of the arguments is summarized by the discussion presented in the above.

First-Order Phase Transition Induced by the M-NM Transition

One unsatisfactory aspect about those attempts mentioned above is that

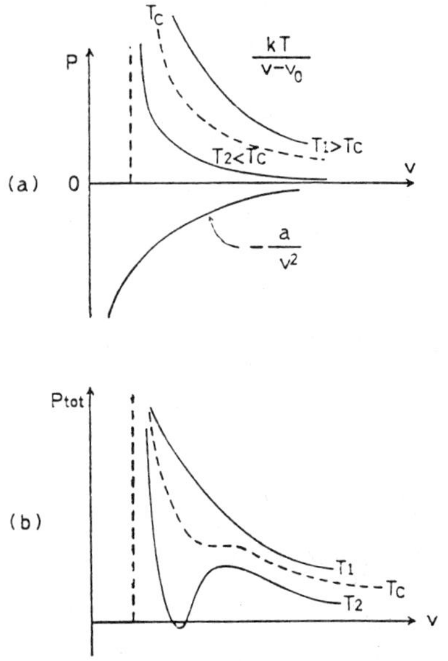

Fig.3. Van der Waals equation of state on the P-v plane.
(a) contributions from the repulsive and attractive terms
(b) the total pressure

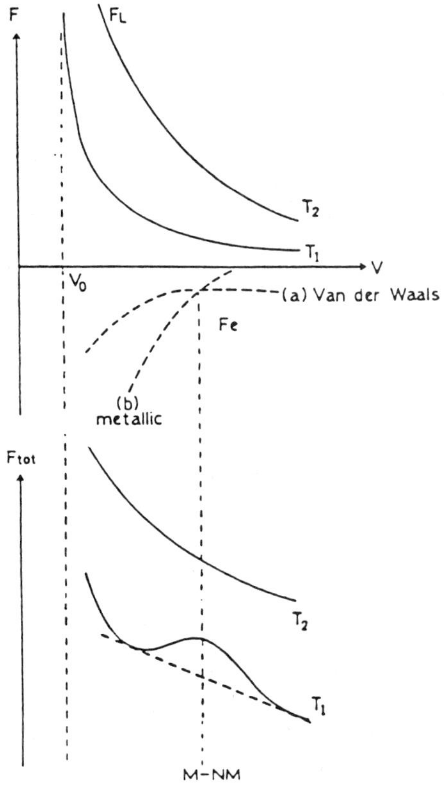

Fig.4. F vs v.
$(T_1<T_c<T_2)$

they do not give any theoretical ground for explaining that the occurrence of an M-NM transition exactly coincides with a point where the first-order transition takes place. On the other hand, a discussion such as given in connection with Fig.1 yields a direct relationship between two transitions.

The situation can be illustrated more clearly in Fig.4 where the solid curves denoted by T_1 and T_2 in the upper half of the figure represent the lattice part of the van der Waals free energy $F_L=-kT\ln(v-v_o)$+constant. For T=0, the broken curve (a) depicts van der Waals cohesion and (b) depicts the Hartree-Fock energy of the metallic configuration. The total free energy curve F_{tot}(on the lower half of Fig.3) has a region of negative convexity for low enough temperatures (T_1) and no such region for high enough temperatures (T_2). The region of negative convexity for low temperatures indicate the appearance of the first-order phase transition since the tangent of common contact can be drawn in this case. The resulting two-phase region covers the volume v_{MNM} at which the crossover takes place from the metallic branch to the nonmetallic branch (denoted by a dotted straight line in the figure). In this analysis, it is obvious that the first-order phase transition is induced by the M-NM transition.

Two Kinds of First-Order Phase Transitions and Two Critical Points

Now, let us recollect the fact that a first-order transition does take place even if there is no M-NM transition and accordingly even when only nonmetallic branch exists. This is obvious from the van der Waals case. For low enough temperatures, the total free energy naturally has the region of negative convexity leading to the appearance of the van der Waals unstable region in the P-v plane. This fact together with the assertion in the preceding subsection indicates the existence of two critical points, one coming from a gas-liquid condensation for which electrons do not play an essential role and one coming from the appearance of two-phase region as a consequence of the change in the cohesive energy where of course the role of electrons is vital.

A possible existence of two critical points was suggested as early as in 1934 by Landau and Zeldovich[11] who predicted what the phase diagram would look like in the P-T plane. The corresponding phase diagrams in the T-v plane may behave as shown in Fig.5 where we have introduced two kinds of liquid states denoted by liquid I(L_I) and liquid II (L_{II}). The latter is identified as a usual liquid state while the former is a new phase. This phase is definitely different from an ordinary gas state because it is a condensed phase — a condensed metallic phase in case of (a) and a condensed nonmetallic phase in case of (b).

It is also interesting to ask how the boundary between metallic and nonmetallic states behaves above the critical point. When the two kinds of the first-order phase transitions happen to coincide or overlap, the M-NM boundary would follow the line (1) in Fig. 6; or the boundary could also behave like(1') as well as like (1") since there is no reason why the M-NM boundary for different temperatures always occurs at the same volume. When the M-NM transition takes place off the critical point of the ordinary gas-liquid condensation, either (2) or (3) in Fig.6 is the possiblity. When parameters fulfill appropriate conditions, the electron-induced two-phase region would be hidden within the gas-liquid two-phase region as illustrated by Fig.6(4) or (5). Just as in the case of (1') and (1"), we may have (2'),(2"), etc., but for the sake of simplicity we have omitted these possibilities from the figure.

No agreement has so far been obtained concerning how the M-NM boundary would terminate at still higher temperatures (and possibly under still higher pressures). Some kinds of plasma phases are expected to set in. A few theo-

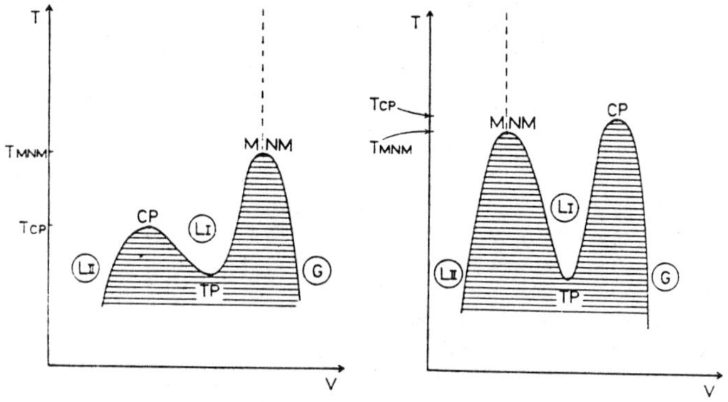

Fig.5. Schematic phase diagrams in the T-v plane.

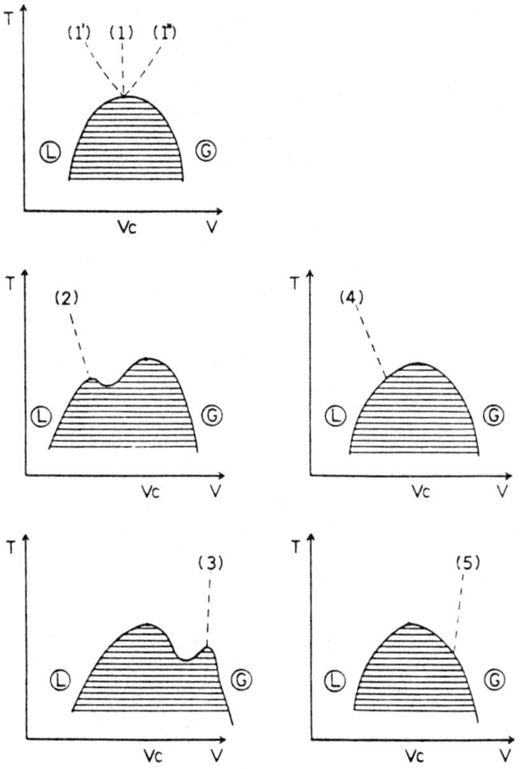

Fig.6 Possible ways the M-NM boundary would behave in the T-v plane where v_c denotes the critical atomic volume. The solid curves show the boundary of the coexistence region which are expressed by hatches. The dotted curves represent the M-NM boundary for each case.

retical attempts[12~14] have been developed to evaluate phase diagrams including metallic, nonmetallic and plasma phases, but the problem is still open for future study.

ELECTRON-INDUCED FIRST-ORDER PHASE TRANSITION

Now, we give from first principles a more detailed analysis of the relationship between the electronic and thermodynamic properties of expanded fluid metals. We treat a system composed of ions and electrons. We assume that the ion part of the Helmholtz free energy, $F_{ion}(v,T)$, is regular as a function of atomic volume v. Then, our special interest lies in the analytic behaviour of the electron part of the free energy, $F_e(v,T)$. That is to say; is $F_e(v,T)$ regular as a function of v? or does it show a singularity such as indicated by Figs.1 and 4 at the volume where the M-NM transition occurs? And if it does, then what kind of singularity is it to be? As for the effects of the singularity reflected on the thermodynamic properties, we defer the discussion until a succeeding section.

Total Hamiltonian

We study a system composed of N ions and N_e electrons described by the total Hamiltonian

$$H_{tot} = H_e + H_{ion} \cdot \tag{6}$$

The electron Hamiltonian H_e for valence electrons and the ionic Hamiltonian H_{ion} are respectively defined by

$$H_e = \sum_{m=1}^{N_e} H_o(\vec{r}_m) + \frac{1}{2} \sum_{m \neq n} v(\vec{r}_m - \vec{r}_n) , \tag{7}$$

and

$$H_{ion} = \sum_{i=1}^{N} K_{ion}(\vec{R}_i) + \frac{1}{2} \sum_{i \neq j} V(\vec{R}_i - \vec{R}_j) , \tag{8}$$

where $\vec{r}_m$ denotes the position of the m-th electron, $\vec{R}_i$ the position of the i-th ion, $v(\vec{r}_m-\vec{r}_n)$ the interaction between the m-th and n-th electrons, $K_{ion}(\vec{R}_i)$ the kinetic energy of the i-th ion, $V(\vec{R}_i-\vec{R}_j)$ the interaction between the i-th and j-th ions, and $H_o(\vec{r}_m)$ is the one-electron Hamiltonian

$$H_o(\vec{r}_m) = - \frac{\hbar^2}{2m} \nabla_m^2 + \sum_{i=1}^{N} u(\vec{r}_m - \vec{R}_i) , \tag{9}$$

m being the electron mass and $u(\vec{r}_m-\vec{R}_i)$ the interaction between the m-th electron and i-th ion. For the sake of simplicity in the formulation, we confine ourselves, in the following, to the case where all valence electrons are s-like. The whole theme of our argument of course remains unchanged even when this restriction to s electrons alone is removed.

Contributions from Ions

For the present purpose, it suffices to treat the ion Hamiltonian for the average number of electrons Z/2 for each spin where Z is the number of valence electrons per atom. Then, we have

$$H_{ion} = \sum_i K(\vec{R}_i) + \frac{1}{2} \sum_{i \neq j} \mathcal{V} (\vec{R}_{ij}) , \tag{10}$$

where $\mathcal{V}(\vec{R}_{ij})$ is the effective potential between ions at $\vec{R}_i$ and $\vec{R}_j$ (for a detailed expression and ita derivation, see ref.1).

Since we are interested in the electron Hamiltonian, the internal energy E_{ion} of the ion system can be approximated by taking the average of (10); ie

$$E_{ion} = \text{const} + \frac{d}{2} \int V(\vec{R}_{ij}) g(\vec{R}_{ij}) d\vec{R}_{ij} \equiv \text{const} - \frac{a}{v} , \quad (11)$$

where d is the number density defined by d=1/v, and $g(\vec{R}_{ij})$ is the pair distribition function. The constant a is positive unless v is unreasonably small since $V(\vec{R}_{ij})$ is most strongly negative at $\vec{R}_{ij}$ where $g(\vec{R}_{ij})$ is dominant. Equation (11) is equivalent to the internal energy term of the van der Waals free energy. The approximate treatment leading to Eq.(11) does not give the strong repulsion for small enough values of v. In the total free energy, however, the strongrepulsion for $v \leqq v_o$ (v_o being the van der Waals' volume) is taken care of in the entropy term.

It is clear from physical consideration that the characteristic features which the ion free energy F_{ion} should retain are (1) the strong repulsion for $v \leqq v_o$; and (2) the attractive part for $v > v_o$. In ref.1, the explicit forms of entropy are obtained for the cases of the van der Waals model, a single-occupancy cell model, the Percus-Yevick equation of state for a hard-sphere system and a lattica gas model. It was shown in ref.1 that the choice of a model does not influence the essential qualitative behaviours of the ion free energy as a function of v. Accordingly, the pressure and all other thermodynamic properties including the critical behaviour of a system do not depend on the kinds of models employed for the ion system.

Contributions from Electrons

Since our purpose is to see the singularites in the electron free energy introduced by the M-NM transition, it would be appropriate to express the electron free energy as a function v and T in the form

$$F_e(v,T) = F_{eo}(v,T) + \Delta F_e(v,T)\Theta(v_t - v) , \quad (12)$$

where $F_{eo}(v,T)$ is a part which is regular through the whole region of v of interest. On the other hand, $\Delta F_e(v,T)$ is an extra term added to the system as a result of the change in the cohesion from a nonmetallic to metallic type. The step function is denoted by Θ and v_t stands for the atomic volume at which the M-NM transition takes place.

In the study of thermodynamic properties, the most natural procedure for evaluating the electron free energy is to calculate Eq.(3) for a given H_e. This procedure, however, is not always easy to follow and very often some approximations are required in order to obtain any useful results at all. Moreover, the approximations employed for this purpose sometimes obscure the physical meanings of the obtained results. In view of this fact, we take an alternative procedure in which we start from the study of thermodynamic properties in the ground state, ie, at absolute zero and take into account the effects of finite temperatures later. Then, what we should be concerned with is, instead of Eq.(12),

$$E_e(v) = E_{eo}(v) + \Delta E_e(v)\Theta(v_t - v) , \quad (13)$$

where $E_e(v)$ is the internal energy of an electron system in the ground state, $E_{eo}(v)$ and $\Delta E_e(v)$ are the regular part and the additional term respectively. A detailed evaluation of $\Delta E_e(v)$ is given in ref.1, and this term is shown to have the form

$$E_e(v) \propto \frac{A v_t}{\lambda} \left(\frac{v_t - v}{v_t} \right)^{\lambda} + \text{higher-order terms.} \quad (14)$$

Table 1. Signs of A and values of λ (for detail, see ref.1)

	A	λ
crystalline model	-	5/2
Hubbard model		
(1) function integral method	-	5/2
(2) Gutzwiller approximation	-	2
(3) alloy analogy approximation	+	5/2
A binary alloy	+	7/2
One electron model defined by a single Green's function[1]	+	

The explicit values of λ and the sigens of A were obtained[1] for various cases. The results are listed in Table 1.

THERMODYNAMIC PROPERTIES NEAR M-NM TRANSITIONS

By making use of the information obtained in the above, we discuss in this section the possible behaviours of some thermodynamic properties near M-NM transitions.

General Discussion

The equation of state is for instance written as

$$P = -\left(\frac{\partial F_{ion}}{\partial v}\right)_T - \left(\frac{\partial F_e}{\partial v}\right)_T$$

$$= \left(\frac{\partial S_{ion}}{\partial v}\right)_T T - \left\{\left(\frac{\partial E_{ion}}{\partial v}\right)_T + \left(\frac{\partial E_{eo}}{\partial v}\right)_T + \left(\frac{\partial \Delta E_e}{\partial v}\right)_T \Theta(v_t - v)\right\}, \quad (15)$$

where we have neglected the term due to the electron entropy because it is small compared to the other terms. It is obvious that the sign of A and the value of λ in ΔE_e determine the behaviours of thermodynamic properties at an M-NM transition point.

The influence of the nonzero additional term, $\Delta E_e(v) \neq 0$, may show up in the following manner.

(1) When the M-NM transition takes place in the pressure-volume region where there is no phase transition otherwise, the additional term may introduce a new first-order phase transition (see Figs.6(1) and (2)).

(2) Even when the effects of the additional term are not so dramatic as to introduce a new phase-transition, it definitely changes the behaviour of the two-phase region in the pressure-volume plane. The way how the change in the behaviour of the two-phase region is caused depends on the sign of A and the magnitude of λ.

(3) The additional terms results in giving singular terms to thermodynamic properties defined by higher-order derivatives of the free energy. These singular terms may be quite noticeable and therefore experimentally detectable even when the singularity expressed by $\Delta E_e(v)$ itself is not overwhelming. Moreover, the singularities in these thermodynamic properties are maintained in the region above the critical temperature even when, at low temperatues, the singularity being covered by the two-phase region does not show any explicit influence to the phase diagram.

Isothermal Compressibility

The isothermal compressibility is one of the most familiar examples which is defined by higher-order derivatives of the free energy. Since the second derivative of $\Delta E_e(v)$ with respect to v behaves as

$$v\left(\frac{\partial^2 \Delta E_e(v)}{\partial v^2}\right)_T = -(\lambda-1)A\left(\frac{v_t - v}{v_t}\right)^{\lambda-2} + \text{higher-order terms}, \quad (16)$$

just below v_t, the isothermal compressibility κ_T has the form

$$\kappa_T = -\frac{1}{v}\left(\frac{\partial v}{\partial P}\right)_T = \frac{\alpha}{f(v)+(\lambda-1)A\left(\frac{v_t-v}{v_t}\right)^{\lambda-2}\Theta(v_t-v)}, \quad (17)$$

where α is a constant and f(v) is a regular function of v. For the alloy analogy solution of the Hubbard model, $\lambda=5/2$ and the behaviour of the resulting κ_T is schematically shown in Fig.7. Note that the singularity in κ_T at $v \lesssim v_t$ is the form of a square root. This means that $\partial\kappa_T/\partial v$ is discontinuous at v_t, and divergent for $v \lesssim v_t$.

Schematic behaviours of κ_T^{-1} as a function of v has previously been discussed[1,15] for various signs of A and for various values of λ. Here, we rephrase the argument in the language of κ_T itself on the basis of (17) for various values of λ.

(a) $\lambda=2$: The isothermal compressibility κ_T is discontinuous at v_t as shown in Fig.8(a) where, for $v \lesssim v_t$, κ_T increases for A<0 and decreases for A>0.

(b) $2<\lambda<3$: κ_T is continuous at v_t but its derivative at $v \lesssim v_t$ is $-\infty$ for A<0 and $+\infty$ for A>0 as shown in Fig.8(b). The above-described example of the Hubbard model is included in this category.

(c) $\lambda=3$: κ_T is continuous at v_t, but its derivative is not, showing a linear slope for $v \lesssim v_t$ which is negative for A<0 and positive for A>0 as shown in Fig.8(c).

(d) $\lambda>3$: κ_T and its first derivative are both continuous at v_t, but its higher derivatives are discontinuous as shown in Fig.8(d).

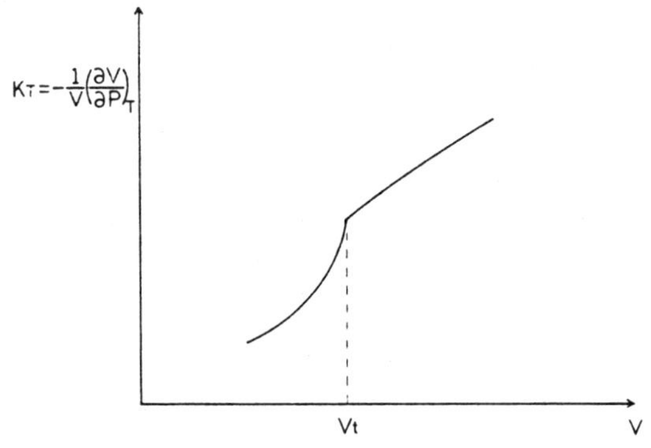

Fig.7. A schematic behaviour of the theoretical isothermal compressibility κ_T for a lattice gas model of the Hubbard Hamiltonian. Note that there exists a singularity at v_t.

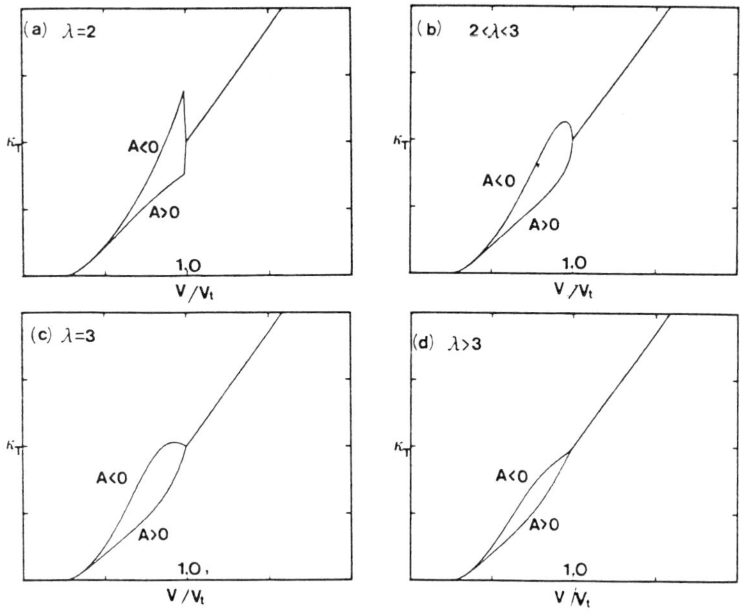

Fig.8. Isothermal compressibility κ_T as a function of v for various values of λ and for the both signs of A. (a) λ=2; (b) 2<λ<3; (c) λ=3; and (d) λ>3.

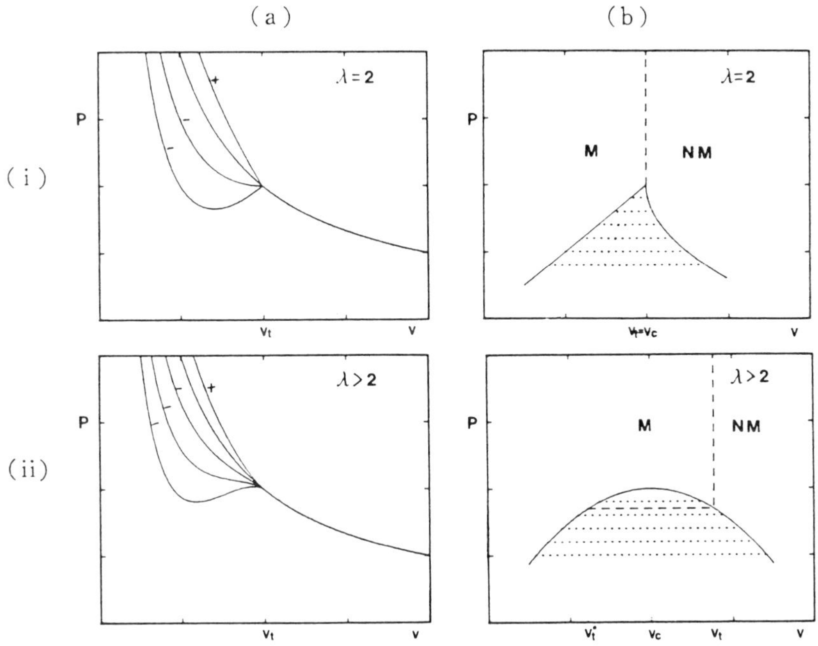

Fig.9. Schematic behaviours of the phase diagram in the P-v plane; (i) λ= 2 and (ii) λ >2
(a) the isotherm; (b) the coexistence region:
Signs + and - denote cases A>0 and A<0 respectively.

In all these cases, the influence of electrons to the isothermal compressibility κ_T as a result of the band-overlap is such that, for $v \lesssim v_t$, it becomes easier to compress the material if A<0 and more difficult if A>0. It is noted that this comparison is done between the cases with and without the singular additional term.

Phase Diagram for an Electron-Induced First-Order Phase Transition

In the present subsection, we study the possible bahaviours of the phase diagram of a first-order phase transition induced by an M-NM transition in the volume region in which there were no phase transitions if it were not for the M-NM transition.

(i) $\lambda=2$: In this case, the first derivative of the additional term $\Delta E_e(v)$ is linear in v and therefore the slope of P as a function of v is discontinuous (see Fig.9(i-a)). As can be seen from this figure, the critical volume coincides with v_t. The starting of the coexistence curve at the critical point (v_c, P_c) is singular (see Fig.9(i-b)); not quadratic as in the van der Waals case but linear on the metallic side and square-root-type on the nonmetallic side, ie

$$\frac{P_c - P}{P_c} \propto \begin{cases} (v_c - v)/v_c & \text{for } v \lesssim v_c \ , \\ \{(v_c - v)/v_c\}^{\frac{1}{2}} & \text{for } v \gtrsim v_c \ . \end{cases} \tag{18}$$

For A>0, the way how the isotherms are modified depends on the actual form of $\Delta E_e(v)$ as a function of v.

(ii) $\lambda>2$: In this case, an isotherm in the P-v plane behaves like the curves as shown in Fig.9(ii-a). Whichever the sign A may take, the critical volume v_c is smaller than v_t, and accordingly falls in the metallic region as shown in Fig.9(ii-b). Near the critical point, the boundary of the coexistence region is of the form

$$\frac{P_c - P}{P_c} \propto \left(\frac{v - v_c}{v_c}\right)^2 \tag{19}$$

for both $v \gtrsim v_c$ and $v \lesssim v_c$. This behaviour of the critical point is the same as that for the van der Waals model.

It is also interesting to note that in this case the boundary of the coexistence curve, which is described by pressure $P_b(v)$ as a function of v, has a singularity, respectively of power $\lambda-1$ and λ at v_t and v_t^*, the latter being the atomic volume of the phase which coexists with the phase having v_t (see Fig.9(ii-b)).

(iii) $1 \leqq \lambda \leqq 2$: The singularity is so strong that the isotherms themselves in the $\bar{\bar{P}}$-v plane are discontinuous at v_t and therefore no critical points exist. It is expected that singularities of this kind would not appear in physical systems. In this connection, it would be worth noting that, literally speaking, the situation in which the intersection between metallic and nonmetaliic branches forms a finite angle as illustrated by Fig.1 does not occur because this corresponds to the case where $\lambda=1$.

Possibility of Two Critical Points

Another interesting effect of the additional term $\Delta E_e(v)$ would be detected when the M-NM transition occurs near the critical point (P_{co}, T_{co}, v_{co}) of a two-phase region caused by other origins. The gas-liquid two-phase region is an example. Let us study the expected modifications of the phase diagrams due to this additional term.

(i) The case where $A<0$

(a) $v_t>v_{co}$: The part of the P-v isotherms where $v< v_t$ is modified and the slope of the isotherm of $T=T_{co}$ is now positive at $v=v_{co}$. Therefore $T_c>T_{co}$ and $P_c>P_{co}$ can be concluded. This critical point is in the metallic region. Generally speaking, another condensation which terminates at a critical point in the metallic region near v_t can take place when some combination of parameters are adequately chosen. This case corresponds to Fig. 6(3) or (5).

(b) When the effect of the additional term is not strong enough, this does not introduce a new critical point and consequently the phase diagram behaves as that shown in Fig.6(4). When the effect is reasonably strong, there appear two critical points and two coexistence regions as shown by Fig.6(2). When the effect is extremely strong, there is a possibility that the original coexistence region is hidden in the two-phase region of the newly introduced condensation and the resulting phase diagram looks like that shown in Fig.6(5).

(ii) The case where $A>0$

(a) $v_t< v_{co}$: The gross feature of the phase diagram is as in Fig.6(4).

(b) $v_t> v_{co}$: The critical point moves in the direction of $P_c< P_{co}$ and $T_c < T_{co}$. The phase diagram is like Fig.6(5).

In the above investigation, λ is considered to be larger than 2. The case $\lambda=2$ has been discussed in the preceeding subsection.

CONCLUSION

In this paper, we have studied how the behaviours of electrons give effects on the thermodynamic properties of a material, specifically in the vicinity of the M-NM transitions. For this purpose, we have chosen expanded fluids as our objects since, in these systems, M-NM transitions are observed to take place and the experimental data of both electronic and thermodynamic properties have reasonably been accumulated. The experimental results of the electronic properties of an expanded fluid such as Hg and alkali metals show rather clearly the transition from a metallic to nonmetallic behaviour when the density of the system is decreased. The experiments for the thermodynamic properties of the same system indicate that the effects of the M-NM transitions are reflected on the thermodynamic properties.[1]

Theoretically, we have given the outline of a method to treat in a systematic manner the relationship between the electronic and thermodynamic properties of a system. Our theoty predicts a possible existence of two kinds of coexistence regions and two critical points corresponding to the gas-to-liquid condensation and the electron-induced transition, respectively. The results of our formulation also suggest that the outcomes of the singularity in the free energy at the M-NM transition should be observed in higher derivatives of the free energy such as the compressibility and the specific heat. The available experimental data seem to be too delicate to be compared to the theory and a further study both from experimental and theoretical sides are desired.

REFERENCES

1. F.Yonezawa and T.Ogawa, Metal-Nonmetal Transitions in Expanded Fluids —— Electronic and Thermodynamic Properties ——, Progr.Theor.Phys.

Supplement 72;1 (1982).
2. N.F.Mott, Metal-Insulator Transitions, Taylor and Francis, London (1974). See also a series of papers by Mott cited therein.
3. N.F.Mott and E.A.Davis, Electronic Processen in Non-Crystalline Materials, Oxford University Press, Oxford, (1971,1979).
4. I.M.Lifshitz, Soviet Phys.JETP11;1130 (1960).
5. I.M.Lifshitz and K.I.Kaganov, Soviet Phys. Uspekhi 5;878 (1963).
6. J.A.Krumhansl, in Physics of Solids at High Pressures, ed.C.Tomizika and R.Emrick, Academic Press, New York, p.425 (1965).
7. M.Cyrot, J.de Phys. 33,125 (1972).
8. M.Cyrot and P.Lacour-Gayet, Solid State Comm. 11;1767 (1972).
9. A.Kawabata, Progr.Theor.Phys. 56;717 (1976).
10. I.Tamura and Y.Nagaoka, Bussei Kenkyu, 26;133 (1976).
11. L.D.Landau and J.Zeldovich, Acta.Phys. chim. USSR 18;194 (1943).
12. G.E.Norman, Soviet Phys. JETP 33;912 (1971)
13. P.P.Edwards and K.J.Sienko, J.Am.Chem.Soc. 103;2969 (1981).
14. F.Hensel, Proceedings of the Eighth Symposium on Thermophysical Properties (1981).
15. T.Ogawa, S.Nara and T.Matsubara, J.de Phys. 41;C8-77 (1980).

METAL-INSULATOR TRANSITION AND LANDAU FERMI LIQUID THEORY

Claudio Castellani(*,#) and Carlo Di Castro(#)

Istituto di Fisica, Università dell'Aquila, L'Aquila Italy(*)
Dipartimento di Fisica, Università degli Studi di Roma "La Sapienza", Roma, Italy(#)

1. INTRODUCTION

The discovery that interaction effects with disorder give anomalous correction terms[1] to the normal Fermi liquid behaviour of an electron system has changed in the last few years our views on the possible interpretations of the metal-insulator transition in the various cases. One cannot advocate any more either the Anderson or the Mott transition, but correlation and disorder appear strictly interconnected giving rise to a wider spectrum of transitions.

In the pure disorder case the first quantum correction term due to the diffusion propagator in the particle-particle channel (Cooperon) leads[2,3] to a value equal to the unity for the index μ specifying the power law behaviour of the conductivity in terms of the deviation from the critical impurity concentration. In the presence of either an external magnetic field or of magnetic impurities, this term is absent and the leading higher order correction term due to the diffusion propagator in the particle-hole channel (diffuson) leads[3] to $\mu=1/2$.

One could simply try to explain the experimental results[4,5] leading to $\mu=1$ for the amorphous alloys of a semiconductor and a metal like $Nb_x Si_{1-x}$ by means of the pure localization phenomena due to the Cooperon. However for example in a-Si:Au $\mu=1$ is experimentally independent of the presence of the external magnetic field[6]. This would not be the case for pure localization. Many other features like the vanishing of the single particle density of states or the temperature dependence of the conductivity would also remain unexplained.

On the other hand $\mu=1/2$ is obtained for uncompensated doped semiconductor like Si:P[7]. One could try to explain this value of μ by supposing that the correlations among the electrons only produce a quasi static spin motion, so that spin scattering suppresses the Cooperon contribution leaving the diffuson term. Many other experiments[8] would remain unexplained like the magnetoresistance, the temperature dependence of the conductivity and so on.

Starting from a generalized non linear σ-model[9-12] or equivalently from a renormalized perturbation theory[13-17] the interaction combined with disorder has been analyzed in many different situations by renormalization group procedure.

The singlet interaction amplitude leads to a universal contribution in the conductivity equation identical in value but completely different in nature from the localization contribution due to the Cooperon. This by itself would lead to an index $\mu=1$ unaffected by the external magnetic field. The triplet contribution instead produces an increase in the conductivity related to an enhancement of the spin susceptibility, thus forbidding a localization transition by overcoming the singlet contribution.

New relevant features appear in the spin susceptibility, in the specific heat and in the single particle density of states. All the renormalization parameters coming from the non linear σ-model can be reinserted in these quantities. One ends up with a "renormalized" Landau theory of Fermi liquids in which the Landau parameters are expressed in terms of the renormalized effective couplings of the theory (inverse conductance, singlet and triplet interaction amplitude). According to the renormalization group equations, in the general case the renormalized Landau parameters would be divergent as the temperature approaches zero. However much before this happens the flow equations are out of their region of validity. The anomalous increase of the spin susceptibility and the corresponding decrease of the spin diffusion constant has suggested[11,14] the formation of pseudo local magnetic moments. The further increase of the triplet contribution would be stopped and the system would be driven to the class of universality where only the singlet contribution drives the transition. The index μ is again equal to unity but the transition would be different from the Anderson transition. In the short range forces case the same situation would bring to a value 1/2 for μ[13]. The existing experiments on the spin susceptibility and on the specific heat for the various systems are not sufficiently detailed to decide on the validity of this mechanism for the enhancement of the spin susceptibility and the specific heat as compared to the one due to pure correlation[5].

2. NON INTERACTING ELECTRONS IN DISORDERED MEDIA

In the absence of the electron-electron interaction, disorder is well known to produce localization at zero temperature. Starting from the metallic side the inverse conductance acts as the only effective coupling. The scaling theory[2] of the localization transition has been fully developed by mapping the problem into a non linear σ-model[3]. The $2+\varepsilon$ expansion for the critical indices leads to at least three classes of universality[3].

We briefly recall here for further reference some properties of an electron gas in the presence of an impurity random potential u(r) with local gaussian correlations

$$\overline{u(r)\, u(r')} = (2\pi N_0 \tau_0)^{-1}\, \delta(r-r')$$

τ_0 being the scattering time in the Born approximation and N_0 the single particle density of states for the free system.

In the extremely weak disorder limit ($\varepsilon_F \tau_0 \gg 1$) the conductivity is given by the Drude formula $\sigma_0 = ne^2 \tau_0/m$.

A diffusive behaviour in the density-density response function K_{oo} is then represented by:

$$K_{oo}(q,\omega)=\frac{-N_oD_oq^2}{-i\omega+D_oq^2} \tag{2.1}$$

where D_o is the diffusion constant $D_o=v_F{}^2\tau_o/d$ such that

$$\sigma_o=-\lim\frac{\omega}{q^2}\ \mathrm{Im}\ K_{oo}(q,\omega)=N_oD_o$$

K_{oo} can be splitted in the static part $(K_{oo})_{++}+(K_{oo})_{--}=-N_o$ and the "dynamic" part K_{oo}^{+-} which include the resummation of the impurity scattering effects

$$K_{oo}^{+-}=\frac{-i\omega N_o}{-i\omega+D_oq^2}= \tag{2.2}$$

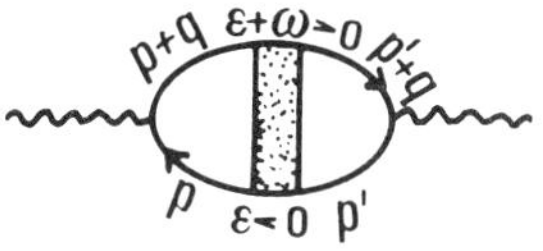

where the superscript +- indicates that the energy of the electron Green functions are on the opposite sides of the Fermi surface. The insertion in the bubble is the particle-hole diffusion propagator (diffuson) defined by a ladder summation of impurity scattering lines:

$$L_d^o\ (q,\omega)=\frac{1}{2\pi N_o\ \tau_o^2}\qquad\frac{1}{-i\omega+D_oq^2}\ ;\ \omega\ ,D_oq^2<\ \tau_o^{-1} \tag{2.3}$$

The amount of disorder is controlled by the effective coupling $t_o=1/(2\pi)^2\sigma_o$, which is small in the weak disorder limit. t_o is marginal at d=2 while at $d=2+\varepsilon$ it acquires a bare dimensionality $-\varepsilon$ when expressed in terms of an inverse length.

At d=2 and at the order t_o a logarithmic correction to the conductivity arises[18] by integrating over the momentum of the diffusion propagator in the particle-particle channel L_c which in presence of time reversal symmetry is given by the same expression (2.3) with q substituted by p+p'+q. One obtains:

$$\sigma=\sigma_o\ (1+t_o\ \ln\frac{\omega}{\Lambda^2}) \tag{2.4}$$

where Λ^2 is an ultraviolet cut-off of the order of τ_o^{-1}.
At the same order in t_o many more correction terms appears in K_{oo}[18,19]. All of them can be recombined in such a way that the expression (2.1) and (2.2) for K_{oo} and K_{oo}^{+-} are still valid with D_o being replaced by $D=Do\ (1+t_o\ \ln\frac{\omega}{\Lambda^2})$ and σ_o by $\sigma=N_oD$.

This general structure is implied by gauge invariance[19] and is preserved in higher order in t_o. Only one renormalization is then required to take care of all the logarithmic singularities. At $d=2+\varepsilon$ the group equation for $t=\lambda^{\varepsilon/2}/(2\pi)^2\sigma(\omega=\lambda\Lambda^2)$ is easily obtained from eq.(2.4)

$$-\frac{\partial t}{\partial\ln\lambda}=-\frac{\varepsilon}{2}\ t+t^2 \tag{2.5}$$

where λ is the rescaling parameter of the cut-off ($\Lambda'^2=\lambda\Lambda^2$) and approaches zero under iteration of the renormalization transformation. In terms of the non linear σ- model this situation corresponds to the general orthogonal case. At the fixed point $t^*=\varepsilon/2$, a one parameter scaling theory follows. The correlation length ξ and the conductivity σ behave as

$$\sigma \sim \xi^{-\mu/\nu} \quad , \qquad \xi \sim (\sigma_0 - \sigma_c)^{-\nu} \quad , \qquad \nu = \frac{1}{\varepsilon} \tag{2.6}$$

where σ_c is the value of σ_0 corresponding to the critical concentration of impurities. The following scaling laws hold[3]

$$x_\sigma = \frac{\mu}{\nu} = \varepsilon \; , \quad x_\omega = 2 + \varepsilon \; , \; x_{\varepsilon_1} = -\frac{3}{\nu} = x_\sigma + 2 - x_\omega - y = -y \tag{2.7}$$

where x_σ, x_ω and x_{ε_1} , are the indices of the conductivity σ , the frequency ω and the dielectric susceptibility ε_1 in terms of an inverse length respectively. y=d-1 specifies the power of the Coulomb interaction in q space.

The presence of magnetic impurities breaks down the time reversal symmetry[3]. An additional scattering time introduces a finite "mass" in the Cooperon. The term t^2 is suppressed in eq.(2.5) and as a first correction a term t^3 appears due to the diffusons. The index μ acquires then the value 1/2 (unitary case).

The spin-orbit coupling in the impurity scattering introduces[3] an antilocalization term $-t^2/2$ in eq.(2.5) thus forbidding at this stage the existence of a fixed point and of a localization transition in the absence of interaction (symplectic case).

The single particle density of states N, the specific heat c_V and the spin susceptibility χ_s are unaffected by pure disorder.

Attempts to explain the experimental results with the non interacting model encounter serious difficulties. Measurements close to the transition as 0.1% of the relative electronic density on uncompensated phosphorus-doped silicon lead to $\mu=1/2$[7]. Compensation tends to increase the value of the index μ from 1/2 to unity[8]. The dielectric susceptibility divergency on the insulating side of the transition in Si:P gives an index value compatible with the related scaling law in eq.(2.7). On the other hand a similar scaling law will hold in the presence of interaction, whereas both the temperature dependence of the conductivity and the magnetoresistance in the metallic side indicate[5,8] that the interaction effects are of importance in various physical quantities, possibly including the single particle density of states[20].

The amorphous alloys[4] of a semiconductor and a metal like Nb_xSi_{1-x} and $Au_x\,Ge_{1-x}$ seem to belong to a different class of universality with $\mu=1$. This value of μ in a-Si:Au is not affected by the presence of an external magnetic field[6]. An external field, as already recalled, would instead suppress the term t^2 appearing in eq.(2.5). The exponent $\mu=1$ appears also in a case (GdS:y) in which the transition is driven by a large applied magnetic field[21]. The single particle density of states in the armorphous systems vanishes at the transition.

Even from this very incomplete list of experiments, it is clear that we need a theory including both correlation and disorder effects in various different symmetry situations before we can attempt any sensible comparison.

3. CORRELATION EFFECTS IN NON DISORDERED ELECTRON GAS

We start by recalling that in the absence of disorder the physical description of an interacting electronic system can be given in terms of the Landau theory[22] with the specific heat c_v, the spin susceptibility χ_s and the thermodynamic density of states $\partial n/\partial\mu$ per spin given by

$$c_v^L = \frac{m^*}{m} c_v^o; \quad \chi_s^L = \frac{m^*/m}{1+F_o^a}; \quad \frac{\partial n}{\partial \mu} = \frac{m^*}{m} \frac{N_o}{1+F_o^s}$$

c_v^o ; χ_s^o and N_o are the corresponding bare values for the free electron gas:

$$c_v^o = \frac{\pi^2}{3} N_o T = \gamma_o T; \quad \chi_s^o = \mu_B^2 N_o; \quad N_o = \frac{1}{(2\pi)^d} S_d \frac{m}{\hbar} k_F^{d-2}$$

Besides F_o^a and F_o^s a third Landau parameter appears in the effective mass ratio $m^*/m = 1+F_1^s/3$.

When rephrased in terms of the triplet and singlet interaction scattering amplitude which appear in the spin and in the density response functions respectively, one obtains

$$c_v^L = \frac{\nu}{N_o} c_v^o = \gamma T, \qquad \chi_s^L = \frac{\nu}{N_o}(1+2\nu\Gamma_t^o)\chi_s^o = \frac{\nu}{N_o} z_2^o \chi_s^o$$

$$\frac{\partial n^L}{\partial \mu} = \nu\left[1-2\nu(\Gamma_s^o - \Gamma_o^o)\right] = \nu z_1^o \qquad (3.1)$$

with $\nu = m^*/m\, N_o$ and $\gamma = m^*/m\, \gamma_o$. Γ_o^c is the static Coulomb scattering amplitude which has been subtracted by the total singlet scattering amplitude Γ_s^c since the screening bubble and therefore $\partial n/\partial\mu$ must be irreducible for cutting a Coulomb line.

When the interaction is assumed to be of the Hubbard type characterized by a strong local repulsive interaction U, the Brinkman and Rice analysis of the Gutzwiller variational approach for one electron per site has been summarized[23] as

$$\frac{m^*}{m} = \left(1-\frac{U^2}{U_c^2}\right)^{-1}; \quad F_o^s(U) = \frac{UN_o}{2}\left(2-\frac{U}{U_c}\right)\left(1-\frac{U}{U_c}\right)^{-2}; \quad F_o^a = F_o^s(-U) \qquad (3.2)$$

At $U = U_c$ the average number of doubly occupied sites vanishes and only localized one electron per site with unpaired spin appears.

This theory does not take magnetic ordering effects into account and gives an oversemplified picture of the Mott-Hubbard transition. From eq.(3.2) it follows that the localization is driven by an infinite effective mass, which, since the factor $1+F_o^a(U_c)$ is finite, controls both the susceptibility and the specific heat enhancement. Even if the previous results are obtained for a system on an ordered lattice, the enhancement of the spin susceptibility for U near to U_c has been taken as an indication of how localized quasi static spins can be obtained by pure correlation.

The hypothesis has been put forward[5] that a similar mechanism could be present in uncompensated Si:P. The Coulomb interaction would enter only to make the collective spin motion quasi static. These magnetic moments would play the role of a random magnetic scattering and therefore they would drive the system into the unitary (non interacting) case with $\mu = 1/2$. In this context measurements of the specific heat[24], spin susceptibility and nuclear spin relaxation time[25] have been alleged[5] which show an anomalous enhancement of c_v and χ_s at low temperature. According to eq.(3.2) an enhancement of the specific heat and of the spin susceptibility, induced in both cases by an increase of the effective mass, should be observed. No quantitative stringent comparison between theory and experiments is however possible at the present to check the suggested hypotesis. The above picture was also strongly based on the fact that uncompensated Si:P could be associated with a model system having one electron per site. The latter property seems to be too stringent for such a complicated system, where hybridization with continuum states should make it difficult to establish this kind of symmetry. We shall show in the next sections that a completely different mechanism, due to combined interaction and disorder, will introduce similar enhancement effects, for which however the specific heat and the spin susceptibility behave differently.

4. INTERACTION AND DISORDER

On the theoretical side the interaction combined with disorder is well known to introduce[1a] corrections to the conductivity which are of the same order as the localization contribution. Anomalous corrections are now present also in the single particle density of states N[1a], in the specific heat c_v[1b] and in the spin susceptibility χ_s[1b,1c]. This was first discovered[1] in the weak localization regime by evaluating the perturbative corrections to the ordinary metallic behaviour both in t_0 and in the potential. These anomalous corrections to the normal Fermi liquid theory (which are logarithmically singular in two dimensions) come out from integrals over the energy and the momentum appearing in the diffusive pole of the diffusion propagators introduced in section 2, thus confining the relevant electron energies in a shell $|\varepsilon| < \tau_0^{-1}$ around the Fermi surface.

On the other hand all the many body effects coming from energies $|\varepsilon| > \tau_0^{-1}$ due to the electron-electron interaction in the absence of disorder can be included by substituting the initial bare potential with the standard Landau Fermi liquid interaction amplitudes appearing in eq.(3.1) of section 3.

Inadequacy of the standard perturbative approach to the problem was found[26] by evaluating the contributions of the order t_0^2 to the single particle density of states. A strong coupling behaviour in the effective interactions was evident at this order thus implying the necessity of an infinite resummation in the interaction parameters.

A relevant progress in this direction was achieved by mapping[9] the problem into a generalized non linear σ- model, whose renormalization group analysis was carried out at first order in t_0, which turns to be the only expansion parameter of the theory. A non trivial a posteriori analysis[13] of the perturbation theory in the presence of disorder allows to obtain the same results. As it is also indicated in reference (15), the natural effective couplings of the theory turn out to be the spin independent interaction scattering amplitude (the singlet Γ_s^0) and the spin dependent part (the triplet Γ_t^0) in the particle-hole channel, plus an interaction amplitude Γ_s^0 associated with the Cooperons. The singlet and the triplet interaction amplitudes can be expressed in terms of the small and large angle scattering amplitudes Γ^0 and Γ_2^0 as $\Gamma_s^0 = \Gamma^0 - \frac{1}{2}\Gamma_2^0$, $\Gamma_t^0 = \frac{1}{2}\Gamma_2^0$

In order to simplify our discussion we suppress the Cooperon contribution and discuss the "pure interaction" model dressed by disorder via the diffusons. The general conclusions we shall derive will not depend on this simplifying assumption.

As already indicated besides the conductivity, the single particle density of states N, the spin susceptibility χ_s and the specific heat c_v are affected by the interaction in the presence of disorder and more than one renormalization parameter has to be introduced.

As consequence of the interaction at first order in t_0 instead of eq.(2.3), a dressed diffusion propagator appears

$$L_d(q,\omega) \sim \frac{\zeta^2}{-iz\omega + Dq^2} \tag{4.1}$$

ζ plays the role of the wave function renormalization and z, whose bare value z° is the unity, allows for an independent frequency renormalization. It was introduced[9] in the context of the generalized non linear σ- model as an additional coupling related to the frequency acting as an external field. D is the dressed diffusion constant at the order considered.

Singular correction terms renormalize also Γ_s^0 and Γ_t^0 into new Γ_s and Γ_t. The singularities of ζ, z, D, Γ_s and Γ_t, which are all logarithmic in two dimensions, are expressed in terms of ω at zero temperature, of T itself at finite temperature and in terms of the rescaling parameter λ in the group equations.

When we use eq.(4.1) together with the interaction amplitudes Γ_s and Γ_t the density-density response function K_{oo} is again the sum of a static term and a dynamic term[13,27]

$$K_{oo}^{st} = -\frac{\partial n}{\partial \mu} = -\nu z_1; \quad K_{oo}^{din} = \Lambda_o K_{oo}^{+-} = \frac{-i\omega\,\nu\zeta^2\Lambda_o^2}{-i\,z_1\omega + Dq^2} \tag{4.2}$$

where the last expression in eq.(4.2) replaces eq.(2.2) and Λ_0 is the vertex that when multiplies the advanced and retarded part K_{oo}^{+-} of the density response function gives the total dynamic part of K_{oo} and z_1 is given by

$$z_1 = z - 2\nu\,\zeta^2\,(\Gamma_s - \Gamma_o) \tag{4.3}$$

Several considerations are in order:

i) $\partial n/\partial\mu$ does not acquire singular corrections[28,26,9,13] and therefore z_1 concides with its bare value $z_1 = 1 - 2(\Gamma_s^0 - \Gamma_o^0)$ appearing in eq.(3.1).

ii)The sum of the two terms in eq.(4.2) must vanish at q=0 for any $\omega \neq 0$ in order to conserve the number of particles and therefore K_{oo} acquires the standard diffusive form

$$K_{oo}(q,\omega) = -\frac{\partial n}{\partial \mu}\frac{D'\,q^2}{-i\omega + D'\,q^2}, \qquad D' = D/z_1 \tag{4.4}$$

since the condition

$$\frac{\nu \zeta^2 \Lambda_0^2}{z_1} = \frac{\partial n}{\partial \mu}$$

has to be satisfied.

iii) This last equation together with the fact that the static Coulomb scattering amplitude Γ_0 as $q \to 0$ becomes $\Gamma_0 = \Lambda_0^2 / 2\frac{\partial n}{\partial \mu}$ implies that[9]

$$z = 2\nu \zeta^2 \Gamma_s \tag{4.3'}$$

so that only z or Γ_s acts as indipendent coupling.

iv) A Ward identity connecting the electromagnetic vertex to the single particle Green function implies that the single particle density of states can be expressed as the limit as $q \to 0$ first and then $\omega \to 0$ of $K_{00}^{+-}(q,\omega)$. This leads to the identification of the wave function renormalization parameter ζ [27,13,16]

$$\zeta = \frac{N}{\nu} \tag{4.5}$$

As usual the wave function renormalization can be reabsorbed into a redefinition of the renormalized couplings $\tilde{\Gamma}_s = \zeta^2 \Gamma_s$ and $\tilde{\Gamma}_t = \zeta^2 \Gamma_t$.

Similar considerations can be carried out for the spin susceptibility leading to[11,14]

$$\chi(q,\omega) = \chi_s \frac{D_s q^2}{-i\omega + D_s q^2}; \qquad D_s = \frac{D}{z_2} \tag{4.6}$$

where

$$\chi_s = \nu z_2, \qquad z_2 = z + 2\nu \tilde{\Gamma}_t \tag{4.7}$$

χ_s has been quite generally connected [16] to z_2 by means of a generalized Ward identity relating the dynamic part of χ to the single particle density of states.

The leading corrections to the specific heat were evalated in perturbation theory[1b,15]; they have been recently reconsidered and in $2+\varepsilon$ dimensions, at least, they have been shown[17] to coincide with the expression for z so that

$$c_v = z \gamma T \tag{4.8}$$

This last result is in agreement with a very simple scaling argument[17]. If the frequency ω is rescaled with the parameter z, the temperature will be scaled in the same way. We consider as usual the logarithm of the partition function to be an invariant of the renormalization group. The termodynamic potential Ω, being multiplied by T will scale as z^{-1}. The specific heat per unit volume $c_v = -T \frac{1}{V} \frac{\partial^2 \Omega}{\partial T^2}$ will have in this case the same equation as the thermodynamic potential per unit volume in the ordinary critical phenomena

$$c_v(T;\lambda) = (\lambda'/\lambda)^{d/2}\, c_v(T';\lambda') \tag{4.9}$$

where T and T' are the rescaled temperature at two different values λ and λ' of the rescaling parameter. If a fixed point of the group transformation exist as $\lambda'/\lambda \rightarrow 0$ then

$$T' = zT\,(\lambda/\lambda')^{\frac{2+\epsilon}{2}} \sim (\lambda/\lambda')^{\frac{d-x_z}{2}}\,T$$

By putting $T'\sim 1$ in eq.(4.9) we obtain

$$c_v \sim T^x \;; \qquad x = \frac{d}{x_T} = \frac{d}{d-x_z} \tag{4.10}$$

For the non interacting case ($x_z=0$) from $x_T=x_\omega=d$ we find the well knonwn result x=1. In the general case $x_\omega=x_T=d-x_z$ and

$$\frac{c_v}{T} = \gamma \sim T^{x-1} = T^{x_z/x_T} \sim z \tag{4.11}$$

In conclusion the general structure of the theory of combined disorder and interaction leads to a "renormalized" Landau theory with eq.(3.1) replaced by

$$c_v = z\,c_v^L \;; \qquad \chi_s = \frac{z_2}{z_2^o}\chi_s^L \;; \qquad \frac{\partial n}{\partial\mu} = \frac{\partial n^L}{\partial\mu} \tag{4.12}$$

where the renormalized parameters z and z_2 are determined in terms of the renormalized couplings $\tilde{\Gamma}_s(\lambda)$, and $\tilde{\Gamma}_t(\lambda)$ via eq's (4.3') and (4.7).

5. DISCUSSION AND CONCLUSIONS

The group equations for the couplings have been derived in different physical situations[9-14].

In the case of non magnetic impurity scattering, but without Cooperon contributions present, they are

$$-\frac{dt}{d\ln\lambda} = -\frac{\epsilon}{2}t + t^2\left[\frac{2\tilde{\Gamma}_s}{z} + 3\left(1 - \frac{z+2\tilde{\Gamma}_t}{2\tilde{\Gamma}_t}\ln\frac{z+2\tilde{\Gamma}_t}{z}\right)\right] \tag{5.1}$$

$$-\frac{d\tilde{\Gamma}_s}{d\ln\lambda} = t\left[-\frac{1}{2}\tilde{\Gamma}_s + \frac{3}{2}\tilde{\Gamma}_t\right]; \quad -\frac{d\tilde{\Gamma}_t}{d\ln\lambda} = \frac{1}{2}t\left[\tilde{\Gamma}_s + \tilde{\Gamma}_t + \frac{8\tilde{\Gamma}_t^2}{z}\right] \tag{5.2}$$

The relation (4.3') leads to the universal singlet contribution[15] $t^2\,2\tilde{\Gamma}_s/z = t^2$ in the equation for t, which is exactly the same as the term that appears in eq.(2.5) due to Cooperon in pure localization. The second term in the square bracket of eq.(5.1) is the triplet contribution.

By solving the group equations in two dimensions one obtains[11,14] that the effective interaction couplings $\tilde{\Gamma}_s$ and $\tilde{\Gamma}_t$ diverge at a finite step λ_c of the renormalization procedure. The triplet contribution in eq.(5.1) eventually overcomes the singlet one and makes σ to increase thus forbidding localization. The same results are obtained in $2+\epsilon$ dimensions except for a region ($\sim \frac{\epsilon}{t_0}(1-\Gamma_t^0/\Gamma_s)>1$) in the initial values of the couplings where λ_c does not exist and the system scales as $\lambda\rightarrow 0$ to a conducting behaviour.

According to Ref. (11) the same divergent behaviour of $\tilde{\Gamma}_s$ and $\tilde{\Gamma}_t$ appears also when Cooperons are considered.

The previous results cannot be taken as conclusive since these divergences drive the group equations out of their region of validity. At present therefore no scaling theory of the metal-insulator transition is available for the general case, which should be the analog of the orthogonal case for the non interacting disordered system.

Whenever by changing the symmetry of the problem we can suppress part of the effects of the triplet contribution we recover a phase transition[10,13,12,15].

A. Spin flip impurity scattering[10,13,15]

When magnetic impurity scattering is present the localization term of eq.(2.5) is absent. In fact an additional scattering time introduces a finite mass in the particle-particle diffusion propagator, which is no more divergent in the limit of zero frequency and momentum. It is therefore justified to neglect the contributions due to the interaction amplitude Γ_3 associated with the Cooperons. Moreover a finite mass appears also in the triplet part of the diffusion propagator in the particle-hole channel so that the contributions coming from Γ_t are no more singular in $2+\varepsilon$ dimensions and disappear from the group equations (5.1) and (5.2). For this system, which is the analog of the unitary case for the non interacting problem the interaction contributes to the conductivity equation only via the universal term, identical to the one due to pure disorder in the orthogonal case. A fixed point is then obtained with an index $\mu=1$. Since the singlet contribution in eq.(5.1) is not affected by the presence of a magnetic field, the value $\mu=1$ for this system is left unchanged by overimposing a magnetic field. This is not the case for the Anderson transition.

Contrary to the non interacting case the single particle density of states approaches zero and a pseudogap appears at the transition.

Since eq.(5.2) gives $z=2\tilde{\Gamma}_s\sim\lambda^{\varepsilon/4}$ as $\lambda\to 0$ it follows[13] $x_\omega = x_T = d-\frac{\varepsilon}{2} = 2+\frac{\varepsilon}{2}$. Nearby the transition the temperature or the frequency dependence of the conductivity can be then derived by the scaling behavior $\sigma\sim\lambda^{\varepsilon/2}$ to be:

$$\sigma \sim \omega^{\varepsilon/2x_\omega}\ ,T=0; \qquad \sigma \sim T^{\varepsilon/2x_T}\ ,\ \omega=0 \qquad (5.3)$$

Eq.(5.3) leads to a power index 2/5 at d=3 to be compared with the value 1/3 obtained in the free case[3]. From eq.(4.11), the specific heat coefficient γ goes as:

$$\gamma \sim T^{\varepsilon/2(2+\varepsilon/2)} \qquad (5.4)$$

The index of the dielectric susceptibility is now related to μ via the modified scaling law $\zeta=\frac{\mu}{\varepsilon}\left[y-\frac{\varepsilon}{2}\right]$ which in three dimensions becomes $\zeta=\frac{3}{2}\mu$.

In the case of short range forces at least at order t^2, the singlet contribution disappears at the fixed point and the pure diffusive localization (unitary case) with $\mu=1/2$ is recovered[13], with no further implications on the termodynamic quantities.

B. External magnetic field[10,13]

The effects of triplet contributions can be reduced also by an external magnetic field large enough ($g\,\mu_B H \gg KT$) to introduce a Zeeman splitting. In this case in addition to the singlet term only the triplet contribution from total zero spin component survives. A line of fixed points appears in $2+\xi$ dimensions with $\mu=1$. Since z stays finite at the fixed point $\gamma = c_v/T = z^* \gamma^L$ acquires no temperature dependance as $T \rightarrow 0$ at the transition. $\zeta = 2\mu$ at d=3 and a pseudo gap is present in the single particle density of states.

C. Spin-orbit impurity scattering[12,15]

When spin-orbit impurity scattering is present only the singlet spin components are diffusive and survive in the particle-hole and particle-particle channels. Pure disorder gives an antilocalization term which is compensated by the singlet universal contribution bringing back the localization phenomena with $\mu=1$. In this case, as well as in case A, the presence of an external magnetic field does not modify the value of the index μ . Eventhough in the spin orbit case the fixed point structure and the approach to criticality is indeed affected by the magnetic field.

D. Spin fluctuation enhancement[11,14,16]

Going back to the general case, with both singlet and triplet contributions, even if $\tilde{\Gamma}_s$ and $\tilde{\Gamma}_t$ tend to diverge, we can still try to make a likely scenario out of these results. From general considerations on gauge invariance $\tilde{\Gamma}_t$ has been associated to the spin susceptibility, which acquires the modified Landau form in eq.(4.12). The group equations give $z \sim (\lambda - \lambda_c)^{-3}$, $\tilde{\Gamma}_t \sim z_2 \sim (\lambda - \lambda_c)^{-4}$. These divergences can be expressed in terms of T by using the relation between the rescaling factor λ and the temperature $(\lambda - \lambda_c) \sim T^{\alpha}$ where $\alpha = 1/3$ or $1/4$ depending on the way we stop the renormalization group procedure at finite temperature[14,16]. The spin susceptibility χ_s is then diverging roughly as $T^{-4\alpha}$ while the spin diffusion constant is approaching zero with the same inverse power. At $d=2+\xi$ the softer divergency of $z=2\,\tilde{\Gamma}_s$ is instead connected with the enhancement of the specific heat constant: $\gamma = c_v/T \sim T^{-3\alpha}$ with a rough estimate of the power law index.

The behaviour of the spin susceptibility and of the spin diffusion constant has suggested the formation of local magnetic moments (of size $\sim (\tau_0 D_0/\lambda_c)^{1/2}$) as $T \rightarrow 0$. The system should eventually crossover into either the high magnetic field (B) case even for very small external magnetic field due to the enhancement of the susceptibility, or to the spin-flip singlet-only (A) case due to the formation of the local magnetic moments. The index μ would be equal to 1 in the first case, while in the second case it would be equal to 1 or 1/2 according to the range of the interaction (long o short range forces respectively).

The above mechanism for the enhancement of the spin susceptibility and the specific heat derives from the combined effect of interaction and disorder. It is therefore totally different from the Brinkman and Rice phenomenon recalled in section 3, where the same enhancement of the spin susceptibility and of the specific heat is driven by an enhancement of the effective mass due to pure correlation. We want here to mention that the presence of localized spins due to single electron occupied states near the mobility edge as a result of a Hubbard repulsion should in any case drive the sistem in the universality classes related to the singlet only cases (A), which run into the unitary case of the non interacting sistem for short range forces only.

Anomalous specif heat[24] and nuclear spin relaxation time[25] are measured in Si:P. Even if invoking crossover effects proved to be an oversimplified picture for the explanation of the value $\mu=1/2$ in this system, it is very important in our opinion to make simultaneous measurements of spin susceptibility and specific heat and individuate region where the enhancement of the two quanties can be possibly ascribed to either pure correlation or disorder and correlation effects.

REFERENCES

1a. B.L.Altshuler, A.G.Aronov and P.A.Lee, Phys. Rev. Lett. 44, 1288 (1980).
H.Fukuyama, J. Phys. Soc. Japan 48, 2169 (1980).
B.L.Altshuler and A.G.Aronov, Solid State Comm. 30, 115 (1979) and Zh. Eksp. Teor. Fiz., 77, 2028 (1979) (Sov. Phys. JETP 50, 968 (1979));
For a review on the metal-insulator transition see: N.F.Mott, "The Metal-Insulator Transition" (Taylor and Francis, London, 1974); "Anderson localization" edited by Y.Nagaoka and H.Fukuyama (Springer, New York 1982); P.A.Lee and T.V.Ramakrishnan, preprint (1984).

1b. B.L.Altshuler, A.G.Aronov and A.Yu.Zuzin, Zh. Eksp. Theor. Fiz. 84, 1525 (1983) (Sov. Phys. JETP 57, 889 (1983).

1c. H.Fukuyama J. Phys. Soc. Japan 50, 3407 (1981).

2. E.Abrahams, P.W.Anderson, D.C.Licciardello and T.V.Ramakrishnan, Phys. Rev. Lett. 42, 673 (1979).

3. F.J.Wegner, Z. Phys. B35, 207 (1979)
S.Hikami, A.I.Larkin and Y.Nagaoka, Progr. Theor. Phys. 63, 707 (1980)
K.B.Efetov, A.I.Larkin and D.E.Khemlnitskii, Zh. Eksp. Teor. Fiz. 79, 1120 (1980) (Sov. Phys. JETP 52 (3), 568 (1980)).
S.Hikami, Prog. Theor. Phys. 64, 1466 (1980); Phys. Rev. B24 2671 (1981).

4. G.Hertel, D.J.Bishop, E.G.Spencer, J.M.Rowell and R.C.Dynes, Phys. Rev. Lett. 50 743 (1983).
W.L.McMillan and J.Mochel Phys. Rev. Lett. 46, 556 (1981).

5. For a review on recent developments in the Metal-Insulator Transition see G.A.Thomas and M.A.Paalanen, preprint 1984.

6. N.Nishida et al., Solid State Electronics, Proceedings of the Santa Cruz Conference (1984).

7. G.A.Thomas, M.Paalanen and T.F.Rosenbaum Phys. Rev. B27, 3897 (1983).

8. For a review on doped semiconductors see T.F.Rosenbaum R.F.Milligan, M.A.Paalanen, G.A.Thomas and R.N.Bhatt, Phys. Rev. B27, 7509 (1983).

9. A.M.Finkel'stein, Zh. Eksp. Teor. Fiz. 84, 168 (1983), (Sov. Phys. JETP 57, 97 (1983)).

10. A.M.Finkel'stein, Pis'ma Zh. Eksp. Teor. Fiz. 37, 436 (1983) (JETP Lett. 37, 517 (1983)).

11. A.M.Finkel'stein, Zeitschrift fur Physik B56, 189 (1984)

12. C.Castellani, C.Di Castro, G.Forgacs and S.Sorella, Solid State Comm. 52 261 (1984).

13. C.Castellani, C.Di Castro, P.A.Lee, and M.Ma, Phys. Rev. B30 527 (1984)

14. C.Castellani, C.Di Castro, P.A.Lee, M.Ma, S.Sorella and E.Tabet, Phys. Rev. B, Rap. Comm. 30, 1596 (1984)
15. B.L.Altshuler and A.G.Aronov, Solid State Comm. 46, 429 (1983)
16. C.Castellani, C.Di Castro, P.A.Lee, M.Ma, S.Sorella and E.Tabet, preprint (1985)
17. C.Castellani, C.Di Castro, unpublished
18. L.P.Gorkov, A.I.Larkin and D.E.Khmel'nitskii, Pis'ma Zh.Eksp. Teor. Fiz. 30 248 (1979) (JETP Lett. 30, 228 (1979))
E.Abrahams and T.V.Ramakrishnan, J.Non-Crist. Sol. 35, 15 (1980)
19. D.Vollhart and P.Wölfle, Phys. Rev. Lett. 45, 842 (1980).
C.Castellani, C.Di Castro, G.Forgacs and E.Tabet, J.Phys. C16, 159 (1983).
20. R.C.Dynes, private comm. (1984).
21. S.von Molnar, A.Briggs, J.Flouquet and G.Remenyi, Phys. Rev. Lett. 51, 706 (1983).
22. See for instance P.Nozière, Theory of interacting Fermi systems (W.A.Benjamin, N.Y. 1964).
23. See for instance: D.Vollhardt, Rev. Mod. Phys. 56, 99 (1984) and references therein.
24. G.A.Thomas, Y.Ootuka, S.Kobayashi and W.Sasaki, Phys. Rev. B24, 4886 (1981).
25. M.A.Paalanen, A.E.Ruckenstein and G.A.Thomas, Solid State Electronic Proceedings of the Santa Cruz Conference (1984).
26. C.Castellani, C.Di Castro, G.Forgacs and E.Tabet, Nucl. Phys. B225, FS9, 441 (1983).
27. C.Castellani and C.Di Castro, to appear in the Reports of the LXXXIX Varenna School (1983) edited by F.Bassani, F.Fumi and M.Tosi (Accademic Press).
28. P.A.Lee, Phys. Rev. B26, 5882 (1982).

LONG-RANGE COULOMB INTERACTION VERSUS CHEMICAL BONDING EFFECTS IN THE THEORY OF METAL-INSULATOR TRANSITIONS

N.H. March

Theoretical Chemistry Department
University of Oxford, 1 South Parks Road
Oxford, OX1 3TG, England

1. INTRODUCTION

In this work, we shall consider two aspects of the metal-insulator transition: (i) The jellium model of electrons with interelectronic Coulomb repulsion e^2/r_{ij}, but no account of electron-ion interaction apart from a uniform, non-responsive neutralizing background of positive charge; (ii) the competition between long-range Coulomb interelectronic repulsion and electron-ion interaction. However, more specifically, we shall focus on the chemical treatment of the metal-insulator transition in H, which is more than simply academic in view of the claims of Hawke et al. (1978) to have seen the transition from molecular H_2 to metallic hydrogen. Then, more briefly, we take up the problem of possible metal-insulator behaviour of alkali atoms. Again, the chemical picture plays a predominant role in current theories.

2. ELECTRON CRYSTALLIZATION IN JELLIUM: A FIRST-ORDER METAL-INSULATOR PHASE TRANSITION

Since the pioneering work of Wigner (1934, 1938), it has been known that, at sufficiently low densities, in the model known as jellium, in which interacting electrons move in a non-responsive uniform neutralizing background of positive charge, Coulomb repulsions will eventually dominate kinetic energy contributions, and one then minimizes the Madelung energy by allowing electron localization on the sites of a crystal lattice. The lowest energy configuration is found to be body-centred cubic.

Though all the above has been well established for about 50 years, Care and March (1975), when reviewing electron crystallization much later, noted the enormous spread in the theoretical estimates of the mean interelectronic spacing r_s, related to the jellium density ρ_o by $\rho_o = 1/\frac{4}{3}\pi r_s^3$. Subsequently, by a combination of theory and experiment, a consensus for a value for the density at which the electron crystal formed had emerged at around $r_s = 70a_o$ (see the lectures of the present writer (March, 1978) in the Proceedings of the Scottish Universities Summer School on 'The Metal-Insulator Transition in Disordered Systems'), a_o being the Bohr radius for hydrogen.

However, the decisive step on this important point for jellium theory came with the quantal Monte Carlo computer simulation by Ceperley and Alder

(1980). This work established that the critical value of r_s for Wigner electron crystallization, say r_c, was near to $r_c = 100\ a_o$, with admittedly substantial error of around $\pm 20\ a_o$.

Much use has been made of the numerical results of this study, especially in the local density version of density functional theory (see for example, the review of Callaway and March, 1984). We shall focus below on those aspects explored in the calculations of Herman and March (1984). In particular, total energies for the paramagnetic and ferromagnetic metallic phases of jellium will be referred to, as well as simple models of the momentum distribution in various phases of jellium.

2.1 GROUND-STATE OF WIGNER INSULATOR

In the case of the electron crystal, Herman and the writer noted that the theoretical expression

$$E(r_s) = -\ 1.79186/r_s + 2.65/r_s^{3/2} - 0.73/r_s^2\ \mathrm{Ry} \tag{1}$$

without adjustment reproduces the Ceperley-Alder energies at $r_s = 50$, 100, 130 and 200 to within 0.124, 0.003, 0.002 and 0.006m Ry respectively. The corresponding computer simulation uncertainties are 0.010, 0.003, 0.002 and 0.001m Ry.

Using eqn. (1) in conjunction with the virial theorem (March, 1958; Argyres, 1967) in the form

$$K + E = -\ r_s \frac{dE}{dr_s}, \tag{2}$$

the kinetic energy K is found to be

$$K(r_s) = 1.325/r_s^{3/2} - 0.73/r_s^2. \tag{3}$$

In the Wigner crystal, the localized electronic orbitals can be described by non-overlapping Gaussians centred on the body-centred-cubic lattice sites at sufficiently low densities ($r_s > \sim 10 a_o$). The single-particle occupation probability of plane wave state $\underline{k}$ can then be approximated as

$$n(k/k_f) = \frac{3\pi^{1/2}}{r_s^{3/4}}\ \exp\,[-(9\pi/4)^{2/3}\ r_s^{-1/2}\ (k/k_f)^2] \tag{4}$$

where k_f is the Fermi wave number. The corresponding kinetic energy is $K(r_s) = 1.5/r_s^{3/2}$, which is to be viewed as a zeroth order approximation to eqn. (3).

2.2 LOW-DENSITY METALLIC PHASES

The ground-state energies for the paramagnetic (P) and ferromagnetic (F) metallic phases of jellium have been calculated by Ceperley and Alder between $1.0 \leqq r_s/a_o \leqq 100$ and $2.0 \leqq r_s/a_o \leqq 100$, respectively, at selected values of r_s. These results have been carefully fitted by Vosko, Wilk and Nusair (1980: VWN) and by Perdew and Zunger (1981: PZ). In spite of their very different analytical forms, the VWN and PZ interpolation formulae both provide good overall fits to the Ceperley-Alder data over the entire range of r_s. On close examination, however, the PZ fit is closer than the VWN fit for $(r_s/a_o) \leqq 20$, while for $(r_s/a_o) \geqq 50$, the PZ fitting errors (though quite small) are about five times larger than their VWN counterparts overall, and in some cases are of opposite sign. Accordingly,

Herman and March (1984) base their results on a weighted average of $E(r_s)$ given by the VWN and PZ interpolation formulae, with VWN being given five times as much weight as PZ.

The kinetic energies for P and F were found from the weighted average of the VWN and PZ fits plus the virial theorem (2), while $K(r_s)$ for the Wigner electron crystal (W) was obtained from eqn (3). As is to be expected, K is higher for the ferromagnetic than for the paramagnetic phases, since the Fermi sphere has radius $2^{1/3}$ times that in the doubly occupied spin-degenerate case of the paramagnetic.

Herman and the writer note that at $(r_s/a_o) = 79 \pm 1$ where the paramagnetic and ferromagnetic total ground-state energies become equal, there is a discontinuity in K, of magnitude $\Delta K = K(P)-K(F) = 0.08$m RY. On the other hand, if one were to adopt the full Fermi sphere approximation, with $r_s = 79$, the numerical result would be highly inaccurate (0.22m Ry), showing that the contribution of electron-hole correlation excitations to the discontinuity is -0.14m Ry. This discontinuity in K is a reflection of the fact that a first-order transition is occurring at around $r_s/a_o = 79$ between paramagnetic and ferromagnetic phases. Bloch (1929), in pioneering work using the Hartree-Fock approximation, found a similar transition at much lower $(r_s/a_o) \sim 6$, but Wigner pointed out that electron correlation would decrease the stability of the ferromagnetic state from the Hartree-Fock prediction.

If one accepts eqn. (1) as an accurate description of the total energy for the Wigner crystal in the range $75 \leqq (r_s/a_o) \leqq 100$, and one then makes suitable allowance for computer simulation and interpolation uncertainties, Herman and the writer find that the cross-over between the paramagnetic metal and the Wigner insulating phase occurs at $(r_s/a_o) = 82 \pm 3$ while the ferromagnetic metal to Wigner insulator transition occurs at $r_s = 84 \pm 4$. Moreover, the discontinuity ΔK for these metal-insulator transitions is found to be somewhat larger than that for the paramagnetic to ferromagnetic transition discussed earlier.

With the aid of eqn. (1) and the PZ and VWN interpolation formula, Herman and the writer are able therefore to pin-point the values of r_s at which the various phase transitions occur. Their estimates are consistent with the earlier values of Ceperley and Alder, but the uncertainties are now substantially reduced. In fact, according to Herman and the writer, the transition densities come so close together that the ferromagnetic phase may just barely emerge as the lowest energy phase or may not emerge at all. This question deserves somewhat further study, as does the question of the possible occurrence of partially polarized metallic phases at intermediate densities.

While the single-particle occupation probability n(k) is a continuous function of k for the Wigner crystal according to eqn. (4), there is a discontinuity in n(k) at the Fermi surface in normal metals, as is well known, and also in low density metallic phases, as has been demonstrated, for example, by the computer simulation studies of Ceperley and Alder (1980). It is of interest to note that March et al. (1979) used this discontinuity in n(k) as an order parameter for a metal-insulator transition at T = 0. The form of n(k) is discussed in some detail by Herman and March (1984).

It is clear then that, particularly following the Ceperley-Alder studies, much is now established at a fully quantitative level about the first-order metal-insulator transition induced at sufficiently low density in jellium by the long-range interelectronic repulsions e^2/r_{ij}.

The metal-insulator transition at T = 0 in jellium is physically a liquid-solid first-order transition. March and Tosi (1985) have compared melting criteria for quantal and classical Wigner crystals.

Of course, allowing electron-ion interaction brings in either band theory, or chemical bonds, to add richness to the already fascinating phases in jellium. We may note, in this context, that it is intuitively clear that antiferromagnetism (see also section 4.1) in the Wigner crystal is favoured at the transition density r_c. This is because, as described above, the Madelung energy favours a body-centred-cubic electron crystal, and a long-range ordered Néel antiferromagnetic insulator at absolute zero is quite natural for this structure, the upward spin electrons being located on the sites of one of the two interpenetrating simple cubic lattices, and the downward spins on the other.

3. METAL-INSULATOR TRANSITION IN HYDROGEN AND IN EXPANDED ALKALI METALS

The transition from the molecular phase of hydrogen to a metallic state was first considered theoretically by Wigner and Huntington (1935). Experiments on the transition to the metallic state of hydrogen at low temperatures have subsequently been reported by Hawke et al. (1978); see, however, the critical comments by Ross (1985). Much theoretical work has been done on the nature of the ground state of metallic hydrogen; see, for instance, the study by Wood and Ashcroft (1982), but it is still not certain whether this has long-range or merely short-range ionic order.

We shall summarize below the main findings of Ferraz et al. (1984) who consider primarily the absolute zero transition in hydrogen, coming from the high density end, i.e. from the metallic phase.

The main achievement of this work is to exhibit by two models the condition for bound state formation around two centres. One approach is via a one-electron framework, exemplified by the Kohn-Sham method in density functional theory (see Lundqvist and March, 1983). Their more refined model, which will be treated below, built from the Heitler-London approach to the free space H_2 molecule, introduces the Thomas-Fermi screening length q_{TF}^{-1} into both nuclear-electron and electron-electron interactions.

3.1 HEITLER-LONDON THEORY OF H_2 MOLECULE WITH SCREENED INTERACTIONS

Their starting point therefore is the Hamiltonian H for two protons at distance R apart in jellium, and two electrons forming a neutral H_2 molecule. The stability of this molecule is investigated taking H as

$$H = \frac{-\hbar^2}{2m} \nabla_1^2 - \frac{\hbar^2}{2m} \nabla_2^2 + V(\underset{\sim}{r}\ R)$$

$$+ V(r_{12}) + \frac{e^2}{|\underset{\sim}{R}_A - \underset{\sim}{R}_B|} \exp(-q_{TF}|\underset{\sim}{R}_A - \underset{\sim}{R}_B|) \qquad (5)$$

where electrons 1 and 2 are attracted by one screened proton at the origin and another at position $\underset{\sim}{R}$ by the potential

$$V(\underset{\sim}{r},\underset{\sim}{R}) = \frac{-e^2}{r} \exp(-q_{TF} r) - \frac{e^2}{|\underset{\sim}{r}-\underset{\sim}{R}|} \exp(-q_{TF}|\underset{\sim}{r}-\underset{\sim}{R}|), \qquad (6)$$

while both electrons repel each other according to the screened interaction:

$$V(r_{12}) = \frac{e^2}{|\underset{\sim}{r}_1 - \underset{\sim}{r}_2|} \exp(-q_{TF}|\underset{\sim}{r}_1 - \underset{\sim}{r}_2|). \qquad (7)$$

Ferraz et al. then follow the Heitler-London approach as modified by Wang (1928) and they therefore write for the electronic wave function of the ground state:

$$\psi(\underset{\sim}{r}_1\underset{\sim}{r}_2) = \psi_A(\underset{\sim}{r}_1)\ \psi_B(\underset{\sim}{r}_2) + \psi_B(\underset{\sim}{r}_1)\psi_A(\underset{\sim}{r}_2) \tag{8}$$

where $\psi_A(\underset{\sim}{r}_1)$ is the trial function $(\alpha|\pi)^{3/2}\exp(-\alpha|\underset{\sim}{r}-\underset{\sim}{R}_A|)$ and similar expressions have been used for other wave functions.

Ferraz et al. find α for each value of R by minimizing the ground state energy E. This allows the determination of E(R) as a function of R for a given density. Their main conclusion is that the molecule becomes unstable for

$$r_s = 1.79a_o. \tag{9}$$

However, they then reconsider the critical densities which correspond to having a bound state around a single He or a single H nucleus. In a Thomas-Fermi calculation they find for the 'united atom' He and for H

$$\begin{aligned} r_s^{\text{critical}}(\text{He}) &= 0.61a_o \\ r_s^{\text{critical}}(\text{H}) &= 2.44a_o \end{aligned} \tag{10}$$

whereas more refined local density calculations yield

$$\begin{aligned} r_s^{\text{critical}} &= 0.44a_o \\ r_s^{\text{critical}} &= 1.9a_o \end{aligned} \tag{11}$$

These eqns. (10) and (11) show that the Thomas-Fermi calculations give a value of r_s^{critical} which is too large by a factor of 1.3. This suggests that the value in eqn. (9) will be too great by the same factor, and that a good approximation to the critical density allowing the formation of bound states in H_2 would be

$$r_s^{\text{critical}}(H_2) \cong \frac{1.79a_o}{1.3} = 1.38a_o. \tag{12}$$

Accordingly, Ferraz et al. expect for $r_s > 1.38a_o$ to have an insulating molecular state. Though one should, no doubt, heed the cautionary remarks of Ross (1985), this value is in remarkable agreement with Hawke et al. (1978).

It is of interest to note here that Raman measurements on solid insulating hydrogen by Sharma et al. (1980) and by Wijngaarden et al. (1982) reveal a maximum in the vibrational frequency of a H_2 molecule as a function of pressure, around 350k bar. Though the metal-insulator transition occurs at around 3 Megabar, Pucci and March (1985) have argued that the maximum is, in fact, a precursor of the metal-insulator transition.

3.2 Expanded alkalis and the metal-insulator transition

It has sometimes been stressed that the major difference between the stability of molecular hydrogen crystal and the metallic alkalis resides in the much larger binding energy of the H_2 molecule in free space compared with the alkali metal dimers. The fact that, for hydrogen, the Heitler-London model of the previous section led back to a criterion for the metal-insulator transition quite near to the one-centre criterion for a bound state, encourages Ferraz et al. to make some observations on transitions in the alkalis, which are briefly summarized below.

The work of Meyer et al. (1967), in which they studied when bound states formed round pseudo-ions, led to the critical values of r_s for the alkalis collected in Table 1. Then, according to the arguments of Ferraz et al. (1984) the critical densities for the metal-insulator transition in the alkalis are in the range going from the densities given by Meyer et al. to the ones obtained by modifying their r_s-values by the factor 1/1.35,* as collected in Table 1 also.

These densities at which bound states occur round pseudo-ions are also compared with the equilibrium values, given in the second row of Table 1. Ferraz et al. argue that the one-centre bound state condition should be corrected by reference to the two-centre condition, the r_s values for bound state formation round one and two centres respectively differing by $(1.16)^2 \cong 1.35$* for hydrogen. Since the dissociation energy for H_2 is substantially greater than for alkali dimers, it seems not unreasonable to assume that this factor 1.35 will be an upper bound.

The scaled r_s values obtained in this way are recorded in the third row of Table 1 and are seen to be quite close to the experimental density except for Li. The conjecture of Ferraz et al. that the factor 1. 35 will be somewhat too large is consistent with the fact that for Rb the r_s value thereby found is a little smaller than the equilibrium r_s for this metal.

It seems therefore that one would not require much expansion of these heavier alkali metals to bring about a transition to an insulating phase. Unfortunately, two ways of causing such expansion, either to take the metals up towards the critical point, or to study metal-ammonia solutions, result in a new ingredient in the problem, namely disorder or inhomogeneity, with a corresponding tendency to localize electrons.

Table 1

Critical values of r_s for the alkalis

	Li	Na	K	Rb	Cs
(r_s/a_o)	7.16	5.49	6.86	6.71	8.03
(r_s^{equil}/a_o)	3.38	3.99	4.86	5.21	5.63
Corrected critical r_s	5.30	4.07	5.08	4.97	5.95

*In section 3.1, the value $r_s = 1.79a_o$ given in eqn. (9), at which the two-centre bound state is lost, corresponds to q_{TF} in eqn. (5) equal to 1.16. The corresponding one-centre bound state criterion is easily found to be $q_{TF} = 1$. Since $q_{TF}^2 \propto r_s^{-1}$, the r_s values for bound state formation round one and two centres respectively differ by $(1.16)^2 \cong 1.35$ for hydrogen.

4. VALENCE BOND PICTURE AND METALLIC COHESION

Poshusta and Klein (1982) have given a valence bond description of H_2 which has been extended to the alkalis by Malrieu et al. (1984). The theory is characterized by the $^1\Sigma_g^+$ and $^3\Sigma_u^+$ potential curves. These have to be known not only near the equilibrium distance r_e of the ground state of the diatom, but also at the larger interatomic distance r of the metal $[r/r_e \cong 1.2]$ and at the second neighbour distance $r' = r\sqrt{2}$ or $r(2/\sqrt{3})$ since these further neighbours play an important role. Even for the ground state, Morse curves or polynomials fitted to the spectroscopic constants would not give realistic values for these larger distances and accurate ab initio curves are to be preferred when available. For the triplet excited state, ab initio calculations are necessary to obtain some information on the very important repulsive interaction. As a consistent choice, Malrieu et al. (1984) use the curves proposed by Konowalow et al. (1980) for Li_2 and for Na_2. For Li_2 the agreement with experiment is excellent (for $^1\Sigma_g^+$, $r_e = 2.69$Å, experiment gives 2.67Å) $D_e = 1.04$eV (experiment: 1.06eV). For Na the core valence correlation begins to play a role, and the calculations did not include this effect, resulting in a somewhat too long ($r_e = 3.172$Å experiment: 3.079Å) and too weak ($D_e = 0.715$eV experiment: 0.748eV) bond. As a result the calculations on Na solid should yield a slight overestimate of the lattice parameter and a slight underestimate of the cohesive energy with respect to that one would obtain from the use of the exact diatomic potential curves.

For K and heavier atoms, lack of realistic potential curves seem to prohibit calculations for the present.

4.1 CRITIQUE AND MERITS OF VALENCE BOND THEORY

We conclude this section with some comments on the validity of valence bond theory in relation to the metal-insulator transition*:

(i) The exchange terms included by Malrieu et al. are just nearest-neighbour contributions. If higher orders are incorporated in the theory, the anti-ferromagnetic nature of the ground-state solution could change. We note that the next-nearest neighbour interactions should come in with a sign which tends to frustrate antiferromagnetic ordering. Such corrections should be especially dramatic for the alkalis treated by Malrieu et al. at ordinary densities. Potential effects of 'frustration' are quantitatively illustrated by Klein (1982).

(ii) Work on the half-filled linear chain Hubbard model (cf Lieb and Wu, 1968) suggests that the associated Heisenberg model, with effective Hamiltonian H_{eff} say, can yield reasonable energies all the way from the atomic limit through to the metallic state. The scheme of Poshusta and Klein (1982) appears to give reasonable results over the full range. Also Dasgupta and Pfeuty (1981) find in a first-order real-space renormalization group treatment that the parameter flow over this entire range (excluding the limit when the Hubbard U = 0) is to the atomic (or valence bond) limit. But the qualification should be made that this model might be rather special. Some of the results of Lieb and Wu (1968) can be interpreted to indicate that the ground state is metallic only when U = 0: firstly because the ground-state energy has a zero radius of convergence about the U = 0 limit and secondly because of the results of Lieb and Wu on ground-state electrical conductivity.

---

*Much of this section has emerged from very helpful correspondence with Professor D.J. Klein, to whom the writer is greatly indebted.

(iii) It might seem, at first sight, that taking a combination of a Néel state ϕ and its spin reversal $\overline{\phi}$ could solve some problems. However, this is disproved by the work of Klein (1977) who demonstrates that even taking suitable combinations over all rotations $R_{\alpha\beta\gamma}\phi$ does not really help to resolve the problems of long-range magnetic order, say in Li, which is predicted, but not observed.

(iv) Metal-insulator transitions conceivably need not involve drastic sudden changes in the electron ground-state wave functions when the nuclear coordinates are constrained to move slowly and smoothly. Rather, the sudden changes in the wave function for an actual metal-insulator phase transition might be associated with a shift between two minima in the potential-energy hypersurface. Indeed this is compatible with the work of Ferraz et al. (1984), discussed in section 3, where one locates an instability for dimerization of two protons in an electron gas, setting in at a certain density. If the electronic wave function can vary smoothly from separated atoms to metal, then it may be possible for an effective Hamiltonian H_{eff} of Heisenberg type to be useful even for the metal.

(v) The use of H_{eff} for other than the ground-state energy raises new problems. The ground state to H_{eff} is just a small component (or a transformation of a small component of the full wave function); indeed the overlap with the true ground state should go as $\sim S^N$, with $|S|< 1$ and 'effective' one-electron overlap'. There should (at metallic densities) be extensive inter-penetration of the H_{eff}-manifold of states by states associated to excited (ionic) 0-order manifolds, and as a consequence H_{eff} by itself should not be of much use for finite temperature properties.

(vi) We want to comment here that Malrieu et al. quote the work of Anderson (1973) and of Pauling (1953). However, Anderson was concerned with the solution of a triangular-lattice Heisenberg model; evidently not intended to describe a metal. It is interesting that Anderson's approximate solution is of a resonance theory form, involving Kekulé structures only. Pauling's work, in contrast, was designed to describe metals, but ionic structures are dominant in his (qualitative) work.

5. SUMMARY

In jellium, as the density is lowered, long-range Coulomb repulsion between electrons eventually results in the freezing of the homogeneous electron liquid; evidently a first-order metal-insulator phase transition. Fermi surface and magnetic ordering effects are briefly discussed as the transition is approached.

Electron-ion effects are then considered in addition to electron-electron correlation, in relation to the metal-insulator transition in (a) hydrogen and (b) alkalis. Case (a) is discussed by means of a Heitler-London treatment of the H_2 molecule in which electron-ion and electron-electron interactions are progressively screened. Finally, for case (b), one-centre bound states round pseudo atoms are briefly discussed, as is a valence-bond treatment by Malrieu et al. of the cohesion of the alkali metals. The importance of the chemical picture in incorporating both electron-ion and electron-electron interactions is quite apparent.

REFERENCES

Anderson, P.W. , 1973, Mater. Res. Bull. 8, 153.
Argyres, P.N. 1967, Phys. Rev. 154, 410.
Bloch, F., 1929, Zeits für Physik 57, 545.
Callaway, J. and March N.H., 1984, Solid State Physics 38, 135 (Academic: New York) Eds: Ehrenreich, H. and Turnbull, D.

Care, C.M. and March, N.H., 1975, Adv. Physics. 24, 101.
Ceperley, D.M. and Alder, B.J., 1980, Phys. Rev. Lett. 45, 566.
Dasgupta, C. and Pfeuty, P., 1981, J. Phys. C.14, 717.
Ferraz, A., March, N.H. and Flores, F., 1984, J. Phys. Chem. Solids, 45, 627.
Hawke, P.S., Burgess, T.J., Duerre, D.E., Huebel, J.G., Keeler, R.N., Klapper, H. and Wallace, W.C., 1978, Phys. Rev. Lett. 41, 994.
Herman, F. and March N.H., 1984, Solid State Commun. 50, 725.
Klein, D.J., 1977, Int. J. Quantum Chem. 12, 291.
Klein, D.J., 1982, J. Phys. A15, 661.
Konowalow, D.D., Rosenkrantz, M.E. and Olson, M.L., 1980, J. Chem. Phys. 72, 2612.
Lieb, E.H. and Wu, F.Y., 1968, Phys. Rev. Lett. 20, 1445.
Lundqvist, S. and March, N.H., 1983, Theory of the inhomogeneous electron gas (Plenum: New York).
Malrieu, J.P., Maynau, D. and Daudey, J.P., 1984, Phys. Rev. B30, 1817.
March, N.H., 1958, Phys. Rev. 110, 604.
March, N.H., 1978, in 'The metal non-metal transition in disordered systems: Proc. Nineteenth Scottish Universities Summer School in Physics: Eds. L.R. Friedman and D.P. Tunstall, p. 1.
March, N.H., Suzuki, M. and Parrinello, M., 1979, Phys. Rev. B19, 2027.
March, N.H., and Tosi, M.P., 1985, Phys. Chem. Liquids, in the press.
Meyer, A., Nestor, C.W. and Young, W.H., 1967, Proc. Phys. Soc. 92, 446.
Pauling, L., 1953, Proc. Natl. Acad. Sci. 39, 551.
Perdew, J.P. and Zunger, A. 1981, Phys. Rev. B23, 5048.
Poshusta, R.D. and Klein, D.J., 1982, Phys. Rev. Lett. 48, 1555.
Pucci, R. and March, N.H., 1985, to be published.
Ross, M., 1985, Rep. Prog. Physics, in the press.
Sharma, S.K., Mao, H.K. and Bell, P.M., 1980, Phys. Rev. Lett. 44, 886.
Vosko, S.H., Wilk, L. and Nusair, M., 1980, Can. J. Physics, 58, 1200.
Wang, S.C., 1928, Phys. Rev. 31, 579.
Wigner, E.P., 1934, Phys. Rev. 46, 1002.
Wigner, E.P., 1938, Trans. Faraday Soc. 34, 678.
Wigner, E.P. and Huntingdon, H.B., 1935, J. Chem. Phys. 3, 764
Wijngaarden, R.J. Lagendijk, A. and Silvera, I.F., 1982, Phys. Rev. B26, 4957.
Wood, D.M. and Ashcroft, N.W., 1982, Phys. Rev. B25, 2532.

Note added after completion of manuscript

My colleague, Dr. David Cooper, has kindly drawn my attention to the very accurate potential energy curve for Na_2 by G. Jeung (J. Phys. B16, 4289, 1983) which is relevant to the discussion of section 4.

THE METAL-INSULATOR TRANSITION IN LIQUID, DOPED CRYSTALLINE AND AMORPHOUS SEMICONDUCTORS: THE EFFECT OF ELECTRON-ELECTRON INTERACTION

A.A.Andreyev and I.S.Shlimak

A.F.Ioffe Physical-Technical Institute
of the Academy of Sciences of the USSR
Leningrad, USSR

ABSTRACT

A review is given of the changes in the temperature behaviour of electric conductivity in the cross-over of the Fermi level from the localized to delocalized states in liquid, doped crystalline, and amorphous semiconductors. Major emphasis is placed on the need of taking into account electron-electron interaction in both the insulating and metallic states in the interpretation of the experimental relationships observed.

Studies of the charge transfer processes in such disordered systems as the amorphous, liquid, and heavily doped semiconductors, "dirty" metals and alloys, are at present at the focus of attention of physicists engaged in condensed state research. Variation of a critical parameter of such a system (or a cross-over from one system to another) may initiate in these materials a transition from semiconducting (or insulating) to metallic state, or vice versa. It is the presence of the disordering factor, namely, the lack of translational symmetry and long-range order in both the amorphous and liquid semiconductors and the doped subsystem of crystalline semiconductors that determines the specific features of the metal-insulator (MI) transition in disordered systems.

This short review should be considered as a modest

token of our respect for Professor N.F.Mott whose ideas have greatly helped to shape the major stages in our understanding of the MI transition,[1,2] in particular, the formulation of an expression for the critical concentration N_c required for the transition to occur:

$$N_c a_0^3 \approx 0{,}2 \tag{1}$$

(where a_0 is the radius of electron localization at an isolated impurity atom in the case of doped crystalline semiconductors, whereas for the amorphous and liquid semiconductors a_0 should be replaced by a quantity ℓ which is on the order of, or greater than the interatomic separation); the concept of the minimum metallic conductivity σ_{min}:

$$\sigma_{min} = C \frac{e^2}{h \ell} \tag{2}$$

(where C is a numerical coefficient of order $10^{-1} - 10^{-2}$, depending on the actual model chosen); the concept of variable range hopping conduction which manifests itself as a continuous decrease of the conduction activation energy with decreasing temperature, the $\sigma(T)$ relationship being expressed analytically as

$$\sigma(T) = \sigma_0 \exp[-(\frac{T_0}{T})^n] \tag{3}$$

In the case of a constant density of states in the vicinity of Fermi level $N(E) = const = N(E_F)$, Mott showed that

$$n = 1/4 \;, \quad T_0 = \frac{C_0}{N(E_F)\, a^3} \tag{4}$$

(where C_0 is a numerical coefficient on the order of 20).

Mott was also the first to suggest the possibility of unactivated hopping conduction in strong electric fields which was subsequently confirmed experimentally[3], and many other ideas.

Particular attention of the theorists and experimenters studying the MI transition in disordered systems was directed at the behaviour of electronic states and the kinetics near the mobility edge E_c. By Anderson[4], in the insulating state and at energies E corresponding to localized states the envelopes of the electronic wave functions fall off exponentially with distance within the localization radius a. Note that as one approaches the MI transition, i.e. as $E \longrightarrow E_c$, the quantity a exhibits a power law divergence

$$a = a_0 \left| \frac{E - E_c}{E_c} \right|^{-\nu} \tag{5}$$

(where a_0 is the "initial" value of the Bohr radius far from the MI transition, ν is the critical exponent).

Relationship (5) resembles the divergence of the correlation radius of fluctuations at the phase transition of the second kind, which provided a fertile ground for the development of the scaling theory of localization.[5-7] This theory uses the analogy with the problem of phase transitions with critical exponents depending on the dimension of space.

As shown by the scaling theory, the dielectric constant $\varkappa$ should exhibit a similar divergence

$$\varkappa = \varkappa_0 \left| \frac{E - E_c}{E_c} \right|^{-\zeta} \tag{6}$$

with the critical exponent ratio $\zeta / \nu = 2$. Various models yield for ν values within 0.5 - 1.0. Another conclusion of this theory is that at $T \longrightarrow 0$ the quantity $\sigma(0)$ may become arbitrarily small since in Eq.(2) ℓ should be replaced by $a \longrightarrow \infty$ from Eq.(5). Nevertheless, σ_{min} has a certain physical meaning of the magnitude of conductivity starting from which $\sigma(0) \longrightarrow 0$.

Another possibility to enlarge our understanding of the MI transition lies in taking into account electron-electron interaction in both the metallic and insulating state.[8-12] The inclusion of Coulomb interaction between the localized carriers brought forward the idea of the presence in the density of states spectrum near Fermi level of a "soft" parabolic gap. As a result, the variable length hopping conductivity will now be described by Eq.(3) with different values of n and T_0:[8,9]

$$n = 1/2 \ , \quad T_0 = C_0' \frac{e^2}{\varkappa a} \tag{7}$$

(here C_0' is a numerical coefficient equal to 2.8). The Coulomb gap model was used to calculate the magnetoresistivity in the context of the variable length hopping conduction mechanism.

Taking into account the electron-electron interaction in the metallic state yielded the temperature correction to conductivity for "dirty" metals and heavily doped semiconductors

$$\sigma(T) = \sigma(0) + AT^x \tag{8}$$

i.e. the sign and magnitude of coefficient A and the values of x which turned out to depend strongly on the band structure of the semiconductor under study, the symmetry of the electronic wave functions, the elastic and inelastic collision mechanisms, etc.

Since the above theoretical studies are still in the beginning stages, any experimental data on the MI transition in disordered systems acquire particular importance.

We are going to discuss below the results of some experimental works carried out with the participation of the present authors and bearing on the various features of the MI transition in liquid, doped crystalline and amorphous semiconductors.

Figs. 1 - 4 display the temperature dependence of the electric conductivity of molten chalcogenides measured in the temperature range from melting point T_m up to 2000°C.

Already the very first work of Ioffe and Regel dealing with liquid semiconductors[14] revealed that the semiconductors retaining short-range order under melting retain in the liquid phase also the semiconducting properties, the electric conductivity of such melts growing exponentially with increasing temperature. Obviously enough, this growth of electric conductivity cannot continue without limit. Indeed, in studies[15-19] extended to higher temperatures the electric conductivity of semiconductor melts was shown to tend to satu-

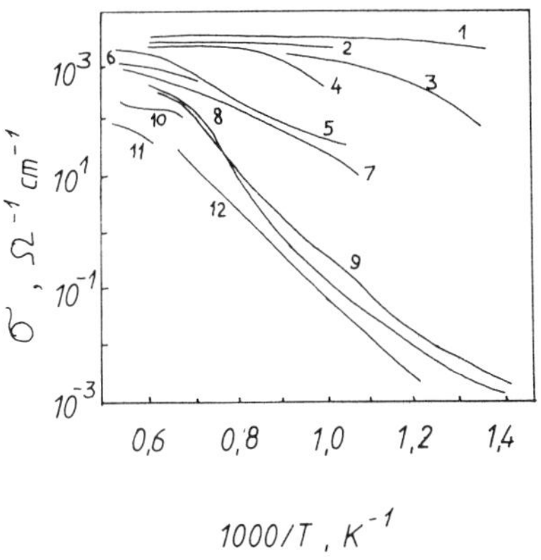

Fig. 1. Temperature dependence of the electric conductivity of semiconductor melts.
1 - Te, 2 - GeTe, 3 - $Si_{0.2}Te_{0.8}$,
4 - InTe, 5 - GeSe, 6 - PbSe, 7 - InSe,
8 - $Se_{0.91}Te_{0.09}$, 9 - As_2Se_3,
10 - CdSe, 11 - ZnTe, 12 - Se.

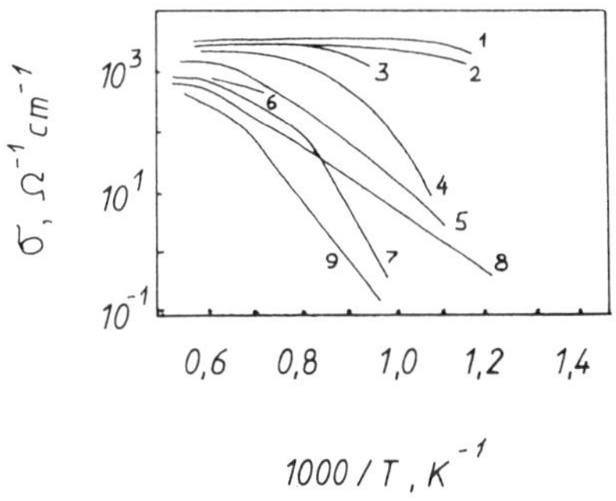

Fig. 2. Temperature dependence of the electric conductivity of semiconductor melts.
1 - Bi_2Te_3, 2 - Sb_2Te_3, 3 - SnTe,
4 - In_2Te_3, 5 - Sb_2Se_3, 6 - CdTe,
7 - $GeSe_2$, 8 - Sb_2S_3, 9 - $Se_{0.9}Ge_{0.1}$.

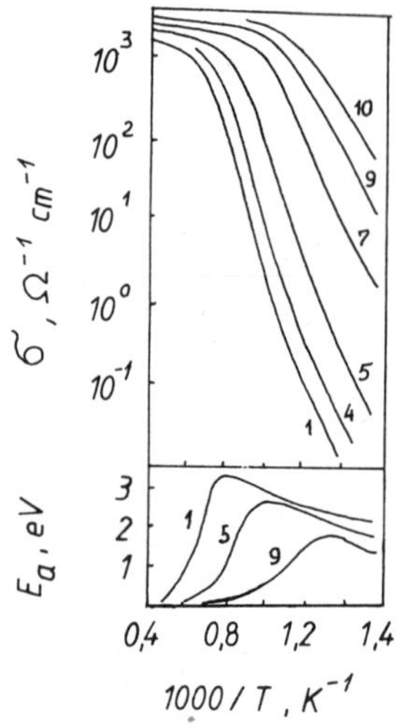

Fig. 3. Temperature dependence of the conductivity and its activation energy for the melts of glassy semiconductors. 1 - $AsSe_{1.5}$, 4 - $AsSe_{1.4}Te_{0.1}$, 5 - $AsSe_{1.2}Te_{0.3}$, 7 - $AsSe_{0.25}Te_{0.75}$, 9 - $AsSe_{0.3}Te_{1.2}$, 10 - $AsTe_{1.5}$.

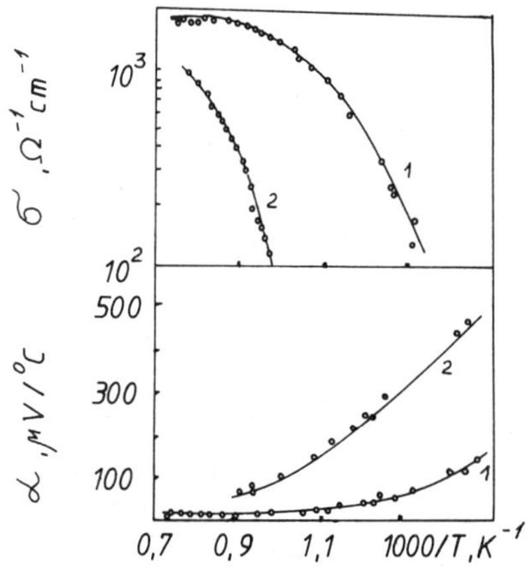

Fig. 4. Electric conductivity and thermo-emf of liquid As_2Te_3 (1) and As_2Te_3 As_2Se_3 (2)

ration at a level characteristic of metals. Within the plateau region the electric conductivity reaches a value of (1 - 3) x10^3 Ohm^{-1}/cm. The typically metallic magnitude of σ, as well as the absence of a significant temperature dependence evidence metallization of the melts.

The most characteristic feature of the electric conductivity of semiconductor melts at saturation is the magnitude of σ_{sat} which for all the studied tellurides and some selenides lies within 2500-3000 Ohm^{-1}/cm (Table 1). Such values of electric conductivity are in accord with the theoretical estimate $\sigma = e^2/3\hbar\ell$ = 2700 Ohm^{-1}/cm made within the model of nearly free electrons under the assumption of only two free electrons per atom participating in conduction, the average interatomic separation being ℓ = 3×10^{-8} cm. Such an agreement which apparently cannot be accidental suggests that only the higher valence subbands are "destroyed" by the fluctuation potential. In this case all the electrons of these subbands should form a Fermi sphere with a radius limited by the values of π/ℓ. The Hall constant calculated for $p_F = \frac{\pi}{\ell}$ turns out to be approximately a factor three above the estimate derived from a consideration of the valence shell structure of the atoms making up the characteristic telluride group and supplemented with an assumption that all the valence electrons participate in conduction (Table 2). Thus the available experimental data permit a conclusion that

Table 1.

Material	Temperature range, K	σ_{sat}, Ohm^{-1}/cm	Material	Temperature range, K	σ_{sat}, Ohm^{-1}/cm
Te	1000-1700	3300	Sb_2Te_3	1360-1700	2800
PbTe	1500-1850	2700	GeSe	1780-2000	2000
SnTe	1430-1740	2800	InSe	1850-2000	1000
GeTe	1430-1780	3000	Sb_2Se_3	1800-2000	1450
In_2Te_3	1540-1750	2500	Sb_2S_3	1880-2000	800
InTe	1450-1750	2500	PbSe	1830-2000	1300
Bi_2Te_3	950-1780	3300			

Table 2. Electric Properties of Semiconductor Melts

Semiconductor	σ Ohm^{-1}/cm	Temp. coeff. of conductivity (sign)	R_H, cm^3/C	R/R_o	α, $\mu V/K$
		Metallic melts of semiconductors			
Ge	15200	-	$-3.6\ 10^{-5}$	1.06	-1.0
Si	12860	-			
InSb	9350	-	$-6.9\ 10^{-5}$	1.4	
InAs	7000	-			
		Metal-like melts			
CdSb	5200	+	$-8.3\ 10^{-5}$	1.5	0
ZnSb	4200	+	$-5.3\ 10^{-5}$	1.2	+2.6
$AuTe_2$	3460	+	$-8.3\ 10^{-5}$	2.1	
Bi_2Te_3	2600	+	$-8.7\ 10^{-5}$	2.2	+1.0
GeTe	2600	+	$-15\ 10^{-5}$	3.6	+20
CuTe	1900	+	$-16\ 10^{-5}$	4.0	
Sb_2Te_3	1850	+	$-17\ 10^{-5}$	3.9	+10
SnTe	1870	+	$-12\ 10^{-5}$	2.3	+42
PlTe	1510	+	$-14\ 10^{-5}$		-3.7
Te	1700	+	$-12\ 10^{-5}$	3.0	+26
$Te_{0.9}Sb_{0.1}$	1250	+	$-17\ 10^{-5}$		+50
AgTe	600	+	$-3.2\ 10^{-4}$		
Tl_2Te_3	500	+	$-7.4\ 10^{-4}$	15	
InTe	500	+	$-5.0\ 10^{-4}$		
In_2Te_3	250	+	$-(5-7)10^{-4}$		+10
TlTe	290	+	$-13.8\ 10^{-4}$	25	
GaTe	250	+	$-3.2\ 10^{-3}$		
Ag_2S	120	+	$-2.8\ 10^{-3}$		+190
		Liquid semiconductors			
Tl_2Te	67	+	$-1.6\ 10^{-3}$		-150
GeSe	51	+	$-3.6\ 10^{-2}$		+50
As_2Te_3	25	+	$-5.0\ 10^{-2}$		+200
$CuSbSe_2$	16	+	$-7.3\ 10^{-3}$		+180
Tl_2Se	10	+	$-2.5\ 10^{-2}$		+240
Tl_2Se + 0.5at% In	40	+	$-2.5\ 10^{-3}$		
$Te_{0.7}Se_{0.3}$	27	+	$-2.6\ 10^{-3}$		
$Te_{0.5}Se_{0.5}$	1.0	+	$-3.0\ 10^{-2}$		+690

Table 2. (Continued)

Semiconductor	σ, Ohm^{-1}/cm	Temp. coeff. of conductivity (sign)	R_H, cm^3/C	R/R_0	α $\mu V/K$
		(Liquid semiconductors)			
InSe	3.0	+	$-1.0\ 10^{-4}$		-220
Sb_2Se_3	2.4	+	$-1.5\ 10^{-3}$		+3
Sb_2Se_3 + 2.5% Sb	3.1	+	$-2.1\ 10^{-2}$		-20
Sb_2Se_3 + 5% Te	3.0	+	$-2.2\ 10^{-2}$		+36
Sb_2S_3	$1\ 10^{-1}$	+	$-3.5\ 10^{-1}$		-100
$60As_2Te_3\ 40As_2Se_3$	$7\ 10^{-1}$	+	-1.0		
V_2O_5	$7\ 10^{-2}$	+	$-3.0\ 10^{-1}$		
$AsTlSe_2$	$1\ 10^{-3}$	+	-10^{-2}		

chemical bonding in melts remains sufficiently strong even at fairly high temperatures. Despite the metallic type of conduction, telluride melts cannot apparently be considered as a system of atoms with fully ionized valence shells, so that the free electron model as it was applied to liquid metals cannot be used for semiconductor melts even in the region where the temperature dependence of electric conductivity saturates.

The available experimental data for electronic conduction melts is summarized in Table 2.[20] When combined with electric conductivity and thermo-emf data, Hall effect measurements permit one to reveal the principal relationships for liquid semiconductors. The melts are arranged in the Table in the order of decreasing absolute magnitude of their electric conductivity at the temperature 20-50°C above T_m. The melts are classified into three groups (this classification was proposed earlier, cf. e.g. ref. 2): 1 - metallic, 2 - metal-like, and 3 - semiconducting melts.

As follows from the definition, the first group is made up by melts of the semiconductors which lose under melting the typical semiconducting properties and exhibit in the liquid state a behaviour, typical for liquid metals: $\sigma = (5\text{-}10)\times10^3\ Ohm^{-1}/cm$, a negative temperature coefficient of electric conductivity, a negative Hall coefficient which coincides with a good accuracy with the value of R_0 predicted by the free electron theory. Melting affects here the short-range order.

Falling into the second group are materials with contradictory properties: indeed, from Hall effect and thermo-emf measurements and the absence of a doping effect by a third component one could conclude that they likewise are metallic melts, were it not for the fact that they have a positive temperature coefficient of electric conductivity, just as semiconductors do. This is believed to be due to a minimum in the density of states spectrum at Fermi level and to a degradation of this minimum with increasing temperature.

The melts of the third group ($\sigma \lesssim (1-2)\times 10^2$ Ohm^{-1}/cm) are characterized by semiconducting properties: indeed, the temperature behaviour of electric conductivity is exponential, the thermo-emf coefficient may become as high as a few hundreds of $\mu V/K$ or even higher, and the Hall coefficient exceeds substantially the values predicted by the free electron theory. A specific feature appears in the form of a p-n anomaly consisting in that the thermo-emf and Hall coefficients have opposite signs, namely, the Hall effect exhibits usually a negative sign with a positive thermo-emf. These properties of liquid semiconductors can be properly understood if we assume that the energy gap in these materials is essentially a mobility gap (just as in amorphous semiconductors) filled by a quasicontinuum of localized states.

The above analysis permits an important conclusion that variation of the electric conductivity of a melt correlates with that of the melt properties, namely, semiconducting properties become revealed progressively more as σ decreases. The boundary values of σ separating the metallic from semiconducting states in a melt are 100-300 Ohm^{-1}/cm which is close to the estimate derived from Eq.(2).

We are turning now to the results obtained on samples of heavily doped crystalline and amorphous semiconductors. Here the MI transition is reached by properly varying the doping level, the temperature behaviour of conductivity being measured at low (helium) and ultralow temperatures. Note that crystalline semiconductors have until recently been the major object of studies. The reason for this lies in that because of the large Bohr radius a_0, the values of N_0 in such crystalline semiconductors as GaAs, Ge, Si are comparatively small (on the order of 10^{16}, 10^{17} and 10^{18} cm^{-3}, respectively) which is substantially below the solubility limit. At such concentrations the doping concept is applicable since a variation of N affects neither the lattice constant of a crystalline semiconductor nor its structure. Apart from this, the technology of introducing impurities in desired concentrations has been well developed for these crystals. In amorphous semiconductors the situation is much more complicated: indeed, the absence of translational symmetry and the inapplicability of the effective mass approximation result in a strong localization of the impurity states which, according to Eq.(1), requires a very high doping level to reach the critical concentration. Means have recently been devised to introduce into amorphous silicon large amounts of phosphorus (donor) or boron (acceptor) by adding to silane during its high frequency decomposition either phosphine or diborane; another efficient method is the ion implantation of impurities into amorphous silicon films. In neither case, however, can one reach doping levels high enough to be able to bring Fermi level to the corresponding mobility edges closer than by 0.2-0.3 eV, and accordingly the growth of electric conductivity with increasing doping level reveals saturation at 10^{-2}-10^{-1} Ohm^{-1}/cm.[21] This indicates that introducing high concentrations of impurities into amorphous semiconductors is not enough, and one has in addition to create a large number of sites in the amorphous network where the impurity could reveal its properties. For group III and V impurities in amorphous silicon or germanium, these should be sites with tetrahedral coordination, i.e. with four nearest neighbours. The number of such sites in real films with a high density of structural defects may turn out to be comparatively small. Apart from this, introduction of an impurity may result in a rearrangement of the local environment involving formation of three- or five-coordinate sites to saturate all covalent bonds. As a result, the metal-insulator transition in amorphous films was observed primarily in the amorphous silicon-metal or amorphous germanium-metal compositions (a review of these works may be found in ref. 22). In such compositions the MI transition is found to occur at metal concentrations of 12-17 at. %. While at such high concentrations one cannot clearly use the concept of doping any more, "modification" appears to be appropriate since the variation of metal content changes the structure of the film proper making it each time essentially a new substance.

Only recently has one succeeded in producing films of amorphous silicon doped heavily with phosphorus and studying the specific features of low temperature conductivity, magnetoresistivity and metal-insulator transition in this material.[23] As far as we know, this is the first work of this kind, and we will dwell on it in more detail. First, however, we are going to present experimental data relating to doped crystalline semiconductors, after which we shall compare them with the results obtained on films of heavily doped amorphous silicon.

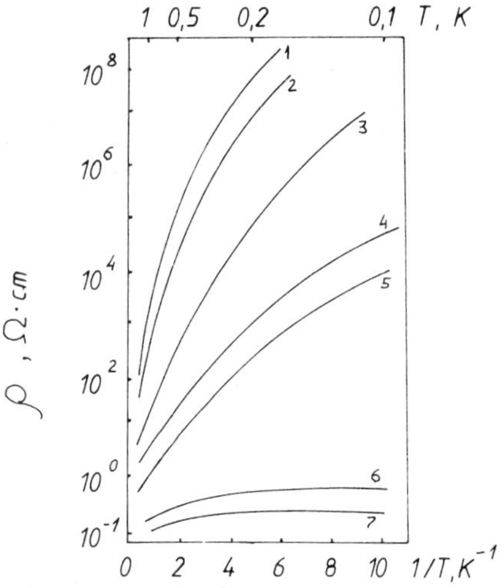

Fig. 5. Temperature dependence of the resistivity of neutron-doped p-Ge <Ga,As> samples with intermediate compensation. 1-7: gallium concentration increases from 3.5×10^{16} to 2×10^{17} cm^{-3}

The low temperature conductivity of doped crystalline semiconductors (with germanium investigated in most detail) was studied under the assumption that up to the MI transition the conduction mechanism can be described by a constant activation energy $\mathcal{E}$ („$\mathcal{E}_2$" for the case of weak, and „$\mathcal{E}_3$" for intermediate and strong compensation). It has, however, been shown[24] that in the insulating state near the MI transition the conduction exhibits variable rather than constant activation energy, which is clearly revealed when extending measurements down into the ultralow temperature domain (Fig. 5). Using the differentiation technique[25] for data treatment showed the curves to be well fitted by Eq.(3) for variable length hopping conduction, but with $n \approx 0.5$. This implies the presence in the density of states spectrum near Fermi level of a parabolic quasi-gap, probably of Coulomb nature.

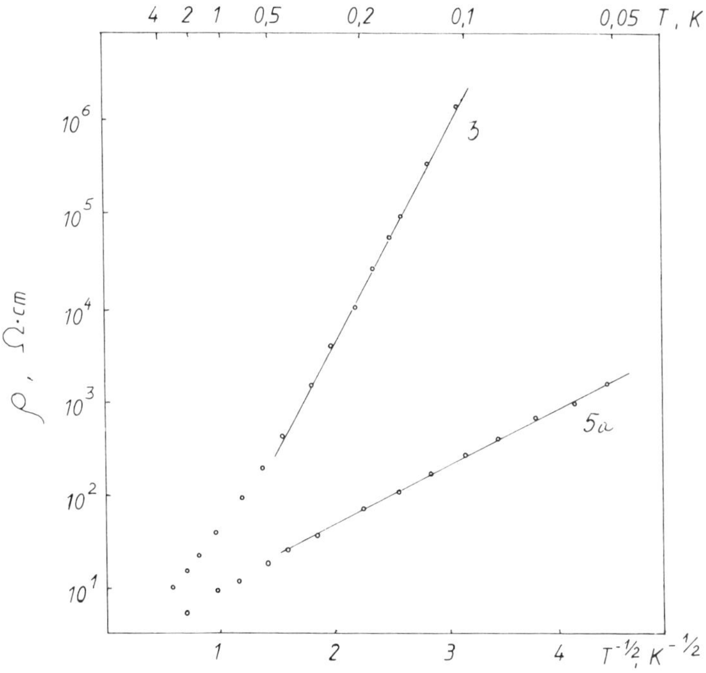

Fig. 6. log ρ vs. $T^{-1/2}$ plots for two of the samples shown in Fig. 5.

Fig. 6 shows log resistivity ρ vs. $T^{-1/2}$ plots for two neutron doped p-germanium samples with intermediate compensation. (The neutron doping technique is characterized by a high spatial homogeneity of doping which is very essential for studies near the MI transition). The slope of the curves T_o is seen to decrease with increasing N. In the context of the Coulomb gap model this may be accounted for, in accordance with Eqs. (5), (6), and (7), by the growth of and as N N_c. (Since $E \longrightarrow E_c$ as $N \longrightarrow N_c$, $|1 - N/N_c|$ in Eqs. (5) and (6) can be substituted for $(E - E_c)/E_c$).

The presence of the parabolic quasigap (and the validity of the $T^{-1/2}$ law for conduction) is apparently characteristic not only for the case of intermediate compensation. Studies of the conductivity of n-Ge <As> samples with weak compensation ($K < 3\%$)[26] have likewise revealed that in the vicinity of the MI transition the $\sigma(T)$ curves, rather than having a constant activation energy, are described by Eq.(3) with n = 0.5 (Figs. 7,8). (Note that these samples possessed also a high homogeneity since ^{75}As was introduced by neutron doping into germanium enriched preliminarily by the ^{74}Ge isotope).

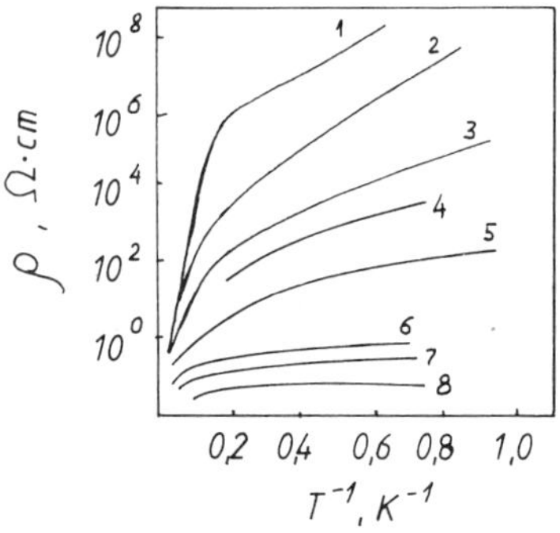

Fig. 7. Temperature dependence of the resistivity of neutron doped n-^{74}Ge<As> samples with weak compensation. 1-8: arsenic concentration increases from 1.1×10^{16} to 7.6×10^{17} cm^{-3}

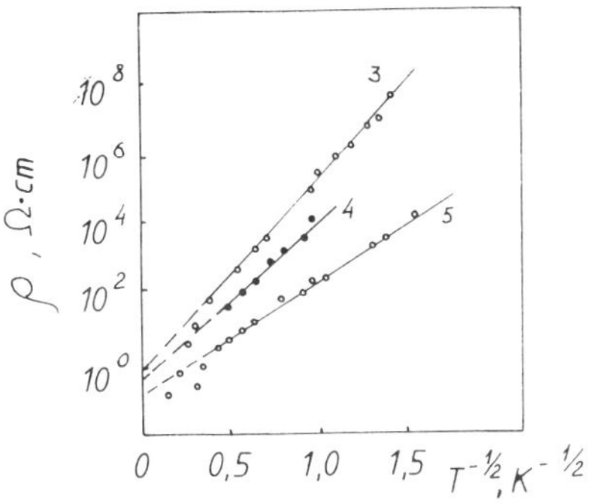

Fig. 8. log ρ vs. $T^{-1/2}$ plots for three of the samples shown in Fig. 7.

As seen from Eq.(7) for T_0, one can find the value of $\varkappa$ if the value of a is determined from independent experiments. This possibility is offered by magnetoresistivity (MR) measurements. Within the temperature domain where the conductivity is described by the $T^{-1/2}$ law the expression for MR has the form[13]

$$\ln \frac{\rho(H)}{\rho(0)} = t \left(\frac{a}{\lambda}\right)^4 \left(\frac{T_0}{T}\right)^{3/2} \qquad (9)$$

(where t = 0.015 and $\lambda = (ch/eH)^{1/2}$ is the magnetic length).

As follows from Eq.(9), MR is proportional to H^2 at constant temperature, and to $T^{-3/2}$ at a fixed magnetic field. Note that from the coefficient of proportionality one can derive the magnitude of the localization radius a. These experiments are reported on in ref. 27. Fig. 9 displays the dependence of a and $\varkappa$ on T_0 which was taken as a parameter showing how close one approaches the MI transition, since $T_0 \rightarrow 0$ when $N \rightarrow N_c$. As seen from the figure, both quantities grow by a power law as the MI transition is approached, the ratio of the critical exponents being 2.3 ± 0.2. It was found also[27] that $T_0 \sim |1 - N/N_c|^2$, which permits separate determination of the critical exponents: ν = 0.60 ± 0.04 and ζ = 1.38 ± 0.07.

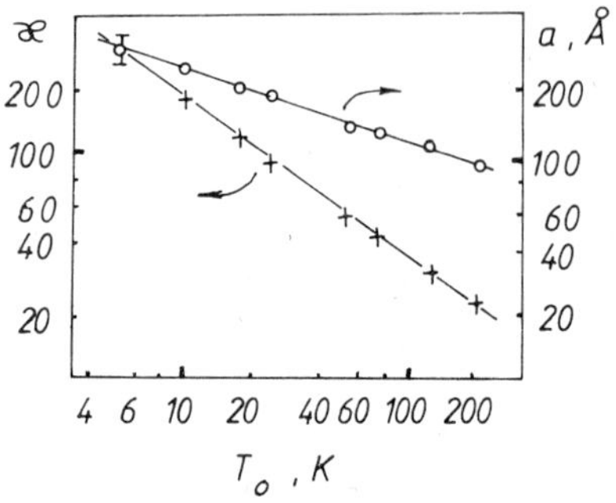

Fig. 9. Dependence of æ and a on T_o as one approaches the MI transition

We are turning now to studies of the low temperature conductivity and MI transition in amorphous silicon films doped heavily with phosphorus.[23] The samples were prepared by the ion implantation of P^+ into a-Si films obtained by vacuum deposition. The implantation dose for all the samples constituted $6x10^{16}$ cm^{-2} with the ion energy of 100 keV. As already mentioned, a simple increase of the dose did not produce the effect of heavy doping, the maximum value of σ at 300 K being not in excess of 0.1 Ohm/cm since the phosphorus atoms were predominantly in other than tetrahedrally bonded environment. To increase the electric activity of phosphorus, the a-Si⟨P⟩ samples were subjected to a heat treatment in glow discharge in a hydrogen atmosphere at different substrate temperatures. In this way films were prepared with an order-of-magnitude higher conductivity, i.e. 1-20 Ohm/cm. The amorphism of the structure was checked by electron diffraction measurements in reflection. One cannot, naturally, exclude the possibility that this increase of electric conductivity obtained by means of a heat treatment is associated with microrearrangement of the structure, i.e. with creation at scale lengths of a few tens of Å of an "intermediate order" structure.[28] Such changes would not be revealed by diffraction methods because of the smallness of size, and at the same time they could result in a substantial increase of localization radius and a decrease of the density of states tail in the vicinity of the corresponding

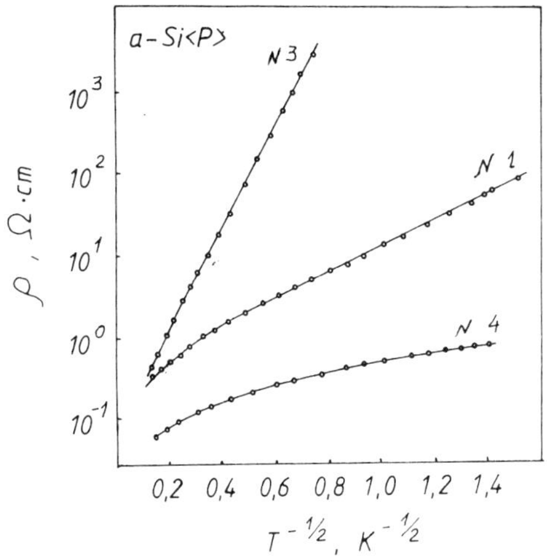

Fig. 10. log ρ vs. $T^{-1/2}$ dependence for amorphous films of heavily doped a-Si<P>.

mobility edge. Both these factors favour enhancement of the doping level and a closer approach to the MI transition.

Figs. 10 and 11 display log ρ vs. $T^{-1/2}$ plots for the a-Si<P> films studied, as well as a temperature dependence of MR for one of the samples. The fact that the MR is quadratic in field and positive permitted determination of a and $æ$ in the way this was done for crystalline n-Ge<As>. The values obtained are shown in Fig. 9. It is remarkable that they turned out to be quite close to the corresponding values in the crystalline semiconductor for the same T_0, although it would be only natural to expect the "initial" values of a_0 and $æ_0$ to be essentially different.

One could present also other experimental findings which remain unexplained; namely, whereas from Eq. (9) it follows that MR $\sim T^{-1.5}$ at a fixed magnetic field, for crystalline germanium the exponent turns out to be smaller ($1.2 \div 1.3$) and for amorphous silicon, somewhat greater ($1.7 \div 1.8$) than 1.5. In refs. [23,27] this difference was deliberately disregarded since the value of a was found by taking fourth root of the slope of the straight line, thus justifying the neglect

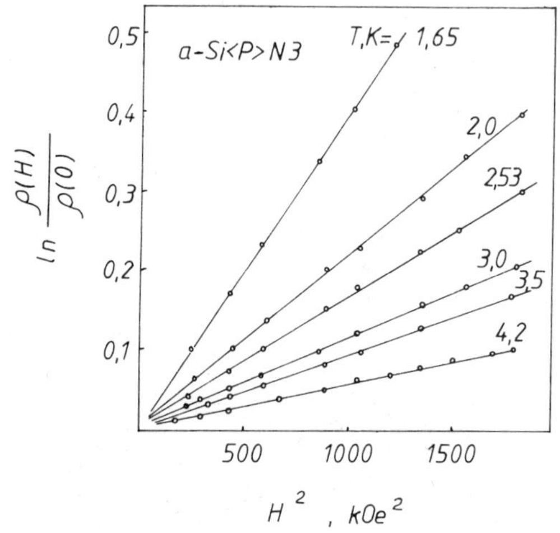

Fig. 11. Magnetoresistivity of sample N° 3.

of a possible error, however this difference may turn out to be fundamental.

Discussion of the theoretical and experimental work dealing with the effect of electron-electron interaction in the metallic state would be beyond the framework of the present review. As for the processes occurring in the insulating state close to the MI transition, we believe that although substantial progress has been achieved in the understanding of the charge transfer mechanisms, a lot remains to be done here, and a contribution of Prof. Mott may again turn out to be decisive.

REFERENCES

1. N.F.Mott, "Metal-Insulator Transitions", Taylor and Francis, London (1974).
2. N.F.Mott and L.A.Davis, "Electronic Processes in Non-Crystalline Materials", Oxford University Press (1978).
3. A.N.Ionov and I.S.Shlimak, Pisma v Zh.E.T.F. 16:374 (1972).
4. P.W.Anderson, Proc. Nat. Acad. Sci. USA 69:1097 (1972).

5. E.Abrahams, P.W.Anderson, D.C.Licciardello and T.V.Ramakrishnan, Phys. Rev. Lett. 42:673 (1979).
6. M.V.Sadovskii, Fiz. Tv. Tela 21:743 (1979).
7. Y.Ymry, Y.Gefen and D.J.Bergman, in: "Proc. Tahiguchi Symposium", ed. Y.Nagaoka, Springer-Verlag, N.Y. (1982).
8. A.L.Efros and B.I.Shklovskii, J. Phys. C 8:149 (1975); A.L.Efros and B.I.Shklovskii, "Electronic Properties of Doped Semiconductors", Springer-Verlag (1984).
9. L.Fleishman, D.C.Licciardello and P.W.Anderson, Phys. Rev. Lett. 40:1340 (1978).
10. M.Pollak, Phil. Mag. 42:781 (1980).
11. A.G.Aronov and B.L.Altshuler, Zh.E.T.F. 77:2028 (1979).
12. B.L.Altshuler and A.G.Aronov, in: "Electron-Electron Interaction in Disordered Systems", ed. A.L.Efros and M.Pollak, North-Holland Publ. Comp. (1985).
13. I.S.Shlimak, A.N.Ionov and B.I.Shklovskii, Fiz. Tekh. Polupr. 17:503 (1983).
14. A.F.Ioffe and A.R.Regel, Progr. Semicond. 4:237 (1960).
15. J.C.Perron, Adv. Phys. 16:657 (1967).
16. V.M.Glazov, S.G.Chizhevskaya and N.N.Glagoleva, "Liquid Semiconductors" (in Russian), Nauka, Moscow (1967).
17. J.T.Edmond, Brit. J. Appl. Phys. 17:979 (1966).
18. A.A.Andreyev, T.Turgunov and V.A.Alekseyev, Fiz. Tv. Tela 16:3660 (1974).
19. A.A.Andreyev, B.T.Melekh and T.Turgunov, Fiz. Tv. Tela 18:244 (1976).
20. V.A.Alekseyev, A.A.Andreyev and M.V.Sadovskii, Usp. Fiz. Nauk 132:47 (1980).
21. W.E.Spear and P.G. Le Comber, Solid St. Comm. 17:1193 (1975); A.V.Dvurechenskii and I.A.Ryazantsev, Phys. St. Sol. 69:K117 (1982).
22. K.Morigaki, Phil. Mag. 42:979 (1980).
23. A.N.Aleshin, A.V.Dvurechenskii, A.N.Ionov, I.A.Ryazantsev and I.S.Shlimak, in: "Meeting on Amorph. Silicon and its Use in Solar Power Technology (Abstracts)", Leningrad (1984).
24. I.S.Shlimak and E.I.Nikulin, Pisma v Zh.E.T.F. 15:30 (1972).
25. A.G.Zabrodskii and I.S.Shlimak, Fiz. Tekh. Polupr. 9:587 (1975).

26. I.S.Shlimak et al. Pisma v Zh.E.T.F. 9:877 (1983).
27. A.N.Ionov, I.S.Shlimak and M.N.Matveyev, Solid St. Comm. 47:763 (1983).
28. V.I.Bonch-Bruyevich and V.D.Istra, Phys. St. Sol. B68:369 (1975).

EXCITON CONDENSATION AND THE MOTT TRANSITION

Leonid A. Turkevich

The Standard Oil Company
Corporate Research Center
4440 Warrensville Center Road
Cleveland, Ohio 44128

INTRODUCTION

Mott has long been preoccupied with electronically driven metal-insulator transitions.[1,2] Mott concentrated on the failure in the atomic limit of the independent electron model for the alkali metals.[3] While Mott ascribed the metal-nonmetal transition in the divalents to disorder,[4] equally dramatic electron correlation effects are also manifested for low density divalent metals.

The most studied of metal-insulator transitions in divalent systems has been that in fluid mercury (Hg).[5] This is undoubtedly due to its relatively accessible liquid-gas critical point ($\rho_c \sim 5.75$ g/cm^3), thereby making feasible a continuous expansion[6] of the condensed metallic phase to the atomic limit--a classic realization of the Mott gedanken expansion experiment.[3] As liquid Hg is expanded, a metal-nonmetal (MN) transition occurs at $\rho_{MN} \sim 9$ g/cm^3. The origin of this transition had been controversial[1,2] until the recent discovery[7,8] by Hensel and coworkers, at Marburg, of a companion dielectric transition at even lower densities in the vapor, $\rho_M \sim 3$ g/cm^3 ("the Marburg line").

In a series of papers, Turkevich and Cohen[9-12] analyzed the phenomonology of expanded fluid Hg and identified the high-pressure, high-temperature phase at the intermediate densities $\rho_M < \rho < \rho_{MN}$ as an excitonic insulator[13] (EI), thus emphasizing the paramount importance of electron-hole correlation in this expanded divalent metal. Exciton condensation[14] at ρ_M results in the observed dielectric anomaly in the vapor; a Mott exciton unbinding[15] is responsible for the sharp metal-nonmetal transition that occurs in the liquid.

While Mott's discussion[15] of the Coulomb driven metal-nonmetal transition was couched in terms of the unbinding of large,hydrogenic (Mott-Wannier) excitons, the strong, on-site electron-hole correlation in divalent Hg makes a Frenkel atomic exciton picture more appropriate. It also changes the energy scale from the typical 10 meV for excitons in amorphous semiconductors to the more dramatic eV of atomic excitations. This obviates the need to worry about competing instabilites (e.g. electron-hole liquid drop) that might occur in the lower energy regime. This paper explicitly relates the exciton condensation instability and the Mott metal-nonmetal transition within a formalism for treating the strong on-site electron-hole correlations in these closed-shell systems. For specificity, we have in mind the parameters for fluid Hg, but our arguments are equally applicable[16] to the other expanded (or chemically diluted) divalent metals (Mg, Zn, Cd) and to the rare-gas solids under extreme pressure (where metallization is expected to occur at $p \sim$ Mbar).

PHENOMENOLOGY OF THE TRANSITIONS IN FLUID MERCURY

Atomic Hg is divalent with a $6s^2$ configuration. As the density of a hypothetical perfect crystal is increased, the 6s and 6p levels broaden into a filled 6s band and an empty 6p band. As the density is further increased, a Wilson metal-nonmetal transition should occur when the two Bloch bands overlap. However, one-electron band-structure calculations[17] all indicate that the single-particle band-gap closes near ρ_c and not at the metal-nonmetal ρ_{MN}. These band-structure calculations are consistent with an extrapolation[18] to zero near ρ_c of the σ_{dc} activation energy in the vapor.

Mott envisages[4] the insulating liquid ($\rho < \rho_{MN}$) as possessing bands which overlap (consistent with the one-electron band-structure calculations) but only via localized states (Mott pseudogap). However, the Knight shift, which measures directly the density of s states at the Fermi level, irrespective of whetehr the states are localized or extended, vanishes[19] for $\rho < \rho_{MN}$, thereby precluding any Fermi level s states. Analyses[20,8] of $\sigma(\omega)$, as deduced from optical reflectivity, also establishes a real gap for $\rho < \rho_{MN}$. Thus the single-particle picture, even augmented by disorder, cannot explain the MN transition in the liquid.

Hensel and colleagues have reported[7,8] a second transition at the lower density ρ_M , as the dense vapor is compressed. The dielectric constant ε (at $\omega \sim 0.6$ eV) follows Clausius-Mosotti behavior for $\rho < \rho_M$, but at $\rho \sim \rho_M$ discontinuously rises from $\varepsilon \sim 2$ to $\varepsilon \sim 10$, with a gradual decrease for $\rho > \rho_M$. The discontinuous increase in ε is also observed[21] at lower frequencies. The optical absorption also qualitatively changes[22] at ρ_M: for $\rho < \rho_M$ the optical absorption is well-characterized by a steep edge, which shifts to lower frequency upon compression--however, at ρ_M the optical absorption suddenly acquires a low frequency plateau below the edge. It is clear that the origin of these remarkable optical anomalies must reside in some change in the electronic structure of Hg at ρ_M.

We have examined in some detail[9-12] the evolution of the excitation spectrum of normal insulating Hg vapor from the sharp atomic transitions to broad exciton bands. The optical absorption edge is identified as the bottom of the singlet $6\ ^1P_1$ Frenkel exciton band. These excitons are bound relative to the single-particle 6p band by $E_B \sim 4$ eV, which is relatively unchanged as the vapor is compressed from infinite dilution through the Marburg line. We have estimated[12] the ρ-dependence of the bottom of this exciton band, where the exciton transfer matrix elements are extracted from calculated Hg_2^* excimer spectra. For a variety of hypothetical crystal structures, the (k=0) bottom of the singlet exciton band becomes degenerate with the atomic ground state $(6\ ^1S_0)$ at $\rho \sim 3$ g/cm^3. The dielectric susceptibility diverges in this model, presaging an instability at ρ_M to the new excitonic insulator ground state, with correlated 6s-6p mixing. The effect of disorder in the fluid is not expected[11] to qualitatively alter this mean-field description of the exciton condensation, although with the density as a thermodynamic variable, the transition may be converted from second-order to first-order.

THEORY OF CONDENSATION AND UNBINDING OF FRENKEL EXCITONS

We now turn to a theoretical discussion of this exciton condensation and its relation to the Mott metal-nonmetal transition. Our treatment is similar to earlier work[23] on condensation of Mott-Wannier excitons. We utilize the model Hamiltonian

$$H = \sum_{k\sigma} \left(\varepsilon_s(k)\, a^+_{k\sigma} a_{k\sigma} + \varepsilon_p(k)\, b^+_{k\sigma} b_{k\sigma} \right) +$$

$$\sum_i \left(U_s\, a^+_{i\uparrow} a_{i\uparrow} a^+_{i\downarrow} a_{i\downarrow} + U_p\, b^+_{i\uparrow} b_{i\uparrow} b^+_{i\downarrow} b_{i\downarrow} + \right. \qquad (1)$$

$$\left. U_{vc} \sum_{\sigma\sigma'} a^+_{i\sigma} a_{i\sigma} b^+_{i\sigma'} b_{i\sigma'} \right)$$

Here $a^+_{k\sigma}$ ($a_{k\sigma}$) creates (destroys) valence (6s) electrons with single-particle energies $\varepsilon_s(k)$ and spin σ; similarly $b^+_{k\sigma}$ ($b_{k\sigma}$) creates (destroys) conduction (6p) electrons with single-particle energies $\varepsilon_p(k)$ and spin σ. This Hamiltonian ascribes all the exciton banding to the banding of the parent single-particle states, and it neglects the smaller[12] dipole-dipole contribution to the banding of the singlet. The single-particle piece of (1) is augmented by repulsive on-site Hubbard U terms. For simplicity we have neglected the spin-orbit splitting between singlet 6 $^1P_1$ and triplet 6 $^3P_1$ excitons. The parameters U_s and U_p are quickly absorbed into Hartree-Fock modifications to the single-particle energies. Thus it is only the competition between the valence-conduction correlation U_{vc} and the single-particle bands that drives the transitions.

As there are no spin flip terms and as we have neglected spin-orbit splitting, we focus only on spin-diagonal correlation functions, suppressing from now on all spin indices. We define the usual temperature single-particle Green's functions

$$G^a_{ij}(\tau-\tau') = - \langle T\, [a_i(\tau)\, a^+_j(\tau')] \rangle$$
$$G^b_{ij}(\tau-\tau') = - \langle T\, [b_i(\tau)\, b^+_j(\tau')] \rangle \qquad (2a)$$

and the anomalous (Gor'kov) Green's functions

$$F^+_{ij}(\tau-\tau') = - \langle T\, [a_i(\tau)\, b^+_j(\tau')] \rangle$$
$$F_{ij}(\tau-\tau') = - \langle T\, [b_i(\tau)\, a^+_j(\tau')] \rangle \qquad (2b)$$

expressing the possible correlation between a conduction electron and a valence hole.

We spatially Fourier transform and utilize the Matsubara (odd) frequency representation for the Green's functions. The equations of motion may be decoupled in the usual generalized Hartree-Fock manner. This identifies the Hartree-Fock single-particle energies as

$$\begin{aligned} \tilde{\varepsilon}_s(k) &= \varepsilon_s(k) + U_s\, G_{ii}^{a}(\eta) + 2U_{vc}\, G_{ii}^{b}(\eta) \\ \tilde{\varepsilon}_p(k) &= \varepsilon_p(k) + U_p\, G_{ii}^{b}(\eta) + 2U_{vc}\, G_{ii}^{a}(\eta) \end{aligned} \tag{3}$$

where η is an infinitesimal. In our subsequent discussion, we will assume a direct gap situation

$$\tilde{\varepsilon}_p(k) - \tilde{\varepsilon}_s(k) \sim 2d + \alpha k^2 \tag{4}$$

where $d > 0$ corresponds to a single-particle (Wilson) insulator and $d < 0$ corresponds to a single-particle (Wilson) metal. Measuring these energies with respect to the chemical potential μ, defines

$$\xi_s = \tilde{\varepsilon}_s - \mu \quad , \quad \xi_p = \tilde{\varepsilon}_p - \mu \tag{5}$$

Defining the correlation gap functions

$$\Delta = U_{vc}\, F_{ii}(\eta) \quad , \quad \Delta^{+} = U_{vc}\, F_{ii}^{+}(\eta) \tag{6}$$

the single-particle Green's functions may be obtained

$$\begin{aligned} G^{a}(k, \omega_n) &= \frac{i\omega_n - \xi_p/\hbar}{(i\omega_n - E_s/\hbar)(i\omega_n - E_p/\hbar)} \\ G^{b}(k, \omega_n) &= \frac{i\omega_n - \xi_s/\hbar}{(i\omega_n - E_s/\hbar)(i\omega_n - E_p/\hbar)} \\ F^{+}(k, \omega_n) &= \frac{-\Delta^{+}/\hbar}{(i\omega_n - E_s/\hbar)(i\omega_n - E_p/\hbar)} \\ F(k, \omega_n) &= \frac{-\Delta/\hbar}{(i\omega_n - E_s/\hbar)(i\omega_n - E_p/\hbar)} \end{aligned} \tag{7}$$

where the renormalized single-particle energies are given by

$$E_s = \tfrac{1}{2} (\xi_p + \xi_s) - \tfrac{1}{2} [(\xi_p - \xi_s)^2 + 4 \Delta^2]^{\frac{1}{2}}$$
$$E_p = \tfrac{1}{2} (\xi_p + \xi_s) + \tfrac{1}{2} [(\xi_p - \xi_s)^2 + 4 \Delta^2]^{\frac{1}{2}} \tag{8}$$

Self-consistency of the decoupling is achieved by the correlation gap equation, which, for low temperatures (compared to eV !) takes the form

$$1 \sim \int \frac{d^3k}{(2\pi)^3} \quad [E_p(k) - E_s(k)]^{-1} \tag{9}$$

The self-consistency condition (9) cannot be satisfied for large single-particle gaps 2d. However, when the system is sufficiently compressed that the gap drops below

$$d_c \sim \frac{\pi}{12} n U_{vc} \tag{10}$$

where n is the atomic density, a self-consistent solution is possible. This is precisely where the (exciton) bound state drops below the top of the filled valence band. The single-particle energies (8) distort from (5), the gap acquiring a Coulomb correlation component Δ. As the single-particle gap $d \rightarrow 0$, the correlation component of the gap grows to a maximum of

$$\Delta_{max} \sim \frac{\pi}{12} n U_{vc} \tag{11}$$

and then decreases as the single-particle bands overlap ($d < 0$), ultimately being driven to zero at

$$d_c \sim - \frac{\pi}{12} n U_{vc} \tag{12}$$

where the exciton bound state is lost as the electron and hole become uncorrelated in a Mott unbinding transition. The symmetry between the exciton condensation threshold (10) and the Mott unbinding (12) is not manifested in the earlier treatments for large (Mott-Wannier) excitons,[24] where, at low temperatures the excitonic insulator state is present for arbitrarily large overlap $d < 0$. This feature was certainly an unsatisfying aspect of that theory, and is superceded by onset of the electron-hole liquid drop instability. It is fortunate that this feature disappears in the treatment of Mott unbinding of Frenkel excitons. The symmetry between (10) and (12) is, in fact, evident in the phase diagram for fluid Hg, where $\rho_{MN} - \rho_c \sim \rho_c - \rho_M$.

A final remark on the order of these transitions. The (T=0) condensation free energy $F_{EI} - F_N \sim - \Delta^2$, both at the condensation instability (10) and at the Mott transition (12). Since the correlation gap $\Delta \to 0$ continuously, both the exciton condensation and the Mott metal-nonmetal transition are second-order within the simple Hubbard-like model (1). The inclusion of density fluctuations, present in the liquid divalent metals near their liquid-gas critical points, is expected[12] to drive both of these transitions first-order. There is, however, reason to believe that the Mott metal-nonmetal transition in this model will be driven first-order, <u>even in the absence of density fluctuations</u>--this last point will become especially relevant for the metallization of the rare-gas solids. As Mott observed,[15] one must also include the effects of screening of the Coulomb interaction in any discussion of the exciton unbinding. In the context of our Frenkel model, we must compare the correlation energy contributions (i.e. the RPA sum of the Coulomb interaction bubble diagrams) for the excitonic insulator and single-particle metallic states. Since we are altering the dielectric function in the excitonic insulator state, we expect these correlation contributions to differ. Preliminary results indicate[25] that for $\Delta > 0$, the excitonic insulator correlation energy is finite order by order, and to lowest approximation may be neglected. Thus when the correlation energy of the electron gas balances the condensation energy of the excitonic insulator, the lowest free energy state switches from the excitonic insulator to the metal, the gap being driven discontinuously to zero. If borne out by detailed calculation, this would provide a nice confirmation of Mott's earliest intuitions on the electronically driven metal-insulator transition.

REFERENCES

1. N. F. Mott, E. A. Davis, Electronic Processes in Noncrystalline Materials (Clarendon, Oxford, 1979).
2. N. F. Mott, Metal-Insulator Transitions (Taylor & Francis, London, 1974).
3. N. F. Mott, Proc. Phys. Soc. (London) A 62, 416 (1949).
4. N. F. Mott, Philos. Mag. 13, 989 (1966), 26, 505 (1972); ref. 2, p.228.
5. For recent reviews, see F. Yonezawa, T. Ogawa, Progr. Theor. Phys. (Japan) Suppl. 72, 1 (1982); H. Endo, ibid., 100.
6. F. Hensel, E. U. Franck, Rev. Mod. Phys. 40, 697 (1968).
7. W. Hefner, F. Hensel, Phys. Rev. Lett. 48, 1026 (1982); W. Hefner, thesis, Univ. of Marburg (1980), unpublished.
8. W.Hefner, R. W. Schmutzler, F. Hensel, J. Phys. (Paris) Colloq. 41, C8-62 (1980).
9. L. A. Turkevich, M. H. Cohen, J. Phys. Chem. 88, 3751 (1984).
10. L. A. Turkevich, M. H. Cohen, J. Non-Cryst. Solids 61-2, 13 (1984).
11. L. A. Turkevich, M. H. Cohen, Ber. Bunsenges. Phys. Chem. 88, 292 (1984).
12. L.A. Turkevich, M. H. Cohen, Phys. Rev. Lett. 53, 2323 (1984).
13. For a review, see B. I. Halperin, T. M. Rice, Solid State Phys. 21, 115 (1968).
14. R. S. Knox, Solid State Phys. Suppl. 5, 100 (1963).
15. N. F. Mott, Philos. Mag. 6, 287 (1961).
16. L. A. Turkevich, M. H. Cohen, unpublished, Bull. Am. Phys. Soc. 30, 273 (1985).
17. H. Overhof, H. Uchtmann, F. Hensel, J. Phys. F 6, 523 (1976); P. Fritzson, K.-F. Berggren, Solid State Commun. 19, 385 (1976); L. F. Mattheiss, W. W. Warren, Jr., Phys. Rev. B 16, 624 (1977).
18. R. Schmutzler, F. Hensel, E. U. Franck, Ber. Bunsenges. Phys. Chem. 72, 1194 (1968).
19. U. El-Hanany, W. W. Warren, Jr., Phys. Rev. Lett. 34, 1276 (1975); W. W. Warren, Jr., F. Hensel, Phys. Rev. B 26, 5980 (1982).
20. H. Ikezi, K. Scwarzenegger, A. L. Simons, A. L. Passner, S. L. McCall, Phys. Rev. B 18, 2494 (1978); C. E. Krohn, J. C. Thompson, Phys. Rev. B 21, 2619 (1980).
21. F. Hensel, G. Schönnherr, private communication.
22. F. Hensel, Phys. Lett. 31A, 88 (1970), Ber. Bunsenges. Phys. Chem. 75, 847 (1971); J. Popielawski, H. Uchtmann, F. Hensel, Ber. Bunsenges. Phys. Chem. 83, 123 (1975); H. Uchtmann, F. Hensel, H. Overhof, Philos. Mag. B 42, 583 (1980).

23. L. V. Keldysh, Yu. V. Kopaev, Sov. Phys.--Solid State 6, 2219 (1965); Fiz. Tverd. Tela 6, 2791 (1964); J. des Cloizeaux, J. Phys. Chem. Solids 26, 259 (1965); D. Jerome, T. M. Rice, W. Kohn, Phys. Rev. 158, 462 (1967).

24. A. N. Kozlov, L. A. Maksimov, Sov. Phys.--JETP 21, 790 (1965); Zh. Eksp. Teor. Fiz. 48, 1184 (1965).

25. L. A. Turkevich, unpublished.

METAL - INSULATOR TRANSITION IN DOPED SEMICONDUCTORS

E.N. Economou and A.C. Fertis

Department of Physics and Research Center of Crete
University of Crete, Heraklio, Crete, Greece

I. INTRODUCTION

Heavily doped semiconductors exhibit, as the concentration of impurities is changed, the phenomenon of metal-insulator transition. Mott[1-4] pointed out that Coulomb repulsion, if large enough in comparison with the transfer matrix element, inhibits electronic propagation in a half-filled band. Subsequently, Hubbard[5] introduced a simple periodic model incorporating this competition between the transfer matrix element V (which facilitates propagation) and the on-site Coulomb repulsion U (which opposes propagation). The metal-insulator transition occurs in this model when

$$U = \alpha Z V, \qquad (1.1)$$

where Z is the coordination number and α is of order unity (recent work[6] finds α=1.34). Taking into account that

$$U = \frac{5}{8} \frac{e^2}{\varepsilon \alpha_B}, \qquad (1.2)$$

where ε is the dielectric constant of the host crystal, and α_B is the effective Bohr radius (α_B = 15.2 ± 1.7Å for Si:P according to Edwards and Sienko[7]); and expressing V in terms of R

$$V(R) = -\frac{e^2}{\varepsilon \alpha_B} f(x)\exp(-x), \qquad (1.3)$$

we can rewrite Eq. (1.1) as

$$n_c^{\frac{1}{3}} \alpha_B = b, \qquad (1.4)$$

where $x = R/\alpha_B$, R is the distance between two nearest neighbor sites ($n=R^{-3}$), f(x) is a polynomial discussed by Fertis et al[8], n_c is the critical concentration, and b is a dimensionless constant, which, using $\alpha = 1.34$, turns out to be 0.176. A similar argument developed by Mott and Davis[4], which is based on a periodic Hubbard model, gives b = 0.172, while Bergren[9] obtains $b \simeq 0.18$; if the host semiconductor has four or six isotropic conduction minima Bergren[9] finds $b \simeq 0.22$.

Edwards and Sienko[7] have documented that Eq. (1.4) is obeyed in a large variety of semiconductors with n_c varying over nine orders of magnitude; the value of b resulting from the experimental analysis[7] is

$$b \simeq 0.26 \pm 0.05 \qquad (1.5)$$

Thus the value of b resulting from Eq. (1.1) is lower than the experimental value. On the other hand, there is an additional mechanism inhibiting propagation : disorder. Anderson, in his pioneering 1958 paper[10], demonstrated that the energy mismatch $\delta\varepsilon$ between neighboring local levels, if large in comparison with V, can force the eigenstates to become localized and thus can trigger a metal-insulator transition. In applying this idea to impurity bands one has to take into account that in the absence of compensation $\delta\varepsilon$ is expected to be small, because the main source of disorder is the random placement of impurities, which affects primarily the transfer matrix element V[11]. Fertis et al[8] have studied systematically the question of metal-insulator transition due to disorder in V. This disorder, which stems from the random placement of the impurities, leads to a critical value of n given again by Eq. (1.4) with $b \simeq 0.16$. On the basis of this result they concluded that both electron repulsion and disorder share responsibility in inducing a metal-insulator transition in doped semiconductors.

There is a third aspect of the problem, which although not capable by itself to trigger the transition, may play a significant quantitative role in various physical quantities. This aspect is the interaction and the final merging of the impurity band with the main band. Ghazali and Serre[12] found that the impurity band has been merged well with the main band when $n^{\frac{1}{3}}\alpha_B \simeq 0.25$ to 0.3. Thus one cannot obtain a satisfactory quantitative picture (at least for n significantly larger that n_c) without taking into account the proximity and the eventual merging of the impurity band with the main band, which is already modified by the merging of the excited impurity states.

There exist extensive experimental studies of various physical quantities near the critical concentration (see e.g. Alexander and Holcomb[13] and Fritzsche[14]). More recently, as a result of theoretical progress in understanding the critical region (see, e.g. ref. 15) and experimental advances in approaching closer to the critical point at very low temperatures[16-19], there is a revitalized interest in the question of metal insulator transition in heavily doped semiconductors.

In the present work we examined the question of metal-insulator transition in doped semiconductors by incorporating on an equal footing both disorder and electron-electron repulsion. However, in our treatment we have ommited the presence of the main band. The effect of this ommision on the value of b seems to be minor. We think that the reason for this fortunate situation is an approximate mutual cancellation of two opposing effects associated with the mixing of our states with the tail of the main band. One effect, which tends to promote metallic behavior, is the opening up of additional channels of propagation due to states in the main band; the other effect, which tends to promote insulating behavior, is the increased disorder due to energy mismatch between impurity levels and host atomic levels.

Other quantities of interest, such as the conductivity and the density of states at the Fermi level, are influenced rather strongly by the presence of the main band.

II. FORMALISM

To describe the impurity band we employ a random, tight binding Hubbard model, with one orthogonalized s-like orbital per impurity atom. The corresponding Hamiltonian has the form

$$H = \sum_{i\sigma} \bar{\varepsilon}_{i\sigma} n_{i\sigma} + \sum_{ij\sigma} V_{ij} c^{+}_{i\sigma} c_{j\sigma} + U \sum_{i} n_{i\sigma} n_{i-\sigma} , \qquad (2.1)$$

where the sites $\{i\}$ refer to the impurity atoms; σ is taken +1 for an electron with spin up and -1 for an electron with spin down; $\bar{\varepsilon}_{i\sigma}$ is the energy level of an isolated impurity; each $\bar{\varepsilon}_{i\sigma}$ is taken as zero in the present work (no diagonal disorder); $c^{+}_{i\sigma}$, $c_{i\sigma}$ are the creation and annihilation operators for a σ-spin electron located at the site i; $n_{i\sigma} = c^{+}_{i\sigma} c_{i\sigma}$ is the number operator for the electrons at the site i; U is the intratomic Coulomb repulsion, given by Eq. (1.2). Each of the transfer matrix elements V_{ij} depends on the relative distance R_{ij} between the atom i and the atom j according to Eq. (1.3) with f(x) given by[8]

$$f(x) = \frac{5}{48} x^2 + \frac{23}{48} x + \frac{11}{16} , \qquad (2.2)$$

which is practically indistinguishable from the form employed by Edwards and Sienko[7]

$$f(x) = \frac{1}{\sqrt{v}} \left(\frac{1}{6} x^2 + \frac{3}{2} x + \frac{3}{2} \right) \qquad (2.3)$$

for v=6. We denote by Z the number of nearest neighbors of each site i connected through a non-zero V_{ij} (j=1,...,Z). Assuming a random uncorrelated placement of the impurities one determines[8] by employing either Eq. (2.2) or Eq. (2.3) the probability distribution of each V_{ij}.

The last term of the Hamiltonian (2.1) which is known as the Hubbard term and characterizes the on-site repulsion between electrons can be approximated[6] by a random one body term;

$$\sum_i n_{i\sigma} n_{i-\sigma} \simeq \sum_i \varepsilon_{i\sigma} n_{i\sigma} , \tag{2.4}$$

where $\{\varepsilon_{i\sigma}\}$ are random variables obeying a binary alloy probability distribution[6]

$$p(\varepsilon_{i\sigma}) = \frac{1}{2} \delta(\varepsilon_{i\sigma} - \varepsilon_A) + \frac{1}{2} \delta(\varepsilon_{i\sigma} - \varepsilon_B), \tag{2.5}$$

with

$$\varepsilon_A = \frac{1}{2} U(n_0 + \mu_i) \tag{2.6}$$

$$\varepsilon_B = \frac{1}{2} U(n_0 - \mu_i) \tag{2.7}$$

The quantity n_0 is the average number of electrons per site and in our present case of no compensation (half-filled band) equals to 1 (n_0=1); the quantity μ_i characterizes the size of the local magnetic moment[6] and is assumed to be a function of U/B_i, where B_i is the local half band width

$$\mu_i = \mu(\frac{U}{B_i}) \tag{2.8}$$

The functional dependence of μ_i vs U/B_i is assumed to be the same as the μ_0 vs U/B function displayed in Fig.2a of ref. 6.

Having approximated the Hubbard many body term of Eq. (2.1) by Eq. (2.4), the motion of a σ-spin electron in the impurity system is described by the Hamiltonian

$$H_\sigma = \sum_i \varepsilon_{i\sigma} n_{i\sigma} + \sum_{ij} V_{ij} c^+_{i\sigma} c_{j\sigma} , \tag{2.9}$$

which is a random Hamiltonian incorporating both diagonal and off-diagonal disorder (i.e. both $\varepsilon_{i\sigma}$ and V_{ij} are random variables). The approximate replacement of the Hubbard many body term by a random one body term seems to be reasonable as far as the determination of the critical point is concerned. However, this replacement may produce serious errors in the behavior of the conductivity near the critical point, because the critical exponent for a disorder induced transition is larger or equal to one, while electron-electron interactions seem to produce[16,20] a critical exponent close to 0.5.

To obtain various quantities of interest, such as the density of states, the mean free path, etc, we need to calculate $\langle G \rangle$, i.e. the average of the Green's function $G(E+is) = (E+is-H_\sigma)^{-1}$. To obtain $\langle G \rangle$ we employ a Homomorphic double site Coherent Potential Approximation (HCPA)[21] with a homomorphic partition as that shown in Fig. 3b of ref. 21. This partition in the case of pure diagonal disorder reduces to the ordinary single site CPA. To proceed with the HCPA we need to express the local half bandwidth in terms of a single V_{ij} and we need the Green's function G^0_{ij} for the periodic case where all the non-zero V_{ij} are the same and all ε_i are zero. For reasons presented elsewhere[8] we have choosen the Bethe-lattice form[15] for G^0_{ij}. This form is characterized by Z, and has a half bandwidth given by $2\sqrt{Z-1}\ V$; for B_i we take

$$B_i = \alpha \langle V_{ij} \rangle + \beta\ V_{ij} , \qquad (2.10)$$

where the constants α and β are determined by the obvious requirements $\langle B_i \rangle = 2\sqrt{Z-1} \langle V_{ij} \rangle$ and $\sigma_B^2 = 2\sqrt{Z-1}\ \sigma_V^2$; σ_B and σ_V are the standard deviations of B_i and V_{ij} respectively.

Having obtained $\langle G \rangle$ we can calculate various quantities of interest as follows:

The density of states per site ρ(E) is given by

$$\rho(E) = - \frac{1}{\pi} \text{Im} \langle G_{ii}(E+is) \rangle , \qquad (2.11)$$

the CPA conductivity σ_0 is[15]

$$\sigma_0 = \frac{2e^2\hbar}{3\pi\Omega} \sum_k \upsilon^2 |\text{Im} \langle G(E_F + is,k) \rangle|^2 , \qquad (2.12)$$

where Ω is the volume and υ is the velocity at the Fermi level E_F. Following the usual practice we assume that

$$\upsilon^2 = \upsilon_m^2 \left(1- \frac{(E_F - E_0)^2}{B^2}\right) , \qquad (2.13)$$

where E_0 is the center of the band, $B = 2\sqrt{Z-1}\ V$, V is given by Eq. (1.3) with $R = n^{-\frac{1}{3}}$, and υ_m^2 is

$$\upsilon_m^2 = 12\ V^2\ R^2/\hbar^2 , \qquad (2.14)$$

The area of the Fermi surface S_F can be expressed in terms of $\rho_F = \rho(E_F)$ and υ :

$$S_F = (2\pi)^3 \hbar\, \upsilon\, \rho_F\, n\ , \tag{2.15}$$

Combining Eqs. (2.12,.15) with the equation[15]

$$\sigma_0 = \frac{2}{(2\pi)^3 3} \frac{e^2}{\hbar} S_F\, \ell\ , \tag{2.16}$$

we obtain the mean free path ℓ. To find the corrected conductivity σ we employ the formula

$$\sigma = \sigma_0 \frac{\varphi(\varphi-1)}{\varphi(\varphi-1) + 6}\ , \tag{2.17}$$

based on the potential well analogy[22,23]. The quantity φ is given by

$$\varphi = S_F\, \ell^2/8.96\ , \tag{2.18}$$

The coefficient of the specific heat per mole γ is given by

$$\gamma = \frac{2\pi^2}{3} k_B^2\, N(E_F) \tag{2.19}$$

where the density of states per mole, $N(E_F)$, is

$$N(E_F) = \rho_F\, N_{Av}\, n/n_{Si}\ ; \tag{2.20}$$

N_{Av} is the Avogardo number and n_{Si} is the number of Si atoms per unit volume.

III. RESULTS AND DISCUSSION

In obtaining the results to be presented below we have usually taken Z=6. We also performed calculations for Z=10 or Z=15 and found that our results remain essentially unchanged. For the function f(x) we have used Eq. (2.2) which yields practically identical results with Eq. (2.3) for v=6. Thus the present model has no adjustable parameters.

In Fig. 1 we display the successive changes in the density of states as the parameter $n^{\frac{1}{3}} a_B$ is varied across the critical value. Shaded areas denote localized states yielding zero conductivity at T=0. The localization of the states has been decided by the L(E)-method as in our previous work[8]. The critical concentration n_c is given by $n_c^{\frac{1}{3}} a_B \simeq 0.235$, in good agreement with the experimental value of 0.26 ± 0.05. The explicit theoretical value for Si : P is $3.70 \times 10^{18} cm^{-3}$ in excellent agreement with the value $3.74 \times 10^{18} cm^{-3}$ obtained in recent experiments[19].

Eq. (2.17) allows an alternative determination of the critical concentration which occurs when $\varphi=1$, i.e. when

$$S_F \ell^2 = 8.96 \ , \tag{3.1}$$

Eq. (3.1) yields a critical value satisfying the relation

$$n_c^{\frac{1}{3}} \ \alpha_B = 0.243 \ , \tag{3.2}$$

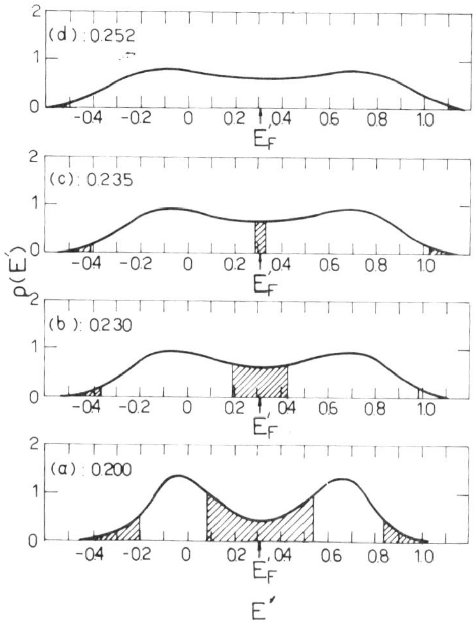

Fig. 1. Density of states ρ vs energy E' ($\equiv E\varepsilon\alpha_B/e^2$) for various values of the parameter $n^{\frac{1}{3}}\ \alpha_B$. Shaded areas denote localized states.

which is in reasonable agreement with the L(E)-method result of 0.235 and again well within the experimental range. For Si : P our explicit result based on Eq. (3.2) is $n_c = 4x10^{18}$ cm^{-3} slightly higher than the recent experimental value. In Fig. 2 we compare our theoretical results with experimental data taken from ref. 17. Although the general features and the

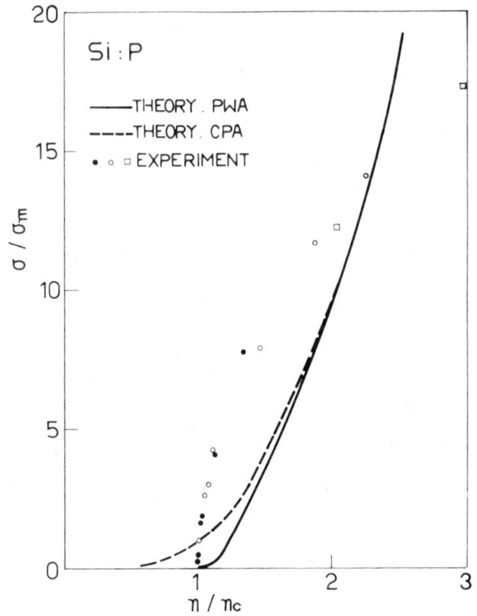

Fig. 2. Conductivity σ vs concentration n; σ_m is Mott's minimum metallic conductivity and n_c is the critical concentration. Solid line is based upon Eq. (2.17) while the dashed line upon Eq. (2.12). Experimental points are from ref. 17.

order of magnitude of the conductivity tend to be reasonable, there are significant discrepancies between theory and experiment both for high n as well as for $n \simeq n_c$. The high n discrepancy can be easily understood as due to our omission of the merging with the main band. For high n the Fermi energy electrons are in the main band and are scattered by the ionized impurities. Rosenbaum et al[19] have shown that this simple picture yields results in good agreement with the experimental data for $n/n_c > 3$. In our model this scattering mechanism has been omitted and thus we overestimate σ for $n/n_c > 2$. The discrepancy near n_c is due to the fact that the critical exponent ν defined by

$$\sigma = \sigma_c \left[(n/n_c)-1\right]^{\nu} , \qquad (3.3)$$

is close to 0.5 for the experimental data, while it is 1 in our approach. The slope of σ as calculated from Eq. (2.17) is approximately 1/6 the slope of σ_0 at n_c; more explicitly we have

$$\frac{\sigma}{\sigma_m} \xrightarrow[n \to n_c^+]{} \left(\frac{n}{n_c} - 1\right) \qquad (3.4)$$

where σ_m, Mott's minimum metallic conductivity, equals to 20 $(\Omega cm)^{-1}$ for Si : P. On the other hand, the experimental data fit Eq. (3.3) well for σ_c=260 $(\Omega cm)^{-1}$ and $\nu \simeq 0.55$. The discrepancy between Eq. (3.4) and experiment is usually attributed to electron-electron interaction effects, which are believed to reduce drastically the critical exponent. It should be pointed out that in the present approach electron-electron repulsion was approximated by a random one body term (Eq. (2.4)); thus our scheme is necessarily predicting the same critical exponent as the localization theory.

In Fig. 3 we plot our preliminary results for the coefficient of the linear term of the specific heat vs concentration. Again the general features and the order of magnitude of our theoretical results are reasonable, although appreciable discrepancies with the experimental data appear both for small and large n. All these discrepancies can be easily attributed to our ommision of the main band, the effect of which is to make the increase of γ with n less pronounced. For large n the Fermi level is well inside the main band and hence γ varies as $n^{\frac{1}{3}}$. This simple model fits the experimental data well[17] down to $n/n_c \simeq 1.2$. For small n ($n/n_c < 1$) there are extra Fermi level states (omitted in our model) coming from the tail of the main band; these states could possibly account for the excess γ observed experimentally in the low n regime.

IV. CONCLUSIONS

Three factors seem to play an important role in the phenomenon of metal-insulator transition in heavily doped semiconductors:

(i) electron-electron repulsion.

(ii) disorder in the off-diagonal (and possibly the diagonal) matrix elements.

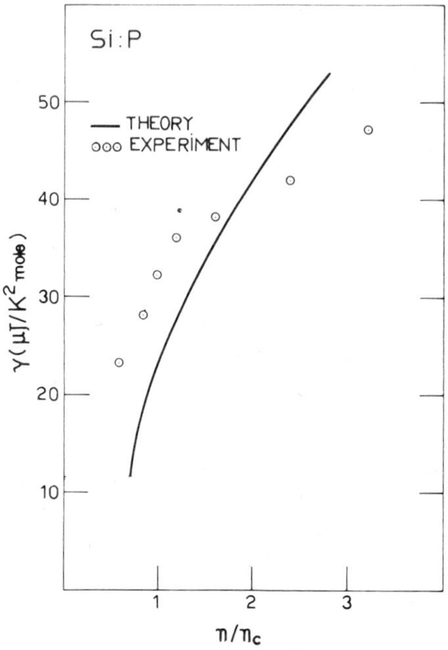

Fig. 3. Coefficient of the linear term of the specific heat γ vs concentration n. Experimental data from ref. 17.

(iii) merging of the impurity band with the main band.

For uncompensated samples electron-electron repulsion and disorder play a dominant and almost equally important role in determining the value of the critical concentration. The role of the main band becomes progressively more important as n increases above n_c; for the conductivity the properties of the main band dominate the experimental features for $n/n_c > 3$, while the region just above n_c seems to be dominated by poorly understood electron-electron interaction effects; the density of states at the Fermi level seems to be dominated by the main band down to almost the critical concentration and even below the critical concentration the tail of the main band seems to make a non negligible contribution.

For compensated samples disorder (both diagonal and off-diagonal) seems to dominate the phenomenon of metal-insulator transition. This is so because compensation on the one hand increases the amount of diagonal disorder and, on the other hand, it shifts the Fermi level away from the center of the unperturbed impurity band and away from the main band; this shifting reduces the role of electron-electron repulsion (which affects more strongly the center of the band) and delays the entrance of the Fermi level into the main band. Experimental data[23] as well as our unpublished preliminary results provide clear evidence in support of a disorder dominated transition in strongly compensated samples.

ACKNOWLEDGEMENTS

Part of this work was supported by the Foundation "Alexandros Onassis", while one of the authors (ACF) was at the N.R.C. "Demokritos", and by the National Hellenic Research Foundation.

REFERENCES

1. N.F. Mott, Proc. Camb. Phil. Soc. 32: 281 (1949).
2. N.F. Mott, Proc. Phys. Soc. A62: 416 (1956).
3. N.F. Mott, "Metal-Insulator Transitions", Taylor and Francis, London (1974).
4. N.F. Mott and E.A. Davis, "Electronic Processes in Non-Crystalline Materials", Oxford University Press, Oxford (1979).
5. J. Hubbard, Proc. Roy. Soc., A281: 401 (1964).
6. R.R. Demarco, E.N. Economou, and C.T. White, Phys. Rev. B18: 3968 (1978).
7. P.P. Edwards and M.J. Sienko, Phys. Rev. B17: 2575 (1978).
8. A.C. Fertis, A.N. Andriotis, and E.N. Economou, Phys. Rev. B24: 5806 (1981).
9. K.F. Bergren, Phil. Mag. 27: 1027 (1973).
10. P.W. Anderson, Phys. Rev. 109: 1492 (1958).
11. N.F. Mott, Phys. Today 31: 42 (1978).
12. A. Ghazali and J. Serre, Phys. Rev. Lett. 48: 886 (1982).
13. M.H. Alexander and D.F. Holcomb, Rev. Mod. Phys. 40: 815 (1968).
14. H. Fritzsche, in "The Metal-Insulator Transition in Disordered Systems", L.R. Friedman and D.P. Tunstall, eds, SUSSP, Edinburgh (1978).
15. E.N. Economou, "Green's Functions in Quantum Physics", Springer-Verlag, Heidelberg (1983).
16. T.F. Rosenbaum, K. Andres, G.A. Thomas, and R.N. Bhatt, Phys. Rev. Lett. 45: 1723 (1980).
17. G.A. Thomas, Y. Ootuka, S. Kobayashi, and W. Sasaki, Phys. Rev. B24: 4886 (1981).
18. T.F. Rosenbaum, R.F. Milligan, G.A. Thomas, P.A. Lee, T.V. Ramakrishan, R.N. Bhatt, K. De Conde, H. Hess, and T. Perry, Phys. Rev. Lett. 47: 1758 (1981).
19. T.F. Rosenbaum, R.F. Milligan, M.A. Paalanen, G.A. Thomas, R.N. Bhatt, and W. Lin, Phys. Rev. B27: 7509 (1983).
20. W.L. McMillan, Phys. Rev. B24: 2739 (1981).
21. F. Yonezawa and T. Odagaki, J. Phys. Soc. Jap. 47: 388 (1979).
22. E.N. Economou, C.M. Soukoulis, and A.D. Zdetsis, Phys. Rev. B30: 1686 (1984); E.N. Economou, to appear in Phys. Rev.
23. G.A. Thomas, Y. Ootuka, S. Katsumoto, S. Kobayashi, and W. Sasaki, Phys. Rev. B25: 4288 (1982).

FLUX QUANTIZATION IN RINGS, CYLINDERS AND ARRAYS

John P. Carini, Dana A. Browne and Sidney R. Nagel

The James Franck Institute and The Department of Physics
The University of Chicago, Chicago, Illinois 60637

In 1959 Aharonov and Bohm[1] proposed a two-slit diffraction experiment which implied that it was the vector potential, **A**, not the magnetic field, **B**, which was of fundamental importance in quantum mechanics. Their experiment showed that the interference pattern caused by an electron diffracting from the two slits depended upon the amount of magnetic flux enclosed by the two possible paths but not upon the strength of the magnetic field in the region of the electron. Since the line integral of **A** around any loop is equal to φ, the magnetic flux through it, the potential must be non-zero in the region occupied by the electron. Even though the magnetic field in the vicinity of the electron is zero, the vector potential is not. Thus the use of the vector potential would not require that there was "action at a distance" whereas the use of the magnetic field would[2]. In addition the interference pattern will be periodic in the flux with a periodicity φ_0 = hc/e. This quantum of flux induces a 2π phase shift of the electron wavefunction around the loop.

In the case of a superconductor, the current is carried by Cooper pairs which have charge (2e). In this case the natural quantum of flux is $\varphi_0/2$ = hc/2e. Only integral multiples of this flux, $n\varphi_0/2$, can be enclosed in a superconducting ring. In normal, that is non-superconducting, metals one might naively expect that the only periodicity of relevance would be φ_0, the value found in the original Aharonov-Bohm effect. It came as quite a surprise when Al'tshuler, Aronov, and Spivak[3] predicted that in the conductivity of a *normal* metal ring there would be oscillations which have a periodicity of $\varphi_0/2$, the superconducting flux quantum. Another novel aspect of their prediction was that these conductivity oscillations should become larger as the disorder in the metal increased. This is quite unusual since one ordinarily expects that interference effects should be washed out by disorder rather than enhanced by it. As the disorder in a metal increases one expects that the coherence length of the wavefunction should decrease and localization effects should become important. The predictions of half flux quantum periodicity were obtained by a consideration of the maximally crossed (or Langer-Neal[4]) diagrams which give the main quantum corrections to the conductivity caused by localization effects. These diagrams give rise to a *negative* contribution in the magnetoresistance which is a hallmark of localization phenomena.

There has been experimental verification of this prediction. Sharvin and Sharvin[5] first showed that there were oscillations in the magnetoresistance of hollow magnesium cylinders which had a period of $\varphi_0/2$. Subsequently they showed that the effect was present in lithium cylinders as well[6]. In both cases the samples were made by coating a thin quartz filament with the metal. The resistance at 4.2 K was measured in the direction of the magnetic field, i.e., along the length of the cylinder. The most pronounced difference between the two materials was in the sign of the resistance oscillations. The resistance of the lithium film decreased as the magnetic field increased away from zero while that of the magnesium film increased. The Sharvin and Sharvin experiments have been confirmed by several different groups for the same materials. Ladan and Maurer[7] repeated the experiment on lithium cylinders and Gijs, Van Haesendonck and Bruynseraede[8] investigated the temperature dependence of the oscillations in magnesium cylinders.

More recently a similar type of oscillation has been observed in more complicated, multiply connected, geometries. In these experiments an *array* of wires was deposited in either a honeycomb or a square network. The magnetic field was oriented perpendicular to the plane of the array and the resistance was measured across it from one side to the other. Conductivity oscillations were observed with a period of $\varphi_0/2$ where, now, the flux is measured through each individual plaquette. Pannetier, Chaussy, Rammal and Gandit[9] observed the effect in honeycomb arrays of magnesium, copper and gold. Bishop and Dolan[10] observed the effect in square arrays. One additional experiment is of interest which did not show this effect. Umbach, Washburn, Laibowitz and Webb[11] studied the conductivity of a single ring which was constructed from a very thin, one-dimensional "wire". Not only were there no clear oscillations with periodicity $\varphi_0/2$ but neither were there any with periodicity φ_0.

The maximally crossed diagrams have been given[12] the name "Cooperons" because of their strong resemblance in zero magnetic field to the Cooper-pair propagator in superconductivity. Several attempts have been made to give a physical picture for the new periodicity which is based on these diagrams. In their first experimental paper on this subject, Sharvin and Sharvin[5] gave an interpretation in terms of a superposition of the two possible trajectories, going in opposite directions, that an electron could take before returning to its original position. Using similar arguments, Bergmann[13] has given a clarification of the meaning of the diagrams in several situations. Basically, an electron can have quantum mechanical interference with itself if it travels around the ring in both a clockwise and a counter-clockwise manner. The scattering from impurities will be identical along the two paths but will occur in the opposite order. In the presence of a magnetic field the electron in each path will have a phase change given by the integral of the vector potential around the cylinder. Constructive interference will again occur when the phase shift between the two paths is 2π so that each path separately has a phase shift of $+\pi$ and $-\pi$. This corresponds to a flux of $\varphi_0/2 = hc/2e$ through the center of the cylinder. This approach gives a very satisfactory picture of the physics of the conductivity oscillations in a cylinder in the single electron approximation. It is difficult to extend this argument to the case where the geometry of the conductor is multiply connected such as in the two-dimensional arrays. It is also not clear how to incorporate electron-electron interactions which must certainly be important in these localization problems. Finally it was not shown in the Bergmann paper why the conductivity oscillations must be strictly periodic with flux $\varphi_0/2$ rather than simply having a large Fourier component at that value.

If there are magnetoresistance oscillations with the new periodicity, $\varphi_0/2$, can one see oscillations in other quantities, such as the eigenvalues or eigenfunctions obtained by solving the Schrodinger equation on a ring? Büttiker, Imry and Landauer[14] investigated this possibility by mapping the problem of a ring with a magnetic flux through its center onto the problem of an electron in a one dimensional periodic lattice. They showed that the vector potential is replaced by the wavevector **k** in the band structure problem. From their analysis they observed that there would *not* be a periodicity with a half flux quantum, $\varphi_0/2$, but only one with φ_0. This periodicity is simply that of the original Aharonov and Bohm[1] paper and is due to the fact that the electron wavefunctions must be single valued around the ring. This same result, a φ_0 periodicity, was found in several studies of the transmission coefficient through a one dimensional ring[15,16]. In those studies, Gefen, Imry and Azbel found that only under very special circumstances could there be a large harmonic at a half flux quantum which would dominate the oscillations. They again concluded that this effect was different from that observed in the experiments mentioned above which found a periodicity of $\varphi_0/2$ in the magnetoresistance oscillations. Büttiker[17] has studied the connection between the two cases mentioned above: that of a single ring penetrated by a magnetic flux[14] and that of a ring connected by current leads[15,16] to reservoirs.

If the new periodicity with flux hc/2e is indeed a single particle effect and is not due to the presence of electron interactions, one would suppose that it could be understood in terms of the solutions to the single particle Schrodinger equation. Along these lines we have developed a picture of the Aharonov-Bohm effect in a disordered normal metal by examining a tight-binding model with nearest neighbor hopping. This approach has several advantages over the methods mentioned above because it can be extended to a variety of situations which are difficult to understand by other approaches. For example, the k-space perturbation theory used by Al'tshuler, Aronov and Spivak[3] which requires a continuum model, and the interpretation of Bergmann[13] via multiple paths do not easily imply half flux periodicity in the case of a regular array of wires or other multiply connected topologies. We are also able to include the effect of electron-electron interactions in a simple manner.

Our work rests on the examination of the symmetries of the single particle Hamiltonian. The tight-binding Hamiltonian we employ is given by $H_{i,j} = V_i \delta_{i,j} + T_{i,j}$ where r_i is the position of the ith site. V_i represents the on-site potential and the nearest neighbor hopping terms in a magnetic field are given by $T_{i,j} = t_{i,j} \exp(2\pi i A_{i,j} e/hc)$, where $t_{i,j} = t_{j,i}$ is the coefficient in zero field and $A_{i,j}$ is the line integral of the vector potential along the link from site i to site j. This model has the correct continuum limit and has an explicit gauge invariance: if $\psi(r_i) \to \psi(r_i) \exp[i\,\chi(r_i)]$ and $A_{i,j} \to A_{i,j} + (2\pi e/hc)[\chi(r_j) - \chi(r_i)]$, then the Hamiltonian has exactly the same form as before. In the following discussion for a single ring of sites with an enclosed flux we will pick the gauge so that $A_{i,j}$ is zero except between site 1 and site N. In this case, $A_{1,N}$ is equal to the enclosed flux φ. The matrix for the tight binding Hamiltonian is:

$$H(\varphi) = \begin{pmatrix} V_1 & -t_{12} & 0 & \cdots & -t_{1,N}\exp(i2\pi\varphi/\varphi_0) \\ -t_{12} & V_2 & -t_{23} & \cdots & 0 \\ 0 & -t_{23} & V_3 & \cdots & 0 \\ \vdots & \vdots & \vdots & & \vdots \\ -t_{1,N}\exp(-i2\pi\varphi/\varphi_0) & 0 & 0 & & V_N \end{pmatrix}$$

In the above form, the Hamiltonian shows an explicit periodicity with period φ_0=hc/e since the phase factor returns to its original value every φ_0. At a flux of hc/2e the terms $T_{1,N}$ and $T_{N,1}$ are the negative of their values at zero flux. This corresponds to anti-periodic boundary conditions around the ring. At multiples of hc/2e the Hamiltonian is real and thus time-reversal invariant. The eigenvalues for a clean system of three sites are shown in Figure 1a and those for a four site system are shown in Figure 1b. The level crossings at multiples of hc/2e are due both to the time reversal invariance and translational symmetry and represent the possibility of forming two different standing waves at these points. The principle effect of disorder (that is, random values of V_i or $t_{i,j}$) on the eigenvalue spectrum is to remove the degeneracies

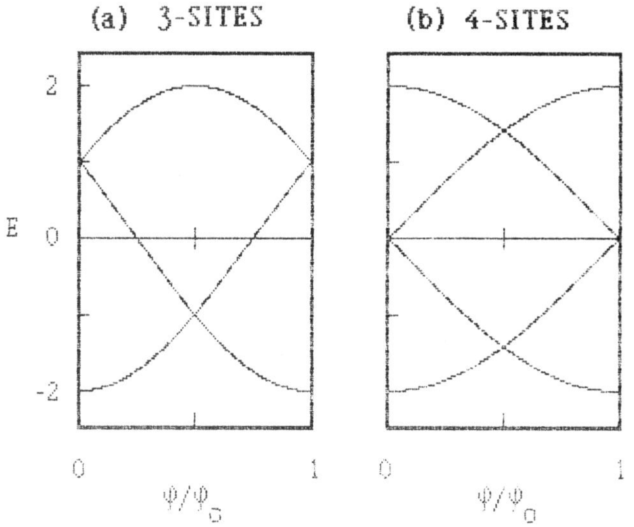

Figure 1 The energy level spectrum as a function of flux of a) an ordered 3-site ring and b) an ordered 4-site ring.

mentioned above by forming standing waves out of the plane waves of the clean system. (The reality of the Hamiltonian at these points implies that the eigenfunctions can always be chosen to be real. Once the degenracies are broken the states can no longer be plane waves.)

The tight-binding model has an additional symmetry that we may exploit[18]. This is a particle-hole symmetry where we change the sign of the wavefunction on alternate sites. This corresponds to a transformation which, because it is unitary, will not change the eigenvalue spectrum of the Hamiltonian. If we were to consider an ordered linear chain of sites rather than a ring, this transformation takes $t_{i,j}$ into $-t_{i,j}$. (The matrix for this situation is the same as given above except that the elements in the top right and lower left corners of the matrix are zero.) By inspection one can see that this will change each eigenvalue into its negative so that the eigenenergies are symmetrically distributed about zero. The same symmetry can also be seen in any ordered ring of an even number of sites. We see this in Figure 1b where for a four site ring the eigenenergies at the top of the band at any value of flux can be mapped into those at the bottom of the band *at the same value of the flux*. For an odd number of sites it is clear that changing the sign of the wavefunction on alternate sites cannot be done without leaving one pair of sites unchanged. Increasing the flux by $\varphi_0/2$ will introduce a phase shift of π which is the same as changing the sign of the wavefunction between one pair of sites. Thus for an odd number of sites on an ordered ring, changing the sign of the wavefunction on alternate sites plus *increasing the flux by* $\varphi_0/2$ maps the spectrum of eigenvalues into itself. This "glide plane" symmetry for the odd-numbered ring can be seen in the eigenvalues of the three site system of Figure 1a.

These symmetries will be broken by diagonal disorder (i.e., random values of V_i) but not by off-diagonal disorder (random values of $t_{i,j}$). We will show how to restore these symmetries by taking an appropriate average over an ensemble of disordered rings. We will restrict ourselves to the case where the distribution of $\{V_i\}$ is symmetric about zero. For a given member of the ensemble of possible random potentials specified by a set $\{V_i\}$, we also consider a second member of the ensemble with Hamiltonian $H'(\varphi)$ with a set of potentials given by $\{V'_i = -V_i\}$. The particle-hole unitary transformation mentioned above is of the form:

$$U^+ = U = \begin{pmatrix} -1 & 0 & 0 & . & . & . & 0 \\ 0 & 1 & 0 & . & . & . & 0 \\ 0 & 0 & -1 & . & . & . & 0 \\ . & . & . & & & & . \\ . & . & . & & & & . \\ 0 & 0 & 0 & . & . & . & (-1)^N \end{pmatrix}$$

For a ring of N sites we find that $U^+ H(\varphi) U = -H'(\varphi + N\varphi_0/2)$. This equation is true even in the presence of off-diagonal disorder. Therefore the symmetry transformation maps $H(\varphi)$ into $-H'(\varphi + N\varphi_0/2)$. It is clear that any quantity that is invariant under U will have the appropriate glide-plane or particle hole symmetry when the two systems $H(\varphi)$ and $H'(\varphi)$ are averaged together. In the calculation of transport coefficients like the conductivity or dielectric constant, we encounter a combination of four wave functions of the form $\psi_\alpha(r_i)\psi_\beta(r_{i+n})\psi_\gamma(r_j)\psi_\delta(r_{j+n})$ which pick up a factor $(-1)^{2n} = 1$ under the transformation U. Thus this quantity is invariant under U and will show the symmetry of the clean system when averaged over the impurity ensemble.

In the limit of a large disordered system, it is permissible to average over the ensemble of impurities. Provided that we look over a narrow band of energies in the large system limit, we do not expect the physical properties to depend upon whether the ring has an even or an odd number of sites. A large odd-numbered ring will have exact glide plane and nearly exact particle-hole symmetries. Likewise a large even numbered ring will have exact particle- hole and approximate glide plane symmetries. The approximate symmetries will become exact as the system size is increased without limit. The contribution of the states in a narrow energy band centered at energy E to the averaged transport coefficients for a flux φ can be mapped into a band centered at (-E) for the same flux by the particle-hole symmetry and to a band centered at (-E) with a flux changed by $\varphi_0/2$ by the glide plane symmetry. Therefore the contribution to the transport coefficient is periodic with period hc/2e, not hc/e as for the clean system. Although the above discussion concerned a single ring, the case of a thin-walled cylinder with a field along the axis of the cylinder can be considered as a stack of rings, so the argument is equally valid in this case as well.

So far, we have only considered a non-interacting system of electrons. The leading corrections due to electron-electron interactions can be included via a random phase approximation to the dielectric constant, ϵ. In the RPA, ϵ is evaluated in the non-interacting limit. We have just shown that in this limit that the electron density and therefore ϵ is periodic with period hc/2e. Therefore the screened electron-electron interactions will also have this periodicity and our conclusions about the periodicity of the conductivity will remain valid even in the presence of interactions.

The above argument can be extended to the case of a two-dimensional array of wires with a magnetic field perpendicular to the plane[19]. For the clean system the above arguments about the "glide plane" or particle-hole symmetry hold when we simply replace the flux through the single ring in the above argument with the flux per elementary plaquette in the array. The number of atoms, N, around each plaquette determines the particular symmetry present. If N is even we have particle-hole symmetry and if it is odd we have "glide plane" symmetry. For the disordered system, the same mapping between the two ensembles can be made. Suppose we consider a square array of wires and place one tight binding site at each vertex so that the symmetry is particle-hole. If we now decorate the lattice by placing 5 sites around each plaquette as in Figure 2, the number N is odd and we have "glide plane" symmetry. We may continue to decorate the lattice; each time we add one site to each plaquette we change the type of symmetry between particle-hole and "glide plane". Continuing in this fashion we fill in each link with a large number of sites so as to approximate the many atoms that constitute the wire. It should not be important whether we stop the process when the number of sites around each plaquette is even or odd so that both particle-hole and "glide plane" symmetries are present and the periodicity with hc/2e is proved as before. This argument works equally well for a triangular or honeycomb lattice, and of course in the presence of electron-electron interactions.

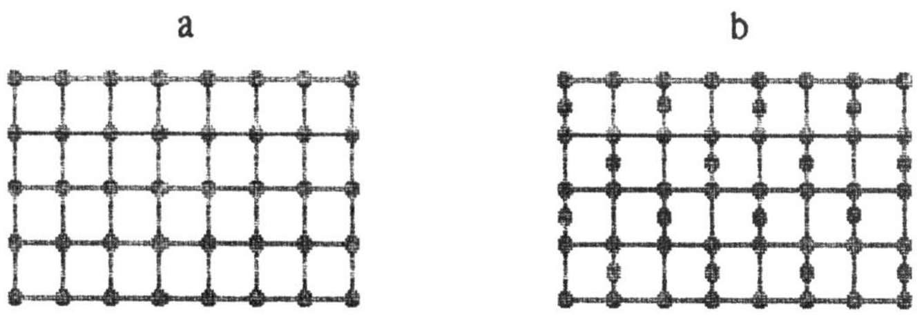

Figure 2 a) The square lattice with a site located at every vertex.
b) The same lattice decorated with five sites per plaquette.

We have now proved that the new periodicity, hc/2e, is quite general in dirty metals and can be seen in a variety of geometries. However we would like to understand what property of the electron wavefunctions is responsible for the variation of the conductivity. An important clue to what is happening can be seen in the case of very small disorder. As mentioned above, the disorder removes the extra degeneracies that occurred in the clean system at multiples of $\varphi_0/2$. These degenerate states may be described either as standing waves or as plane waves at these points. In the presence of disorder, standing waves are the correct wavefunctions since non-degenerate eigenfunctions of a real matrix may always be chosen to be real. At these points the shift in the eigenvalues must be treated by degenerate perturbation theory and is linear in the amount of disorder. At other values of flux the wavefunctions are complex and the effect of disorder is diminished. From non-degenerate perturbation theory we know that the relative shifts of the energies is quadratic in the amount of disorder. Thus we see that the disorder has a larger effect when the wavefunctions are real than when they are complex.

In order to probe the localization of the wavefunctions we[20] have studied the participation ratio, P, which is sensitive to how localized or extended an eigenfunction is and yet is still one of the most simple properties to calculate: $P = [N^{-1}\Sigma|\psi(r_i)|^2]^2/[N^{-1}\Sigma|\psi(r_i)|^4]$. One can easily verify that P varies between P=1 for a totally extended state and $P=N^{-1}$ for a very well localized one. In Figure 3 we show how P varies as a function of flux for a one dimensional ring with a moderate amount of disorder. In the case of very weak disorder, P is 2/3 for the standing wave states near $\varphi = n\varphi_0/2$ and P is 1 for the plane wave states at the other values of flux. The drop in the participation ratio at multiples of hc/2e is due to the fact that the real wavefunctions (standing waves) must have nodes in order to be orthogonal to each other whereas for intermediate values of the flux, the wavefunctions are complex and nodes are unnecessary for orthogonality. Hence we see that the reality of the Hamiltonian every $\varphi_0/2$ is the driving force behind the periodicity with half the normal flux quantum. The degree of localization is very sensitive to the presence or absence of time reversal invariance: the standing waves occurring at multiples of hc/2e are much easier to localize than are the plane waves found at other flux values. This is the familiar result that the effect of a magnetic field on a disordered metal is to destroy the localization and produce the negative magnetoresistance unique to the localization problem.

The simplest experimental system that shows the half flux periodicity is the resistance measured along the length of a hollow cylinder[5-8]. As already mentioned, our proof that $\varphi_0/2$ is the fundamental periodicity of the transport coefficients in the ring is valid for this case as well. Do our arguments about the nature of the wavefunctions in the ring also apply to the more complicated geometry?

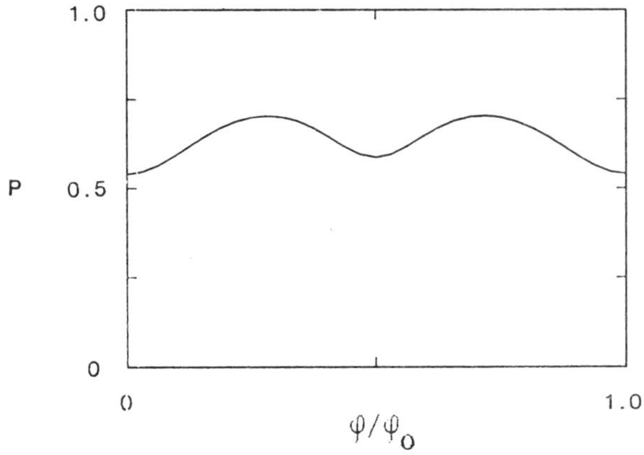

Figure 3 The participation ratio for a 32-site ring with an intermediate amount of disorder ($\Delta V = 1$).

The wavefunctions of the ordered cylinder are plane waves in the direction transverse to the axis of the cylinder while they are standing waves in the direction parallel to the axis (if non-periodic boundary conditions are used). Any degeneracies due to motion along the length of the cylinder depend critically on the boundary conditions and can occur at any value of the flux as shown in Figure 4. Disorder breaks every degeneracy and the participation ratio has structure at the place where each one is broken. In spite of this, the structure at $n\varphi_0/2$ continues to dominate the behavior of the transport properties. First, almost all the states have the degeneracies at $n\varphi_0/2$ due to time reversal invariance, while only a few are degenerate at any other value of the flux. These accidental degeneracies do not contribute coherently to the behavior of the cylinder as do those due to time reversal invariance since the latter always occur at the same values of φ. A calculation of the conductivity contains an average over a band of energies of width k_BT which consists of many states most of which are more localized at $n\varphi_0/2$ than elsewhere. Second, at the accidental degeneracies, the wavefunctions have a plane wave component while those at $n\varphi_0/2$ are pure standing waves. Only the structure at half flux quanta survives as the disorder is increased. This can be seen in Figure 5 where we see that at moderate disorders the participation ratios of even individual states resemble those of one-dimensional rings.

We have seen that transport around a cylinder behaves in a similar manner to that around a ring. However we are also interested in the experimental situation where the resistivity is measured along the axis of the cylinder. In order to study this we have calculated several quantities which are sensitive to the extent of the wavefunction in that direction. The first of these

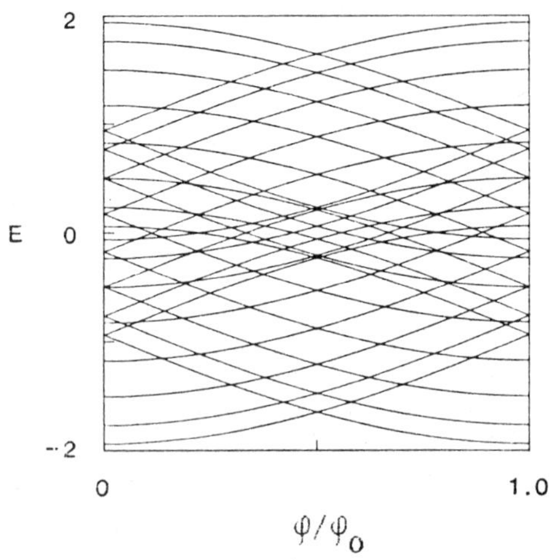

Figure 4 The energy spectrum as a function of flux for an ordered cylinder. The cylinder consists of 8 rings 4 sites each.

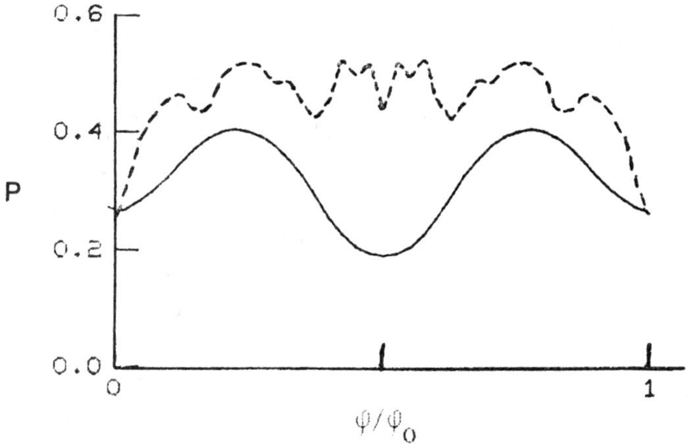

Figure 5 The participation ratio for a single state in a cylinder for two disorders: dashed line is $\Delta V = 2$ and solid line is for $\Delta V=6$.

is the "longitudinal partial participation ratio", P_L, which sums the modulus squared of the wavefunction around each individual ring, squares the result and then sums over all the rings in the cylinder:

$$P_L = \left(N_R^{-1} \sum_{j=1}^{N_R} |\psi_j|^2 \right)^2 / \left(N_R^{-1} \sum_{j=1}^{N_R} |\psi_j|^4 \right) \quad \text{and} \quad |\psi_j|^2 = \sum_{i=1}^{N_C} |\psi_{j,i}|^2 .$$

Here N_R is the number of rings in the cylinder, N_C is the number of sites per ring and $\psi_{j,i}$ is the wavefunction on the ith site in the jth ring. P_L is N_R^{-1} for a wavefunction localized on a single ring and is equal to 1 for completely extended wavefunction. A partial participation ratio, P_R, can also be defined to measure the localization around the cylinder by interchanging N_R and N_C in the above equation.

In a clean system, or one with little disorder, the wavefunctions are standing waves in the direction parallel to the axis so that $P_L \approx 2/3$; there is structure in $P_L(\varphi)$ only at accidental degeneracies and none at $\varphi = n\varphi_0/2$. The flux dependence of P_R looks like that of P itself. Structure at multiple values of $\varphi_0/2$ appear in P_L only when the disorder is increased and the longitudinal and transverse motions of the electrons (i.e., those parallel and perpendicular to the axis of the cylinder) are coupled by scattering. This is seen in Figure 6.

The oscillations in the longitudinal localization length can also be seen in very long cylinders using quasi one-dimensional recursive matrix methods developed by MacKinnon[21]. The measured quantity is the transmission coefficient, T, from one end of the cylinder to the other for an electron of energy, E. From this we can obtain an effective localization length, λ.

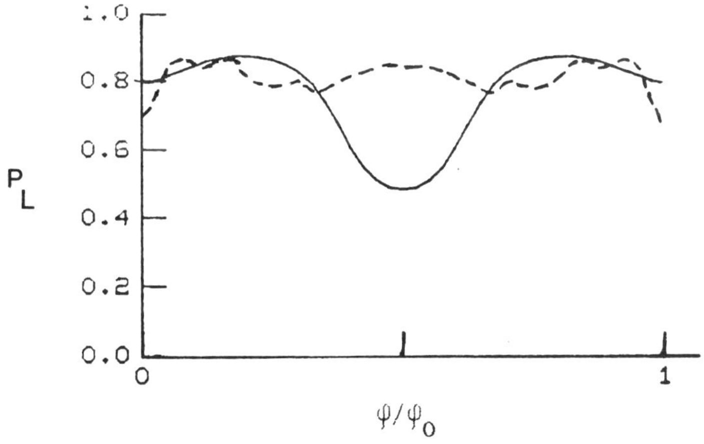

Figure 6 The partial participation ratio, P_L, for one state in a cylinder for two different disorders: the dashed line has $\Delta V = 2$ and the solid line has $\Delta V = 6$.

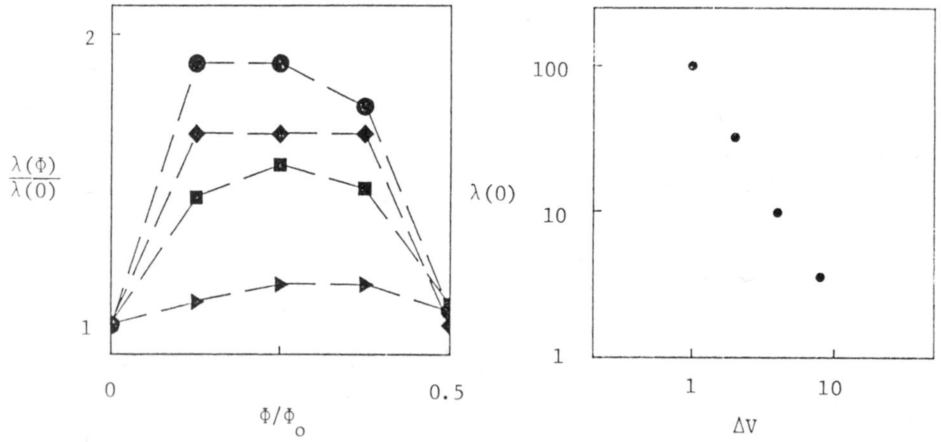

Figure 7 a) The localization length as a function of flux for a long cylinder at various amounts of disorder. $\Delta V = 1$ (circles), 2 (diamonds), 4 (squares), and 8 (triangles). λ is normalized to its value at $\varphi = 0$.
b) $\lambda(\varphi = 0)$ as a function of disorder, ΔV.

$$\lambda^{-1} = \lim_{N_R \to \infty} \{(-2N_R)^{-1} \ln T(E,N)\} = \lim_{N_R \to \infty} \{(-2N_R)^{-1} \ln \mathrm{Tr} \,|\langle 1 | \, G(E) \, | N_R \rangle|^2\},$$

where G(E) is the Green's function from the first to the N_Rth ring. If the length of the cylinder is much longer than λ the oscillations at $\varphi = n\varphi_0/2$ become apparent even without averaging over a band of energies. These oscillations grow stronger as the disorder is decreased as seen in Figure 7.

The two-dimensional cylinder is certainly a richer system than our starting point of the ring, but the dominant effect remains the existence of degenerate time-reversed states and the breaking of those degeneracies by the disorder and by the flux.

In conclusion we have demonstrated that the new periodicity in the flux quantization is quite general. It appears in all the transport properties for all levels of disorder and is unaffected by electron-electron interactions. The half-flux quantum periodicity is visible in many different geometries as well, including wire arrays which was a situation beyond the scope of the original calculation[3]. Our proof of this periodicity relies on various symmetries of the Hamiltonian. In the large system limit, when an ensemble average is taken over the disorder, the transport properties of our tight-binding model have this $\varphi_0/2$ periodicity. The question still remains whether the conductivity of an individual system will be periodic since no average over an ensemble is taken in that case. In other words, is a typical result for a single system similar to the average of many different ones? In the case of cylinders and arrays one expects that the answer is yes and that the system self-averages. For one dimensional rings the answer is not as clear. MacKinnon[22] has shown for one-dimensional wires that the conductivity of any member of the ensemble does approach the average value, albeit slowly. The experimental results of Umbach[11] which show no strong $\varphi_0/2$ periodicity in the conductivity of single rings does suggest that averaging may be important. Our studies so far have been on rings that were too small to examine the large N limit. However, for small rings we have found that as N increases the deviations about the average value do decrease.

ADKNOWLEDGMENTS

We wish to thank K.A. Muttalib for his collaboration at the beginning of this work. We are also grateful to L.P. Kadanoff, A. Libchaber and Y. Nagel for many stimulating discussions and suggestions. J.P.C. and S.R.N. were supported by NSF grant DMR 82-01182 and DAB was supported by NSF-MRL grant DMR 82-16892 and by an Exxon Postdoctoral Fellowship.

REFERENCES

1) Y. Aharonov and D. Bohm, Phys. Rev. 115, 485 (1959).
2) R.P. Feynmann, **The Feynmann Lectures on Physics,** Vol. 2 (Addison-Wesley, New York, 1965) Chapter 15.
3) B.L. Al'tshuler, A.G. Aronov and B.Z. Spivak, Pis'ma Zh. Eksp. Teor. Fiz. 33, 101 (1981) [JETP Lett. 33, 94 (1981)].
4) J.S. Langer and T. Neal, Phys. Rev. Lett. 16, 984 (1966).
5) D. Yu. Sharvin and Yu. V. Sharvin, Pis'ma Zh. Eksp. Teor. Fiz. 34, 285 (1981) [JETP Lett. 34, 272 (1981)].
6) B.L. Al'tshuler, A.G. Aronov, B.Z. Spivak, D. Yu. Sharvin and Yu. V. Sharvin, Pis'ma Zh. Eksp. Teor. Fiz. 35, 476 (1982) [JETP Lett. 35, 588 (1982)].
7) F.R. Ladan and J. Maurer, C. R. Acad. Sci. 297, 227 (1983).
8) M. Gijs, C. Van Haesendonck, and Y. Bruynseraede, Phys. Rev. Lett. 52, 2069 (1984).
9) B. Pannetier, J. Chaussy, R. Rammal and P. Gandit, Phys. Rev. Lett. 53, 718 (1984).
10) D.J. Bishop and G. Dolan (unpublished).
11) C.P. Umbach, S. Washburn, R.B. Laibowitz and R.A. Webb, Phys. Rev. B 30, 4048 (1984)..
12) P.W. Anderson, Physica 109&110B, 1830 (1982).
13) G. Bergmann, Phys. Rev. B 28, 2914 (1983).
14) M. Büttiker, Y. Imry and R. Landauer, Phys. Lett. 96A, 365 (1983).
15) Y. Gefen, Y. Imry and M. Ya. Azbel, Phys. Rev. Lett. 52, 129 (1984).
16) Y. Gefen, Y. Imry and M. Ya. Azbel, Surface Science 142, 203 (1984).
17) M. Büttiker, in **Localization, Interaction and Transport in Impure**

Metals, edited by L. Schweitzer and B. Kramer, (PTB, Braunschweig, 1984) p. 107.
18) D.A. Browne, J.P. Carini, K.A. Muttalib and S.R. Nagel, Phys. Rev. B 30, 6798 (1984).
19) D.A. Browne, J.P. Carini and S.R. Nagel (in press).
20) J.P. Carini, K.A. Muttalib and S.R. Nagel, Phys. Rev. Lett. 53, 102 (1984).
21) A. MacKinnon and B. Kramer, Phys. Rev. Lett. 47, 1546 (1981).
22) A. MacKinnon, J. Phys. C 13, L1031 (1980).

LOCALIZATION AND HEAVY FERMIONS

M. Cyrot

Laboratoire Louis Néel
CNRS, 166X
38042 Grenoble Cedex, France

ABSTRACT

Starting from an idea due to Mott, we propose an explanation for heavy fermions. They would stem from the almost localized f electrons. Consequences are drawn and it is shown that Wilson ratio smaller than one can be obtained. We suggest that superconductivity can be found in this low Wilson ratio heavy fermions compounds.

Localization due to correlation is an idea that Mott[1] tried to keep alive at a time where band theory seemed to answer any questions of solid state physicists. The idea that an half filled band could be insulating was difficult to handle. The mathematical description of the Mott transition was first put forwards by Hubbard[2]. The alloy analogy was improuved[3] and permitted to handle simultaneously the magnetism of the different phases. A very important step in the understanding of the transition is due to Brinkman and Rice[4] using Gutzwiller's[5] variational approach. Thy showed that before localization, the effective mass of the carriers was increasing.

This concept of nearly localized Fermi liquid was first applied to electrons in solids. Recently Brinkman and Anderson[6] realised that such a concept could also be relevant to liquid Helium three. A new type of description of this liquid is now appearing and Vollhardt[7] has recently remforced this point and showed how Gutzwiller's variational approach can be casted into a Fermi liquid theory. Such a new approach permits to understand the behavior of the specific heat and susceptibility as a function of pressure.

To my knowledge, Mott[9] was the first to suggest that the mass enhancement of a nearly localized Fermi liquid could be linked with the properties of a Kondo lattice. He interpreted the Kondo frequency as describing the jump of f electrons into the Fermi sea. In that case the number of f electrons is not exactly an integral number. If the f states form a band of highly correlated electrons, this number of holes determines the effective mass of the current carriers. He proposed[9] this interpretation for $CeAl_3$ and $FeZn_{20}$.

In the following we want to develop the idea that heavy fermions correspond to nearly localized f electrons. We consider a lattice of f states and a Fermi sea and want to describe the transition from the virtual bound state with a non integral number of f electrons to the state giving the Kondo lattice where the number of electrons is just one. In between the f electrons are nearly localized and we want to argue that such a situation can explain some behaviors of the heavy fermions compounds.

We consider an Anderson lattice with U infinite i.e. we study only the lower band as a function of position into the Fermi sea. We consider the case where the hybridization tail of the f-state resonance is still finite at the Fermi level but small. The Luttinger[8] sum rule must apply. This sum rule says that the Fermi surfaces volume is equal to the number of conduction electrons per unit cell. In the case of mixed valence compounds, the number of electrons must include the f electrons. In the opposit limit, the Kondo lattice, the number of electrons excludes the f electrons. We are in a similar situation as in the Hubbard model with one electron per atom . For a critical value of U where the metal insulator transition happens, the Fermi surface disapperars. Brinkman and Rice have shown that the effective mass of the carriers increases. Mott argue that if $N\xi$ is the number of current carriers, in order that Luttinger's theorem is not violated, the effective mass of these carriers must be

$$\frac{m^*}{m} = \frac{1}{\xi}$$

In our case, the volume of the Fermi surface must change discontinuously. If ξ represents now the probability of a f site to be non occupied, the mass enhancement will be given by ξ^{-1}.

This could be the origin of heavy fermions. Another explanation is that heavy fermions exist in the Kondo lattice because of the resonance of width T_K at the Fermi level ; the mass enhancement being $1/T_K$. Although both explanation lead finally to the same order of magnitude as Mott links T_K and ξ, the consequences are somewhat different and the physics too. Nozieres[10] argues that the ground state of a Kondo lattice can only be magnetic, the non magnetic state with a Kondo resonance being impossible because one needs one electron per spin to obtain a non magnetic case. Thus within T_K of the Fermi surface, it is not possible to obtain enough electrons. Although Mott didn't consider this possibility in his book, it is not contradictory with his view. In order to be non magnetic we must have T_K greater than the RKKY interaction. Linking T_K to the fraction of the rare earth ions which have a valency different from the majority, the Kondo metal in a sense have intermediate valence.

Thus if we consider the lowering of a f level compared to the Fermi energy, when the f level is closed to E_F we have the intermediate valence case. As the f level decrease, the tail of the virtual level band arrive near E_F i.e. in the case $|E_F-\varepsilon_f| \sim \Delta$, we arrive in the region of localisations of the f electrons and obtain heavy fermions. If $E_f-\varepsilon_f$ still increases, we obtain the Kondo lattice.

As magnetism is concerned, as argued by Nozières, the Kondo lattice is probably magnetic. In the region of heavy fermions, we must have a transition to non magnetism. In fact this region $(E_F-\varepsilon_f) \sim \Delta$ is the region where T_K is of the order of the RKKY interaction. This is probably the reason why heavy fermions compounds can be magnetic or not.

We now want to describe the physical consequences of this model. As in the Brinkman and Rice Fermi liquid, both specific heat and susceptibility are enhanced. The local spin fluctuation are enhanced because opposite spin electron cannot occupy the same site. The enhancement is due to the mass. Thus contrary to the nearly ferromagnetic model, it is the same for susceptibility and specific heat. Castaing[11] describes Helium three as a nearly localized liquid where each spin feels a local field which have a distribution f(ε) of width T_K. Such an approach can be used for the contribution of the f electrons. One can also introduce an antiferromagnetic molecular field θ which must tend to the Néel temperature of the Kondo lattice. In that case, f electron can be described by a Landau theory with Landau parameter F_o^a such that

$$1 + F_o^a = \frac{1}{4}(1 + \frac{\theta}{T_K})$$

If $\theta << T_K$ we have $F_o^a = -\frac{3}{4}$ as in liquid helium.

However θ can be large and we can consider the opposite case $\frac{\theta}{T_K} >> 1$. The Wilson ratio which is the inverse, is now smaller than one. This could be an explanation of why heavy fermions compounds can have a Wilson ratio smaller than one. This situation can occur when antiferromagnetism disappears.

One puzzling problem when we compared the properties of $UBe_{13}CeCuSi_2$ and Upt_3 is the different behavior of the resistivity in the last compounds. If a single model must explain the occurrence of heavy fermions, one must explain this difference. At T=0 the resistivity is zero for a perfect crystal and for T≠0 it should rise. The maximum in resistivity can be explain by the fact that of $T > T_K$ only a function $\frac{T_K}{T}$ of the electrons within a range kT of the Fermi energy can be scattered into the region of high density of states. Thus if the scattering is due to impurities or any other mechanism that does not increase with T we expect a decrease in ρ. However if the scattering is due to phonon then it is proportional to T and we expect a resistivity practically independent of T. This is probably the case of UPt_3. If impurities are added one can expect that a maximum will be obtained as in $CeCuSi_2$. It seems to be the case[12].

Within this model, some problems remain unsolved. Anderson[13] emphasized that if Kondo lattice can be described by a Fermi liquid theory, triplet superconductivity would occur. In an experimental point of view, no definitive conclusions can be drawn. Each point in favor of triplet superconductivity is not conclusive. Moreover some points are against. The effect of impurities as not so drastic as triplet superconductivity would suggest. In Upt_3, the pressure effect[14] on the spin fluctuation temperature and the superconductivity temperature is different although simple explanation would predict the same behaviour for triplet superconductivity. If this type of superconductivity is ruled out, another problem remains. Does the attractive interaction is through phonons in this type of materials or can we think to an attractive electron-electron interaction which could lead to singlet superconductivity. There is probably an argument against an extrapolation of results on a simple Fermi liquid. In our case there exist heavy fermions and normal fermions contrary to the Helium case. An antiferromagnetic coupling between heavy fermions is perhaps possible through the normal ones.

If we consider Castaing's result in the limit $\theta >> T_K$ the Landau parameter F_o^a is large. If we can apply to our heavy fermions an hamiltonian similar to the Hubbard one, a large positive value of F_o^s gives a large negative value of F_o^s. Thus we can reach the instability

$$1 + F_o^s < 0$$

which can be the superconducting instability. Thus superconductivity would be found in heavy fermions compounds with very low Wilson ratio. This seems to be the case[15].

REFERENCES

1. N.F. MOTT, Proc. Phys. Soc. A 62, 416 (1949)

2. J. HUBBARD, Proc. Roy. Soc. A 227 (1964) ; ibid 281, 401 (1964)

3. M. CYROT, Physica, 91 B+C, 141 (1977)

4. W.F. BRINKMAN and T.M. RICE, Phys. Rev. B2, 1324 (1970)

5. M.C. GUTZWILLER, Phys. Rev. Lett. 10, 159 (1963) ; Phys. Rev. A 134, 923 (1964)

6. P.W. ANDERSON and W.F. BRINKMAN, In the Helium liquids, J.G.M. Armitage and I.E. Farqhar eds ; Academic Press New York (1975)

7. VOLLHARDT, Rev. Mod. Phys. 56, 99 (1984)

8. J.M. LUTTINGER, Phys. Rev. 119, 1153 (1960)

9. N.F. MOTT, Metal insulator transitions. Taylor and Francis LTD (1974)

10. P. NOZIERES, to be published.

11. B. CASTAING, J. Phys. Lettres 41, L333 (1980)

12. J.J.M. FRANSE, Private communication.

13. P.W. ANDERSON, Phys. Rev. B 30, 1549 (1984)

14. J.O. WILLIS, J.D. THOMPSON, Z. FISK, A. de VISSER; J.J.M. FRANSE and A. MENOVSKY, to be published.

15. G.R. STEWARDT, to be published.

ANDERSON LOCALIZATION

B. Kramer

Physikalisch-Technische Bundesanstalt
Bundesallee 100
3300 Braunschweig, Federal Republic of Germany

A. MacKinnon

The Blackett Laboratory
Imperial College
Prince Consort Road
London

INTRODUCTION

The electronic properties of disordered systems have been studied quite extensively during the last few decades. There is a variety of reasons for this fact. First of all, there is a fundamental interest in disordered systems because "real" solids contain a certain amount of disorder, due to impurities, defects etc.. Thus, the understanding of their properties requires some knowledge of the quantum mechanical properties of a particle moving in a random potential. In contrast to the usual "textbook" examples of quantum mechanical systems, like the "hydrogen atom" or the "Bloch electron", for instance, where symmetry plays an outstanding role in the solution of the problem, disordered systems provide a particular challenge for the theory because of the complete lack of symmetry due to the random character of the Hamiltonian. In particular, it has turned out that the interplay of quantum mechanics and statistics which is one of the characteristic features of these systems, leads to complications which are so severe that even the simplest problems are forbiddingly difficult to solve. In spite of many efforts, for instance, it is up to now impossible to calculate the density of states, except for special models (Kramer and Weaire, 1979).

Most important, in connection with disordered solids, is the transport problem. Amorphous semiconductors are becoming more and more common in modern device technology. If they are properly prepared, one can control their electronic properties by doping, as in the case of crystalline semiconductors. Therefore, a sound knowledge of the electronic processes which determine the transport is of outstanding importance. One of the "standard books" about disordered solids is devoted to this subject (Mott and Davis, 1971).

The prerequisite for the understanding of the electronic processes is the knowledge of the properties of the single electron states. Their shape, whether, and under what conditions, they may be localized, has been the subject of numerous papers during the last twenty five years (Kramer et al., 1985). The "classical" paper in which this problem was formulated was that of Anderson (1958). Although the qualitative aspects of the problem seemed to be understood it took about a quarter of a century until significant progress was made with respect to the understanding of the "critical behaviour" of the electrical properties near the "Anderson transition", i. e. the critical disorder at which all quantum states become localized, and the system is insulating. This was achieved by reformulating the problem in terms of the renormalization group method which was originally designed for the treatment of phase transitions (Wegner, 1976). One of the most remarkable result of these considerations was that for dimensionality lower than three all one electron eigenstates of a random system are localized (Abrahams et al., 1979). An Anderson transition, which is the disorder induced metal-insulator transition can only exist in three and (possibly) in higher dimensions.

Numerical methods, although having been applied and continuously improved during the course of the years (Kramer et al., 1981), were also unsuccessful until the development of a "finite size scaling method" for this problem (MacKinnon and Kramer, 1981). Using this method some results of the renormalization group theory could be verified, others, such as the critical behaviour of the system near the Anderson transition, not (MacKinnon and Kramer, 1983).

In this paper, we shall review some of the results of this work, and of additional calculations (Bulka et al., 1985) which were done in order to clarify the "phase diagram" for localization in three dimensions. Although the latter are not yet complete, and do not yet fit too well into the scaling picture, we think that they are an instructive contribution to elucidate the interplay between the mechanisms which lead to, and destroy, localization.

MECHANISMS IMPORTANT FOR LOCALIZATION

Let us first consider a classical particle moving in a random potential as, for instance, the one shown in Fig. 1. In this case it is comparatively simple to decide, whether a particle is "localized" in a finite region of space or whether it can move throughout the whole space.

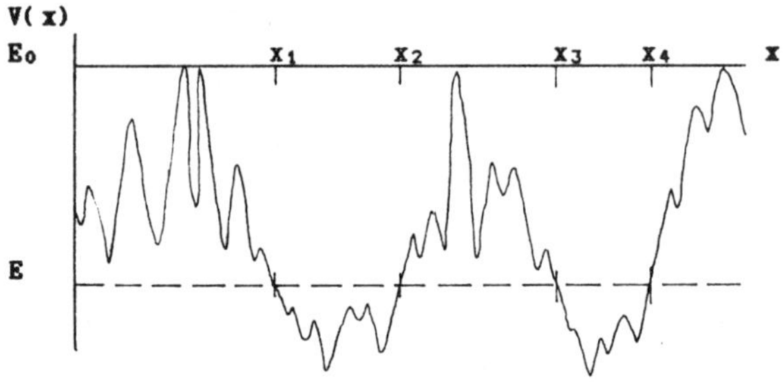

Fig. 1: Classical particle moving in a random potential V(x).

Because of the energy conservation law the allowed intervals on the x-axis are given by the condition that the total energy E is larger than the potential energy, $E > V(x)$. For $E < E_0$ (cf. Fig. 1) these are the intervals $[x_1, x_2]$, $[x_3, x_4]$, etc.. If the particle is within one of these intervals, it will remain there for an infinitely long time. We can say that it is in a "localized state". On the other hand, if $E > E_0$, the particle can move along the whole x-axis. It is in a "delocalized state". E_0 is the energy which separates localized from extended states.

What happens if we consider instead a quantum mechanical particle?

First of all, for $E < E_0$, the particle is no longer restricted to one of the intervals, due to quantum mechanical tunneling. The quantum states may extend along the whole x-axis, though they are certainly statistically varying due to the random nature of the potential (Fig. 2a).

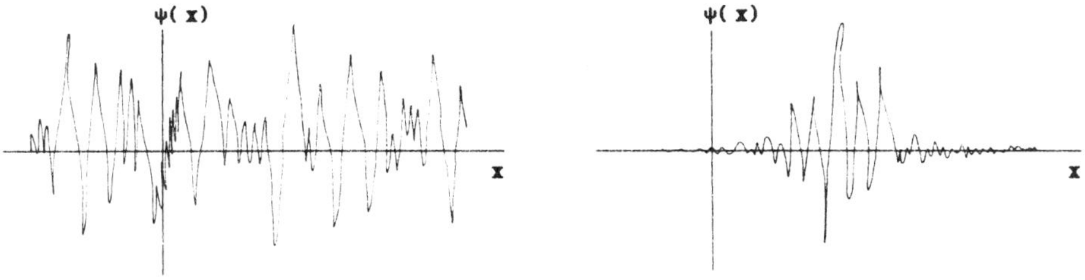

(a) (b)

Fig. 2: Random extended state (a), localized state (b)

On the other hand, for $E > E_0$, one can imagine that the "probability wave" which corresponds to the particle can be constructed from waves which are repeatedly scattered by the spatial fluctuations of the potential. The resulting state may have the form shown in Fig. 2b. For large x the random oscillations die away exponentially, due to destructive quantum interference.

Thus, quantum mechanical tunneling may lead to delocalization of a localized particle, whereas quantum interference may result in localization of a delocalized state. There are examples for both mechanisms. Bloch states in an ideal crystal are delocalized. The quantum states in a one-dimensional random system are all localized. In contrast to the case of a classical particle, where the distinction between localized and delocalized states is trivially possible, the theoretical treatment of quantum mechanical localization is extremely complicated. This is due to the competition between tunneling and quantum interference. Qualitatively, one expects that the states in the tails of a band are more localized than those near the band centre.

For any quantitative theory one needs a definition of localization in terms of a mathematical expression. In analogy to potential-localization one may assume that the localized states in a random potential are also exponentially decaying at infinity.

$$\psi(x) \rightarrow f(x)\, e^{-|x|/\xi} \qquad (1)$$

$f(x)$ is a statistically varying prefactor. As long as ξ is finite the state ψ is localized. ξ is the localization length depending on the strength of the random potential, the energy, and on the dimensionality of the system. When ξ tends to infinity the state is extended. The energy E_c at which ξ diverges for fixed disorder is called the "mobility edge".

Besides the fact that it contains some arbitrariness due to the assumption about the form of the wave function, this definition is difficult to deal with, since one had to calculate the eigenstates of a Hamiltonian in order to be able to distinguish between localized and extended states.

As we are interested in the transport properties of a macroscopic system it would make sense to have a definition in terms of an ensemble average of a quantity, which directly relates to the conductivity. The microscopic calculation of the conductivity is extremely complicated. Therefore, we use a different approach. We consider the probability for a particle to make a transition from the site x to x' which is described by

$$W([\mathbf{H}], E; x, x') = |G(E; x, x')|^2 \tag{2}$$

where $G(E; x, x')$ is the one-particle propagator $\langle x|(E-\mathbf{H})^{-1}|x'\rangle$. We may further define the average transmission probability by carrying out the ensemble average

$$W(E; x, x') = \int D[\mathbf{H}]\ P[\mathbf{H}]\ W[\mathbf{H}] \tag{3}$$

Here, $P[\mathbf{H}]$ is the probability distribution of $\mathbf{H}$ in the ensemble.

The localization length may now be defined as

$$2/\xi = -\lim_{|x-x'|\to\infty} (\ln W(E; x, x'))/|x-x'| \tag{4}$$

This is consistent with eq. (1) except that the localization length is now given by an ensemble average. Since the quantity on the right hand side of eq. (4) is self-averaging, one even may omit the ensemble average in the thermodynamic limit. This makes the definition eq. (4) very convenient for numerical studies in addition to its direct relation to the transport properties.

THE RECURSIVE METHOD

Throughout our calculations we use the following Hamiltonian which is defined on a lattice represented by an orthonormal set of site states $|j\rangle$, $\Sigma_j |j\rangle\langle j| = \mathbf{1}$.

$$\mathbf{H} = \sum_{j,k} |j\rangle\langle k|\ (V + \varepsilon_j \delta_{jk}) \tag{5}$$

V is the transition matrix element between the two sites j and k. It corresponds to the kinetic energy. ε_j is the site energy and represents the potential energy. We take only the diagonal part of $\mathbf{H}$ as random, and distributed according to

$$P(\varepsilon_1 \ldots . \varepsilon_N) = \prod_j^N p(\varepsilon_j) \tag{6}$$

where N is the number of the lattice sites, and $p(\varepsilon)$ the normalized box function of width W. The disorder parameter is then $w = W/V$. For simplicity, we consider a square lattice.

The method of calculation of the localization length can best be demonstrated for a 1-D system. First, the Hamiltonian for N lattice sites, $H^{(N)}$, is decomposed according to

$$H^{(N)} = H^{(N-1)} + |N\rangle\langle N|\,\varepsilon_N + V\,(|N-1\rangle\langle N| + |N\rangle\langle N-1|)$$

$$= H_0 + H' \qquad (7)$$

H_0 contains only N-1 sites plus the Nth site, but the latter is not coupled to the former. $H' = V\,(|N-1\rangle\langle N| + |N\rangle\langle N-1|)$ is the coupling operator. Then one can solve the resolvent equation

$$G^{(N)} = G_0 + G_0\,H'\,G^{(N)} \qquad (8)$$

The matrix element of the resolvent between the first and the Nth site can be calculated recursively from the equations

$$G^{(N)}_{1N} = G^{(N-1)}_{1N-1}\,g_N \qquad (9)$$

$$g_N = (E - \varepsilon_N - g_{N-1})^{-1} \qquad (10)$$

The localization length is

$$2/\xi = -\lim_{N\to\infty} \sum_{i=1}^{N} \log |g_i|^2/N \qquad (11)$$

This enables us to calculate the localization length to arbitrary accuracy, in principle. No computer storage is needed. The statistical error in ξ may also be calculated recursively (MacKinnon 1980, MacKinnon and Kramer 1983).

A remarkable feature of these equations is that they may be generalized to systems with finite cross section. In this case eq.s (9) and (10) become matrix equations the dimension of which are given by the number of sites within the cross section of the system, M^{d-1} (d dimensionality). The localization length is then M-dependent, such that

$$\xi_M = \xi_M(E, W, d) \qquad (12)$$

and the localization length for d>1 is given by the limit of ξ_M when $M\to\infty$.

Furthermore, the general idea may be used to generate recursive equations also for other electronic properties, such as the density of states, and the conductivity (MacKinnon 1985).

SOME RESULTS FOR THE LOCALIZATION LENGTH

The procedure described above is only the first step in the evaluation of the data. In order to explain the second step let us consider the set of raw data shown in Fig. 3. The data were obtained for the Anderson model eq. (5) (square lattice) for E = 0 i.e. in the center of the band, for a variety of disorder parameters. In order to be able to perform the limit of large M we have to investigate the behaviour of the data in this region. Inspection of Fig. 3 shows that it might be possible to scale the data points onto one and the same curve(s) using the scale transformation

$$1/M \to \xi(0, W, d)/M \qquad (13)$$

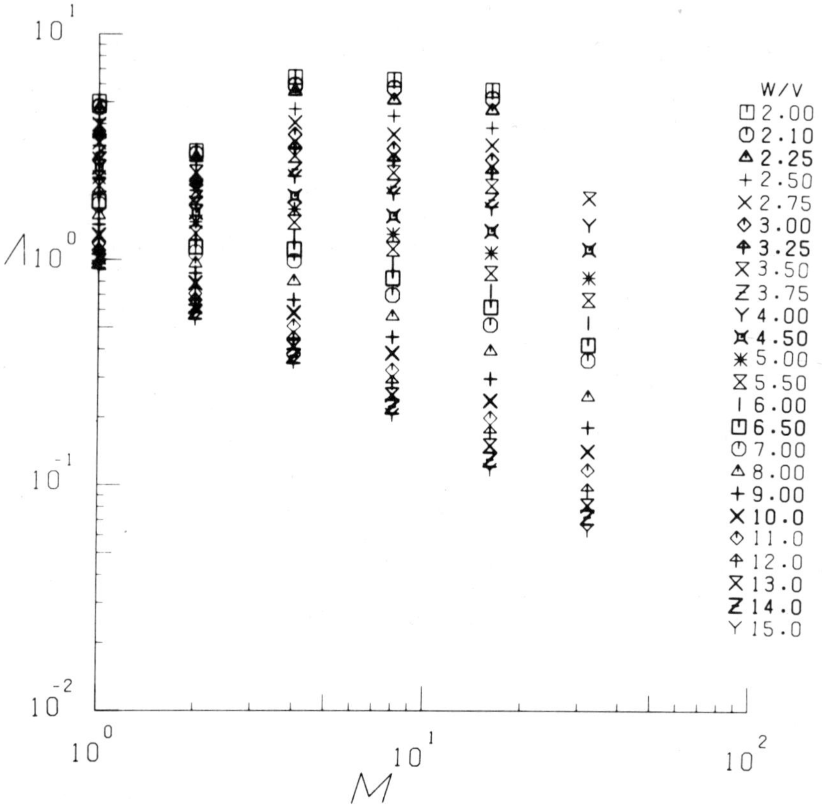

a

Fig. 3: Double logarithmic plot of the renormalized exponential decay length $\Lambda = \xi_M/M$ of the transmission coefficient of strips of width M (a), and of bars of cross section M^2 (b). The values of the disorder parameter $w = W/V$ are indicated, E is taken at the band center of the tight binding band.

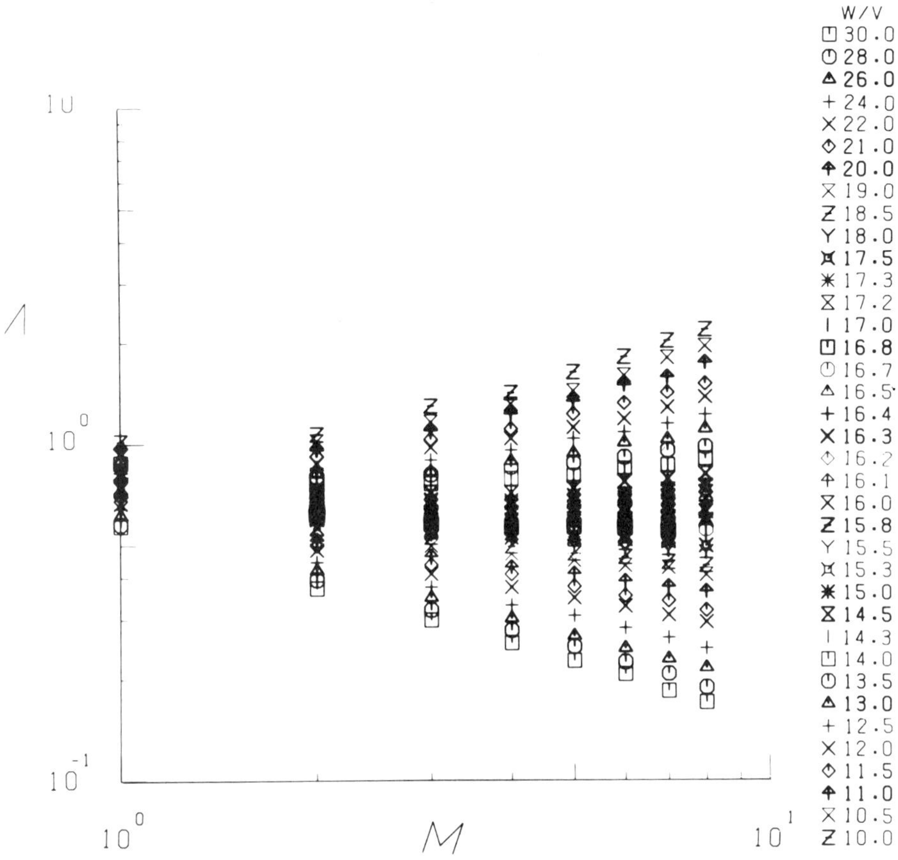

Λ
M
W/V
30.0
28.0
26.0
24.0
22.0
21.0
20.0
19.0
18.5
18.0
17.5
17.3
17.2
17.0
16.8
16.7
16.5
16.4
16.3
16.2
16.1
16.0
15.8
15.5
15.3
15.0
14.5
14.3
14.0
13.5
13.0
12.5
12.0
11.5
11.0
10.5
10.0

b

Fig. 3

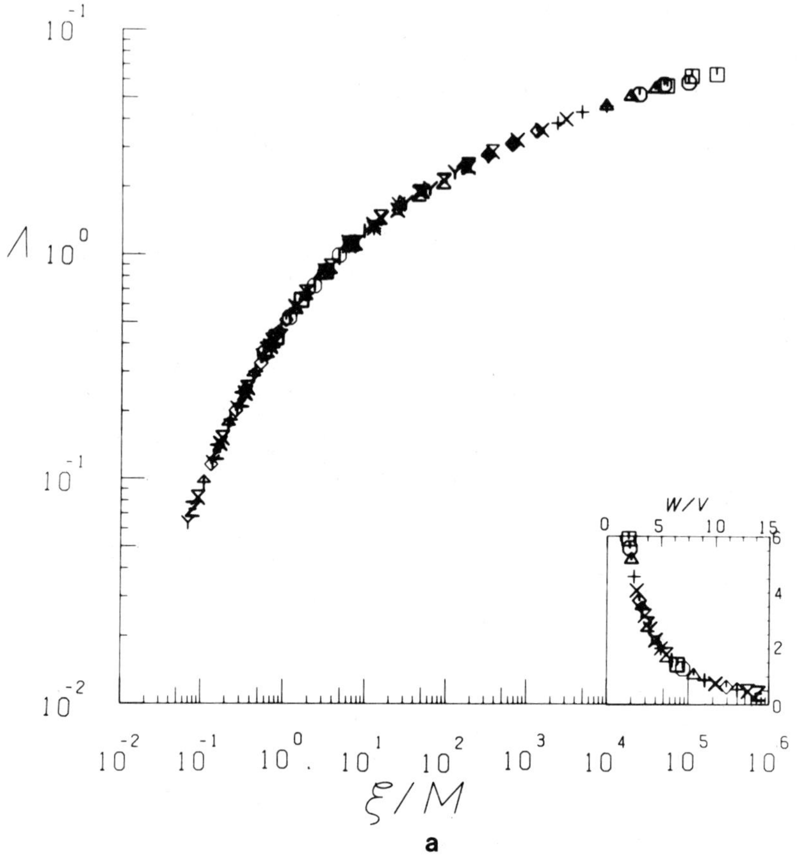

a

Fig. 4: Renormalized decay length Λ of strips of width M (a), and bars of cross section M^2 (b) as a function of ξ/M. The scaling parameter ξ was chosen to fit all data from Fig. 3 onto one and the same curve (scaling functions $f_d(x)$). The inserts show the scaling parameters ξ as a function of the disorder.

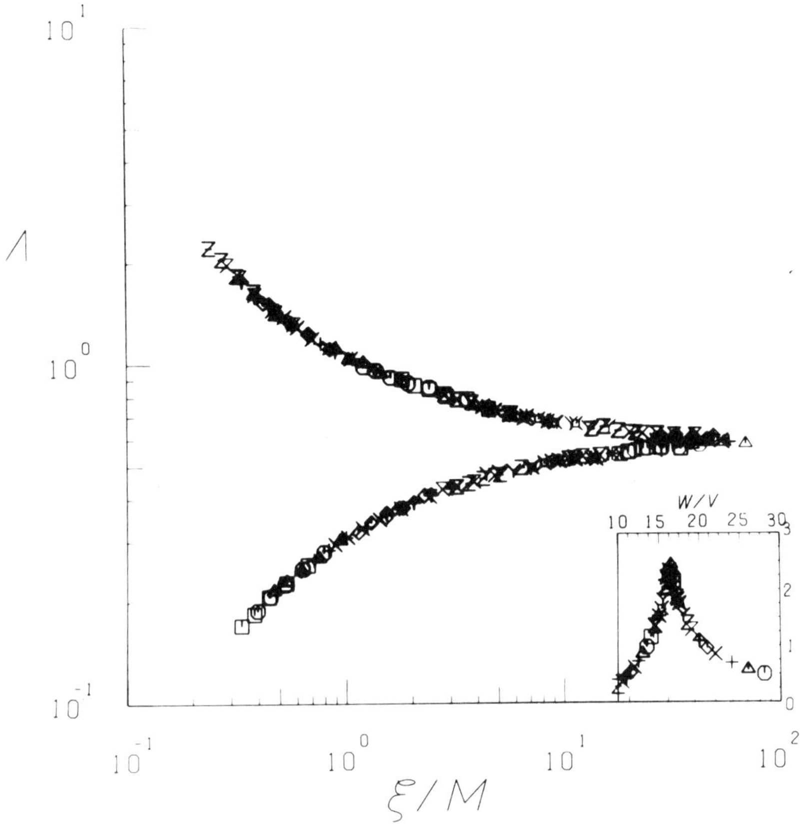

b

Fig. 4

Fig. 4 shows that this is indeed possible, within the statistical error which was fixed at 1% in these calculations (MacKinnon and Kramer, 1983). This means that

$$\xi_M(0,W,d)/M = f_d(\xi(0,W,d)/M) \tag{14}$$

It is important to note that the form of the scaling function $f_d(x)$ depends characteristically on the dimensionality d. For d = 2 the scaling function consists of one branch only whereas for d = 3 there are two branches. Of the latter, one contains the data for $W<W_c = 16.5$ ($f_3>0.6$), the other those corresponding to $W>W_c$ ($f_3<0.6$).

The scaling parameter is independent of M. Next we investigate the question of its physical meaning. We first consider d = 2. Here f_2 is a monotonically increasing function which, for small x tends to

$$f_2(x) = x \tag{15}$$

If W = const $x \to 0$ means $M \to \infty$. In this case the scaling parameter is therefore the localization length in the infinite 2-d system. All states are thus localized in 2d. In 3d the lower branch of the scaling function ($W > W_c$) may be discussed by using the same argument. It corresponds therefore to the localized regime. The upper branch is somewhat more difficult. First of all, from the analysis of the data one observes that

$$f_3(x) \approx 1/x \tag{16}$$

for small x. Since the dimensionless conductance can be related to ξ_M and is proportional to M for large M, we obtain (MacKinnon and Kramer, 1983)

$$\xi = 1/\sigma \tag{17}$$

where σ is the dc-conductivity. Thus, the upper branch of $f_3(x)$ describes the metallic regime. W_c is the critical value of the disorder for localization.

Analysing the curves in more detail one can derive information about the critical behaviour of the system. One obtains that near the transition σ^{-1} and ξ diverge with the same power law, the critical exponents being $s = \nu = 1.5$, respectively. In 2d one obtains an essential singularity at $W = 0$ which is of the form $\xi \approx \exp(1/W^n)$, $n = o(1)$.

In order to investigate the phase diagram for localization, i.e. the dependence of the critical disorder, W_c, on the energy E, one can apply the above described procedure to the data at different energies, $\xi_M(E,W,d)$. This works as long as one stays near the band centre where the density of states is large. As the band edge is approached the numerical calculations become extremely difficult. On the one hand, this is due to the smallness of the density of states. On the other hand near the band edge a second length comes into play: the phase correlation length, which becomes very large at the band edge. Thus, in order to avoid boundary effects the width of the system has to be very large.

The phase diagram shown in Fig. 5 was obtained from data for $\xi_M(E,W,d)$ which were taken with a statistical accuracy of 2% (Bulka et al., 1985). Using such data the scaling analysis described above cannot be performed very reliably. However, as the data for $E \neq 0$ behave qualitatively as those in the band centre (cf. Fig. 3) we have estimated the critical disorder directly from the slope of ξ_M/M as a function of $1/M$.

We believe that the accuracy of the points $W_c(E)$ near the band centre is better than 10%, whereas in the band edge region this might be overestimated. There are two surprising features in Fig. 5: Firstly, the phase separation curve obtained numerically seems to agree extremely well with that obtained from a CPA-calculation (Economou et al., 1984) which is also shown in Fig. 5. Secondly, starting at E = 0, $W_c(E)$ stays almost constant, then, around E = 6V, starts to decrease first slightly, and then more rapidly. At E = 8V we observe that the slope of the curve is infinite. For W_c < 12 the phase separation curve follows the band edge. We note that for E > 6V as a function W/V one has two localization-nonlocalization transitions.

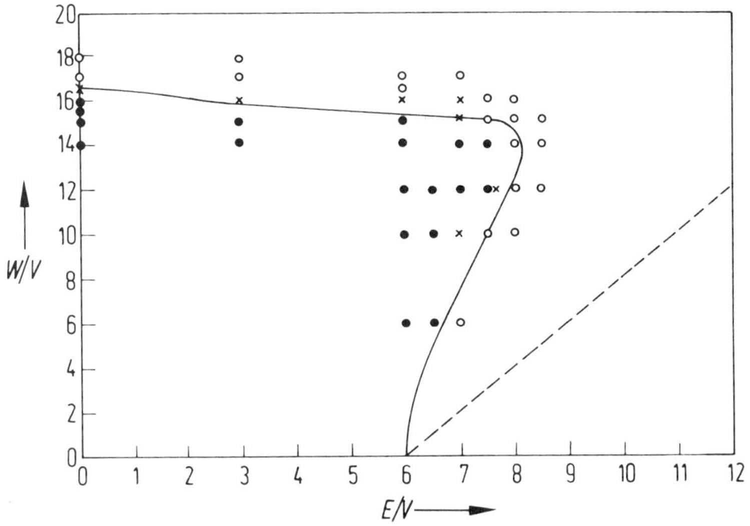

Fig. 5: Phase diagram for localization for the Anderson model (simple cubic lattice)

DISCUSSION OF THE RESULTS

One may understand the behaviour of the data qualitatively by using the mechanisms important for localization which were above.

Near the centre of the band, and for small disorder, $W \ll 1$, the free electron picture is a good starting point. Quantum interference is here the mechanism which is important and may be considered as the precursor of localization. The phenomenon has been discussed extensively by Bergmann (1984) in particular with respect to its experimental consequences. Considering the respective terms in the expression for the conductivity one may calculate the quantum corrections to the classical conductivity quantitatively. These perturbational results are also in quantitative agreement with the results derived from the numerical data shown in Fig.s 3 and 4 (MacKinnon and Kramer, 1983). As the disorder is increased quantum interference should become more destructive, and, simultaneously, less important. Potential effects, and tunneling processes become increasingly important for localization, particularly near the critical disorder W_c.

Near the band edges of the ordered system the free electron picture is certainly not a good starting point. Instead, one may assume that for small disorder the states in the outermost tails of the band are potential-localized within the potential wells which occasionally occur due to the random potential fluctuations. If we increase the disorder (while fixing the energy) the density of states strongly increases due to the exponential shape of the band tail. The distance between the localization centres decreases. Tunneling may lead to an increase of the localization length, and even to complete delocalization of the states. If the disorder is further increased we leave the band tail region, the density of states tends to decrease again as 1/W, as near the band centre. Localization is again enhanced by the combination of interference and potential effects.

It is not yet clear how to fit this interpretation of the phase diagram of localization into the scaling picture.

ACKNOWLEDGEMENTS

This work was supported by the Deutsche Forschungsgemeinschaft, and the British Petroleum Venture Research Unit.

REFERENCES

Abrahams, E., Anderson, P. W., Licciardello, D. C., and Ramakrishnan, T. V., 1979, Scaling Theory of Localization: Absence of Quantum Diffusion in Two Dimensions, Phys. Rev. Letters 42: 673

Anderson, P. W., 1958, The Absence of Diffusion in Certain Random Lattices, Phys. Rev. 109: 1492

Bergmann, G., 1984, Weak Localization in Thin Films, Phys. Reports 107: 3

Bulka, B., Kramer, B., and MacKinnon, A., 1985, Mobility Edge in the Three Dimensional Anderson Model, to be published

Economou, E. N., Soukoulis, C. M., and Zdetis, A. D., 1984, Localized States in Disordered Systems as Bound States in Potential Wells, Phys. Rev. B30: 1686

Kramer, B., Bergmann, G., and Bruynseraede, Y., 1985, "Localization, Interaction, and Transport Phenomena in Impure Metals", Proceedings, Springer Series in Solid State Sciences, to be published

Kramer B., MacKinnon, A., Weaire, D., 1981, Numerical Study of Conductivity for the Anderson Model in Two and Three Dimensions, Phys. Rev. B23: 6357

Kramer B. , Weaire, D. L., 1979, Theory of Electron States in Amorphous Semiconductors, in: "Amorphous Semiconductors", M. H. Brodsky ed., Springer Verlag, Heidelberg

Mott, N. F., and Davis, E. A., 1971, "Electronic Processes in Non-crystalline Solids", Oxford UP, London

MacKinnon, A., 1980, The Conductivity of the One-Dimensional Disordered Anderson Model: A New Numerical Model, J. Phys. C13: 1031

--, 1985, The Calculation of Transport Properties and Density of States of Disordered Solids, Z. Phys., to be published

MacKinnon, A., Kramer, B., 1981, One Parameter Scaling of Localization Length and Conductance in Disordered Systems, Phys. Rev. Letters 47: 1546

--, --, 1983, The Scaling Theory of Electrons in Disordered Solids: Additional Numerical Results, Z. Phys. B53: 1

Wegner, F. J., 1976, Electrons in Disordered Systems. Scaling near the Mobility Edge, Z. Phys. B25: 327

EFFECT OF PHASE CORRELATIONS ON THE ANDERSON TRANSITION

Moshe Kaveh

Department of Physics
Bar-Ilan University
Ramat-Gan, Israel

We study the effect of phase correlations between eigenfunctions on the Anderson transition. Arguments in support of Mott's wavefunctions near the mobility edge are given. Using these strongly correlated wavefunctions for systems with dimensionality $d > 2$, we produce the results of the scaling theory that the conductivity σ drops continuously to zero as $\sigma \sim e^2/\hbar\xi^{d-2}$ or $\sigma \sim e^2/\hbar L^{d-2}$ if $L < \xi$. In two dimensions, $\sigma = 0$ for $L \to \infty$ for any degree of disorder although the nature of localization (power-law or exponential) cannot be determined by the present method. We also discuss the Anderson transition in the presence of a magnetic field. For weak magnetic fields, we give an argument that supports a continuous transition followed by a shift of the mobility edge to lower values. For strong magnetic fields the phase correlations are destroyed and a discontinuous transition may occur. Present experiments indicate a very sharp transition but more experiments are needed to verify whether it is a first order transition.

INTRODUCTION

A non-interacting disordered electron system can be either in a metallic state or an insulating state. It was first pointed out by Anderson (1958) that when a critical degree of disorder is reached, a metal-insulator transtion occurs. This is the Anderson transition. This problem was intensively studied by Mott during the past two decades. In particular, the concepts of hopping transport in the Anderson insulator were developed in detail (Mott 1974, Mott and Davis 1979), and observed in many systems. Mott (1970) was the first to introduce the concept of a mobility edge E_c, which separates exponential localized states from extended states. When the Fermi energy lies below E_c, $\sigma = 0$ at zero temperature and hopping (or variable range hopping) transport is observed at non-zero temperatures. The nature of the wavefunctions below the mobility edge are well-known; they decay as $\exp(-r/\xi)$ from a given point which depends on the energy. For $E \to E_c$ (from below) $\xi \to \infty$. For $E > E_c$, the eigenstates are extended. The exact form of ψ_E for $E > E_c$ is not known. In 1970 Mott advanced the random phase approximation for ψ_E which is given by

$$\psi_E = \sum C_n \exp(i\psi_n)\phi_E(r - R_n) \tag{1}$$

where the sum is over all the random positions of the potential wells (in the Anderson model), ψ_n are random phases and $\phi(r - R_n)$ are atomic wavefunctions. These wavefunctions were used in the Kubo-Greenwood formula to calculate σ, which is proportional to the square of the following matrix element

$$<\psi_E \left|\frac{\partial}{\partial x}\right| \psi_{E'}> \tag{2}$$

where $E' = E = E_F$. It was assumed that the phases $\psi_n(E)$ and $\psi_n(E')$ are uncorrelated. Under these assumptions the matrix element in eq. (2) cannot vanish. This led to the concept of a minimum metallic conductivity. This is achieved for the "most disordered" wavefunction in the random phase approximation, namely that ψ_n is random between any two potential wells over a distance, a. This was correlated to the Ioffe-Regel concept that the phases are random over a distance given by the elastic mean free path ℓ, which cannot be smaller than a. The Kubo-Greenwood formula, when calculated with ψ_E given by (1), leads to the Boltzmann conductivity, $\sigma_B = e^2 K_F{}^2 \ell/3\pi^2\hbar$. In the Ioffe-Regel limit $K_F\ell = K_F a \simeq \pi$ and the minimum Boltzmann conductivity is $\sim 0.3\ e^2/\hbar a$. Another order of magnitude reduction in the conductivity comes from the reduced density of states near the mobility edge. This leads to a minimum metallic conductivity

$$\sigma_{min} \simeq 0.03\ e^2/\hbar a \tag{3}$$

Thus, in the framework of the random phase approximation a discontinuous drop in the conductivity occurs, jumping from the value given by (3) to zero. It was not until the work of Abrahams et al (1979) that a scaling theory of the Anderson transition was developed and a continuous transition was predicted. This important prediction led to a new flood of experiments which conclusively observed conductivities much lower than σ_{min}. The most detailed study (Rosenbaum et al 1982, Thomas 1984) is for Si:P in which σ drops by orders of magnitude below σ_{min}. These results must mean a breakdown of the random phase approximation. Thus, the remaining problem is to construct an extended wavefunction for which $<\psi_E|\frac{\partial}{\partial x}|\psi_{E'}> = 0$ when $E = E' = E_c$. Far above E_c where $K_F\ell >> 1$, Kaveh and Mott (1981, 1982, 1983) have shown that ψ_E can be expanded in terms of plane waves and contains a power-law component for

$d \geqslant 2$. This ψ_E with a power-law component, when inserted in the Kubo-Greenwood formula, leads to a conductivity which is smaller than the Boltzmann conductivity. Thus, the nature of ψ_E for $E >> E_c$ is well understood (for a review see Kaveh 1985), and is given by

$$\psi_E = \psi_{RPA} + \psi_{ext}/r^{d-1} \tag{4}$$

where ψ_{RPA} is ψ_E in the random phase approximation and ψ_{ext} is some extended wavefunction. For $E \to E_c$, $K_F \ell \sim 1$ (an expansion in terms of plane waves is not useful) and eq. (4) is not valid (see a detailed discussion by Kaveh 1985).

Only in 1984 Mott reproposed eq. (1) but pointed out that the assumption that the phases $\psi_n(E)$ and $\psi_n(E')$ are uncorrelated is unjustified (at zero temperature and zero magnetic field).

In this paper, we discuss this phase correlation and show how to determine ψ_E when $E \to E_c$. This is in agreement with the form of ψ_E as proposed by Mott (1984). We then show how this leads to a conductivity which tends continuously to zero as $\sigma \sim e^2/\hbar\, \xi^{d-2}$ in agreement with the scaling theory. The effect of a magnetic field on this phase correlation is discussed and the different possibilities for the nature of the Anderson transition in the presence of magnetic field are pointed out.

DIFFUSION NEAR A MOBILITY EDGE

We expand the eigenstates near the mobility edge $E \gtrsim E_c$ in terms of all the localized eigenstates ψ_E, with $E' \leqslant E_c$. Thus, instead of eq. (1) we write

$$\psi_E = \sum_{E'<E_c} C_{E'} \Phi_{E'} \quad ; \quad E > E_c \tag{5}$$

where Φ_E' is given by

$$\Phi_{E'} = \sum_n a_n^{(E')} \phi_{E'}(r - R_n) \exp[-|\underline{r} - r_o(E')|/\xi(E')] \tag{6}$$

and $\xi(E')$ is the localization length, $\phi_{E'}$ atomic states and $\underline{r}_o(E')$ the centre around which the eigenfunction is localized. The eigenstate $\Phi_{E'}$ is a real wavefunction.

We now show the following points:

(i) The smallest energy E' that contributes to ψ_E is given by

$$E'_{min} = 2E_c - E$$

This means that, if $E = E_c + \Delta E$, then $E'_{min} = E_c - \Delta E$.

(ii) If $\xi^{-1}(E'_{min}) \sim (\Delta E)^\nu$, then for three dimensions,

$$\sigma \sim (\Delta E)^\nu$$

Thus, as first claimed by the scaling theory (Wegner 1979, Abrahams et al 1979), the critical exponent of σ is the same as the exponent of the localization length.

(iii) If the length of the sample L(or any other relevant length scales like the inelastic diffusion length $L_i = (D\tau_{in})^{1/2}$, or the frequency dependent length, $L_\omega = (D/\omega)^{1/2}$) is shorter than $\xi(E'_{min})$, then

$$\sigma(E) \sim e^2/\hbar L \tag{7}$$

To prove these statements we consider two localized states $\Phi_{E'}$ and $\Phi_{E''}$ for which their centres are as close as possible. For a system of dimensionality d,

$$E'' - E' \simeq 1/N(E')\xi(E')^d \tag{8}$$

The distance, R, between the centres $\underline{r}(E')$ and $\underline{r}(E'')$ is given by

$$R = |\underline{r}(E') - \underline{r}(E'')| = \xi(E')\ln\frac{I}{|E'-E'|} \tag{9}$$

where I is the overlap integral. Thus, the distance between the centres is about $\xi(E')$. In order that state with energy between E' and E'' can form an extended state it is required that an electron be able to diffuse from one centre to another. The diffusion time is

$$t_D \simeq \frac{\xi^2(E')}{D(E)} \tag{10}$$

where D(E) is the diffusion constant at energy $E > E_c$. If

$$\frac{\hbar}{t_D} > E'' - E' \tag{11}$$

any two localized states with energy between E' and E'' will contribute to an extended state.

We thus need

$$\frac{\hbar\, D(E)}{\xi(E')^2} > \frac{1}{N(E')\xi(E')^d} \tag{12}$$

D(E) is non-zero and therefore it is easier for states with larger localization length to fulfill this condition. This condition can be rewritten by using the Einstein relation $\sigma(E) = e^2 N(E)D(E)$ as

$$\sigma(E) > \frac{e^2}{\hbar\xi(E')^{d-2}}\frac{N(E)}{N(E')} \tag{13}$$

Since N(E') changes very little near the mobility edge we may approximate $N(E)/N(E') = 1$. We thus have

$$\sigma(E) > \frac{e^2}{\hbar\xi(E')^{d-2}} \tag{14}$$

The smallest localization length, $\xi(E'_{min})$, obeys

$$\sigma(E) \simeq e^2/\hbar\xi(E'_{min})^{d-2} \tag{15}$$

Thus energies $E' < E'_{min}$ will not contribute to the extended state ψ_E in eq. (5). If we now suppose that

$$\sigma(E) = \sigma_o[(E - E_c)/E_c]^{\mu} \tag{16}$$

and

$$\xi^{-1}(E'_{min}) = \xi_o^{-1}[(E_c - E'_{min})/E_c]^{\nu} \quad , \tag{17}$$

it follows from equating eq. (15) with eq. (16) that

$$\mu = (d - 2)\nu \tag{18}$$

and

$$E'_{min} = 2E_c - E \tag{19}$$

We see that states for which $E' < E_c - \Delta E$, where $\Delta E = E - E_c$ do not contribute to the diffusion constant D(E) and hence do not appear in eq. (5) for ψ_E. The critical exponent of the conductivity is given by eq. (18). This simple argument produces the main result of the scaling theory. Moreover, it also produces the third point. If $L < \xi(E'_{min})$, it must follow that two localized states at energy E' are now separated by a distance which is smaller than $\xi(E')$. Thus,

$$E' - E'' = 1/N(E')L^d \tag{20}$$

In order that these two states should cause a finite diffusion constant at energy E we need

$$\hbar D(E)/L^2 \simeq 1/N(E')L^d \tag{21}$$

or

$$\sigma(E) \simeq e^2/\hbar L^{d-2} \; ; \qquad L < \xi(E'_{min}) \tag{22}$$

NATURE OF WAVEFUNCTIONS NEAR THE MOBILITY EDGE

From the previous arguments, we see that ψ_E is constructed from localized states in a range ΔE below E_c. Every state $\Phi_{E'}$, for $E_c - \Delta E < E' < E_c$ with $\Delta E = E - E_c$, is a real function but ψ_E need not be real because C(E') can be complex. ψ_E will contain many maxima corresponding to all the centres $r_o(E')$. However, for $E' \simeq E_c$, $\xi(E' \simeq E_c)$ is so large that ψ_E is not affected much by these states. The dominant structure of ψ_E is produced by states $\Phi_{E'}$ with the smallest localization length, i.e., $E' \simeq E'_{min}$. This argument leads to the form of ψ_E that was recently proposed by Mott (1984). Mott's ψ_E consists of peaks separated by a distance R which is given by

$$R = \xi(E'_{min}) \ell n \frac{I}{|E'_{min} - E'|} \tag{23}$$

ψ_E is real around the peaks but complex around the minima. Mott's argument is based on a hypothesis, first pointed out by Imry (1980), that for $L < \xi(E_c - \Delta E)$ the state ψ_E for $E = E_c + \Delta E$ and the state $\Phi_{E'}$ for $E' = E_c - \Delta E$ is indistinguishable. Indeed, both states lead to the same conduction, $e^2/\hbar L^{d-2}$ for $L < \xi(E'_{min})$ as shown in the previous section. Thus,

$$\psi_E(r) \simeq \Phi_{E'_{min}}(r) \quad ; \qquad L < \xi(E'_{min}) \tag{24}$$

States with $E' > E'_{min}$ do not affect this property much since $\xi(E' > E'_{min}) > \xi(E'_{min})$. From this property one can argue (Mott 1984, Mott and Kaveh 1985) that all the states with energy $E \geqslant E_c$ in a small range of energy δE are very similar and

$$\psi_{E + \delta E} \simeq \psi_E \quad ; \qquad \delta E \simeq 0 \tag{25}$$

over a distance $\xi(E'_{min})$ around every maxima. This means that all the eigenfunctions ψ_{E_1} and ψ_{E_2} for $E_2 \simeq E_1$ will possess the same peaks at the same locations separated by a distance $R = \xi(E'_{min}) \ \ell n I / \ E'_{min} - E') \simeq \xi(E'_{min})$. We shall see that it is this property that reduces the conductivity.

CONDUCTIVITY NEAR THE MOBILITY EDGE: KUBO-GREENWOOD METHOD

We now apply the Kubo-Greenwood method to calculate the conductivity near E_c for a non-interacting disordered electron gas, by using Mott's wavefunctions. The Kubo-Greenwood formula for the conductivity is given by

$$\sigma = c_o L^d |\langle\psi_E| \frac{\partial}{\partial x} |\psi_{E + \delta E}\rangle|^2_{Av} \{N(E)\}^2 \tag{26}$$

where the matrix element is averaged over all states $\psi_{E + \delta E}$ letting $\delta E \to 0$.

We first calculate σ for $L < \xi(E'_{min})$. It is clear that ψ_E and $\psi_{E + \delta E}$ are strongly correlated for $L < \xi(E'_{min})$. We proceed by making the following transformation

$$\langle\psi_E| \frac{\partial}{\partial x} |\psi_{E + \delta E}\rangle = (m\delta E/\hbar^2)\langle\psi_E|x|\psi_{E + \delta E}\rangle \tag{27}$$

and find the dependence of the matrix element on L in the following way. The energy separation between two states, δE, can be written as

$$\delta E \simeq 1/N(E) L^d \tag{28}$$

For strongly correlated functions we must have

$$\langle\psi_E|x|\psi_{E + \delta E}\rangle = c_1 L \tag{29}$$

where c_1 is a constant independent of L (its value however, depends on the actual form of ψ_E). Inserting eqs. (27) - (29) in the Kubo-Greenwood formula for σ we get

$$\sigma = c\, e^2/\hbar L^{d-2} \quad ; \qquad L < \xi(E'_{min}) \tag{30}$$

where c is a constant depending on c_1. We see that for $d > 2$, $\sigma \to 0$ as $L \to \infty$ in agreement with the prediction first made by the scaling theories (Wegner 1979, Abrahams et al 1979). Thus, Mott's form of ψ_E with its properties as discussed in the previous section yield a vanishing conductivity at the mobility edge. It should be noted that the minimum metallic conductivity concept (Mott 1970) was obtained under the assumption that ψ_E and $\psi_{E'}$ are uncorrelated. In this case the matrix element in eq. (27) must depend on the dimensionality of the system

$$<\psi_E| \frac{\partial}{\partial x} |\psi_{E + \delta E}> \propto L^{-d} \cdot L^{d/2} \tag{31}$$

The first factor comes from the normalization factors of ψ_E and the $L^{d/2}$ term comes from the uncorrelated random phases of ψ_E and $\psi_{E + \delta E}$. This random phase approximation leads to a conductivity which is independent of L. Thus, the origin of $\sigma \to 0$ indicates a breakdown of the random phase approximation. The correlated wavefunctions lead to a conductivity which decreases with the size of the system.

We now calculate $\sigma(E)$ for $L \to \infty$. The matrix element on the RHS of eq. (27) is changed since now $L > \xi(E'_{min})$. The phases of ψ_E and $\psi_{E'}$ are now correlated only over a distance $\xi = \xi(E'_{min})$. Thus,

$$<\psi_E|x|\psi_{E + \delta E}> \simeq c_1 \xi (R/\xi)^{d/2} (L/R)^{d/2} = c_1 \xi (L/\xi)^{d/2} \tag{32}$$

Inserting this result in eq. (26) and using eqs. (27) - (28) leads to

$$\sigma = c\, e^2/\hbar \xi^{d-2} \quad ; \qquad L > \xi \tag{33}$$

where ξ is calculated at an energy $E_c - \Delta E$ and $\Delta E = E - E_c$. As $E \to E_c$, $E'_{min} \to E_c$ and $\xi \to \infty$ leading to $\sigma \to 0$. Thus, $\sigma(E = E_c) = 0$ for $L \to \infty$. We see that the form for ψ_E as given by eq. (5) and described in the previous section produces a continuous Anderson transition with a critical index $(d - 2)\nu$. Calculations in $d = 2 + \varepsilon$ (Berezin et al 1980, Hikami 1980), show that $\nu \simeq 1/(d - 2)$ and thus for $d > 2$,

$$\sigma(E) \sim (E - E_c) \tag{34}$$

with a critical exponent about unity.

TWO DIMENSIONS

The above arguments cannot be carried out in two dimensions. If we assume a finite diffusion constant at energy $E > E_c$, we immediately face a difficulty. The kinetic energy an electron needs in order to hop from one localized state with energy E' to another localized state separated in energy by about $1/N(E')\xi(E')^2$ is given by $\hbar D(E)/\xi(E')^2$. Thus, the condition for having an extended state that is given by eq. (12) does not depend any more on $\xi(E')$ and for $E \to E_c$ an extended state with a non-zero diffusion constant cannot be formed. One possibility is that there is no mobility edge in two dimensions and that all eigenstates are exponentially localized. This hypothesis was made by the scaling theory (Wegner 1979, Abrahams et al 1979). Another possibility is that for $L \to \infty, \sigma \to 0$ but there is a mobility edge which separates exponential localized states from power-law localized states. For $r \to \infty$, it was suggested (Kaveh and Mott 1981, Haydock 1981, 1983, Davies et al 1983; for a review see Kaveh 1985) that

$$|\psi_E| \propto \begin{cases} r^{-s} ; & E > E_c \\ \exp(-2/\xi) ; & E < E_c \end{cases} \tag{35}$$

If we again express ψ_E in terms of localized eigenstates Φ_E, we have

$$\psi_E = \sum_{E'<E_c} C_{E'} \Phi_{E'} \tag{36}$$

but now without a minimum value for E'. Thus, for $E > E_c$ all eigenstates with $E' < E_c$ contribute to ψ_E. For $E < E_c$, only one particular $C_{E'=E}$ is non-zero. The most direct calculation of ψ_E for $E > E_c$ is by Haydock (1981, 1983). He used the recursion method to solve the Schredinger equation for a two-dimensional random system. He finds that a mobility edge exists and $|\psi_E| \propto 1/r^s$ for $r \to \infty$ with s increasing as E decreases. Kaveh (1984) has shown that a characteristic length which is given by $L_o = \ell \exp(\pi K_F \ell/2)$ is the relevant length separating the two forms of ψ_E. For $r > L_o$, $|\psi_E| \sim r^{-s}$ and for $r < L_o$, ψ_E is quasi extended and is given by (Kaveh and Mott 1981, 1982, 1983, Kaveh 1984)

$$\psi_E = \psi_{ext}(1) + \psi_{ext}(2)/r \quad ; \qquad r < L_o \tag{37}$$

where $\psi_{ext}(1)$ and $\psi_{ext}(2)$ are extended wavefunctions with random phases. This was shown to lead to the well-known logarithmic correction for the conductivity as first obtained by Abrahams et al (1979). The two possibilities for the nature of localization of ψ_E for $r > L_o$ (exponential or power-law) leads to very different transport mechanisms at low temperatures. This was recently studied in detail (Kaveh 1984, 1985). It was shown (Kaveh 1985) that all the available data in inversion layers support power-law localization.

EFFECT OF MAGNETIC FIELD ON THE CRITICAL BEHAVIOUR OF σ

We have seen that in three dimensions for a non-interacting disordered system

$$\sigma \sim (E - E_c)^{\nu \simeq 1} \tag{38}$$

We now discuss the effect of a magnetic field. We expect that if the cyclotron length $L_H = \sqrt{ch/eH}$ is larger than $\xi(E'_{min})$, the conductivity $\sigma(E)$ is little affected. ψ_E remains the same function being constructed from all the states in the range $\Delta E = E_c - E'_{min}$. But for larger magnetic fields where $L_H < \xi(E'_{min})$, we expect that σ will be affected. We may, however, still expand ψ_E as in eq. (5) but with localized states, $\Phi_{E'}(r,H)$, that depend on L_H. For $L_H > \xi(E')$, we may take $\Phi_{E'}$ as before, independent of H. L_H is effectively the new localization length for $L_H < \xi(E')$; E'_{min} must be shifted. The diffusion constant and hence the conductivity may be determined from $\hbar D(E)/L_H^2 \simeq 1/N(E')L_H^3$, which leads to

$$\sigma(E) \simeq C\, e^2/\hbar L_H \tag{39}$$

This comes from states <u>below</u> $E_c - \Delta E$ (where $\Delta E = E - E_c$), since $L_H < \xi(E'_{min})$. Thus, $\sigma(E)$ will not depend on $\xi(\Delta E)$, once $L_H < \xi(E'_{min})$. We now ask what is $\sigma(E)$ for E just below E_c in the presence of a magnetic field. Shapiro (1984) argued that a weak magnetic field pushes E_c to lower values, namely $E_c(H) < E_c(o)$. The above argument suggests that at $E = E_c(o)$, $\sigma \simeq e^2/\hbar L_H$. Can σ vanish at the new mobility edge? Before answering this question, we give an estimate of the shift in the mobility edge due to the magnetic field. We have already pointed out that E'_{min} is shifted to a lower value which we denote by E_o. We must have $\xi(E_o) = L_H$. Writing $\xi(E')^{-1} = \xi_o^{-1}(E_c - E')/E_c$, we find that the shift in E'_{min}, $\delta E = E'_{min} - E_o$ for small magnetic fields is given by

$$\delta E \simeq E_c(o)\xi_o/L_H \tag{40}$$

If we suppose that this is also the shift in the mobility edge, we get $E_c(H) - E_c(o) \propto H^{1/2}$. In this case, states in the range $E_c(H) < E < E_c(o)$ are <u>extended</u> and the conductivity is given by

$$\sigma \simeq C\, e^2\, \hbar\xi(E') \tag{41}$$

where $E' = E_c(H) - \Delta E$ and $\Delta E = E - E_c(H)$. The localization length in the presence of a magnetic field may be written as

$$\xi^{-1} = \xi_o^{-1}(E_c(H) - E)^{\nu(H)} \tag{42}$$

Thus, in this picture we still get a continuous drop of σ at the new mobility edge with an index $\nu(H)$, i.e., $\sigma \sim (E_c(H) - E)^{\nu(H)}$ for $E_c(H) \leqslant E < E_c(o)$. Hikami (1981) argued that a continuous decrease of σ to zero occurs in the presence of a weak magnetic field but the critical index changes to $\nu(H) = 1/2$. This result can be seen as follows. By calculating a scaling function which vanishes at G_c, one may expand $\beta(G)$ to give

$$\beta(G \simeq G_c) = \frac{1}{\nu}\left| \frac{G - G_c}{G_c} \right| \tag{43}$$

The exponent ν is the same as the exponent of the localization length $\xi^{-1} \sim (E_c - E')^{\nu}$ and due to eq. (18) is also the critical exponent of σ (Abrahams et al 1979). Calculating $\beta(G)$ in the perturbation region (large G) leads to (Hikami 1981, Wegner 1981)

$$\beta(G) = 1 - \frac{c_1}{G} - \frac{c_2}{G^2} \tag{44}$$

For zero magnetic field, it was shown (Gorkov et al 1979, Hikami 1981) that $c_2 = o$. Extrapolating eq. (44) to $G = G_c$ (where $\beta(G_c) = o$) leads to $\beta(G) = (G - G_c)/G_c$ which implies $\nu = 1$. For non-zero magnetic fields it was shown (Oppermann and Jungling 1980, Hikami 1981, Wegner 1981) that $c_1 = 0$ but $c_2 \neq o$. Extrapolating eq. (44), with this result, to $G = G_c$ leads to $\beta(G) = 1 - (G_c/G)^2 \sim 2(G - G_c)/G_c$, implying $\nu = 1/2$. If however, the magnetic field does not affect the mobility edge we expect (Mott and Kaveh 1985) that σ will jump discontinuously to zero from its value $\sim e^2/\hbar L_H$ at the mobility edge. Thus, an experiment is able to test whether, in the presence of a magnetic field, the transition remains continuous but ν changes from unity into 1/2 or whether a discontinuous transition occurs. No such experiments have yet been performed.

As the magnetic field reaches large values the overlap between atomic wavefunctions decreases. For $L_H < a_B$, where a_B is the atomic Bohr radius, the overlap between two atomic wavefunctions centred at the donor locations becomes so small that an Anderson transition is induced. This transition is achieved even for $E_F >> E_c(o)$, i.e., well in the metallic state (without the magnetic field). Thus, $E_c(H)$ is pushed to higher values. Shapiro (1984) suggested a phase diagram in which, for small magnetic fields, $E_c(H)$ first decreases and then rapidly increases for higher fields due to the "magnetic freezeout". If this phase diagram is correct, it suggests that there exist two critical magnetic fields ($H_c(1)$ and $H_c(2)$) for which $\sigma = o$ <u>outside</u> this region. At these fields $\sigma = o$ and must undergo a <u>maximum</u> as a function of H in between. It is very interesting to verify this prediction. Another problem is the nature of the metal-insulator transition at high magnetic fields. Biskupski (1984) and Long and Pepper (1984) have found experimentally, for large magnetic fields, a very sharp transition and, in fact, have interpreted it as a discontinuous transition. They found that just below the transition $\sigma \simeq 0.03\ e^2/\hbar a$, which coincides with Mott's minimum metallic conductivity. We suggest that further experiments be performed to check whether the possibility of a continuous (although sharp) transition can be ruled out. This transition has not yet been studied theoretically in detail. Nevertheless, arguments for a discontinuous transition were advanced by Mott (1984) and Mott and Kaveh (1985) in which the strong magnetic field destroys the phase correlations between the wavefunctions near the mobility edge leading again to the validity of the random phase approximation and hence to a minimum metallic conductivity.

In summary, we feel that the Anderson transition (for non-interacting electrons) and the nature of the wavefunctions near E_c in a zero magnetic field are now fairly understood. The situation in the presence of a magnetic field is less clear both theoretically and experimentally. The different possibilities of the transition in this case have been pointed out.

REFERENCES

Abrahams E, Anderson P W, Licciardello D C and Ramalerishran T W, 1979, Phys. Rev. Lett. 42, 693

Anderson P W, 1958, Phys. Rev. 109, 1492

Berezin E, Hikami S and Zinn-Justin J, 1980, Nucl. Phys. B165, 528

Biskupski G, 1984, Proceedings of the Braunschweig Conference ibid. 1982, Thesis Lille

Davies R, Pepper M and Kaveh M, 1983, J. Phys. C16, L285

Haydock R V, 1981, Phil. Mag. B43, 203 ibid. 1984, APS March Meeting

Hikami S, 1980, Progr. Theor. Phys. 64, 1466

——— 1981, Phys. Rev. B24, 2671; ibid Phys. Lett. 98B, 208

Imry Y, 1980, Phys. Rev. Lett. 44, 469

Kaveh M and Mott N F, 1981, J. Phys. C 14, L183

——— 1982, J. Phys. C 15, L697

——— 1983, Phil. Mag. B 47, L9 ibid. J. Phys. C 16, L1067

Kaveh M, 1984, J. Phys. C (in press)

Kaveh M, 1985, Phil. Mag. (in press)

Long A P and Pepper M, 1984, J. Phys. C 17, 3391

Oppermann R and Jungling K, 1980, Phys. Lett. 76A, 449

Mott N F, 1970, Phil. Mag. 22, 7

——— 1974, Metal-Insulator Transitions

——— 1984, Phil. Mag. B 49, L75

Mott N F and Davis E A, 1979, Electron Processes in Non-Crystalline Materials

Mott N F and Kaveh M, 1985, Phil. Mag. (in press), ibid Adv. in Phys. (in press)

Rosenbaum T F, Andres K, Thomas G A and Bhatt R N, 1982, Phys. Rev. Lett. 48, 1284

Shapiro B, 1984, Phil. Mag. B 50, 241

Thomas G, 1984, Phil. Mag. B 50, 169

Wegner F, 1979, Z. Phys. B 35, 207

Wegner F, 1981, Nucl. Phys. B 180, 77

ELECTRON-LATTICE-INTERACTION INDUCED LOCALIZATION IN SOLIDS

David Emin

Sandia National Laboratories
Albuquerque, New Mexico 87185, USA

INTRODUCTION

The idea that the electron-lattice interaction could provide a strong driving force for electronic localization in solids originated with Landau in 1933.[1] He suggested that an electron might find itself bound in the potential well established by the displacements of the atoms of a solid from their carrier-free equilibrium positions. The lowering of the electron's energy resulting from its being bound may more than offset the strain energy associated with the atomic displacements. Then, it is energetically favorable for the atoms to displace so as to form such a well with the electron trapped within it. In other words, the atomic displacment pattern is stablized by the trapping of the electron. The atoms then assume "new" equilibrium positions consistent with the presence of the trapped electron. Since the electron is trapped in the atomic displacement pattern which is itself stablized by the presence of the trapped charge, the electron is said to be self-trapped. Reflecting early considerations of self-trapping in polar semiconductors, the unit composed of the self-trapped carrier and its atomic displacment pattern is termed a polaron. This terminology prevails today and is currently applied to self-trapped carriers in both polar and nonpolar materials. The adjectives "small" and "large" affixed to the term polaron denote the spatial extent of the wavefunction of the self-trapped carrier. A large polaron has a length or radius which is large compared with unit-cell dimensions. A small polaron is severely localized on this scale.

The basic physical quantities associated with self-trapping in nonmagnetic crystalline solids are the electron-lattice interaction, the stiffness of the lattice and the transfer energy associated with moving a carrier between adjacent degenerate sites. The electron-lattice interaction denotes the dependence of the energy of an electron on the positions of the atoms surrounding it. The short-range component of the electron-lattice interaction results from the dependence of a localized electron's energy on the positions of the atoms adjacent to it. The short-range component is a general feature of electrons in condensed matter. The long-range component of the electron-lattice interaction arises from the electron's interaction with the electrostatic fields associated with atomic displacements of charged atoms. As such, its presence is not universal.

The purpose of this article is discuss electron-lattice-interaction induced electronic localization in solids for which the electron-lattice interaction is, in the main, of short-range. First, the dependence of the nature of the localization on dimensionality will be addressed. It will be seen that self-trapping in two- and three-dimensions is dichotomous. That is, a carrier is either self-trapped to form a small-polaron or it is not self-trapped at all.[2-6] In most solids the parameters are such that charge carriers are near the borderline between being quasifree and self-trapped.[4] Second, a situation is described in which rising temperature presents a (low-mobility) self-trapped carrier with the possibility of being in a (high-mobility) quasifree state.[4] This leads to a semiconductor-to-semiconductor transition in which the low-temperature transport is via (low-mobility) small-polaron hopping and the high-temperature transport is via (high-mobility) itinerant motion. This suggests a mechanism for a thermally induced insulator-to-metal transition in which the charge carriers on the insulating side of the transition are small polarons. Third, it is shown that defect states also are dichotomous.[5,6,7] They are either of large radius with weak interaction with the lattice or they are severely localized with major electron-induced atomic displacements. In addition, it is shown how other seats of localization, such as defects or disorder, act synergistically with the electron-lattice interaction to produce severe localization. Here, the combination of another source of localization with the electron-lattice interaction produces qualitatively more severe localization than would be produced by either localization mechanism singularly.

The remaining sections of this article describe two distinct examples of the electron-lattice interaction producing severe localization in crystalline solids. In the fourth segment of this paper it is shown how small-polaronic localization of an electronic excitation on the surface of a wide-band semiconductor occurs as an essential step in electronically stimulated atomic desorption.[9] Here again, defects and disorder greatly facilitate localization and thereby enhance desorption. Finally, an example of how the electron-lattice interaction acts in tandem with spin disorder to produce severe small-polaronic localization is considered. In particular, with rising temperature the large-radius donor state of a ferromagnetic semiconductor, e.g. EuO,[10] abruptly shrinks to a severely localized small-polaronic donor. Namely, with increasing excitation of spin waves the free-energy minimum of the system shifts from one associated with a large-radius donor to that of a small-polaronic donor.[11] This problem provides an analytically soluable example of how disorder can act as a trigger for small-polaron formation.

ADIABATIC THEORY OF SELF-TRAPPING

To begin our discussion of self-trapping, let us study an interacting electron and deformable lattice in the "adiabatic" limit. That is, we envision the electron's motion being sufficiently rapid that it is able to adjust to the instantaneous positions of the relatively sluggish atoms. Hence, the electron "sees" a static, albeit deformable, lattice. To find the groundstate of this system, we shall investigate the energy of the coupled electron-lattice system as a function of static atomic displacements. Thus, formally, in the adiabatic limit we ignore the kinetic energy of the atoms of the solid and search for the adiabatic potential.

For simplicity, let us consider an electron added to an isotropic continuum.[3-5] The short-range electron-lattice interaction in this continuum arises from the electron's potential energy depending upon the dilation of the elastic continuum at the electron's position. In the standard situation the electron-lattice interaction is taken to be linear. That is, the potential energy of an electron at position $\mathbf{r}$, $V(\mathbf{r})$, is simply proportional to the continuum's dilatation at that point, $\Delta(\mathbf{r})$: $V(\mathbf{r}) = -Z\Delta(\mathbf{r})$. Here Z is the electron-lattice coupling constant, analogous to the deformation-potential constant in solids. The electron's energy operator is the sum of its kinetic energy operator, T, and its potential energy operator, V(**r**). The electronic energy for a given strain field is then the expectation value of the energy operator with the electron's eigenfunction $\psi(\mathbf{r})$:

$$E_{electron} = \int dr\ \psi^*(\mathbf{r})\ [T + V(\mathbf{r})]\ \psi(\mathbf{r})\ . \qquad (1)$$

In addition to the electron's energy, the total energy of the system includes the strain energy of the deformable continuum. The strain energy of this deformable continuum is

$$E_{strain} = (S/2) \int dr\ \Delta^2(\mathbf{r})\ , \qquad (2)$$

where S is a stiffness constant for the deformable continuum.

To find the adiabatic groundstate, we minimize the total energy of the system with respect to dilatations of the deformable continuum. This yields the minimization condition

$$\Delta(\mathbf{r}) = (Z/S)|\psi(\mathbf{r})|^2\ . \qquad (3)$$

We note that the minimium energy dilatation is proportional to the probability of the electron occupying the location of the dilatation. Furthermore, the minimum energy dilatation is proportional to the electron-lattice interaction constant, Z, and inversely proportional to the stiffness of the continuum, S. The wave equation for the electron in the adiabatic groundstate is then found by utilizing the minimum-energy dilatation, Eq. (3), in the electron's potential energy:

$$[T - (Z^2/S)|\psi(\mathbf{r})|^2]\ \psi(\mathbf{r}) = E\ \psi(\mathbf{r})\ . \qquad (4)$$

This is a nonlinear equation. The greater the degree of the electron's localization, the deeper the strain-induced potential well which fosters the electron's localization. The nonlinearity of the electron's wave equation denotes the feedback aspect of self-trapping.

A scaling argument provides a simple way of discerning the nature of the solutions of this nonlinear wave equation.[4,5] Namely, we note that the groundstate solution must be spherically symmetric. As such, we can investigate various spherically symmetric solutions by varying their characteristic "radius," R. It is found that the groundstate energy is of the form:

$$E(R) = T_e/R^2 - V_{int}/2R^d\ , \qquad (5)$$

where d is the system's dimensionality. The first term of E(R) arises from the electron's kinetic energy and the second term arises from the system's electron-lattice interaction.

As illustrated in Fig. 1, for a one-dimensional system (d = 1) the adiabatic energy possesses a single finite-radius minimum. This corresponds to the formation of a finite-radius polaron. However, in two dimensions (d = 2) the system has the two radically distinct solutions illustrated in Fig. 1. For $T_e > V_{int}/2$ the energy falls monotonically to a minimum at $R = \infty$. This corresponds to the carrier being unbound in a undilated continuum. Alternatively, if $T_e > V_{int}/2$, the system has a solitary minimum at R = 0. This corresponds to the carrier shrinking to the smallest possible size as it is bound in the infinitely deep potential well associated with an infinitely great local strain. This is the continuum version of a carrier forming a small polaron. With small-polaron formation

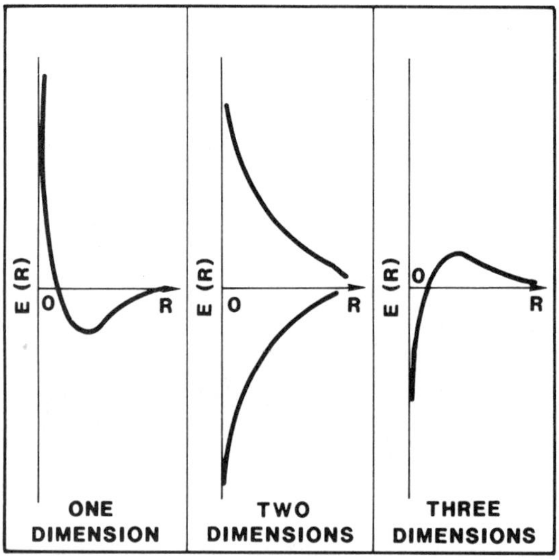

Fig. 1. E(R)-versus-R curves for an excess charge in one-, two-, and three dimensional continuua.

in a real solid, composed of discrete atoms, the carrier, rather than being confined at a point, is simply localized on a single site. Thus, the minimum acceptable value of R, R_{min}, is of the order of an interatomic separation. Thus, we see that self-trapping is dichotomous in two dimensions with a short-range electron-lattice interaction. The carrier either self-traps to form a small polaron or it remains free.

For a three-dimensional system (d = 3), Fig. 1, the kinetic energy term dominates at large radii, while the electron-lattice term dominates as R approaches zero. Thus, E(R) is a peaked function which has minima at R = 0 and as R approaches infinity. The minimum at R = 0 corresponds to the electron being severely localized in an infinitely deep potential well. It is the small-polaron solution. The minimum at $R = \infty$ corresponds to an unbound electron moving in an dilation-free continuum. It is the free-electron solution. Thus, in its groundstate, an electron in a three-dimensional deformable system is either severely localized to form a small polaron or it remains quasifree. The system is dichotomous. The absolute minimum is determined by the position of the small-R cut-off, R_{min}.

It should also be noted that to pass adiabatically from the quasifree solution to the small-polaron solution requires transcending an energy barrier.[3-6] This indicates that for a free-electron to self-trap the system must pass through an energetically unfavorable deformation. In other words, the system must negotiate the "barrier to self-trapping." The time associated with a carrier waiting to transcend this barrier is termed "the time delay for self-trapping."

SEMICONDUCTOR-TO-SEMICONDUCTOR TRANSITION

We have seen that an electron in a three-dimensional deformable solid with a short-range electron-lattice interaction either forms a small polaron or remains quasifree. Based on this realization, one can carry out a variational study of the low-lying eigenstates of a carrier in a discrete deformable solid.[3,4] In addition to taking explicit account of the atomicity of a real solid, such treatments transcend the adiabatic approximation. As such, they provide criteria for small-polaron and quasifree-carrier existence. As an outgrowth, these studies suggest a mechanism for a semiconductor-to-semiconductor transition. In such a transition, as the temperature is raised, charge transport abruptly changes from being due to small polarons to being dominated by itinerant carriers.

Typical results of such studies are schematically indicated in Fig. 2. Here the energy of a charge carrier is plotted against the electron-lattice coupling strength, γ. Increasing γ may be thought of as decreasing the stiffness of the deformable lattice. The electron-lattice coupling strengths γ_{max} and γ_{min} are the limiting coupling strengths for the respective existence of quasifree and small-polaronic charge carriers. Small polarons can exist for values of γ exceeding γ_{min}. Quasifree carriers can exist for values of γ below γ_{max}. Between these two limiting values quasifree and small-polaronic carriers can coexist.

For values of γ below γ_{min} a small polaron, if formed, will spontaneously decompose as the electron lowers its kinetic energy by spreading out. For values of γ in excess of γ_{max} a quasifree carrier will spontaneously self-trap since it can no longer move between sites sufficiently rapidly to preclude inducing significant atomic displacments. Indeed, since the force that a severely localized stationary carrier exerts on the adjacent atoms is typically large (~ 2-3 eV/A), it is only by moving rapidly between sites that an excess charge can exist without producing significant atomic displacements. Thus, the collapse of a quasifree carrier into a small polaron is analogous to a pebble skimming on the surface of a pond sinking as it slows.

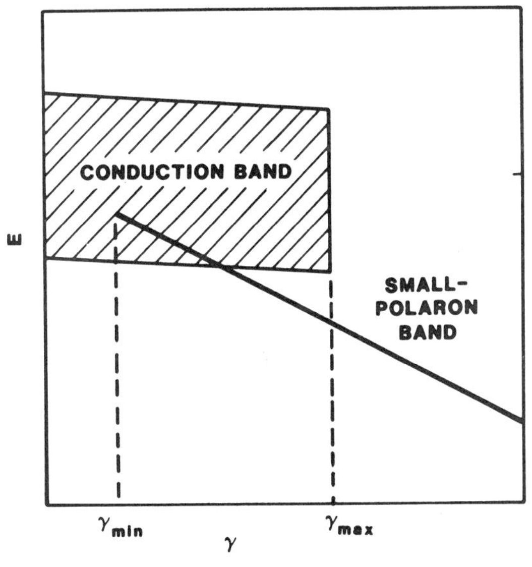

Fig. 2. The energy levels of an excess carrier in a three-dimensional deformable lattice are plotted against the electron-lattice coupling strength.

With rising temperature the coexistence region broadens as q_{min} decreases slowly and γ_{max} rises rapidly.[4] The rise in γ_{max} with temperature occurs because an electron requires an increasingly strong electron-lattice interaction to significantly affect atomic motion as the thermal agitation of the atoms increases. In other words, with rising temperature the average momenta of the vibrating atoms increases.

The rapid increase of γ_{max} with rising temperature provides a mechanism for a semiconductor-to-semiconductor transition.[4] In particular, envision the electron-lattice coupling parameter of a material exceeding γ_{max} at low temperatures. Then, at low temperatures only small polarons exist. The electronic transport is the low-mobility hopping of small polarons. With increasing temperature γ_{max} increases and ultimately γ_{max} exceeds the value of the electron-lattice coupling that characterizes the material. Two types of charge carriers can then coexist: high-mobility quasifree carriers and low-mobility small polarons. If the energy gap between the conduction band and the small-polaron band is sufficiently narrow, the number of high-mobility carriers will be large enough to dominate the electronic transport. This yields a semiconductor-to-semiconductor transition in which the low-temperature electronic transport is that of small polarons and the high-temperature electronic transport is dominated by the motion of itinerant carriers. One might also expect this type of driving mechanism to be operative in semiconductor-to-metal transitions in which the low-temperature electronic transport is that of small polarons.

DEFECTS, DISORDER AND SELF-TRAPPING

We have seen that there is a dichotomy between charge carriers in solids; they are either small polarons or quasifree. Similarly, there is a dichotomy between defect states in semiconductors and insulators.[5 7 8] Namely, defect states are either severely localized small-polaronic defect states or they are weakly localized states for which the lattice displacements are minimal. For example, consider a coulombic defect added to the previously considered deformable continuum.[5,8] With the defect the adiabatic energy of the system garners a term in addition to those of Eq.(5):

$$E(R) = T_e/R^2 - V_{int}/2R^d - V_{def}/R \ . \qquad (6)$$

In the absence of the electron-lattice interaction ($V_{int} = 0$) the adiabatic energy has a solitary finite-radius minimum (curve a of Fig. 3). However, with progressively strong electron-lattice interactions the adiabatic energy curves become those of curves b and c. Curve b depicts two minima: a finite radius minimum and a small-polaronic minimum at $R = 0$. Curve c depicts only a solitary minimum: the small-polaronic minimum. Noting that the E(R)-versus-R curves only have meaning when R exceeds R_{min}, we see that there are four possible situations. With vanishing electron-lattice interaction there is only a finite-radius defect state. As the electron-lattice interaction is increased a metastable small-polaronic defect state (at $R = R_{min}$) occurs. With a further increase of the electron-lattice interaction the small-polaronic defect state becomes stable and the large-radius defect state becomes metastable. Finally, with a sufficiently strong electron-lattice interaction only the small-polaronic defect state is possible. Thus, the electron-lattice interaction can lead to the collapse of defect states to severely localized (small-polaronic) defect states.

An alternative way of viewing the situation is to consider the effect of progressively increasing the strength of the defect potential on the original dichotomy between quasifree and small polaronic carriers. As the

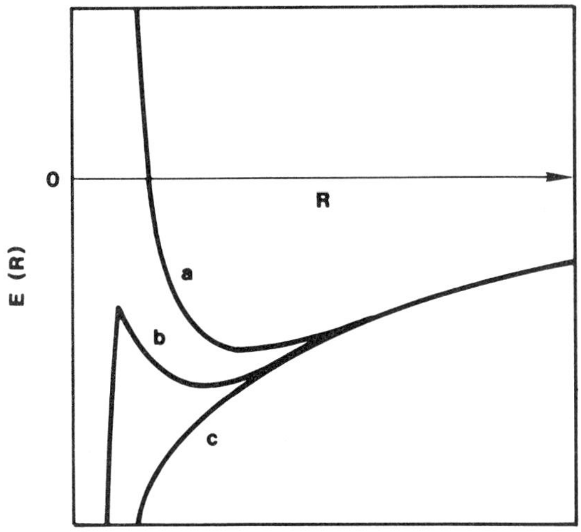

Fig. 3. The E(R)-versus-R curves for an electron in a three-dimensional deformable continuum containing an attractive coulombic defect. The three curves (a, b, and c) correspond to zero, and two successively larger values of the electron-lattice coupling strength.

strength of the defect potential is increased (1) the quasifree state is converted into a large-radius defect state and (2) the barrier to self-trapping is reduced. With a further increase in the strength of the defect potential the barrier to self-trapping is eliminated. Then, only the small-polaronic defect state remains. This approach directly shows how disorder fosters small-polaron formation.

The essential features of self-trapping in disordered systems is clear. Namely, the electron-lattice interaction of a severely localized stationary charge is generally sufficiently strong so as to induce very significant displacements of the atoms surrounding it. It is only the rapid motion of a charge between sites which precludes it becoming self-trapped. Because of the feedback nature of self-trapping (i.e., the nonlinearity of the associated wave equation), with a short-range electron-lattice interaction carriers either remain quasifree or self-trap to form small polarons. The imposition of a localizing potential, associated with defects or disorder, fosters localization. By impeding the rapid intersite motion of a charge carrier such a localizing potential encourages small-polaron formation. Thus, disorder aids self-trapping. Small-polaron formation is therefore more prevalent in disordered materials than in their crystalline counterparts. These findings are general. They do not depend on the detailed form of the defect or disorder potential.

ELECTRONICALLY STIMULATED DESORPTION

The electronically stimulated desorption of atoms from the surface of a semiconductor is a phenomenon in which small-polaron formation plays a crucial role. Imagine a layer of adsorbed atoms on the surface of a semiconductor (e.g., hydrogen adsorbed on the surface of silicon). Under a variety of stimuli (e.g., irradiation) bond-breaking excitations, such as holes, emerge at the surface. Confining such an excitation to a single surface bond severs that bond and expels the associated adsorbed atom from the surface. Thus, while the usual polaron situation has a static localized charge producing large atomic displacements, here (in this surface-physics analogy) a static localized charge actually breaks a bond and ejects an atom from the surface. Furthermore, as in small-polaron formation, the essential problem is that a surface electronic excitation is not generally confined to a single site (bond). Rather, the excitation can frequently move between degenerate surface sites in less time than is required for a significant displacement of a surface atom to occur. Hence, we are led to ask how an electronic surface excitation localizes on a site so as to produce desorption.

This problem can be approached in close analogy with the previously described adiabatic theory of self-trapping.[9] Here we envision surface (two-dimensional) self-trapping in which the presence of a carrier at a solitary surface site not only acts to displace the surface atom but also breaks the bond of that surface atom to the bulk. In essence we now have a more complicated electron-lattice interaction. For pedagogic simplicity, we model the system as a deformable continuum. Then, the potential energy of an electronic excitation at position **r** is

$$V(\mathbf{r}) = -Fx(\mathbf{r}) - (k/2)x^2(\mathbf{r}) , \qquad (7)$$

where $x(\mathbf{r})$ is the bond displacement at surface position $\mathbf{r}$; F is the force exerted by the electronic excitation to expel the surface atom; k is the stiffness constant of the electronic-excitation-free surface bond. The strain energy of the surface bonds is given by:

$$E_{strain} = (k/2) \int d\mathbf{r}\, x^2(\mathbf{r}) . \qquad (8)$$

The electronic energy of the system is the expectation value of the sum of the excitation's kinetic and potential energies. $T + V(\mathbf{r})$:

$$E_{electronic} = \int d\mathbf{r}\, \psi^*(\mathbf{r}) \,[\, T + V(\mathbf{r}) \,]\, \psi(\mathbf{r}) , \qquad (9)$$

where $\psi(\mathbf{r})$ is the electronic eigenfunction for the deformation pattern $x(\mathbf{r})$.

In analogy with our adiabatic treatment of the groundstate of the polaron system, we now minimize the total energy of the system with respect to deformations of surface bonds. This yields, in analogy with Eq.(3), an expression for the groundstate deformation pattern in terms of the groundstate electronic wave function

$$x(\mathbf{r}) = F\, |\psi(\mathbf{r})|^2/k\, [\, 1 - |\psi(\mathbf{r})|^2] . \qquad (10)$$

Here the surface strain at position **r** is proportional to both the probability of the electronic excitation being at position $\mathbf{r}$, $|\psi(\mathbf{r})|^2$, and the expelling force that such an excitation exerts, F. The surface strain is also inversely proportional to the stiffness of the occupied bonds; the dependence of the denominator on the occupation probability of the electronic excitation reflects the weakening of the surface bonds induced by the presence of the electronic excitation. When the excitation is completely delocalized, with the excitation's probability amplitude being negligibly small (inversely proportional to the area of the surface), the denominator of Eq.(10) is simply k. Conversely, when the excitation is completely localized on a single surface bond, the denominator of Eq.(10) vanishes since the occupied bond is then broken. Clearly, with the deformation field at its adiabatic minimum, given by Eq.(10), the wave equation associated with tne potential, Eq.(7),is highly nonlinear.

As in the polaron problem, we may use a scaling argument to ascertain the nature of the solutions of this highly nonlinear equation. We note that the groundstate eigenfunctions of the surface problem are circularly symmetric (d=2). We can then consider eigenfunctions of varying radii. The resulting total adiabatic energy is of the form:

$$E(R) = T_e R^{-2} - (F^2/2k)R^{-2}\int dr\ |\psi(r)|^4/[1 - |\psi(r)|^2 R^{-2}] \ . \quad (11)$$

The E(R)-versus-R curves are of the two types shown in Fig. 4. The upper curve corresponds to $T_e > F^2/2k$ and the lower curve to $T_e < F^2/2k$. The situation at arbitrarily large R corresponds to an electronic excitation spread out on an infinite undeformed surface. Conversely, when the excitation is localized at a single site, desorption occurs and E(R) approaches negative infinity. The desorption process corresponds to shrinking from infinite R to the value of R corresponding to desorption. The upper curve possesses a barrier to desorption. Since an energy barrier must be transcended before desorption can occur, it corresponds to delayed desportion. Namely, the atoms about the desorbing site must achieve a minimum deformation before desortpion will occur. The lower curve of Fig. 4 depicts spontaneous desorption. Figure 5 provides a physical picture of electronically stimulated desorption. Successive subfigures correspond to decreasing values of R. As the electronic excitation becomes increasingly localized at the desorption site, the bond at that site progressively weakens. The bond is broken when the excitation is completely localized on the desorption site.

Just as defects foster self-trapping in the bulk, so they encourage desorption on a surface. The addition of a defect potential to the upper curve will reduce or eliminate the barrier to desorption. Similarly, a defect potential will steepen the spontaneous desorption curve. This corresponds to accelerating the desorption process. In addition, limiting the surface coverage, setting a maximum value of R, also fosters desorption. In particular, it reduces the barrier experienced on the upper curve and eliminates the most gently falling portion of the lower desorption curve.

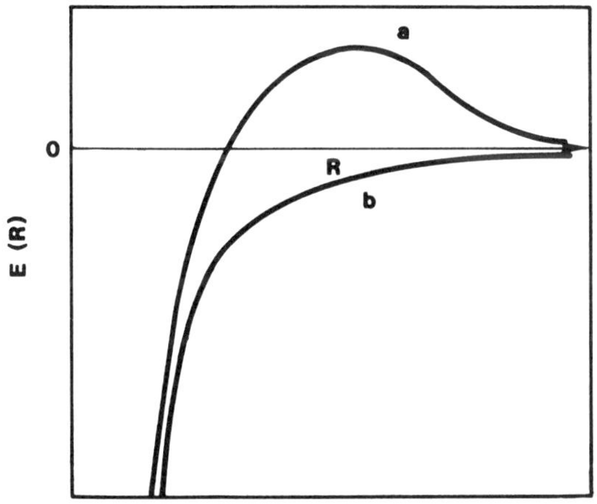

Fig. 4. The E(R)-versus-R curves for electronically stimulated desorption. Curve a corresponds to time-delayed desorption while curve b denotes spontaneous desorption. E(R) approaches negative infinity as a surface atom is desorbed.

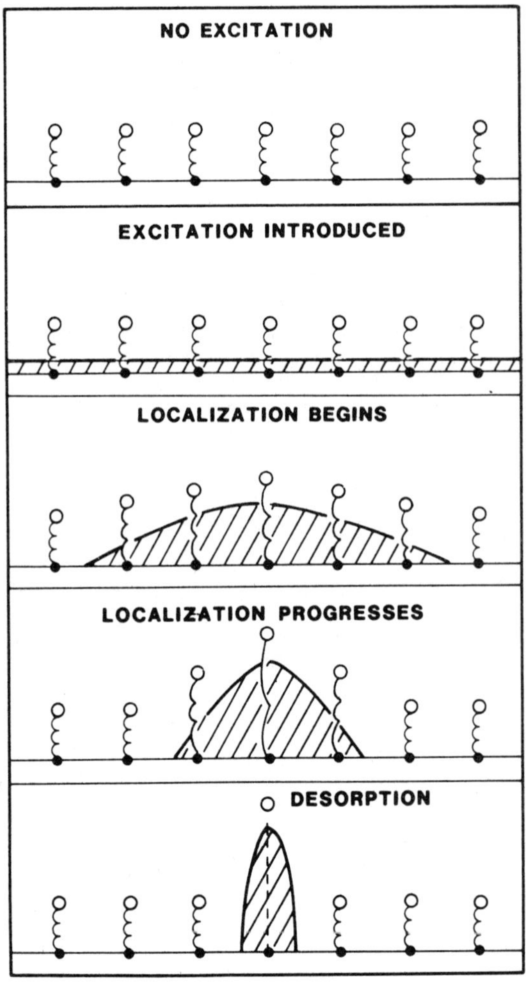

Fig. 5. The electronically stimulated desorption of one of a line of adsorbed atoms is schematically depicted. The number of links in the springs coupling the adsorbed atoms to the bulk denotes the stiffness of the bonding of the adsorbed atom to the host surface atoms. The cross-hatched area depicts the wavefunction of the electronic excitation. As it localizes desorption occurs.

Thus, electronically stimulated desorption is associated with the self-trapping of a surface electronic excitation. Here, as in the bulk, sources of localization, defects and limited coverage, facilitate self-trapping. As such, they aid electronically stimulated desorption.

SPIN-DISORDER-INDUCED SELF-TRAPPING IN FERROMAGNETIC SEMICONDUCTORS

A major difficulty in studying the effect of disorder on self-trapping lies in our inability to represent disorder in an analytically tenable manner. Indeed, in practice the disorder of a semiconductor when it is "amorphous" is neither well specified nor well controlled. It is therefore of interest to examine systems for which disorder can be imposed in a controlled and reversible manner. Ferromagnetic semiconductors, such as EuO, are such systems. As the temperature is raised toward the Curie temperature the degree of spin disorder experienced by a charge carrier increases. Here one finds evidence that is consistent with the idea that disorder facilitates the formation of small-polaronic states. In particular, in these ferromagnetic semiconductors, large-radii donors abruptly shrink to severely localized states as the temperature is raised toward the Curie temperature.[10] In this section a recent theoretical study of the thermally induced abrupt collapse of a donor state in a ferromagnetic semiconductor[11] will be summarized.

The energy of the magnetic interaction between a donor electron, of spin σ and wavefunction $\psi(\mathbf{r})$, with a local spin at position $\mathbf{r}$, $S(\mathbf{r})$, is simply that due to intraatomic exchange with interaction constant I:

$$E_{int} = -I \int d\mathbf{r}\, \sigma\cdot S(\mathbf{r})\, |\psi(\mathbf{r})|^2 . \qquad (12)$$

One may treat the deviations of the local spins from the magnetization direction in terms of the creation of spin waves. Then, the interaction energy for a donor of radius R and spin in the magnetization direction is

$$E_{int} = -IS + (I/N)\sum_{q} \Sigma_{q'}\, c^*_{q}\, c_{q'}(1 + |q - q'|^2R^2/4)^{-2}, \qquad (13)$$

where S is the magnitude of a local spin of the ferromagnet, N is the number of sites in the crystal, and c^*_q and c_q are, respectively, creation and annihilation operators for spin waves of wavevector $\mathbf{q}$. The interaction of the donor electron with the spin waves leads to an alteration of the magnon energies. Calculated perturbatively through second order one finds the change of the magnon energies to be

$$\hbar\Delta\omega_q = I/N - (I/N)^2\sum_{q'} \frac{(1 + |q - q'|^2R^2)^{-2}}{\hbar\omega_{q'} - \hbar\omega_q} . \qquad (14)$$

The donor-induced change of the magnon frequencies gives rise to a donor-induced change of the magnon free-energy. In particular, the second-order contribution to the shift of the magnon frequencies leads to an R-dependent change of the magnon free energy: $-(108/\pi^{4/3})k_BT(I/k_BT_c)^2R^{-2}$. The R-dependence of the free-energy of a donor in a ferromagnetic semiconductor is then the sum of the nonmagnetic terms, given in Eq.(6), and the above contribution. It is clear that the magnetic contribution provides increasing localization as the temperature is raised. In fact, as illustrated in Fig. 6, with increasing temperature the donor state of lowest free energy shifts from being one of large radius to being a severely localized small-polaronic donor. The parameters are chosen to model the donor-state collapse observed in EuO.[10] Thus, we see an explicit illustration of how the imposition of disorder can induce small-polaron formation.

SUMMARY

The electron-lattice interaction plays a central role in situations in which excess electronic charge is severely localized. In these instances the presence of the added charge induces significant atomic displacements of

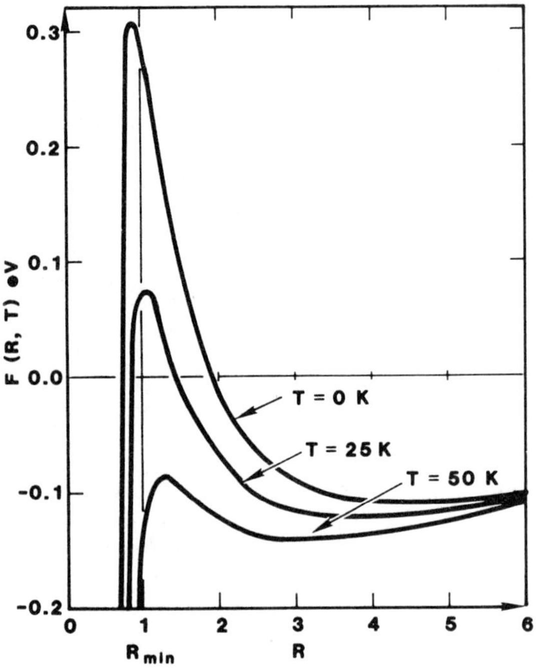

Fig. 6. The free-energy of a donor in a ferromagnetic semiconductor is shown at three temperatures. Due to the discreteness of the lattice the minimum acceptable value of R is R_{min}. In the units of this plot $R_{min} = 1$. At T = 0 K, only the large-radius donor can exist. At T = 25 K, the small-polaronic donor state is metastable. Above T = 50 K the severely localized small-polaronic donor becomes stable. The parameters for this plot are chosen to model the donor state in EuO.

the surrounding atoms. This occurs because in solids the forces exerted by a severely localized electronic charge on the surrounding atoms are typically quite large. It is only the ability of a charge to move rapidly between atomic sites which enables it to avoid small-polaron formation.

When an excess charge interacts with the atomic system via a short-range electron-lattice interaction the eigenstates of the coupled electron-lattice system are dichotomous. The charge either induces minimal atomic displacements or it forms a small-polaron. This dichotomy accounts for the sharp distinction between systems with itinerant carriers and small-polaron transport. This dichotomy also leads to a mechanism for a semiconductor-to-semiconductor transition. Here, low-temperature transport is via small-polaron motion and high-temperature transport is dominated by the motion of itinerant carriers. Another manifestation of this dichotomy is the qualitative distinction between large-radius defect and impurity states (such as the phosphorus donor in crystalline silicon) and severely localized small-polaron defects (such as F-centers). Analysis of electronically stimulated desorption of atoms from the surface of a solid also manifests this dichotomy. Namely, the desorption can occur spontaneously or only upon transcending an energy barrier (time-delayed desorption).

A central theme of these studies is that disorder and the electron-lattice interaction act synergistically in inducing small-polaron formation. Here, disorder and the electron-lattice interaction can produce qualitatively more severe localization acting in concert than would arise from each effect acting individually. This occurs because of the nonlinear (feedback) nature of self-trapping. An example of the dramatic effect of disorder on self-trapping is found in the thermally induced collapse of a shallow donor state in a ferromagnetic semiconductor. In this instance, with increasing temperature the electron of a large-radius donor encounters greater spin disorder. Ultimately, the donor collapses to a severely localized small-polaronic donor state. Thus, one expects small-polaron formation to be much more prevalent in disordered materials than in crystalline solids.

ACKNOWLEDGEMENT

This work is supported by the United States Department of Energy under contract DE-AC04-756DP00789.

REFERENCES

1. L. Landau, On the Motion of Electrons in A Crystal Lattice, Phys. Z. Sowjetunion 3:664 (1933).
2. E. I. Rashba, Theory of Strong Interaction of Electron Excitation with Lattice Vibrations in Molecular Crystal, Opt. Specktrosk. 2:75 (1957).
3. Y. Toyozawa, Self-Trapping of an Electron by the Acoustical Mode of Lattice Vibration. I, Prog. Theor. Phys. 26:29 (1961).
4. D. Emin, On the Existence of Free and Self-Trapped Carriers in Insulators: An Abrupt Temperature-Dependent Conductivity Transition, Adv. Phys. 22:57 (1973).
5. D. Emin and T. Holstein, Adiabatic Theory of an Electron in a Deformable Continuum, Phys. Rev. Lett. 33:303 (1976).
6. N. F. Mott and A. M. Stoneham, The Lifetime of Electrons, Holes and Excitons before Self-Trapping, J. Phys. C 10:3391 (1977).
7. P. W. Anderson, Effect of Franck-Condon Displacements on the Mobility Edge and the Energy Gap in Disordered Materials, Nature 235:163 (1975).
8. D. Emin, Comments on the Theory of Localized States in Semiconducting Noncrystalline Solids, in "Physics of Structurally Disordered Solids," S. S. Mitra, ed., Plenum, New York (1976).
9. D. R. Jennison and D. Emin, Strain-Induced Localization and Electronically Stimulated Desorption and Dissociation, Phys. Rev. Lett. 51:1390 (1983).
10. J. B. Torrence, M. W. Schaefer and F. R. McGuire, Bound Magnetic Polarons and the Insulator-Metal Transition in EuO, Phys. Rev. Lett. 29:1168 (1972).
11. D. Emin, M. Hillery and N. H. Liu, Thermally Induced Abrupt Collapse of a Shallow Donor State in a Ferromagnetic Semiconductor, Phys. Rev. Lett. (in press).

DENSITY CORRELATIONS NEAR THE MOBILITY EDGE

Franz Wegner

Institut für Theoretische Physik
Ruprecht-Karls-Universität
D-6900 Heidelberg, F.R. Germany

ABSTRACT

The correlations of the eigenfunctions of a particle in a spinindependent time-reversal invariant random potential near the mobility edge in d = 2+ε dimensions are determined. The formulation in terms of the nonlinear sigma model is used and the previously employed technique to derive the participation ratio near criticality is extended to correlations by means of the operator-product expansion.

INTRODUCTION

In 1958 Anderson[1] argued that strong disorder would prevent noninteracting particles (originally the magnetization) from diffusing. In the sixties the idea of a mobility edge separating localized from extended eigenstates in noninteracting systems with weak disorder was advocated by Banyai[2], Mott[3], and Cohen, Fritzsche, and Ovshinsky[4]. In the early seventies the transition from the extended to the localized states was considered a critical phenomenon[5] governed by power laws and critical exponents[6-9]. Although it became clear that a field theoretic formulation[10-12] with an n=0 component field S could be given similarly as for the random walk and the polymer problem the usual Landau-Ginsburg theory did not yield correct results.

First scaling theories were developed in the middle of the seventies by Thouless et al.[13] and the present author[14]. The scaling law

$$s = (d - 2)\nu \tag{1}$$

which connects the critical exponent of the residual conductivity $\sigma \sim (E - E_c)^s$ and the localization length $\xi \sim (E_c - E)^{-\nu}$ seemingly favoured Mott's idea of a minimum metallic conductivity[3] in d=2 dimensions and it supported Cohen's and Jortner's[15] and the author's[14] arguments for a continuously vanishing conductivity at the metal-insulator transition in d=3 dimensions. In 1978 numerical calculations by Licciardello and Thouless[16] raised doubts on the existence of a mobility edge in two dimensions.

The ends of 1978 and 1979 finally brought the important breakthrough for the solution of the mobility edge problem for noninteracting particles. Four approaches were put forward:

(i) the selfconsistent approach by Götze[17] based on a mode coupling formalism, subsequently modified by Vollhardt and Wölfle[18] for time-reversal invariant potentials.

(ii) the scaling approach by Abrahams, Anderson, Licciardello, and Ramakrishnan[19] based on the scaling ideas by Edwards, Licciardello, and Thouless[13]. The authors went beyond the coherent potential approximation taking into account contributions to the conductivity originally considered by Langer and Neal[20].

(iii) a systematic $1/n$ expansion by Oppermann and the author[21] for a model with n orbitals per lattice site[22]. This is a generalization to d dimensions of the zero-dimensional model introduced by Wigner[23] for a statistical description of nuclei.

(iv) a nonlinear sigma by the author[24] and Schäfer[25] where it became apparent that the diffusion modes are the Goldstone modes of this model. It turned out that the composite fields $Q = SS$ suggested by Aharony and Imry[26] are the proper fields for this problem, despite the fact that the order parameter $< Q >$ does not vanish at the mobility edge.

All these approaches agree that d=2 is the lower critical dimensionality and that as long as spin-orbit effects can be neglected there are no extended states in two dimensions. However, the localization length may become extremely large yielding a quasimetallic behaviour. For small $\varepsilon = d-2$ the conductivity exponent s approaches one if time-reversal invariance is conserved (spin-orbit coupling neglected), and $1/2$ if it is broken.

It is Nevill Mott's merit to have stimulated by his work the interest in the metal-insulator transition over decades and to have contributed many extremely useful ideas, concepts and results.

After this 1979 breakthrough the field has seen rapid progress, see e.g. the Proceedings of the Taniguchi Symposium[27] and of the LITPIM conference in Braunschweig[28]. But there are still important questions to be answered, two of which are:

(i) Although the exponent s is known for the time-reversal invariant (orthogonal) case up to four-loop order, $s = 1 + O(\varepsilon^4)$,[29] it is not clear whether it stays at s=1. There is no estimate available for s in dimension three whose precision is comparable (for estimates see refs. 30, 31) to that of estimates of critical exponents from high-temperature series expansions in usual critical phenomena. The situation is even worse if time-reversal invariance is broken by spin-flip scattering (unitary case), $s = 1/2 + O(\varepsilon)$, and in the time-reversal invariant spin-orbit potential (symplectic) case for which there is not even an ε-expansion exponent available.

(ii) It is not known whether there is an upper critical dimensionality d_c for these systems. The selfconsistent approximation yields $d_c=4$. Kunz and Souillard[32] have collected more arguments for this value of d_c. Strayley[33] argues that $d_c=6$. Other observations that four and six are special (not necessarily critical) dimensions are found in refs. 34-36. None of these arguments have paved the way for an ε-expansion around d_c. An ε-expansion around $d_c=8$, [37] violates exact results[38].

After this sketchy historical introduction to the modern theory of the mobility edge I turn to a rather special problem: the correlations of the eigenfunctions in a random potential in $d = 2+\varepsilon$ dimensions close to the mobility edge. The formulation in terms of the nonlinear sigma model[25] is used and the technique to determine the participation ratio near criticality[39] is extended to determine the correlations by means of the operator-product expansion[40-43]. Restriction is made to the case of a spinindependent potential obeying time-reversal invariance whose mobility edge behaviour is thus governed by the orthogonal universality class. Some results derived here were reported earlier[44].

In ref. 39 the critical behaviour of

$$\hat{\rho}\, P^{(k)}(E) = \overline{\sum_i |\psi_i(r)|^{2k} \delta(E - E_i)} \sim (E_c - E)^{\pi_k} \tag{2}$$

in the region of the localized states, and of

$$\hat{\rho}^k / p^{(k)}(E) = \lim_{\eta \to 0} \overline{\rho_\eta(r)^k} \sim (E - E_c)^{-\mu_k} \tag{3}$$

in the region of the extended states were determined with

$$\rho_\eta(r) = \sum_i |\psi_i(r)|^2 \, \delta_\eta(E - E_i) \tag{4}$$

and the smeared δ-function

$$\delta_\eta(E - E_i) = \frac{\eta/\pi}{(E-E_i)^2+\eta^2} \, . \tag{5}$$

The sums run over all eigenstates ψ_i (energy E_i) of a tight-binding model with random potential. $\hat{\rho}$ is the density of states per energy and orbital. The bar indicates the ensemble average. $P^{(2)}$ is the inverse participation ratio, $p^{(2)}$ a "smeared" participation ratio. The critical exponents are related by

$$\pi_k = (k - 1)d\nu - \mu_k \, . \tag{6}$$

For a tight-binding model with real matrix elements an ε-expansion for a $d = 2+\varepsilon$ dimensional system predicts in one-loop order

$$\mu_k = k(k - 1) + O(\varepsilon), \tag{7}$$

$$\pi_k = (k - 1)(2\varepsilon^{-1} + 1 - k) + O(\varepsilon). \tag{8}$$

Here we consider the corresponding density correlations. For localized states we define

$$\hat{\rho}\, \Gamma^{loc}_{k_1,k_2}(r_1 - r_2, E) = \overline{\sum_i |\psi_i(r_1)|^{2k_1} |\psi_i(r_2)|^{2k_2} \, \delta(E - E_i)} \, . \tag{9}$$

Since the nonlinear σ-model does not allow a direct determination of Γ we consider

$$A_{k_1,k_2}(r_1 - r_2, E, \eta) = \overline{\rho_\eta(r_1)^{k_1} \rho_\eta(r_2)^{k_2}} \, . \tag{10}$$

In terms of the spectral function

$$S_{k_1,k_2}(r_1 - r_2,\varepsilon_1,\ldots\varepsilon_{k_1+k_2}) =$$

$$= \prod_{b=1}^{k_1} < r_1|\delta(\varepsilon_b - H)|r_1 > \prod_{b=1}^{k_2} < r_2|\delta(\varepsilon_{k_1+b} - H)|r_2 > \quad (11)$$

A can be written

$$A = \prod_{b=1}^{k_1+k_2} (\int d\varepsilon_b \delta_\eta(E - \varepsilon_b))\, S_{k_1k_2}(r_1 - r_2,\varepsilon_1,\ldots\varepsilon_{k_1+k_2}). \quad (12)$$

For localized states the most singular contribution as a function of the energies ε to S is given by $\hat{\rho}\Gamma$,[45]

$$S = \hat{\rho}\Gamma^{loc}_{k_1,k_2}(r_1 - r_2,\varepsilon_1) \prod_{b=2}^{k_1+k_2} \delta(\varepsilon_b - \varepsilon_1) + \ldots \quad (13)$$

which can be extracted by the limit

$$\hat{\rho}\Gamma^{loc}_{k_1,k_2}(r_1 - r_2,E) = C^{-1}_{k_1+k_2} \lim_{\eta\to 0} (\eta^{k_1+k_2-1} A_{k_1,k_2}(E,\eta)) \quad (14)$$

where

$$C_k = \eta^{k-1} \int (\delta_\eta(x))^k dx = (2\pi)^{1-k} \frac{(2k-3)!}{(k-1)!} \quad (15)$$

In the region of the extended states no such δ-function contribution as in (13) exists and it has to be expected that for fixed r_1 and r_2 the function S varies smoothly as a function of the energies ε as long as the energy differences are small in comparison to $D/|r_1 - r_2|^2$, where D is the diffusion constant. Thus we expect the limit $\eta\to 0$ of A to exist and introduce

$$\hat{\rho}^{k_1+k_2}\Gamma^{ext}_{k_1,k_2}(r_1 - r_2,E) = \lim_{\eta\to 0} A_{k_1,k_2}(r_1 - r_2,E,\eta). \quad (16)$$

MOBILITY EDGE BEHAVIOUR

We now map the problem of an electron in a random potential on the nonlinear sigma-model which is governed by

$$\mathcal{H}_o = \frac{\hbar\pi\rho}{4} \int d^dr\, tr(\frac{D}{2}\nabla Q(r)\nabla Q(r) + 2\eta s Q(r)). \quad (17)$$

The matrices Q(r) are composed of matrix elements $Q^{pp'}_{bb'}(r)$ with the energy indices p, p' which assume the values 1 and 2 corresponding to energies

$$z_p = E + s_p\eta, \quad s_1 = i, \quad s_2 = -i. \quad (18)$$

The matrix s is diagonal,

$$s^{pp'}_{bb'} = s_p\, \delta_{pp'}\, \delta_{bb'}. \quad (19)$$

The replica indices b, b' run from 1 to m, where m=0 due to the replica trick. The independent fields are the elements of $Q^{12}(r)$, which are real,

$$Q^{21}_{bb'} = Q^{12}_{b'b} , \quad Q^{11} = -i(1 + Q^{12}Q^{21})^{1/2} , \quad Q^{22} = i(1 + Q^{21}Q^{12})^{1/2} . \tag{20}$$

Here we normalize Q so that its eigenvalues are $\pm i$, and we incorporate the hyperbolic symmetry[25] which accordingly yields a different sign convention from that in refs. 24, 39, 46 for the elements of Q and t below. D is the microscopic diffusion constant, ρ the density of states per energy and volume,

$$\hat{\rho} = v\rho, \; v = a^d, \tag{21}$$

where v is the volume per electronic orbital, a the lattice spacing (in the case of a simple (hyper) cubic lattice). The conventional dimensional coupling constant t is given by

$$\hbar\rho D a^{d-2} = (2^d \pi^{d/2+1}\Gamma(d/2)t)^{-1}, \tag{22}$$

which in d=2 dimensions reduces to

$$\hbar\rho D = (4\pi^2 t)^{-1}. \tag{23}$$

The dimensionless conductance (per spin degree of freedom) is given by

$$g = \hbar e^{-2} a^{d-2} \sigma = \hbar\rho D a^{d-2} \tag{24}$$

according to the Einstein relation and is thus directly expressed by t, (22). Thus g,[19] is the proper quantity to be renormalized. The quantity A can be expressed as

$$A_{k_1 k_2} = \hat{\rho}^{k_1+k_2} < \prod_{b=1}^{k_1} \hat{Q}_b(r_1) \prod_{b=1}^{k_2} \hat{Q}_{k_1+b}(r_2) > \tag{25}$$

where the expectation value is taken with respect to $e^{-\mathcal{H}_o}$ and

$$\hat{Q}_b(r) = \frac{i}{2} (\hat{Q}^{11}_{bb}(r) - \hat{Q}^{22}_{bb}(r)). \tag{26}$$

In order to evaluate (20) we add appropriate source terms to $\mathcal{H}_o$

$$\mathcal{H} = \mathcal{H}_o(t,\eta) - \mu_1 \prod_{b=1}^{k_1} \hat{Q}_b(r_1) - \mu_2 \prod_{b=1}^{k_2} \hat{Q}_{k_1+b}(r_2). \tag{27}$$

Since the tight-binding model is defined on a lattice with spacing a, there is a cut-off of the Fourier components at π/a in (17). Crudely speaking the large Fourier components down to $\pi/(ae^{\ell})$ are integrated over under the renormalization group. Then the length is rescaled by a factor e^{ℓ} so that the cut-off resumes its original value. Then r_i, t, μ_i, and η obey the renormalization group equations

$$\frac{dr_i}{d\ell} = -r_i, \; \frac{dt}{d\ell} = -W(t), \; \frac{d\mu_i}{d\ell} = \zeta_{2k_i}(t)\mu_i, \; \frac{d\eta}{d\ell} = d\cdot\eta \tag{28a-d}$$

where in one-loop order[46,39] one has

$$W(t) = \varepsilon t - 2t^2 + O(t^3), \quad \zeta_{2k}(t) = 2k(k-1)t + O(t^2). \tag{29}$$

More precisely $\Pi \hat{Q}_b$ has to be expanded into eigenoperators which obey equations like (28c). Here we consider only the leading contribution which is the symmetric one of ref. 39.

Similarly as in ref. 43 in a first step we follow the renormalization group procedure until $r_1(\ell) - r_2(\ell) = r(\ell)$ has reached the distance a,

$$re^{-\ell_o} = a. \tag{30}$$

Under this change of length scale the localization length ξ changes by the same factor $e^{-\ell_o}$ to ξ_o,

$$e^{\ell_o} = \frac{r}{a} = \frac{\xi}{\xi_o} \tag{31}$$

and t has reached the value t_o, which from (28a,b) obeys

$$\ell_o = -\int_t^{t_o} \frac{dt}{W(t)} . \tag{32}$$

Since (31)

$$\ell_o = \ln\frac{\xi}{a} - \ln\frac{\xi_o}{a} , \tag{33}$$

comparison of (32) and (33) shows

$$\ln\frac{\xi}{a} = \int^t \frac{dt}{W(t)} \tag{34}$$

which defines (up to a factor) ξ/a as a function of t or t as a function of ξ/a,

$$t = t(\xi/a) \tag{35}$$

and similarly

$$t_o = t(\xi_o/a) = t(\xi/r). \tag{36}$$

For μ_i one obtains from (28)

$$\ln\frac{\mu_{i,o}}{\mu_i} = \int_{\mu_i}^{\mu_{i,o}} \frac{d\mu_i}{\mu_i} = -\int_t^{t_o} \frac{\zeta_{2k_i}(t)dt}{W(t)} = \ln\frac{Z_{2k_o}}{Z_{2k_{i,o}}} \tag{37}$$

where Z_{2k} is a function of t or according to (35) of ξ/a. Thus

$$\mu_{i,o} = \mu_i \frac{Z_{2k_i}(\xi/a)}{Z_{2k_i}(\xi/r)} \tag{38}$$

and

$$\mathcal{H}(\ell_o) = \mathcal{H}_o(t_o,\eta_o) - \mu_{1,o}\prod_{b=1}^{k_1}\hat{Q}_b(r_1e^{-\ell_o}) - \mu_{2,o}\prod_{b=1}^{k_2}\hat{Q}_{k_1+b}(r_2e^{-\ell_o}). \tag{39}$$

The expectation value in (25) is obtained from

$$\hat{\rho}^{-k_1-k_2} A = \frac{\partial^2}{\partial\mu_1\partial\mu_2} < e^{-\mathcal{H}} > = \frac{\partial^2}{\partial\mu_1\partial\mu_2} < e^{-\mathcal{H}(\ell_o)} > =$$

$$= \frac{\partial^2}{\partial\mu_1\partial\mu_2} < e^{-\mathcal{H}'(\ell_o)} > \tag{40}$$

with

$$\mathcal{H}'(\ell_o) = \mathcal{H}_o(t_o,\eta_o) - \mu_{1,o}\mu_{2,o} \prod_{b=1}^{k_1} \hat{Q}_b(r_1 e^{-\ell_o}) \prod_{b=1}^{k_2} \hat{Q}_{k_1+b}(r_2 e^{-\ell_o}). \tag{41}$$

The product of the $\hat{Q}$'s can be expanded as

$$\mathcal{H}'(\ell_o) = \mathcal{H}_o(t_o,\eta_o) - \mu_o \prod_{b=1}^{k_1+k_2} \hat{Q}_b(\frac{r_1+r_2}{2} e^{-\ell_o}) - \dots , \tag{42}$$

where subleading terms have been omitted and

$$\mu_o = \text{const}\cdot\mu_{1,o}\mu_{2,o}. \tag{43}$$

Now we apply the renormalization group to $\mathcal{H}'$ until t reaches a given value $\bar{t}$. The corresponding localization length be $\bar{\xi}$. Then

$$\bar{\mu} = \mu_o \frac{Z_{2k_1+2k_2}(\xi/r)}{Z_{2k_1+2k_2}(\bar{\xi}/a)} \tag{44}$$

and

$$A = \frac{\bar{\mu}}{\mu_1\mu_2} < \prod_{b=1}^{k_1+k_2} \hat{Q}_b >. \tag{45}$$

The expectation value is evaluated for $\mathcal{H}_o(\bar{t},\bar{\eta})$. Since $\bar{t}$ is fixed, it depends only on k_1, k_2 and

$$\bar{\eta} = (\xi/\bar{\xi})^d\cdot\eta \tag{46}$$

according to (28d). Thus

$$A_{k_1,k_2}(r,E,\eta) = \frac{Z_{2k_1}(\xi/a)\; Z_{2k_2}(\xi/a)\; Z_{2k_1+2k_2}(\xi/r)\; f_{k_1k_2}(\bar{\eta})}{Z_{2k_1}(\xi/r)\; Z_{2k_2}(\xi/r)} \tag{47}$$

From (29), (34), (37) one obtains

$$\xi(t) = \begin{cases} a\left|\frac{t}{t-\varepsilon/2}\right|^{1/\varepsilon}, & \varepsilon \neq 0, \\ a\exp(\frac{1}{2t}), & \varepsilon = 0, \end{cases} \tag{48}$$

and

$$Z_{2k}(\xi/a) = \begin{cases} [(\xi/a)^{\varepsilon} \pm 1]^{k(k-1)}, & \varepsilon \neq 0, \\ [\ln(\xi/a)]^{k(k-1)}, & \varepsilon = 0, \end{cases} \tag{49}$$

where the plus (minus) sign holds in the regions of the extended (localized) states. For the extended states one expects from (16) that $f(\bar{\eta})$ approaches a finite limit as $\bar{\eta}$ goes to zero, thus

$$\Gamma^{ext}_{k_1,k_2}(r,E) = \text{const}\cdot[(\frac{\xi}{a})^{\varepsilon} + 1]^{\kappa}[(\frac{\xi}{r})^{\varepsilon} + 1]^{2k_1k_2} \tag{50}$$

valid for $\varepsilon > 0$ and all $r >> a$, with

$$\kappa = k_1(k_1 - 1) + k_2(k_2 - 1). \tag{51}$$

In the region of the localized states the factor $\eta^{k_1+k_2-1}$ appears in front of A. Thus one expects f to diverge like $(\bar{\eta})^{-k_1-k_2+1}$ yielding for $\xi >> r >> a$

$$\Gamma^{loc}_{k_1k_2}(r,E) = \begin{cases} \text{const}\ \xi^{-(k_1+k_2-1)d}[(\frac{\xi}{a})^{\varepsilon} - 1]^{\kappa}[(\frac{\xi}{r})^{\varepsilon} - 1]^{2k_1k_2}, & \varepsilon \neq 0 \\ \text{const}\ \xi^{-(k_1+k_2-1)d}(\ln(\xi/a))^{\kappa}(\ln(\xi/r))^{2k_1k_2}, & \varepsilon = 0. \end{cases} \tag{52}$$

In the calculation given above we used the one-loop result (29). If we put instead close to the mobility edge, $t \approx t_c$

$$W(t) = W'(t_c)(t-t_c) = -(t-t_c)/\nu;\ \zeta_{2k}(t) = \zeta_{2k}(t_c) = \mu_k/\nu \tag{53}$$

where μ_k is the critical exponent of (3), then we obtain for finite ν

$$\Gamma^{ext}_{k_1,k_2}(r,E) = \text{const}\cdot(\frac{\xi}{a})^{(\mu_{k_1}+\mu_{k_2})/\nu}(\frac{\xi}{r})^{(\mu_{k_1+k_2}-\mu_{k_1}-\mu_{k_2})/\nu} \tag{54}$$

for extended states. For localized states it has to be multiplied by $\xi^{-(k_1+k_2-1)d}$. These results agree with (50) and (52) for $(\xi/a)^{\varepsilon} >> 1$, $(\xi/r)^{\varepsilon} >> 1$ for the exponents μ_k, (7) and $\nu = 1/\varepsilon$.

DISCUSSION

The final expressions (50), (52), (54) allow several observations. In the region of extended states the amplitude correlation decays over the distance ξ. For $\xi >> r$ the correlation becomes much larger than the limit $r \to \infty$ which is expected to be the product $1/(p^{(k1)}p^{(k2)})$, (3). Note that one obtains $1/p^{(k)}$ and $P^{(k)}$ by setting $k_1 = k$, $k_2 = 0$. It should be noted that $\Gamma_{2k_1,2k_2}$ is not proportional to $(\Gamma_{k_1,k_2})^2$. This is an indication of the strong fluctuations of the amplitudes of the eigenfunctions near the mobility edge.

For d=2, Γ^{loc} varies logarithmically with r. An effective critical exponent

$$\frac{d \ln \Gamma^{loc}}{d \ln r} = - \frac{2k_1k_2}{\ln(\xi/r)} \qquad (55)$$

may be defined. If $\xi >> r$ and in numerical calculations r can only be varied over approximately one order of magnitude, then this exponent remains nearly constant as a function of r and one may obtain the misleading impression of an exponent varying continuously as a function of ξ or disorder.

For d=3 and $k_1=k_2=1$ the correlation has been investigated by Soukoulis and Economou[47]. We note that the nonzero finite density of states at the mobility edge implies $\pi_1 = 0$, (2), which with (6) yields $\mu_1 = 0$. Thus the exponent D_c in ref. 47 is connected with μ_2/ν in (54) by

$$D_c = 1.7 \pm 0.3 = 3 - \mu_2/\nu. \qquad (56)$$

Prelovšek[48] has estimated the exponent π_2 of the inverse participation ratio which with (6) yields

$$\pi_2 = 1.4 = 3\nu - \mu_2. \qquad (57)$$

These estimates yield $\nu = \pi_2/D_c \approx 0.8$. The numerical errors allow for $\nu = 1$. The one-loop ε-expansion yields $\mu_2 = 2$ and thus $D_c = \pi_2 = \nu = 1$ for d=3.

REFERENCES

1. P.W. Anderson, Phys. Rev. 109, 1492 (1958)
2. L. Banyai, Physique des Semiconducteurs, Hulin, ed., Dunod Paris, 417 (1964)
3. N.F. Mott, Phil. Mag. 13, 989 (1966); Adv. Phys. 16, 49 (1967)
4. M.H. Cohen, H. Fritzsche, S.R. Ovshinsky, Phys. Rev. Lett. 22, 1065 (1969)
5. S.F. Edwards, J. Phys. C3, L30 (1970)
6. P.W. Anderson, Proc. Nat. Acad. Sci. 69, 1097 (1972)
7. T. Lukes, J. Noncryst, Solids 8-10, 470 (1972)
8. R.A. Abraham, S.F. Edwards, J. Phys. C5, 1183, 1196 (1972)
9. K.F. Freed, Phys. Rev. B5, 4802 (1972)
10. S.K. Ma, unpublished note (1972)
11. S.F. Edwards, J. Phys. C8, 1660 (1975)
12. D.J. Thouless, J. Phys. C8, 1803 (1975)
13. J.T. Edwards, D.J. Thouless, J. Phys. C5, 807 (1972)
D.C. Licciardello, D.J. Thouless, Phys. Rev. Lett. 35, 1435 (1975)
14. F.J. Wegner, Z. Phys. B25, 327 (1976)
15. M.H. Cohen, J. Jortner, Phys. Rev. Lett. 30, 699 (1973)
16. D.C. Licciardello, D.J. Thouless, J. Phys. C11, 925 (1978)
17. W. Götze, Sol. St. Comm. 27, 1393 (1978); J. Phys. C12, 1279 (1979)
18. D. Vollhardt, P. Wölfle, Phys. Rev. Lett. 45, 842 (1980); Phys. Rev. B22, 4666 (1980)
19. E. Abrahams, P.W. Anderson, D.C. Licciardello, T.V. Ramakrishnan, Phys. Rev. Lett. 42, 673 (1979)
20. J.S. Langer, T. Neal, Phys. Rev. Lett. 16, 984 (1966)
21. R. Oppermann, F. Wegner, Z. Phys. B34, 327 (1979)
22. F.J. Wegner, Phys. Rev. B19, 783 (1979)
23. E.P. Wigner, Ann. Math. 62, 548 (1955); 67, 325 (1958)
24. F. Wegner, Z. Phys. B35, 207 (1979)
25. L. Schäfer, F. Wegner, Z. Phys. B38, 113 (1980)
26. A. Aharony, Y. Imry, J. Phys. C10, L487 (1977)
27. Anderson Localization, Y. Nagaoka and H. Fukuyama, eds., Springer Series in Solid State Sciences 39

28. Proceedings of the International Conference on Localization, Interaction, and Transport Phenomena in Impure Metals, Braunschweig 1984, L. Schweitzer, B. Kramer, eds., (to appear) and supplement (1984)
29. S. Hikami, Nucl. Phys. B215 [FS7], 555 (1983)
30. S. Sarker, E. Domany, Phys. Rev. B23, 6018 (1981)
31. A. MacKinnon, B. Kramer, Z. Phys. B53, 1 (1983)
32. H. Kunz, B. Souillard, J. Physique-Lett. 44, L503 (1983)
33. J.P. Straley, Phys. Rev. B28, 5393 (1983)
34. S.F. Edwards, M.B. Green, G. Srinivasan, Phil Mag. 35, 1421 (1977)
35. K. Ziegler, Phys. Lett. B22, 4666 (1980)
36. P. Prelovšek, Phys. Rev. B23, 1304 (1981)
37. A.B. Harris, T.C. Lubensky, Sol. St. Comm. 34, 343 (1980); Phys. Rev. B23, 2640 (1981)
38. F. Wegner, Z. Phys. B44, 9 (1981)
39. F. Wegner, Z. Phys. B36, 209 (1980)
40. L.P. Kadanoff, Phys. Rev. Lett. 23, 1430 (1969)
41. A.M. Polyakov, Zh. Eksp. Teor. Fiz. 57, 271 (1969); JETP 30, 151 (1969)
42. K.G. Wilson, Phys. Rev. 179, 1499 (1969)
43. F.J. Wegner, J. Phys. A8, 710 (1975)
44. F.J. Wegner, in ref. 27, p.8
45. E.N. Economou, M.H. Cohen, Phys. Rev. Lett. 25, 1445 (1970)
46. E. Brézin, J. Zinn-Justin, Phys. Rev. Lett. 36, 691 (1976); Phys. Rev. B14, 3110, 4976 (1976)
47. C.M. Soukoulis, E.N. Economou, Phys. Rev. Lett. 52, 565 (1984)
48. P. Prelovšek, Phys. Rev. B18, 3657 (1978)

AN ALTERNATIVE THEORY FOR THERMOELECTRIC POWER IN ANDERSON-MOTT INSULATORS

M. Pollak* and L. Friedman†

*Department of Physics
University of California
Riverside, CA 92521

†GTE Laboratories
Sylvan Road
Waltham MA

ABSTRACT

A new theory is developed for the thermoelectric power due to single-phonon assisted hopping. The theory relates the thermoelectric power to the average site energy on the current-carrying percolation cluster. In addition to providing new insight into this transport property, resulting formulae are easier to use than those in other theories. A calculation of the Seebeck coefficient for a nearly constant density of states, but with a small term linear in energy, yields a temperature dependence in agreement with previous theories.

INTRODUCTION

The Seebeck coefficient S of a material is defined as the electric field E set up by a temperature gradient ∇T, i.e. $S=E/\nabla T$. Onsager relations relate S to the Peltier coefficient Π, which is usually easier to calculate, by $S=\Pi/T$. Π is defined as the average heat E carried by an electron, when the electric current density is $\underset{\sim}{j}$, and the thermal gradient $\Delta T=0$, i.e. $\Pi=E/j$. A method to calculate Π for hopping transport has been developed by Overhof[(1)], and used extensively by many authors. In this paper we propose an alternative method which is easier to use. The basic element in our calculation consists of the evaluation of the average site energy on the current carrying percolation cluster (or, approximately, on the critical percolation cluster). This average is calculated from the distribution $p(E_1)$ of site energies on the percolation cluster. A method to obtain $p(E_i)$ from the density of states $N(E_i)$ has been given by Pollak[2], and more accurately by Friedman and Pollak[3], and will be reviewed and adapted to the present need in the next section. When $p(E_i)$ is known, the Peltier coefficient is calculable from

$$\Pi = \int_{-\infty}^{\infty} E_i p(E_i) dE_i \qquad (1)$$

We now present the justification for the above procedure. Consider two planes, transverse to $\underset{\sim}{j}$, separated by a large distance, i.e. by a distance larger than the correlation length of the current carrying percolation cluster. Say that the current enters the region between planes at the left plane and leaves at the right plane. For sufficiently large "samples", i.e. in the thermodynamic limit, the two planes will intersect the percolation cluster at sites in such a way that their energy distribution is $p(E_i)$. (The fact that the planes may intersect the percolation cluster between sites rather than at sites is unimportant here. When the plane happens to intersect the cluster at a bond, rather than at a site, we can consider the relevant site to be at that end of the bond which is closer to the other plane). The condition div $\underset{\sim}{j}=0$ implies that for every electron entering the volume between the planes at the left plane, there is an electron leaving that volume at the right plane. Moreover for reasons of macroscopic homogeneity, the net number of electrons entering at any E_i at the left surface is equal to the net number of electrons leaving from a site at the same E_i on the right surface. Since electrons are indistinguishable, we can associate an electron entering at E_i with an electron leaving at E_i, which corresponds to a picture where the electron carries an energy E_i through the space between the planes.

It remains to show that E_i should be identified with the heat flow. For that purpose we must show that all of E_i corresponds to heat, and that there is no other heat flux associated with the current density $\underset{\sim}{j}$. The first part seems obvious. As for the second part, the only other heat flux associated with j could be carried by phonons and arise from the electron-phonon interaction. In the Miller-Abrahams mechanism of hopping transport, where transport is effected by electron transitions associated with the absorption or emission of single delocalized phonons no such flux can exist. The reason lies in momentum conservation. Since the initial as well as the final electronic states have no momentum, no phonon momentum can be generated by the transition, [i.e. the phonons involved must be of the form $[\exp(i\underset{\sim}{k}.\underset{\sim}{r}) + \exp(-i\underset{\sim}{k}.\underset{\sim}{r})]$. No net energy flux due to phonons can therefore result from the current. It is important to stress that this is not necessarily true when either localized phonons, or many-phonon processes are involved in the transitions. In the first case, an energy flux can be generated by the "hopping" of localized phonons, in analogy with the energy flux resulting from the hopping transport of localized electrons. In the case of many-phonon transitions, an energy flux can also exist, if the dispersion relation of the relevant phonons is different from the linear relation characteristic of low frequency acoustic phonons. For then the superposition of phonons which has zero momentum may have a non-zero energy flux. The localized phonon and many-phonon mechanisms may be related, since localized phonons can be constructed as super-positions of many delocalized phonons. In conclusion, then, for the single-phonon assisted hopping the entire heat flux is given by $\int E_i p(E_i) dE_i$, as in eq. (1). However, for more complex hopping processes the heat flux due to phonons must also be evaluated. To compare the results of our theory with those of Overhof[1], we shall study only the Miller-Abrahams[4] process of single-phonon assisted hopping, for which there is no current induced energy flux of phonons. The transport theory then can be cast into the form of a random resistance network, with resistances

$$Z_{ij} = Z_o \exp(-2r_{ij}/a) \exp(-E_{ij}/kT) \tag{2}$$

between any sites i,j. The quantities R_{ij} and E_{ij} are random distance and energy variables, respectively, determined by sites i and j. More specifically, r_{ij} is the distance between sites i and j, and E_{ij} is a simple function[2] of the site energies E_i, E_j. A percolation procedure already alluded to is commonly used[2] to calculate the steady state transport properties of the network, and we shall make use of some aspects of such works. In particular we shall make use of results in references (1,2,3).

GENERAL THEORY

We now turn to the calculation of $p(E_i)$. Obviously, $p(E_i)$ must be proportional to the probability that there is a state at E_i, i.e. to the density of states $N(E_i)$. It also must be proportional to the probability that a site at E_i is connected to the backbone of the percolation cluster[5] (i.e. to the current carrying part of the percolation cluster). Such a site must have at least two resistances with values Z_m or less connected to it, i.e., the second smallest resistance emanating from the site must be not larger than Z_m, the maximum resistance on the percolation cluster. The probability for this will be denoted by $p_2(Z_m|E_i)$. The condition that the second smallest resistance be smaller than Z_m is necessary but not sufficient for the site to be connected to the backbone cluster, because the other ends of the two resistances may not be connected to further resistances with values Z_m or less. Thus, not all sites at E_i which satisfy the stated condition are connected to the backbone, so $p_2(Z_m|E_i)$ must be multiplied by another factor $p<1$. This p can depend only weakly on E_i. The reason is that the only correlation between neighboring resistances comes from their common site energy.[2] Thus the probability that the percolation cluster is continued beyond the two resistances accounted for by $p_2(Z_m|E_i)$ depends only indirectly, and thus weakly, on E_i. We then assume that p is independent of E_i. So far $p(E_i) \propto N(E_i)p_2(Z_m|E_i)\ p$. In order that $\int_{-\infty}^{\infty} E_i p(E_i)dE_i$ correspond to a properly averaged E_i, we need to assure that $p(E_i)$ is a correctly normalized probability density, i.e. that $\int_{-\infty}^{\infty} p(E_i)dE_i=1$. This is assured by writing

$$p(E_i) = N(E_i)\ p_2(Z_m|E_i) / \int_{-E_m}^{E_m} N(E_i) \qquad p_2(Z_m|E_i)dE_i \tag{3}$$

Notice that the factor p cancels due to its (approximate) independence on E_i. The limits of integration are contracted, in accordance with $p_2 \equiv 0$ for $|E_i| \geqslant E_m$. The probability $p_2(Z_m|E_i)$ can be expresed in terms of $P(Z_m|E_i)$, the average number of resistances $Z<Z_m$ connected to a site at E_i. The expression is[3]

$$p_2 = 1 - e^{-P}\ (1 + P) \tag{4}$$

The argument $(Z_m|E_i)$ has been left out from p_2 and P above, since no confusion can result.

$P(Z_m|E_i)$ is obtained from the distribution of the random variables r and E of eq. (2),

$$P(Z_m|E_i) = (4\pi/3)(a/2kT)^3\ [(E_m-E_i)^3 \int_0^{E_i} N(E)dE + \int_{E_i}^{E_m} (E_m-E)^3 N(E)dE + \int_{-E_m+E_i}^{0} (E_m-E_i+E)^3 N(E)dE] \quad , \quad E_i>0 \quad , \tag{5a}$$

$$P(Z_m|E_i) = (4\pi/3)(a/2kT)^3\ [(E_m-E_i)^3 \int_0^{-E_i} N(-E)dE + \int_{-E_i}^{E_m} (E_m-E)^3 N(-E)dE + \int_{-E_m+E_i}^{0} (E_m+E_i+E)^3 N(-E)dE] \quad , \quad E_i<0 \tag{5b}$$

The expression has been split into the negative and positive regimes of E_i. Usually eq. (5) is written in a more compact form, but here we wish to better display the symmetry properties of $P(Z_m|E_i)$ with respect to E_i. Comparison of the corresponding terms in the equations for $E_i>0$ and for $E_i<0$ reveal that antisymmetric parts of N(E) generate antisymmetric parts of $P(Z_m|E_i)$, and vice versa.

Combining eqs. (1), (3), and (4) we now have

$$\Pi = \int_{-E_m}^{E_m} E_i N(E_i) \ [1-e^{-P}(1+P)] \ dE_i \Big/ \int_{-E_m}^{E_m} N(E_i) \ [1-e^{-P}(1+P)] \ dE_i \qquad (6)$$

Eq. (6) is the final form for the Peltier coefficient, with P given in eq. (5). While eq. (5) can be integrated explicitly for various densities of states, eq. (6) probably has to be solved numerically for most cases.

It is instructive to separate p_2, P, and N(E) into symmetric and antisymmetric parts with respect to E_i, annotated by by subscripts s and a respectively. Only the antisymmetric part of $N(E_i)[1-e^{-P}(1+P)]$ will contribute to Π. The nonvanishing part of the numerator of eq. (6) is

$$E_i[N_a(E_i) \ p_{2s}(Z_m|E_i) + N_s(E_i)p_{2a}(Z_m|E_i)] \qquad (7a)$$

and the nonvanishing part of the denominator is

$$[N_s(E_i)p_{2s}(Z_m|E_i) + N_a(E_i) \ p_{2a}(Z_m|E_i)] \qquad (7b)$$

where

$$\begin{aligned} p_{2s} &= 1-\exp(-P_s) \cosh P_a \ (1+P_s) + P_a \exp(-P_s) \sinh P_a \\ p_{2a} &= -\exp(-P_s) \cosh P_a \ P_a + (1+P_s) \exp(-P_s) \sinh P_a \end{aligned} \qquad (8)$$

Eq. (7a) shows that Π vanishes when $N(E_i)$ is symmetric, since then $N_a(E_i)=0$, from eq. (5) $P_a=0$, and from eq. (8) p_{2a} is also zero. Thus eq. (7a) gives zero.

The two contributions in eq. (7a) can be ascribed distinct physical meanings. The first term reflects directly the fact that the flow of electrons is easier on the side of the Fermi level where sites are more numerous. The second term reflects the percolative aspects of the problem; namely the fact that it is not only easier to find a site at high densities of states, but from a given site it is also easier to connect resistances to sites at high densities of states.

It is important to note that in addition to the explicit temperature dependence of Π contained in eq. (6) there is also an implicit temperature dependence. This arises from the fact that the energy E_i in all the above expressions is measured from the Fermi level. For asymmetric $N(E_i)$, the Fermi level is generally temperature dependent, which means that for such cases $N(E_i)$ is also implicitly temperature dependent. For strong asymmetries, such an effect must be taken into account.

APPLICATIONS

To compare the present theory with that of Overhof[1], we apply it to the same densities of states. As far as we know, the only density of states for which an analytical result for the temperature dependence of Π is available, using that theory, is a constant density of states with a small linear asymmetry, i.e.

$$N(E) = N_o + \alpha E \quad , \qquad \alpha E_m << N_o \quad . \tag{9}$$

Overhof's result for such a density of states is $\Pi \propto T^{1/2}$. Substituting eq. (9) into eq. (5),

$$P_s = (2\pi/3)(a/2kT)^3(E_m-E_i)^3(E_m+E_i) \; ; \quad P_a = (\alpha E_i/2N_o)P_s \tag{10}$$

Since the condition $\alpha E_m << N_o$ makes $P_a << 1$, eq. (8) can be approximated by

$$\begin{aligned} p_{2s} &= 1 - (1+P_s)\exp(-P_s) \\ p_{2a} &= P_a P_s \exp(-P_s) \end{aligned} \tag{11}$$

Substitution of eqs. (9), (10), and (11) into eq. (7) and then into eq. (6) gives

$$\Pi = kT\left(\frac{3\alpha^2 kT}{2\pi N_o^3 a^3}\right)^{1/2} \frac{\int_{-\varepsilon_m}^{\varepsilon_m} \varepsilon_i^2[1 - (1 + y + y^2/2)\exp(-y)]d\varepsilon_i}{\int_{-\varepsilon_m}^{\varepsilon_m} [1 - (1 + y)\exp(-y)]d\varepsilon_i} \tag{12}$$

$$y = (\varepsilon_m - \varepsilon_i)^3(\varepsilon_m + \varepsilon_i)$$

where we put $\varepsilon_{i,m} = (2\pi/3)^{1/4} N_o^{1/4} (a/kT)^{3/4} E_{i,m}$, which are dimensionless quantities. We know from percolation theory[2] that $E_m/kT \propto (kT)^{1/4}$, which makes ε_m independent of temperature, of a, and of N_o, i.e. a pure number. The integrals are also pure numbers and can be readily evaluated. We obtain

$$\Pi = 0.084\ kT \left(\frac{3\alpha^2 kT}{2\pi N_o^3 a^3} \right)^{1/2} , \tag{13}$$

The Seebeck coefficient then is

$$S = \Pi/T = .058k\left(\frac{\alpha}{N_o}\right)\left(\frac{kT}{N_o a^3}\right)^{1/2} \tag{14}$$

which is in agreement with the $T^{1/2}$ dependence obtained by Overhof.[1]

We note here without proof (to be published elsewhere) that the above method can be extended to densities of states

$$N(E) = \beta|E^n| + \alpha E \quad , \tag{15}$$

where α is again small, i.e. $\alpha|E| << \beta E_m^n$. The result is

$$S \propto T^{(2-n)/(4+n)} \tag{16}$$

SUMMARY

We developed a new theory for the thermoelectric power due to single-phonon assisted hopping. The theory is simpler to use, and provides physical insight into the thermoelectric power. The formula for the Peltier coefficient Π is given in eqs. (6) and (5). The Seebeck coefficient then can be calculated using the Onsager relation $S = \Pi/T$. The theory has been applied to the case of a constant density of states plus a small term linear in energy. The result is in agreement with that previously obtained by Overhof.

The authors are grateful to the editors of this volume for their invitation to make a contribution in honor of Professor Sir Nevill Mott, whose work has inspired us for many years. M.P. would also like to thank JSPS, and Professor H. Kamimura for the opportunity to spend time at the Department of Physics of the Tokyo University. This work is in part due to the stimulating atmosphere there.

REFERENCES

1. H. Overhof, Phys. Stat. Sol.(b) 67, 709 (1975).
2. M. Pollak, The Metal Non-Metal Transition in Disordered Systems, L. Friedman and D. Tunstall, Eds., SUSSP, Edinburgh 1978, p. 238.
3. L. Friedman and M. Pollak, Phil. Mag. B 44, 487 (1981).
4. A. Miller and E. Abrahams, Phys. Rev. 120, 745 (1960).
5. S. Kirkpatrick, Rev. Mod. Phys. 45, 574 (1973).

TRANSPORT PROPERTIES NEAR THE PERCOLATION THRESHOLD OF CONTINUUM SYSTEMS

B.I. Halperin and S. Feng

Department of Physics
Harvard University
Cambridge, MA 02138

P.N. Sen

Schlumberger-Doll Research
Old Quarry Road
Ridgefield, CT 06877

INTRODUCTION

The purpose of this article is to review some recent results on the critical exponents for transport properties near the percolation threshold of several types of continuum models.[1] We have found that the exponents governing the behavior of the electrical conductivity, the elastic constants, or the fluid permeability in such systems can be quite different from the corresponding exponents for conventional discrete lattice percolation models.[2-6] This is in contrast to the exponents for *geometrical* percolation properties, such as the correlation length exponent ν, which are believed to be the same for these models as for ordinary lattice percolation.[7,8]

In several respects, our analysis relates strongly to the work of Sir Nevill Mott on the law of "hopping-conductivity," engendered by the phonon-assisted tunneling of electrons between localized states in the bandgap of a semiconductor.[9] Mott's law, which states that the logarithm of the resistivity varies as the inverse of the fourth root of the temperature, at low temperatures, is the result of a model in which there is a wide distribution in the microscopic hopping rates between sites, arising from the strong exponential dependence of the rate on the randomly distributed site energy and the random distances between neighboring sites. It may be noted that a simple calculation based on the arithmetic mean of the hopping rate between nearest neighbor sites would give too large a conductivity in this model, as the mean is dominated by a few near-neighbor pairs, which have a large hopping rate, but do not form a connected network across the sample. Similarly, an estimate based on the arithmetic mean of the hopping *time* from a random site in the system would lead to an underestimate of the conductivity.[10] Mott was able to obtain a correct estimate of the conductivity, however, by use of a "typical" hopping distance, chosen so as to maximize the rate of transport across the sample.

The model of hopping conduction is equivalent to a resistor network with a very wide variation in the conductance of the individual bonds.[10-12] It was noted by Ambegaokar, Halperin and Langer,[10] and independently by Pollak,[11] that the conductivity of such a network is actually determined by a percolation problem, and that the Mott hopping law could thus be rederived from the threshold value of the conductance just sufficient to form a connected network across the sample. A similar interplay of the ideas of percolation and of optimization of the path across a network with varying bond-strengths is at the heart of the following analysis.

In a recent note,[1] the present authors considered transport properties near the percolation threshold of a class of "Swiss-cheese" continuum models, where spherical holes are randomly placed in a uniform transport medium. In addition to reviewing this model, below, we shall discuss the "inverted Swiss-cheese" model where the randomly placed spheres are the conducting medium, and the spaces between them are insulating. We also discuss another class of models, where the conducting regions are the portions of space where a specified smooth stochastic "potential-function" $V(\vec{r})$ is less than some cutoff value V^*. (We denote this as the "potential model.") A two-dimensional experimental realization of the potential model was studied by Smith and Lobb,[13] who used the intensity of a laser-speckle pattern to generate the function $V(\vec{r})$, and used high contrast film and photolithographic techniques to produce the two-dimensional conducting sample.

Table 1. Estimates of the differences between the transport percolation exponents in the continuum models and the corresponding exponents on a discrete lattice. (See text for definitions.)

	Swiss-Cheese Model			Inverted Swiss-Cheese Model and Potential Model		
	Conductivity ($\bar{t}-t$)	Elasticity ($\bar{f}-f$)	Permeability ($\bar{e}-t$)	Conductivity ($\bar{t}-t$)	Elasticity ($\bar{f}-f$)	Permeability ($\bar{e}-t$)
2D	0	3/2	3/2	0	0	0
3D	1/2	5/2	5/2	0	1/2	1/2

The results of our analysis for the Swiss-cheese models in two and three dimensions are summarized in Table 1. The exponents $\bar{t}$ and $\bar{f}$ are defined by assumption that the macroscopic electrical conductivity Σ and the shear modulus N vanish as the volume fraction q of holes approaches a critical value q_c, according to the power laws $\Sigma \sim (q_c - q)^{\bar{t}}$, and $N \sim (q_c - q)^{\bar{f}}$. The fluid permeability k, defined as the volume-rate of fluid flow through the space between the random spheres, under a unit macroscopic pressure gradient, will clearly vanish at the same value q_c of the sphere-volume-fraction, and it is assumed also to follow a power law, $k \sim (q_c - q)^{\bar{e}}$. We find in two dimensions, that the conductivity exponent $\bar{t}$ is the same as the exponent t for a standard lattice resistor network at percolation, but that the exponent $\bar{f}$, for the Swiss-cheese model is significantly larger than the exponent f of a lattice model with both bond-stretching and bond-bending elastic forces, which was recently studied.[2-5] For the *three*-dimensional Swiss-cheese model, we find that *both* $\bar{t}$ and $\bar{f}$ are larger than the corresponding discrete percolation exponents. Moreover, while the permeability and conductivity exponents are identical to each other in the standard lattice percolation model, we find that $\bar{e}$ is dramatically larger than $\bar{t}$ in our continuum model, in both two and three dimensions.

The results of our analysis for the "potential model" are summarized in the last three columns of Table 1. Note that in this case the conductivity exponent $\bar{t}$ is the same as the lattice exponent t in both two and three dimensions while the exponents $\bar{e}$ and $\bar{f}$ are significantly larger than the lattice values in $d=3$ but not in $d=2$. The results for "inverted Swiss-cheese models" are the same as for the potential model, as summarized in Table 1.

In our analysis, we follow previous authors[7,13-15] in mapping the continuum models onto a type of discrete random network. Unlike the standard discrete percolation problem, however, we must employ a *continuous* distribution of bond strengths, and our analysis shows that in many cases there is a large probability density for finding a small bond strength. It has been known for some time that such a distribution can lead to an increase of the conductivity exponent.[17,18] In reference 1, we examined the elasticity, as well as the conductivity and permeability exponents, by considering the contribution of the "singly connected bonds" in the "nodes-links-blobs" picture of the percolation backbone,[19] similar to the analysis of Kantor and Webman for the lattice elasticity problem.[3] Although our method was different from those in refs. 17 and 18, the various methods lead to rather similar results, at least in the conductivity case. In the discussion below, we concentrate on the electrical conductivity and fluid permeability.

THE SWISS-CHEESE MODEL

The mapping of the Swiss-cheese model onto a discrete random network was described by Elam, Kerstein and Rehr,[7] and is illustrated in Figure 1 for the two-dimensional case of random circular holes punched in a conducting sheet. In higher dimensions, the construction corresponds to a "Voronoi tesselation", and the bonds are the edges of the Voronoi polyhedra. In two dimensions, a bond is present if the two neighboring holes do not overlap, but the "strength" of the bond i depends crucially on the channel width δ_i (Figure 2). It is important to note that δ_i has a continuous probability distribution $p(\delta)$ which approaches a *finite* limit $p(0)$, for $\delta \to 0^+$. Although there will clearly be some short-range correlations between the values of δ_i on nearby bonds, these correlations should not affect the finiteness of the distribution for $\delta \to 0$, and we shall ignore such correlations entirely.

Next, we shall analyze the strength of a bond. In the two-dimensional example, the electrical conductance g_i associated with a narrow neck of width δ_i is given, up to a constant of order unity which we ignore, by

$$g_i \approx \sigma_0 \delta_i^{1/2}/a^{1/2} \tag{1}$$

where σ_0 is the microscopic conductivity of the material, and $\underset{\sim}{a}$ is the hole radius. This result can be understood simply if we approximate the neck by a thin rectangle of width δ_i and length $\ell_i \approx (\delta_i a)^{1/2}$, as illustrated in Figure 2.

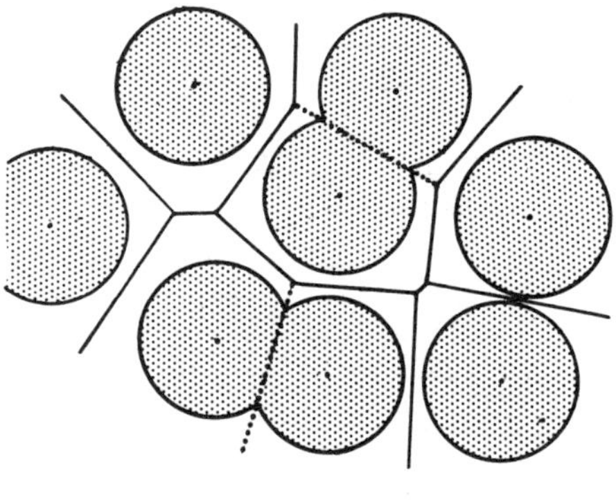

Fig. 1. Swiss-cheese model in two dimensions. Straight lines show the bonds of the superimposed discrete network; dotted lines are the missing bonds.

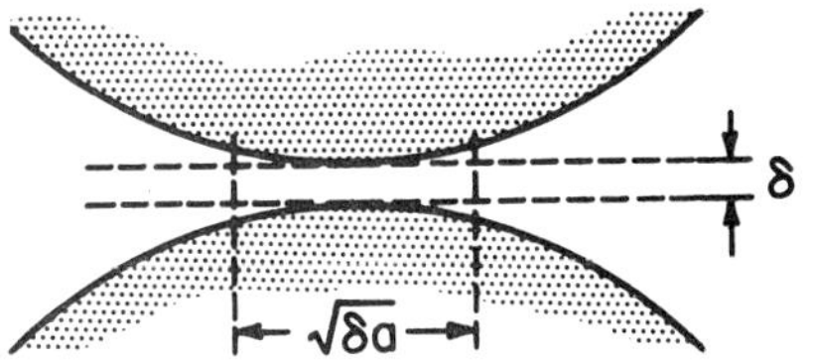

Fig. 2. Narrow neck in the two-dimensional Swiss-cheese model. Dashed lines outline a rectangular approximation to the neck.

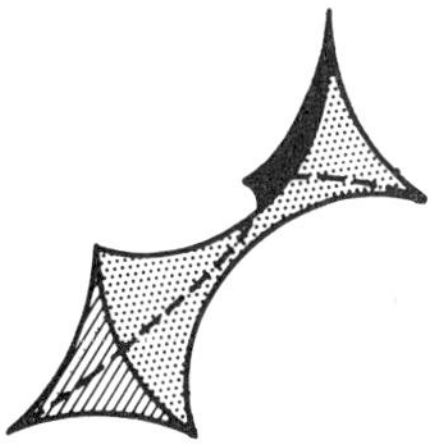

Fig. 3. Narrow portion of a bond passing between three overlapping spherical holes, in the three-dimensional Swiss-cheese model.

In the three-dimensional Swiss-cheese model, the smallest cross-section of a bond has roughly the shape of a triangle, with a side δ_i which again has a finite probability density $p(0)$ in the limit $\delta_i \to 0$. (See Figure 3.) Since this smallest cross-section persists over a distance which is roughly $(\delta_i a)^{1/2}$, we find that the bond has an electrical conductance $g_i \propto \delta_i^{3/2}$.

The difference between the permeability and conductivity problems arises from the different behaviors of the bond strengths, for small δ_i. In three dimensions, the flow of a viscous fluid through a narrow channel like that in Figure 3 is proportional to $\delta_i^4/\ell_i = \delta_i^{7/2}/a^{1/2}$, while in a two-dimensional version of the model, the flow varies as $\delta_i^{5/2}/a^{1/2}$.

In the elasticity problem, we need to find the stiffness constants γ_i that describe the energy cost of bending the bond or twisting the bond through a specified angle. These bond stiffness constants are found to vary as $\delta_i^{5/2}$ and $\delta_i^{7/2}$ in $d=2$ and $d=3$, respectively.[1,20]

THE POTENTIAL MODEL

The discrete network associated with the potential-model was discussed in detail by Weinrib,[14] for the particular case of the two-dimensional laser speckle pattern, used by Smith and Lobb.[13] A similar construction was suggested earlier, by Ziman,[15] for a three-dimensional model. In these constructions, one associates nodes of the discrete network with local minima of the potential function, and one associates bonds with the saddle points connecting two valleys. We assume that in the vicinity of a saddle point, the potential $V(\vec{r})$ may be expanded as

$$V(\vec{r}) = V(\vec{r}_i) + \frac{1}{2}\sum_{\alpha\beta} (\vec{r}-\vec{r}_i)_\alpha (\vec{r}-\vec{r}_i)_\beta \, U_{\alpha\beta} \tag{2}$$

where $\vec{r}_i$ is the position of the saddle, and $U_{\alpha\beta}$, the matrix of second derivatives of V, has one negative eigenvalue λ_1, and (d-1) positive eigenvalues, $\lambda_2,\ldots,\lambda_d$. If $h_i \equiv V^* - V(\vec{r}_i)$ is positive, the bond at $\vec{r}_i$ will be occupied. At its narrowest point, the bond will have an elliptical cross-section with principal diameters $\Delta_2 = (8h_i/\lambda_2)^{1/2}$, $\Delta_3 = (8h_i/\lambda_3)^{1/2}$, etc. (see Figure 4). If h_i is small, the length ℓ of the constricted region of the bond is given by $\ell \approx |8h_i/\lambda_1|^{1/2}$, which is of the same order as the diameter Δ_2, if we assume that all eigenvalues of $U_{\alpha\beta}$ have similar magnitudes.

To compute the transport properties, we replace the bond by a cylinder of length ℓ and diameter Δ_2. Then the electrical conductance g_i is proportional to $h_i^{1/2}$ in $d = 3$, and h_i^0 in $d = 2$. (More accurately $g_i \propto \ell n\, h_i$ in this case.) For the fluid flow through the bond we find $g_i \propto h_i^{3/2}$ in $d = 3$ and $g_i \propto h_i$ in $d = 2$. Similarly, the bond-bending and torsion constants obey $\gamma_i \propto h_i^{3/2}$ in $d = 3$ and $\gamma_i \propto h_i$ in $d = 2$, for this model.

Finally, we note that except in "pathological" cases we may assume that the distribution of saddle point values $V(\vec{r}_i)$ is regular in the vicinity of the critical value, so that the probability distribution $p(h_i)$ has a finite value $p(0)$ in the limit $h_i \to 0$.

THE INVERTED SWISS-CHEESE MODEL

For the inverted Swiss-cheese model, a bond occurs when two spherical conductors overlap, as shown in Figure 5. The diameter Δ_i of the constricted region is related to the overlap distance δ_i, by $\Delta_i = (2\delta_i a)^{1/2}$, where a is the sphere radius. For electrical conduction, or for viscous fluid flow through an aperture, we expect that the region of maximum dissipation will extend for a distance of order $\Delta_i/2$ on either side of the aperture,[20] so that we may again approximate the bond by a cylinder with length ℓ_i equal to its diameter $\Delta_i \propto \delta_i^{1/2}$. It is clear, also, that

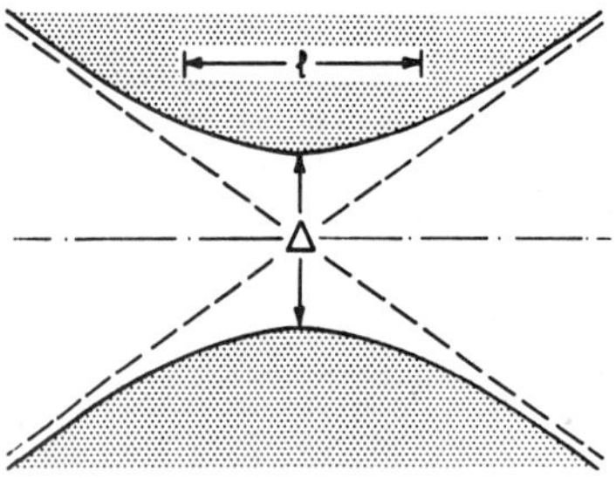

Fig. 4. Geometry of a narrow neck in the potential model.

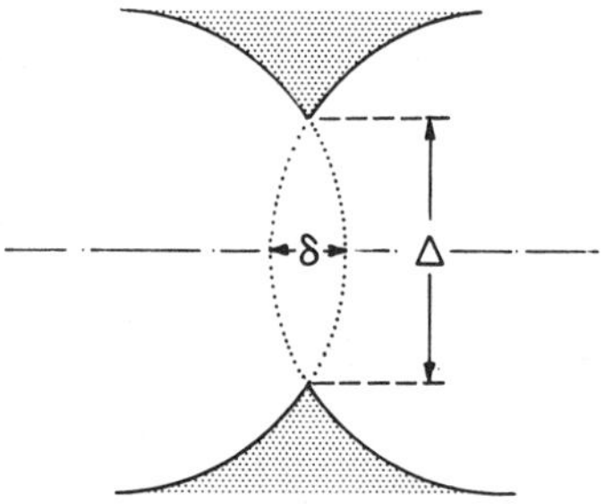

Fig. 5. Neck geometry in the inverted Swiss-cheese model.

the distribution $p(\delta_i)$ is finite at $\delta_i = 0$. Therefore, we find that the probability distribution of the bond strength (either g_i or γ_i) has the same limiting behavior for the inverted Swiss-cheese model as for the potential model discussed above.[20]

PERCOLATION PROPERTIES

Next we consider how bonds are connected in the macroscopic system near the percolation threshold. In the nodes-links-blobs picture the conducting "backbone" of the infinite cluster is imagined to consist of a network of quasi-one-dimensional string segments ("links"), tying together a set of "nodes" whose typical separation is the percolation correlation length $\xi \approx (q_c - q)^{-\nu}$. Each string is supposed to consist of several sequences of singly-connected bonds, in series with thicker regions, or "blob", where there are two or more conducting bonds in parallel.[19]

To estimate the macroscopic conductivity, we ignore the resistance of the blobs, and approximate the conductance G of a string by

$$G^{-1} = \sum_{i=1}^{L_1} g_i^{-1} \tag{3}$$

where the sum is restricted to the L_1 singly connected bonds on the string. It has been shown[19] that the typical value of L_1 is proportional to $(q_c - q)^{-1}$. In the conventional case, where each bond has *unit conductance*, the conductance of a string will be L_1^{-1}, and the conductivity of the network will be $\Sigma \sim \xi^{2-d} L_1^{-1}$, where d is the spatial dimension. Thus this analysis predicts $t \approx 1 + (d-2)\nu \equiv t_1$, a result which slightly underestimates the true value of t in 2 and 3 dimensions.[1,19]

Now we must estimate a *typical value* of the string conductance G for the case, where there is a distribution of bond strengths. If a string contains many singly connected bonds, we should be able to replace the sum in Eq. (3) by an integral over the probability distribution $p(\delta)$, provided that we properly control the contribution of the weakest bonds. In particular, if $g_i = g(\delta_i)$, we replace Eq. (3) by

$$G^{-1} = L_1 \int_{\delta_{min}}^{\infty} [p(\delta)/g(\delta)]\, d\delta \tag{4}$$

where δ_{min} is the minimum value of δ for the singly connected bonds on the string. It may be seen that for large values of L_1, the typical value of δ_{min} is equal to δ_0/L_1, where $\delta_0 \equiv 1/p(0)$. [More generally,

$(1-\int_0^{\varepsilon} p(\delta)\,d\delta)^{L_1} \approx e^{-\varepsilon L_1/\delta_0}$ is the probability that *none* of the L_1 bonds has $\delta_i < \varepsilon$.]

Now, suppose that the function $g(\delta)$ behaves, in the limit $\delta \to 0$, as

$$g(\delta) \propto \delta^x \quad , \tag{5}$$

where x is a constant. Then if $x < 1$, the integral (4) converges, for $\delta_{min} \to 0$, and hence we find simply

$$G^{-1} \approx L_1 \langle g^{-1} \rangle \quad , \tag{6}$$

where $\langle\,\rangle$ denotes an average over the distribution $p(\delta)$. In this case, the resistance of the chain is determined by the mean resistance of a bond in the same way as for a chain of equal bonds. On the other hand, for $x > 1$, we see that the value of G^{-1} is determined by δ_{min}, and that typically

$$G^{-1} \propto L_1^x \; . \tag{7}$$

(The case of $x = 1$ leads to logarithmic corrections, which we ignore throughout this paper.) Thus, by considering only the contributions of the singly connected bonds on the links of the percolating backbone, we are led to the following estimate for conductivity exponent $\bar{t}$:

$$\begin{aligned} \bar{t} &\approx 1 + (d-2)\nu = t_1 \quad , && \text{for } x < 1 \; , \\ \bar{t} &\approx x + (d-2)\nu = t_1 + x - 1 \; , && \text{for } x > 1 \; , \end{aligned} \tag{8}$$

where t_1 is the exponent for the conventional percolation problem, in the same approximation. The estimates for the permeability exponent $\bar{e}$, listed in Table 1, are obtained from the same equation (8), using the values of x appropriate to the fluid flow case.

For the elastic problem, we define a force constant K for a string such that $\frac{1}{2} Ku^2$ is the energy cost to displace one end of the string by a small distance u, when the other end of the string is clamped in a position and orientation. If one only considers the compliance of the singly-connected bonds in the string, one finds[3]

$$K^{-1} = \sum_{i=1}^{L_1} \zeta_i^2/\gamma_i \tag{9}$$

where γ_i is the bending force constant of bond i, and ζ_i, the moment-arm of the i-th bond, is a length of order ξ. If all bonds have the same bending constant, as is the case in the conventional lattice percolation model, we find $K \sim \gamma/L_1\xi^2$. Since the macroscopic elastic constants are proportional to $\xi^{2-d}K$, this implies the relation $f \approx 1 + d\nu \equiv f_1$, a result first obtained by Kantor and Webman.[3]

If the bond strengths γ_i have a distribution with $\gamma_i \propto \delta_i^x$, for $\delta \to 0$, we estimate the sum in (9) by an integral, similar to Eq. (4). Now we find $K^{-1} \propto L_1\xi^2$, for $x < 1$, and $K^{-1} \propto L_1^x\xi^2$, for $x > 1$. The corresponding estimates for the elasticity exponent are

$$\bar{f} \approx 1 + d\nu = f_1 \quad , \qquad \text{for} \quad x < 1 \; ,$$
$$\bar{f} \approx x + d\nu = f_1 + x - 1 \; , \qquad \text{for} \quad x > 1 \; . \tag{10}$$

The use above of the "typical value" of K or G may be justified by the same percolation argument employed by Ambegaokar et al.[10] and Pollak[11] in their rederivations of Mott's law of hopping conductivity.[9] We assume that the conducting backbone will fall apart if a critical fraction f_c of the links are randomly cut, where f_c is not very close to either 0 or 1. Then the transport properties of the network are primarily determined by the threshold value K_c or G_c, such that a fraction f_c of the strings have $K < K_c$, or $G < G_c$. If $f_c = 1/2$, then K_c and G_c are the median value of K and G; but the same dependence on L_1 is obtained for any fixed value of f_c.

The exponents derived above, by considering only the singly connected bonds on the strings (links) of the percolating backbone, are presumably lower bounds to the true exponents of the various models. On the other hand, if the exponents t and f in Tables 1 and 2 are defined as the exact lattice values,[6] then the exponent *differences* $(\bar{t} - t)$, $(\bar{f} - f)$ and $(\bar{e} - t)$ listed in the table turn out to be upper bounds for the Swiss-cheese continuum model. These bounds can be established by an extension of the "variational method" employed in ref. 17, where Kogut and Straley established a lower bound to the electrical conductivity by considering the contribution of a connected subnetwork of strong conductors, optimally chosen.[20] (This variational method, of course, is similar in spirit to the method employed by Mott in the hopping conduction problem.[9])

We may summarize our theoretical results as follows. If we consider a network in which a fraction p of the bonds are randomly present, with bond strengths g_i or γ_i that vary as $\delta_i^{\,x}$, where the δ_i are independent random variables with a probability distribution that remains finite in the limit $\delta \to 0$, and if $\bar{t}$ and $\bar{f}$ are the exponents for the vanishing of the network conductivity and elastic moduli as $p \to p_c$, then for $x > 1$ we have the inequalities

$$t_1 + x - 1 \leqslant \bar{t} \leqslant t + x - 1$$
$$f_1 + x - 1 \leqslant \bar{f} \leqslant f + x - 1 \qquad (11)$$

In so far as the differences $(t - t_1)$ and $(f - f_1)$ are small, these inequalities determine $\bar{t}$ and $\bar{f}$ to a reasonable accuracy. For $x \leqslant 1$ we expect $\bar{t} = t$ and $\bar{f} = f$. In order to obtain the results of Tables 1 for the continuum models, we have simply used the values of x appropriate to the narrow bonds in each model. The fluid permeability differs from the electrical conductivity only because of the different values of x in the two cases.

EXPERIMENTS AND NUMERICAL SIMULATIONS

As was mentioned above, Smith and Lobb[13] have studied experimentally the electrical conductivity of a two-dimensional continuum system falling in the class of the "potential model" described above. Their result for the conductivity exponent $\bar{t} = 1.30$ is in excellent agreement with the lattice value for t, as we would expect from our analysis, summarized in Table 1.

The simplest way to observe experimentally the difference between continuum and discrete percolation exponents might be to measure the elastic modulus of a sheet with randomly located circular holes (two-dimensional Swiss-cheese model). The system studied recently by Benbuigui[5] is not of this type, however, as the holes were centered at random sites on a discrete lattice, and there are consequently no narrow necks in the system. We expect the elasticity exponent of Benguigui's system to be the same as the lattice exponent f, in agreement with his observations. The conductivity experiments of Last and Thouless,[21] which first demonstrated that the two-dimensional conductivity exponent is greater than unity, were also carried out in a geometry without narrow necks.

Roberts and Schwartz[16] have studied numerically the electrical conductivity (and recently the permeability) of a porous rock, using a model similar to our Swiss-cheese model, except that the centers of their interpenetrating insulating spheres were chosen originally from a Bernal distribution, rather than completely at random. We would not expect this additional short-range correlation to affect the critical exponents. As noted by Roberts and Schwartz, however, the volume fraction of conductor in these models is very small (≈ 3%) at the percolation threshold, and we expect that the critical exponent may be observable only for q very close to q_c. Roberts and Schwartz do not investigate this, but study instead a wide range of conducting volume fractions above percolation. We note that the analysis of Roberts and Schwartz involved mapping onto a discrete network, similar to ours, with bond strengths determined by the cross-sectional area of the necks.

Wong, Koplik and Tomanic[22] have studied a network of conducting pipes, in which all bonds on a regular network are present, but there is a wide distribution of pipe radii. The exponents of their model depend on parameters in the distribution, but they find typically a permeability exponent about twice the conductivity exponent, which is not far from our finding.

ACKNOWLEDGMENTS

The authors are grateful for very helpful discussions with L.M. Schwartz, C.J. Lobb and Y. Kantor. Work at Harvard was supported in part by the NSF, through the Harvard Materials Research Laboratory, and grant DMR 82-07431.

REFERENCES

1. B.I. Halperin, S. Feng, and P.N. Sen, preprint, 1985.
2. See, for example, S. Feng, P.N. Sen, B.I. Halperin, and C.J. Lobb, *Phys. Rev.* B30:5386 (1984), and references therein.
3. Y. Kantor and I. Webman, *Phys. Rev. Lett.* 52:1891 (1984); Y. Kantor, *J. Phys.* A17:L843 (1984).
4. D.J. Bergman, *Phys. Rev.* B31:1696 (1985).
5. L. Benguigui, *Phys. Rev. Lett.* 53:2028 (1984).
6. Numerical values of discrete lattice exponents defined in the text are: for $d=2$, $t_1=1$, $t\approx 1.3$, and $f\approx f_1\approx 3.7$; for $d=3$, $t_1\approx t\approx 1.9$, and $f\approx f_1\approx 3.6$.
7. W.T. Elam, A.R. Kerstein, and J.J. Rehr, *Phys. Rev. Lett.* 52:1516 (1984).
8. E.T. Gwalinski and H.E. Stanley, *J. Phys.* A14:L291 (1981).
9. N.F. Mott, *Phil. Mag.* 19-835 (1969).
10. V. Ambegaokar, B.I. Halperin, and J.S. Langer, *Phys. Rev.* B4:2612 (1971). See also M. Pollak and B.I. Halperin, *Solid State Comm.* 13:869 (1973).
11. M. Pollak, *J. Non-Cryst. Solids* 11:1 (1972).
12. See also W. Brenig, P. Wölfe, and G. Döhler, *Phys. Lett.* A35:77 (1971); R. Jones and W. Schaich, *J. Phys.* C5:43 (1972).
13. L.N. Smith and C.J. Lobb, *Phys. Rev.* B20:3653 (1979).
14. A. Weinrib, *Phys. Rev.* B26:1352 (1982).
15. J.M. Ziman, *J. Phys.* C1:1532 (1968).
16. J.N. Roberts and L.M. Schwartz, *Phys. Rev.* B (in press) and private communications.
17. P.M. Kogut and J.P. Straley, *J. Phys.* C12:2151 (1979).
18. A. Ben-Mizrahi and D.J. Bergman, *J. Phys.* C14:909 (1981).
19. A. Coniglio, *Phys. Rev. Lett.* 46:250 (1981); R. Pike and H.E. Stanley, *J. Phys.* A14:L169 (1981).

20. Details will be given elsewhere.
21. B.J. Last and D.J. Thouless, *Phys. Rev. Lett.* 27:1719 (1971).
22. P.Z. Wong, J. Koplik, and J. Tomanic, *Phys. Rev.* B30:6606 (1984).

FIRST-ORDER PHASE TRANSITION TO THE METALLIC STATE IN DOPED POLYACETYLENE: SOLITONS AT HIGH DENSITY

J. Chen, T.-C. Chung, F. Moraes and A.J. Heeger

Institute for Polymers and Organic Solids
Department of Physics
University of California
Santa Barbara, CA 93106

INTRODUCTION

The insulator-metal transition in conventional three dimensional semi-conductors is conceptually understood in the context of the Mott transition. The localized donor or acceptor states of the dilute limit evolve into "metallic" impurity bands when the wave-function overlap is sufficient to give effective screening of the localizing electron-electron interaction.

In quasi-one-dimensional semi-conductors, and in particular in conducting polymers, the evolution of the metallic state at high doping levels is not yet understood. In the case of a two-fold degenerate ground state, for example trans-$(CH)_x$, there is considerable evidence that in the dilute limit charge is stored in the form of spinless charged solitons,[1-3] and that at higher doping levels (above ∿6 mole%) a truly metallic state is achieved.[1-7] There is, however, at best an incomplete understanding of the transition. In a series of recent experimental studies, we demonstrated that this major change in electronic structure takes place as a first-order phase transition from a soliton lattice to a metal.[2,8] This is the first experimental demonstration of an abrupt first order transition as a function of impurity concentration in any doped semiconductor system.

EXPERIMENTAL TECHNIQUES

Semitransparent polyacetylene films were synthesized on quartz substrates. The films were n-type doped with sodium (Na^+), using either a chemical or an electrochemical method, to obtain $[(Na^+)_y(CH)^{-y}]_x$. Chemical doping was carried out by exposing the polymer film to a solution of sodium benzophenone in tetrahydrofuran, Na^+B^- (THF),

$$(CH)_x + xyNa^+B^- \rightarrow [(Na^+)_y(CH)^{y-}]_x + xyB \quad (1)$$

An apparatus was designed for in situ chemical doping and measurement of the ESR and optical spectra (or the conductivity and optical spectra) in a completely sealed system.[2] Our methodology was to first obtain a visible-ir spectrum at each doping level to assure dopant uniformity and to determine the dopant concentration, using the previously published optical data obtained as a function of dopant concentration. The apparatus was then immediately moved to carry out the ESR or σ measurements. The sealed apparatus design resulted in excellent sample stability. Magnetic resonance data were monitored over several days and found to be fully reproducible. After the conductivity measurements, σ was monitored over a period of several weeks with no significant change in value.

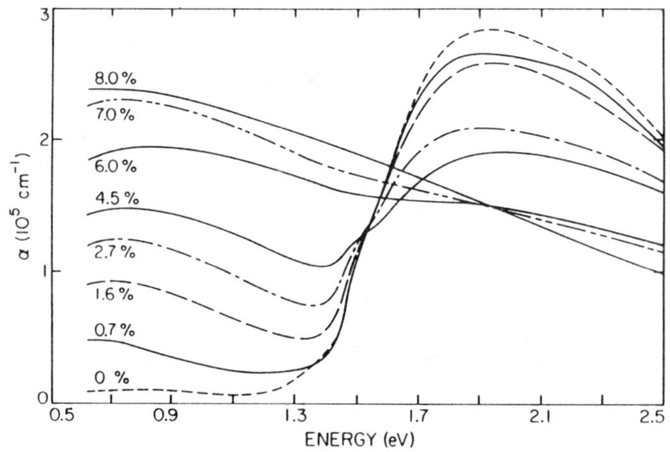

Fig. 1. Optical absorption coefficient (α) vs photon energy for $[Na^+)_y(CH)^{y-}]_x$. The data were obtained from the identical samples used in Fig. 5 (see text).

The y values (Na^+) were obtained from the series of spectra shown in Fig. 1. We found excellent stability; the spectra were reproducible over many days. Note that the soliton midgap transition[7] remains evident even at concentrations as high as 4.5%, implying the presence of solitons at high density.

A separate and independent series of ESR measurements were carried out in situ on $[(Na^+)_y(CH)^{-y}]_x$ during electro-chemical doping.[8] A schematic diagram of the electrochemical cell is sketched in Fig. 2; the $(CH)_x$ sample, electrolyte and counter electrode were assembled inside a 3 mm esr tube. The glass was sealed across the electrical contacts which were connected externally to a variable voltage supply. The esr tube (electrochemical cell) was mounted with the tip inside the microwave cavity of an IBM Instruments (Bruker) E-200D ESR spectrometer. To avoid having the Na metal and Nickel mesh inside the cavity, the counter electrode was mounted above the $(CH)_x$ film. The glass capillary tube (around the platinum wire leading to the $(CH)_x$ electrode) was included to minimize background currents (∿ 0.02 μ amps).

EXPERIMENTAL RESULTS

In Fig. 3, we plot the electron spin contribution to the magnetic susceptibility (χ) of trans-$[Na_y^+(CH)^{-y}]_x$ as a function of its electrochemical potential, μ (referenced to Na metal).[8] Note that both χ and μ

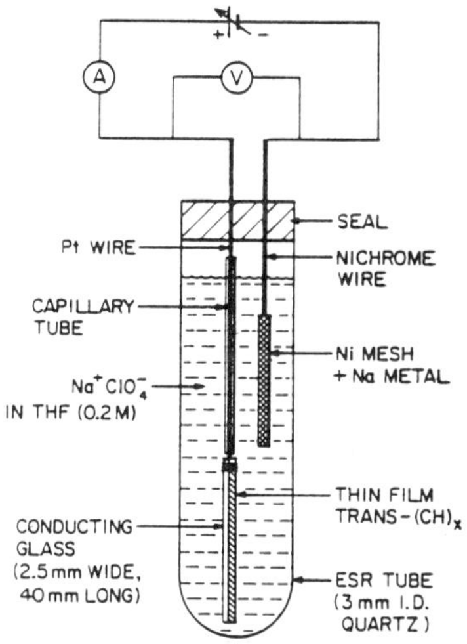

Fig. 2. Diagram of electrochemical cell used for in situ esr measurements.

are intrinsic thermodynamic variables. The abrupt increase in χ at μ = 0.8 eV (relative to Na) and the observation of hysteresis in χ vs μ are indicative of a first order phase transition.

The data of Fig. 3 were obtained using the in situ electrochemical method described above. In order to avoid problems arising from the electrochemical decomposition[9] of ClO_4^-, the cell was used only at voltages $\geq$ 0.70 V.

Data were taken by first setting the external voltage and monitoring the current while allowing the cell to come to equilibrium (typically 12 to 36 hours). After stepping the voltage by ΔV = 0.1 V, the initial current (for voltages beyond the injection threshold, $\cong$ 1.5 V) was 0.5 μ amp falling to < 0.02 μ amp as the cell approached equilibrium. At equilibrium, the measured external voltage is the open-circuit voltage,

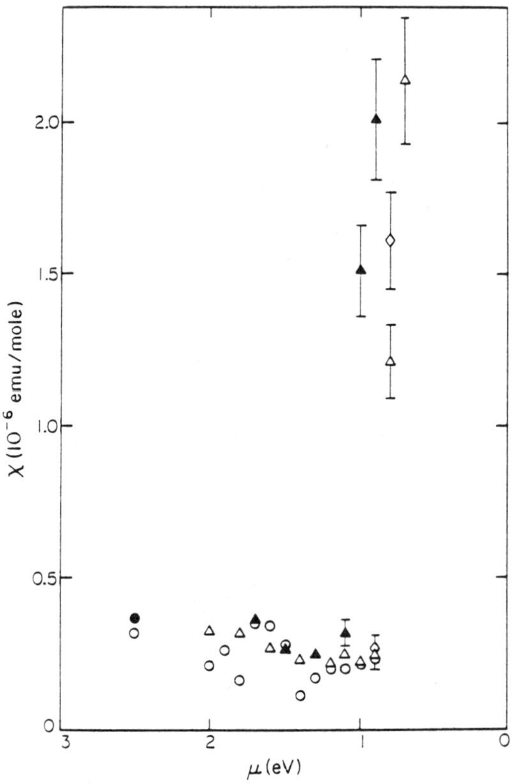

Fig. 3. Spin susceptibility of $[Na_y^+(CH)^{-y}]_x$ asa function of the chemical potential of the doped polymer (referenced to Na metal). The results were obtained by esr measurements carried out in situ during electrochemical doping; o - first cycle doping, o - first cycle undoping; Δ - second cycle doping, Δ - second cycle undoping; □ - third cycle doping.

which is by definition the electrochemical potential of the doped polymer referenced to that of the Na counter electrode. After allowing the cell to come to equilibrium, the esr signal was recorded using signal averaging to obtain the necessary signal-to-noise ratio from the thin film samples. The data were integrated twice to obtain the spin contribution to the magnetic susceptibility (χ). An NBS ruby standard was attached to the outside of the esr tube (near the sample) for calibration of the absolute magnetic susceptibilities. Symmetric esr lines and accurate χ values were obtained since the sample thickness ($\cong$ 0.5 μm) was much less than the microwave skin depth, even at the highest concentrations.

The data shown in Fig. 3 were obtained from several doping and undoping cycles (each cycle is denoted with a different symbol). The cell was stable over a period of approximately three weeks, and χ values were reproducible from cycle to cycle. The error bars denote mean deviations of several independent measurements of χ at that voltage setting. The hysteresis was carefully checked by allowing 36 hours for the sample to come to equilibrium at the appropriate voltages; it is a genuine feature of the data.

The esr linewidth (ΔH) is shown as a function of μ in Fig. 4. The step-like increase and the clear hysteresis are consistent with the corresponding χ data (Fig. 3) and corroborate the first order transition. In all cases the esr spectrum consisted of a single, symmetric line even though the linewidth in the two regimes differed by a factor of 20; a result which demonstrates uniform doping. There is no indication whatsoever of metallic island formation,[10] which would show up as pre-transition increases in χ and ΔH.

The data of Figs. 3 and 4 are qualitatively similar to the results for χ and ΔH vs. y obtained from chemical doping of $[Na_y(CH)^{-y}]_x$ as shown in Fig. 5.[2] In both cases the two regimes are clearly defined with a sharp transition between them. In the chemical doping experiments, detailed measurements of the temperature dependence of the susceptibilities demonstrated that the weak narrow line below the transition obeyed Curie's law ($\chi \sim 1/T$) indicative of a dilute concentration ($\sim 1.5 \times 10^{-4}$ per carbon) of localized magnetic states, whereas the large susceptibility above the transition was temperature independent indicative of the Pauli susceptibility (χ_p) of a metal. Although it was not possible to carry out temperature dependence measurements with the electrochemical cell, the excellent agreement of both χ and ΔH in the two regimes (Figs. 3 and 4) with the corresponding values obtained by chemical doping implies the same behavior.

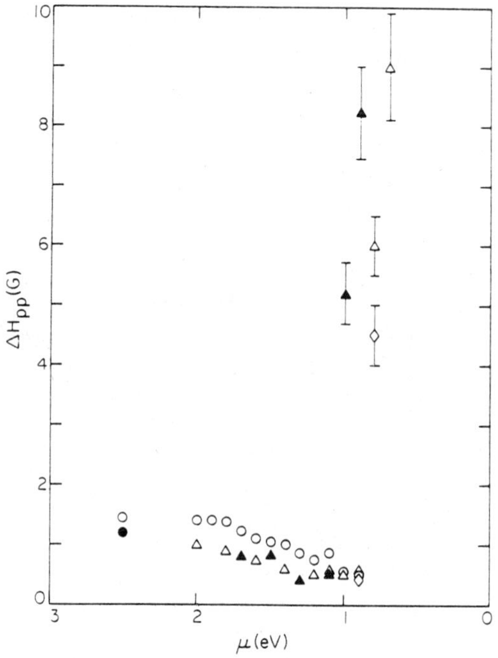

Fig. 4. Peak-to-peak esr linewidth (ΔH_{pp}) of $[Na^+_y(CH)^{-y}]_x$ as a function of the chemical potential of the doped polymer (references to Na metal); o - first cycle doping; o - first cycle undoping; Δ - second cycle doping, Δ - second cycle undoping; □ - third cycle doping.

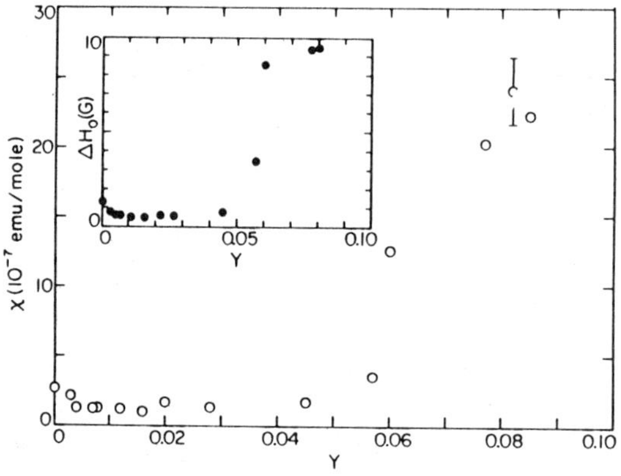

Fig. 5. Magnetic susceptibility (room temperature) for $[(Na^+)_y(CH)^{y-}]_x$ vs. y. The inset shows the concentration dependence on the ESR linewidth.

The initial susceptibility due to neutral solitons decreased with doping. For μ beyond the injection threshold,[11] $\cong$ 1.5 volts with respect to Na, the number of Curie spins was essentially independent of μ (and thus independent of y). The narrow linewidth implies a mobile species, not bound to the Na^+ (Coulomb binding to Na^+ would yield a broad line comparable with that of the metallic state). As suggested for polythiophene,[12] this concentration independent Curie contribution may arise from formation of one kinetically meta-stable spin-½ polaron per two $(CH)_x$ chains, since at any dopant level approximately half the chains will have an odd number of charges. The observed concentration (∿ one per 10^4 carbon atoms) is consistent with the average molecular weight[13] of $(CH)_x$ prepared by the Shirakawa method.

The susceptibility results shown in Figs. 3 and 5 demonstrate without ambiguity that for dopant levels below the first-order transition, charge is stored predominantly in a non-magnetic configuration. This conclusion combined with the results of infrared studies[14] on doped and photoexcited trans-$(CH)_x$ demonstrate that doping occurs via the formation of spinless charged solitons, and that these solitons dominate the physical properties at all concentration below the sharp transition to the truly metallic state (characterized by a finite density of states at the Fermi energy and an associated Pauli spin susceptibility). In the metallic regime, $\chi_p \cong 2\text{x}10^{-6}$ emu/mole in agreement with previous measurements.

DISCUSSION: ORIGIN OF THE FIRST ORDER PHASE TRANSITION

Electrochemical data[11] on $[Na_y^+(CH)^{-y}]_x$ and structural studies of electrochemically prepared samples[15] have identified ordered phases of Na-doped trans-$(CH)_x$. These results together with the susceptibility data (Figs. 3 and 5) suggest to us the possibility of the formation of a soliton lattice. The electrochemical voltage spectroscopy curves of Shacklette et al.[11] and x-ray diffraction results of Baughman et al.[15] demonstrate Na^+ insertion via a sequence of stoichiometric phases. The discontinuous transformation from one phase to the next causes the appearance of plateaus in μ vs. y. For $[Na_y^+(CH)^{-y}]_x$ the first plateau extends to a concentration of y $\cong$ 6 mole% with a transition at $\mu \cong$ 0.8 eV (referenced to Na).[11] This value is in close agreement with the transition observed in χ (Fig. 3). The existence of an ordered phase of the Na-doped polymer in which charge is stored in the form of spinless charged solitons implies the formation of a soliton lattice. Thus, the data of Figs. 3, 4, and 5 indicate a first-order phase transition from soliton lattice to metal.

Within the Peierls model, the one-dimensional metal is unstable to the formation of charge density waves (CDW) with $q = 2k_F$ where k_F is the Fermi wavenumber. In this case, doping would change $2k_F$ and continuously shift the period of the CDW. The addition of commensurability energy causes q to stick at the commensurate value. For the particular case of the two-fold degenerate ground state of trans-$(CH)_x$, the effect is sufficiently strong to dominate the condensation energy. As a result, instead of becoming incommensurate as q deviates from π/a (due to charge transfer doping) the charge is stored in charged solitons.[16,17] The data of Figs. 3 and 5 indicate that the resulting soliton lattice is stable until the concentration is sufficiently high that the corresponding free energy is greater than the free energy of the metal.

The energy of the soliton lattice has been analyzed in considerable detail;[18-22] within the single chain approximation, exact solutions are available. The results can be summarized in terms of the increase in energy above the undoped ground state:

$$\Delta E = N_{s,\bar{s}} (E_s + \frac{16}{\pi} \Delta e^{-d/\xi}) \tag{1}$$

where $N_{s,\bar{s}}$ is the number of kinks (solitons or antisolitons), E_s is the creation energy for a soliton, ξ is the soliton half-width and d is the period of the soliton lattice (in each repeat unit there exists a soliton and an antisoliton so that $N_{s,\bar{s}} = 2/d$). Within this model the cross-over to metallic behavior would occur when the energy δE needed to introduce electrons or holes into the soliton band becomes less than zero. Lin-Liu and Maki[18b] noted that by creating neutral solitons, holes could be inserted into the otherwise full mid-gap-band. The energy cost per neutral soliton (note, however, that they must be made in pairs) is given by the following:[23]

$$\delta E = (\frac{1}{\pi}\Delta - \frac{1}{2} U_{eff}) + \frac{16}{\pi}\Delta \left(1 + \frac{d}{\xi}\right)e^{-d/\xi} - 4\Delta e^{-d/2\xi} \tag{2}$$

where $E_s^N = (2\Delta/\pi - \frac{1}{2} U_{eff})$ is the energy for creation of a neutral soliton and U_{eff} is the repulsive interaction between two electrons (or holes) in a soliton.[24] From numerical estimates based on eqn. (2), one can obtain a critical concentration for $\delta E<0$ of about 6% (i.e. in approximate agreement with the experimental value) provided $U_{eff} \sim \Delta$. A more detailed analysis shows that this novel transition is first order, in agreement with the experimental findings (Figs. 3, 4 and 5).

The above argument ignores many important features of the problem; including interchain transfer, randomness and disorder, and the long-range Coulomb interaction. Although the existence of ordered phase[11,15] implies that the Na^+ doped polymer is relatively well ordered, the effects of residual disorder must be taken into account, especially for other dopants.[7]

SUMMARY AND CONCLUSION

In summary, the spin susceptibility and esr linewidth have been measured as functions of the dopant concentration and, independently, as functions of the chemical potential through in situ measurements on $[Na^+_y(CH)^{-y}]_x$. The results show step-like increases in χ and ΔH, with hysteresis, indicative of a first order phase transition.

The transport and magnetic data are compared in Fig. 6, where we plot χ_p (right-hand scale) and σ(300K) (left-hand logarithmic scale) for direct comparison. We conclude that $\chi_p < 10^{-8}$ emu/mole for $y < 0.05$, whereas σ is quite high, e.g., 1 - 10 $\Omega^{-1}cm^{-1}$ in the dopant range from 2-4 mole%. Transport by mobile charged solitons is consistent with the susceptibility Figs. 3 and 5), the spectroscopic data (Fig. 1, ref. 7) and the doping-induced ir modes,[14] which prove that for $y < 0.05$ charge is stored in the spinless gap states of charged solitons.

Epstein et al.[25] have argued that the conductivity in the non-magnetic regime is not due to mobile solitons, but results from electrons hopping among localized soliton levels spread through the gap by disorder. This cannot be the case for $[Na^+_y(CH)^{-y}]_x$. The sharp first order phase transition and the associated abrupt increase in the density of states by a factor $\geq$ 200 are inconsistent with a model dominated by disorder.

Dong and Schrieffer[26] have shown that a careful treatment of the three-dimensional Coulomb interaction can lead to "melting" and the formation of a highly correlated soliton liquid. Since such a state would exhibit electrical conductivity, the soliton liquid may be responsible for the relatively high electrical conductivity at concentrations in the non-magnetic regime below the first order transition to the metallic state.

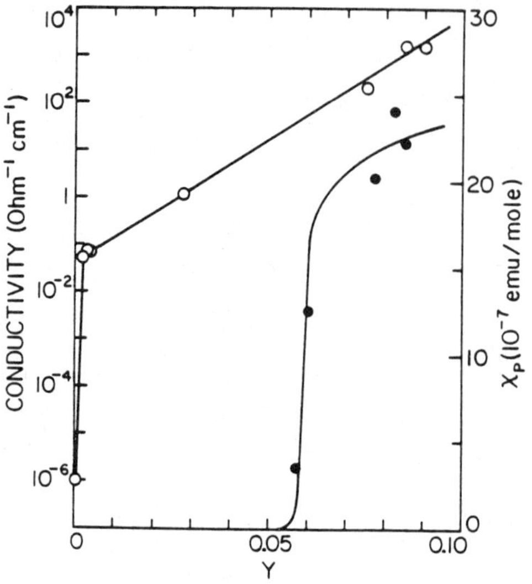

Fig. 6. χ_p vs y (right-hand scale)and σ_{dc} (300K) vs y (left-hand logarithmic scale). For y < 0.5, $\chi_p < 10^{-8}$ emu/mole.

The demonstration of charge storage in spinless solitons in trans-$(CH)_x$ is the first experimental manifestation of fermion charge fractionalization through soliton formation. Although the solitons in trans-$(CH)_x$ have integer charge, the reversed spin-charge relation arises directly from charge fractionalization in the fermion system (spins $\pm\frac{1}{2}$) with commensurability two.[27,28]

The first order phase transition to the metallic state (finite density of states at the Fermi level) has been interpreted as arising from the spontaneous generation of neutral solitons and the formation of a soliton lattice with a partially filled "mid-gap" electronic energy band. This novel metallic state is particularly interesting since it offers an opportunity for detailed studies of solitons at high density.

ACKNOWLEDGEMENTS

These magnetic resonance studies were supported by the National Science Foundation, DMR82-12800. We thank Dr. Y.R. Lin-Liu and Dr. S. Kivelson for important comments and helpful discussions.

REFERENCES

1a. For a review, see S. Etemad, A.J. Heeger and A.G. MacDiarmid, Ann. Rev. Chem. Phys. 33, 443 (1982).

b. Recent results are summarized in the Proceedings of the Los Alamos Workshop on Synthetic Metals; J. Synth. Metals 9, (1984).

2. T.-C. Chung, F. Moraes, J.D. Flood and A.J. Heeger, Phys. Rev. B 29, 2341 (1984) (and references therein).

3. S. Ikehata, J. Kaufer, T. Woerner, A. Pron, M. Druy, A. Sivak, A.J. Heeger and A.G. MacDiarmid, Phys. Rev. Lett 45, 1123 (1980).

4. B.R. Weinberger, J. Kaufer, A.J. Heeger, A. Pron and A.G. MacDiarmid, Phys. Rev. B 20, 223 (1979).

5. Y.W. Park, A.J. Heeger, M.A. Druy and A.G. MacDiarmid, J. Chem. Phys. 73, 946 (1980).

6. A.J. Epstein, H. Rommelman, M.A. Druy, A.J. Heeger and A.G. MacDiarmid, Sol. St. Commun. 38, 683 (1981).

7. A. Fedlblum, J.H. Kaufman, S. Etemad, A.J. Heeger, T.-C. Chung and A.G. MacDiarmid, Phys. Rev. B 26, 815 (1982).

8. J. Chen, T.-C. Chung, F. Moraes and A.J. Heeger, Sol. State Commun. (in press).

9. L. Shacklette, private communication.

10. See, for example, Y. Tomkiewicz et al., Phys. Rev. Lett. 43, 1532 (1979); Phys. Rev. B 29, 4348 (1981).

11. L.W. Shacklette, N.S. Murthy and R.H. Baughman, Proc. of ICSM 84 Abano, Italy; to be published in Mol. Cryst. and Liq. Cryst.

12. F. Moraes, D. Davidov, M. Kobayashi, T.-C. Chung, J. Chen, A.J. Heeger and F. Wudl, J. Synth. Metals (in press).

13. J.C.W. Chien, M. Schen and F. Karasz, Makro. Molec. Chemie, Rapid Commun. 5, 217 (1984).

14. See the following and references therein:

a. G. Blanchet, C.R. Fincher and A.J. Heeger, Phys. Rev. Lett. 50, 1938 (1983); 51, 2132 (1983).

b. Z. Vardeny, J. Orenstein and G.L. Baker, Phys. Rev. Lett. 50, 2032 (1983).

c. Z. Vardeny, E. Ehrenfreund, O. Brafman and B. Horovitz, Phys. Rev. Lett. 51, 2326 (1983).

15. R.H. Baughman, L.W. Shacklette, N.S. Murthy, G.G. Miller and R.L. Elsenbaumer, Proc. of ICSM 84, Abano, Italy; to be published in Molec. Cryst. and Liq. Cryst.

16. W.-P. .Su, J.R. Schrieffer and A.J. Heeger, Phys. Rev. Lett 42, 1698 (1979); Phys. Rev. B 22, 2209 (1980).
17. M.J. Rice, Phys. Lett 71A, 152 (1979).
18a. H. Takayama, Y.R. Lin-Liu and K. Maki, Phys. Rev. B 21, 2388 (1980).
b. Y.R. Lin-Liu and K. Maki, Phys. Rev. B 22, 5754 (1980).
19. M.J. Rice and J. Timonen, Phys. Lett 73A, 368 (1979).
20. J.P. Albert and C. Jouanin, Mol. Cryst. Liq. Cryst. 77, 297 (1981).
21. E.J. Mele and M.J. Rice, Phys. Rev. B 23, 5397 (1981).
22a. B. Horovitz, Phys. Rev. Lett. 46, 742 (1981).
b. S.A. Brazovskii, S. Gordyunin and N.N. Korova, Zh. Eksp. Teor. Fiz. Pis'ma Red. 31, 486 (1980); JETP Lett. 31, 456 (1980).
23. Equation 2 differs from that given by Lin-Liu and Make (ref 18b). The positive term, $(16\Delta/\pi)(d/\xi)\exp(-d/\xi)$, arises from the increase in the inter-soliton repulsion caused by the addition of neutrals; we thank Y.R. Lin-Liu for calling our attention to this term. The negative term, $-\frac{1}{2}U_{eff}$, comes from the inclusion of the short range Coulomb interaction within the Hubbard model (ref. 23).
24. S. Kivelson and D.E. Heim, Phys. Rev. B 26, 4378 (1982).
25. J. Dong and J.R. Schrieffer, to be published. See also J. Dong, PhD thesis, UCSB, 1984.
26. A.J. Epstein, D. Hoffman and D. Tanner, Phys. Rev. Lett. 23, 1866 (1983).
27. R. Jackiw and C. Rebbe, Phys. Rev. D 13, 3398 (1976).
28. R. Jackiw and J.R. Schrieffer, Nucl. Phys. B190, 253 (1981).

THE GERMANIUM GRAIN BOUNDARY: A DISORDERED TWO-DIMENSIONAL ELECTRONIC SYSTEM

G. Landwehr[a], and S. Uchida[b]

a) Physikalisches Institut der Universität Würzburg
D-8700 Würzburg, Federal Republic of Germany
b) Department of Applied Physics, University of Tokyo
Tokyo, Japan

INTRODUCTION

Detailed work on dislocations in semiconductors started in the fifties when it became obvious that they can be electrically active[1] and influence the quality of device material. From early work on plastically deformed germanium it was concluded, that edge dislocations in this semiconductor can behave as acceptors. Because grain boundaries contain arrays of dislocations, a p-type space charge layer can be formed adjacent to a grain boundary. This aspect is of technological importance for the manufacture of solar cells from polycrystalline material, and consequently a considerable amount of work has been done in this field recently. The main concern is ususally the carrier transport across grain boundaries. Also the trapping of carriers is of interest.

After the first artificial germanium bicrystals of large diameter were grown, it became possible to study electrical transport across and along grain boundaries in detail.[2] In particular germanium bicrystals a large p-type conductivity was observed along the grain boundary which was comparable in magnitude to that of a clean germanium surface. This was the motivation for a detailed study of galvano-magnetic properties of medium angle grain boundaries at low temperatures by Landwehr and Handler.[3] Measurements of conductivity and Hall effect of bicrystals with a common (100)plane and a tilt angle of 20° indicated much similarity to heavily doped p-type germanium with a hole concentration of about $5x10^{18}/cm^3$ and a Hall mobility of about 300 cm^2/Vs. The conductivity was almost temperature independent below 20 K as expected for a degenerate electron gas in conjunction with impurity scattering. Whereas the magneto resistance at high temperatures corresponded to the expectations of common transport theory it turned out that at helium temperatures it was several orders of magnitude larger than predicted. Also, the dependence on the magnetic field was unusual. Eventually, the experiments were terminated because no satisfactory explanation of the data could be proposed.

The situation changed, however, when a few years ago detailed theories for transport in two-dimensional systems were developed. The theory of weak localization[4] predicts a logarithmic increase of resistance with decreasing temperature and a negative magneto resistance. Altshuler, Aronov and Lee[5] showed that electron-electron scattering in the presence of impurities should also lead to a logarithmic divergence of the resistance in two-dimensional systems. Subsequently, detailed predictions for the magneto resistance of two-dimensional systems were made by Altshuler et al.[6] and by Fukuyama[7] and Kawabata[8] with the magnetic field oriented perpendicular or parallel to the conducting layer. Under these circumstances it seemed appropriate to resume the transport studies in p-type inversion layers adjacent to grain boundaries in germanium bicrystals. The progress in the experimental techniques in the last 20 years - the availability of high magnetic fields in combination with millikelvin temperatures - was another incentive to have a new look at the problem.

There was a third reason to continue the studies on low temperature transport in germanium-bicrystals. It had become evident quite early, when the thickness of the inversion layer in a bicrystal was estimated employing the Thomas-Fermi method[9], that quantum effects should show up due to the confinement of the holes to a sheet of about 100 Å. The work on p-channel MOSFETs has demonstrated that in spite of the complicated valence-band structure of silicon it has been possible to calculate the electrical subbands. It turned out that the band structure of an inversion layer depends very much on the electric field present in the inversion layer.[10] An extension of the subband calculations to germanium-bicrystals has been made in the meantime.[11] These can serve as a basis for the interpretation of experimental data.

In the following, we shall first discuss the crystallographic properties of the grain boundaries in bicrystals. Subsequently, the theoretical calculations of the electronic structure will be outlined together with the major results. Then we shall show typical data which demonstrate that p-type inversion layers in germanium bicrystals in many respects behave like disordered two-dimensional systems. Finally, we shall compare our results with recent theories.

GERMANIUM BICRYSTALS

A germanium bicrystal can be grown with the Czochralski-technique, this has been demonstrated by Mataré and coworkers.[12] If two seed crystals with (001) orientation are tilted along a (100) axis by an angle θ which may vary between a few degrees and 25°, a regular array of edge dislocations is supposed to arise at the interface with a distance $D = a/2\sin(\theta/2)$ (Fig. 1). This has been checked by electron microscopy for small tilt angles. For medium angle boundaries with $\theta > 5^{o}$ the compression and dilation zones of adjacent dislocations overlap to a large extent so that the lattice distortion is essentially relaxed a few lattice constants away from the grain boundary. But nevertheless a substantial amount of disorder will remain.

In germanium bicrystals derived from (100) seeds the edge dislocations have dangling bonds, as a consequence of which they act as line defects with acceptor character. For tilt angles $< 5^{o}$ the spacecharge cylinders surrounding each edge dislocation do not overlap so that at helium temperatures undoped specimens behave as insulators. If the tilt angle is increased to 7°, hopping conduction parallel to the grain boundary becomes possible. The resistivity of p-type spacecharge layers in (100) bicrystals with tilt angles between 7° and 30° was systematically studied at low temperatures by Vul and Zavaritskaya.[13] For specimens with a tilt angle $< 10^{o}$ activated resistivity was observed. For larger tilt angles the resistivity was not much dependent of temperature, indicating degenerate behaviour. There was a pronounced difference in the measured resistivity, depending on the current direction relative to the dislocations. In specimens in which the current was directed perpendicular to the dislocations the resistivity was larger than in samples in which current and dislocations were parallel. The anisotropy of the resistivity decreased with increasing tilt angle and had almost vanished for $\theta = 30^{o}$. This indicates that a model with independent indivi-

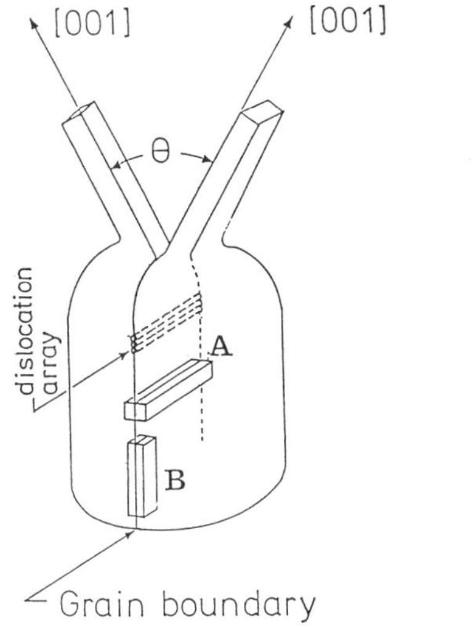

Fig. 1. Germanium bicrystal, with 2 samples A and B of different orientation (schematic). θ = tilt angle.

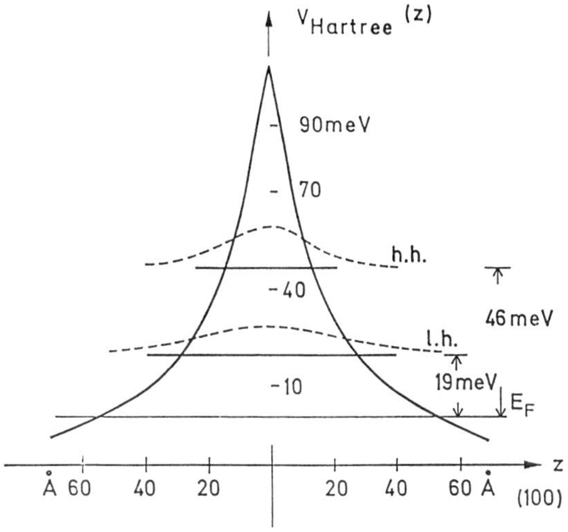

Fig. 2. Hartree potential energy for a (100) p-type inversion layer adjacent to a grain boundary in Ge. Two electric subbands are occupied, which can be attributed to heavy (hh) and light holes (lh). E_F = Fermi energy. Hole concentration: $10^{13}/cm^2$.

dual dislocations looses its significance for nominal dislocation distances of the order of 10 Å. Nevertheless, it seems justified to discuss experimental transport data for bicrystals with tilt angles between 10 and 20° in terms of a built-in linear dislocation lattice.

It should be mentioned that the grain boundary model discussed so far seems to be too simple for the germanium lattice. Many years ago, Hornstra[14] studied the problem in detail and proposed specific models of edge dislocations with a (100) glide plane, with the dislocation axis along [110] and a [110] Burgers vector. In one case, two free bonds per atom are available at the termination of the edge dislocation, and in the other configuration reconstruction has occured with no dangling bonds. Lateron, theoretical calculations were performed for various dislocations in the diamond lattice.[15] Because it has been impossible to obtain self-consistency and to take relaxation effects into account, one cannot expect that the improved models describe accurately the reality.

SUBBAND CALCULATIONS

The interpretation of transport data for p-type materials is usually complicated by the fact that valence bands are degenerate and anisotropic. In bulk germanium the two valence bands - the light and heavy hole band - are degenerate at the center of the Brillouin-zone and a third band is split off by spin-orbit interaction. In two-dimensional systems an additional complication arises. If the de Broglie wavelength of free carriers is larger than the thickness of the conducting layer, quantum effects show up.[16] Whereas the carriers are free to move parallel to the interface in a two-dimensional system, they are bound in the perpendicular direction by the surface potential. The boundary quantization is the origin of the so-called electric subbands.

A p-type inversion layer adjacent to a grain boundary in a germanium bicrystal is similar to that in a p-channel silicon MOSFET except that the potential in a grain boundary is symmetrical with respect to the interface. The problem is relatively complex because in the degenerate valence bands of both germanium and silicon the carrier motions parallel and perpendicular to the interface are strongly coupled, with the coupling depending on the magnitude of the electric field in the grain boundary region. It has been known for some time[10,16] that in p-type silicon inversion layers in MOSFETs the band structure is substantially modified by the electric field in the interface, for instance the effective masses differ significantly from the bulk values and depend on the carrier concentration. Usually the light and heavy hole subbands are well separated. Whereas in n-channel MOSFETs in most cases only one subband is occupied,[17] it has been shown that for p-channel devices of (110) orientation both the light and heavy hole subband can be populated.[10b]

Because usually the transport properties of semiconductors - especially in high magnetic fields - depend strongly on the presence of a second carrier, it seemed desirable to get detailed insight into the electronic band structure of p-type inversion layers in germanium bicrystals. Consequently, the methods which were employed for p-channel MOSFETs were modified.

In order to obtain electrical subbands, constant energy contours and effective masses, a one-dimensional Schroedinger-equation was solved in the effective mass approximation self-consistently in conjunction with Poisson's equation.[11,18] The band structure was incorporated by inserting the well-known Luttinger and Kohn 6x6 Hamiltonian. The Hartree-potential was obtained by solving Poisson's equation taking into account the charge of both fixed impurities and free carriers. In order to make the calculations managable the tilt angle was neglected and the grain boundary was assumed to be (100) plane. The negatively charged dislocations were considered as scattering centers and the possible periodicity of the dislocation potential was neglected. This is a possible drawback which will be discussed with the presentation of experimental data lateron. It should be noted that the electric subbands are twofold spin-degenerate since the germanium grain boundary problem has inversion symmetry, contrary to the MOSFET case. Details of the calculations are contained in ref.[18].

The Hartree-potential as a function of the distance from the grain boundary plane is shown in Fig. 2 for a total free hole concentration of $10^{13}/cm^2$. Two electric subbands are occupied, the upper one 46 meV above the Fermi level and the lower one 19 meV. For a carrier concentration of $4x10^{12}/cm^2$ the energies are 20 meV and 7 meV, respectively. The envelope functions calculated for the two subbands - indicated by dotted lines - allow the identification as heavy (hh) and light (lh) subband. One can recognize that the band bending is quite substantial. The Fermi level is located about 100 meV below the potential maximum. The thickness of the inversion layer is less than 100 Å resulting in a volume hole concentration which is found in heavily doped material.

The constant energy contours for the light and heavy holes for a hole concentration of $10^{13}/cm^2$ is shown in Fig. 3. The E=constant lines for the heavy holes are strongly warped with negative masses around [100] directions. The light holes, on the other hand, are rather isotropic. The effective masses calculated from the constant energy contours are shown in Fig. 4. Whereas at high carrier concentrations the mass difference between light and heavy holes is not large it becomes significant at typical carrier concentrations of $4x10^{12}/cm^2$.

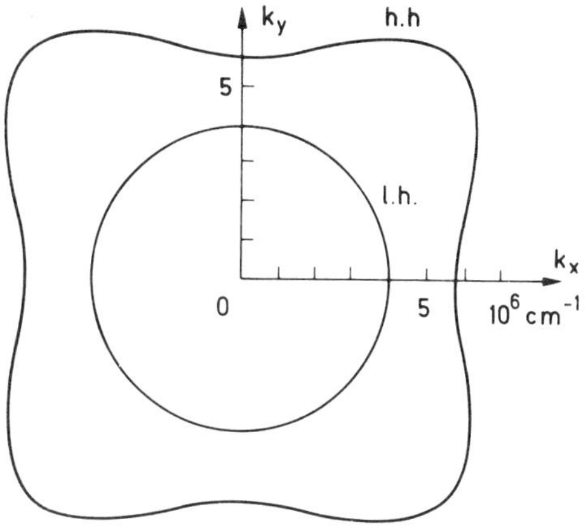

Fig.3. Constant energy contours in a (100)-plane of k_x-k_y space for light (lh) and heavy holes (hh) for a total hole concentration of $10^{13}/cm^2$.

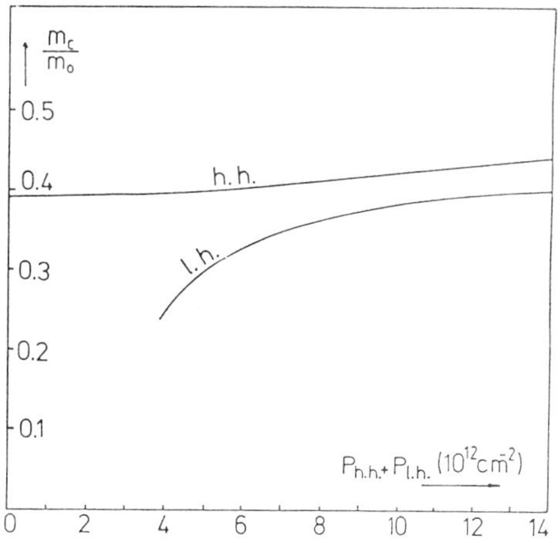

Fig.4. Cyclotron mass for light and heavy holes as a function of the total hole concentration.

The dependence of the Fermi energy on the total free hole concentration is plotted in Fig. 5 (full lines). According to the calculations the light hole subband is populated around $2x10^{12}/cm^2$ and up to hole concentrations of $2x10^{13}/cm^2$ only two subbands are occupied. The dotted line in Fig. 5 represents the ratio of light to heavy holes. One recognizes that for a typical total hole concentration of $5x10^{12}/cm^2$ about 20 % of the carriers are light holes. This means that in an inversion layer the ratio of light to heavy holes is considerably higher than in bulk material.

In our subband calculations many body effects have been neglected because they would have complicated the work too much. This drawback does not seem to be serious, however. From previous work on n-type silicon inversion layers[19] it is known that at low surface electron concentrations many body effects should be taken into account. At carrier concentrations above $10^{12}/cm^2$ the screening is rather effective, however, so that many body effects should play only a minor role. This seems also to apply to p-type inversion layers. A particular bicrystal with common (100) plane and 15° tilt angle showed very high hole mobility of more than 2000 cm^2/Vs. This allowed to observe Shubnikov-de Haas oscillations.[18] Their analysis showed reasonable agreement between experiment and subband calculations.

EXPERIMENTAL DATA

The bicrystals studied were grown from [001] seed crystals and tilted along a [100] axis. The twist angle was nominally zero. The bulk was doped with antimony with a concentration up to $10^{16}/cm^3$. The crystals were grown at the Max-Planck-Institut fuer Festkoerperforschung, Stuttgart. Two kinds of samples were cut from the bicrystals with the dislocations parallel (type A in Fig. 1) or perpendicular (type B) to the long axis of the specimens. Separate current and potential contacts made of indium allowed to measure the electrical properties of the p-type space charge layer adjacent to the grain boundary even if the n-type bulk was conducting at high temperatures.

We investigated samples from bicrystals with tilt angles between 7° to 25°. As expected the 7° specimens showed activated behaviour and had a high

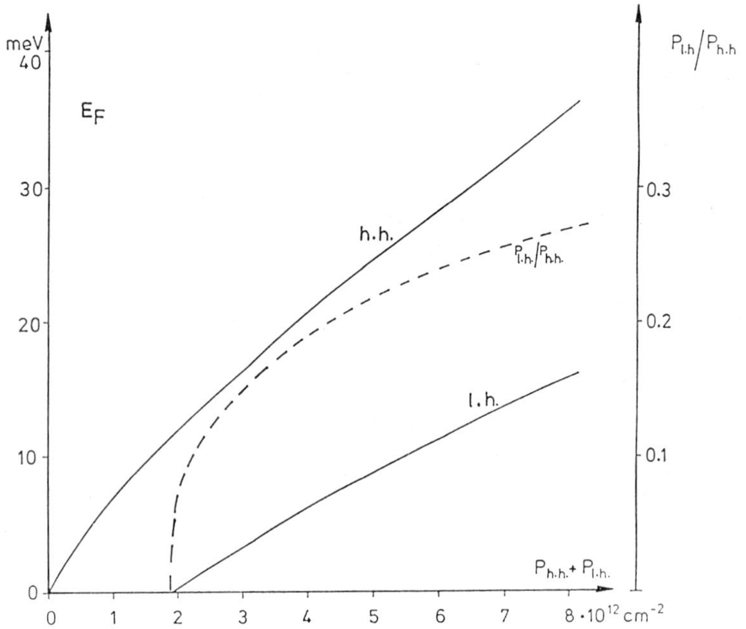

Fig.5. Fermi energy E_F for light and heavy holes and ratio of light to heavy holes P_{lh}/P_{hh} as a function of the total hole concentration.

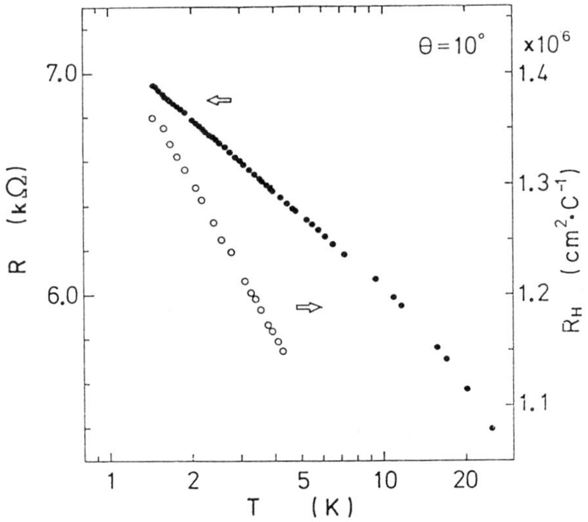

Fig.6: Resistance per unit area R and low field Hall coefficient R_H for a Ge-bicrystal with 10° tilt angle as a function of temperature.

resistance at helium temperatures. Samples with tilt angles of 10°, 11°, 13°, 15°, 20° and 25° were degenerate and showed a logarithmic increase of resistance with decreasing temperature. Except the 15° specimens the resistivity of the samples with perpendicular dislocations was higher than that with parallel dislocations. The specimens behaved qualitatively although not quantitatively alike. We have presented already data for the 10° and 15° bicrystals[20,18] previously and shall report here additional results for samples with 10° tilt angle and perpendicular dislocations. The rest of the data will be discussed elsewhere.

In Fig. 6 resistivity R and reduced Hall constant R_H are shown for a sample with parallel dislocations as a function of temperature. It is evident that both R and R_H increase logarithmically with decreasing temperature. The increase in resistivity per temperature decade is very large and amounts to 19.5 %. Samples of type B also show logarithmic behaviour. The logarithmic slope of the curve representing the Hall data is two times as large as that for the resistivity plot. Behaviour of this kind is characteristic for disordered two-dimensional systems in which many body effects dominate.

The Hall constant R_H was derived from data obtained in relatively low magnetic fieds. R_H showed a pronounced field dependence, it decreased at T = 4.2 K from 1.15×10^6 cm^2 C^{-1} at 0.5 T to about 8×10^5 cm^2 C^{-1}. The Hall mobility calculated from the low field Hall coefficient (corresponding to a hole concentration of 5×10^{12}/cm^2 was around 200 cm^2/Vs.

The magneto resistance of parallel specimens is shown in Fig. 7. The increase in resistance was investigated for magnetic fields B up to 20 T, with B oriented always perpendicular to the current. The parameter ϕ in the figure is the angle between the magnetic field and the normal of the grain

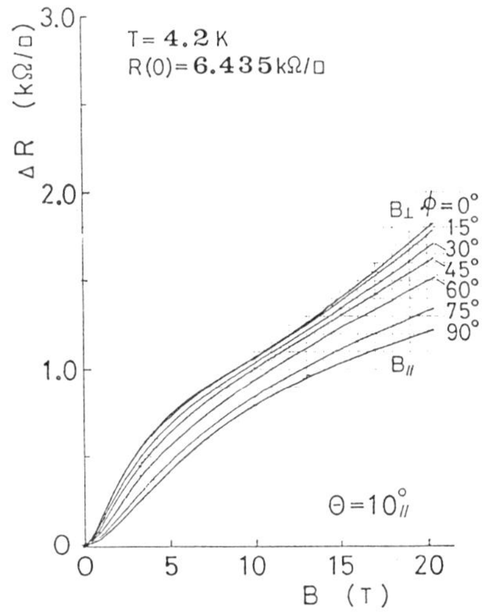

Fig. 7. Magnetoresistance as function of a transverse magnetic field B with the angle ϕ between grain boundary normal and B as parameter for a sample of type A.

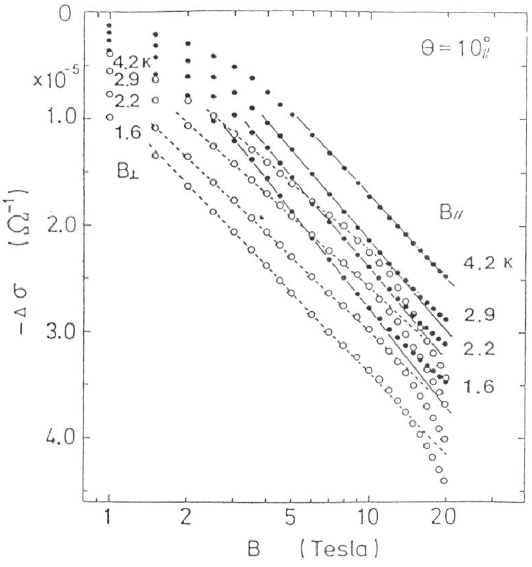

Fig. 8. Negative change of conductivity as function of transverse magnetic field B on a logarithmic scale for B parallel and perpendicular to the grain boundary at four temperatures.

boundary plane. It is evident that for $\phi = 90^o$, when the magnetic field is parallel to the grain boundary, there is a substantial magneto resistance. Also, the effect in the transverse configuration $\phi = 0$ is much larger than in thin metal films or silicon MOSFETs.[21] A plot of the negative change in conductivity against the logarithm of the magnetic fields for B parallel and perpendicular to the grain boundary (Fig. 8) reveals that in rather broad ranges of the magnetic field the magneto conductivity has a logarithmic functional dependence. Data have been plotted for four temperatures between 4.2 K and 1.6 K. One can recognize that the magneto resistivity is increasing with decreasing temperature for both orientations of B. Deviations from logarithmic behaviour can be observed at both low and high magnetic fields. One should note that the slope of the $-\Delta\sigma$ versus ln B curves is increasing with decreasing temperature.

Transverse magneto resistance data for a 10° specimen with perpendicular dislocations have been plotted in Fig. 9. Again the angle between grain boundary and magnetic field is the parameter. Whereas the resistance per unit area tends to saturate in a parallel magnetic field there is a strong almost linear increase of Δ R beyond 10 T. The increase in resistance is quite substantial and amounts to almost 50 % in a field of 20 T. A logarithmic plot of the negative change in conductivity against the magnetic field shows again (Fig. 10) that in certain ranges of the magnetic field dependence of σ is logarithmic. Deviations show up beyond 10 T with opposite sign for parallel and perpendicular magnetic fields. Again the magnitude of the magneto conductivity increases with decreasing temperature.

The magneto resistance data for both parallel and perpendicular specimens scale rather well if they are plotted against the ratio of magnetic field devided by temperature, B/T, for the temperature range between 4.2 and 1.6 K. This is shown in Fig. 11 for a parallel specimen at a magnetic field of perpendicular orientation. As discussed by Kawabata[8], this kind of behaviour suggests that Zeeman-effects are important. The scaling behaviour for a 10° degree sample with parallel dislocations can be seen for a parallel magnetic field in Fig. 12. Drastic deviations from B/T scaling became evident[22] when the temperature was lowered to the millikelvin range. From Fig. 12 one can recognize that at small B/T values a negative differential magneto resistance develops at a temperature of 0.15 K. However, no negative magneto resistance was observed in the millikelvin range for specimens with perpendicular oriented dislocations. This again indicates that grain boundaries behave anisotropic and that they cannot be considered as a system with random disorder.

DISCUSSION

During the last years much attention has been paid to the investigation of the low temperature properties of two-dimensional electronic systems with disorder. The small logarithmic decrease of conductivity with decreasing temperature in thin metal films and semiconductor inversion layers has been explained by the weak localization model proposed by Abrahams et al.[4] The logarithmic increase in resistance is explained in the framework of the one-electron theory by quantum interference effects. Somewhat later, Altshuler et al.[5] showed that electron-electron interaction in conjunction with impurity scattering also leads to a logarithmic decrease in conductivity. There is a fundamental difference between the two mechanisms: Whereas in the weak localization model the conductivity change is caused by a variation of the mobility, in the interaction model the density of states is affected. In order to distinguish between the two theories it is useful to analyze experiments on disordered 2d systems employing magnetic fields. The Hall-effect in a system in which weak localization dominates should be indepent of temperature. For an interacting system one expects a logarithmic increase of the Hall constant with decreasing temperature because of the modification of the density of

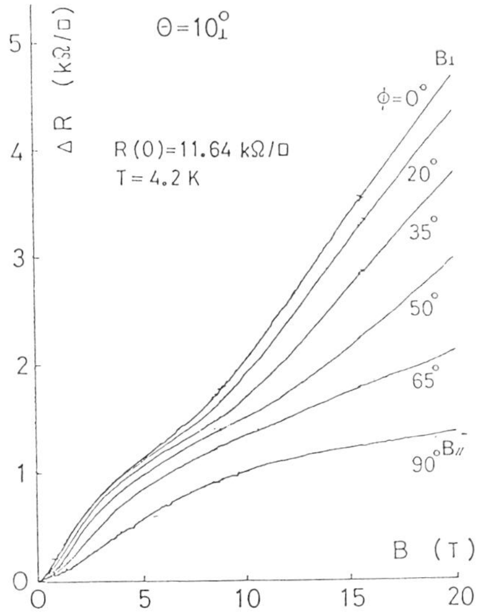

Fig.9. Magneto resistance as a function of a transverse magnetic field B with the angle ϕ between grain boundary normal and field as parameter for a sample of type B.

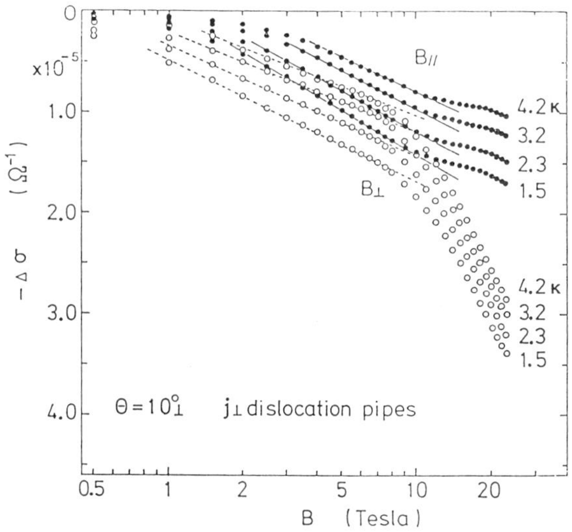

Fig.10. Negative change of conductivity as function of transverse magnetic field B on a log scale for B parallel and perpendicular to the grain boundary at four temperatures.

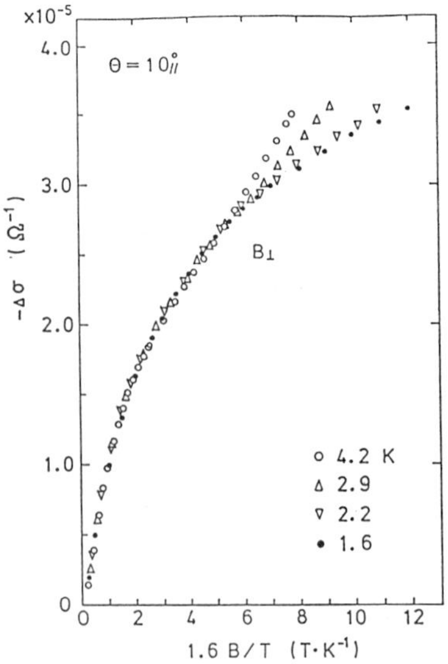

Fig.11. Negative change of conductivity for a bicrystal with θ = 10° as a function of the ratio magnetic field B/temperature T for four temperatures.

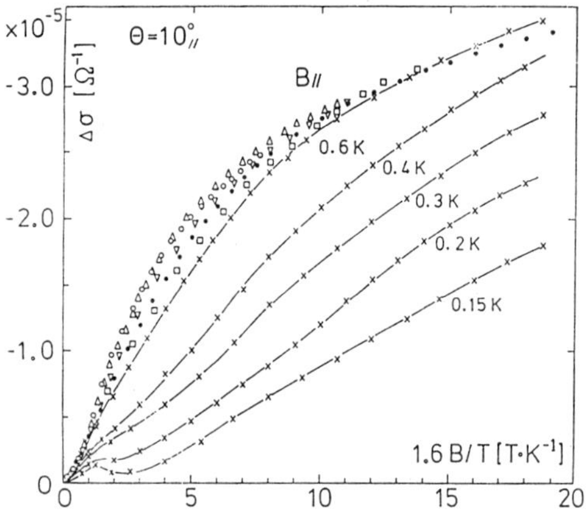

Fig.12. Negative change of conductivity as a function of magnetic field B/temperature T for the temperature intervals 4.2 K - 1.6 K (points) and 0.6 K - 0.15 K (drawn lines).

states by many body effects. There is also a qualitative difference in the transverse magneto resistance to be expected. In a weakly localized sytem, the symmetry of the quantum interference effects is broken by a transverse magnetic field and a negative magneto resistance effect is predicted. In an interacting system the Hartree and exchange effects are influenced by a magnetic field in a very subtle way. The expected magneto resistance is always positive and an increase of resistance is predicted even if the magnetic field is parallel to the plane of the conducting layer. The Zeeman splitting of the energy levels can result in a large logarithmic magneto resistance if the g-factor is sufficiently large.

The theoretical treatment of disordered electronic systems has advanced very rapidly in the recent past. For detailed information the reader is referred to a conference report edited by Nagaoka and Fukuyama.[23] Some of the early models have been revised, see e.g. Finkelstein[24] and Altshuler[25].

In the bicrystals investigated by us we never found a negative magneto resistance except under special conditions in the millikelvin range. The Hall constant of the samples studied increased logarithmically with decreasing temperature. All this indicates that weak localization does not play a significant role in the specimens studied. The physical reason for this has to be sought in the strong spin-orbit interaction in p-type germanium.

The following observations have to be explained:

1. A logarithmic correction is seen in both conductivity and Hall coefficient.
2. The magneto resistance is positive when the magnetic field is applied perpendicular to the inversion layer and it is anomalously large considering the rather low Hall mobility of 200 cm^2/Vs.

3. A positive and very large magneto resistance exists when the field is oriented parallel to the grain boundary.
4. In both configurations of the magnetic field does the magneto resistance increases significantly as the temperature is lowered.
5. The magnetic field dependence of the magneto resistance is logarithmic over wide field ranges for parallel and perpendicular fields and the logarithmic slopes are different for the two configurations.
6. Deviations from the logarithmic behaviour of the magneto resistance are observed at high magnetic fields.
7. Although there is much similarity in the expèrimental data for samples with parallel and perpendicular dislocations pronounced differences exist especially in the magneto resistance.

We have discussed our previous data on 10° bicrystals with parallel dislocations[20] in terms of the interaction constants g_1 - g_4 introduced by Fukuyama.[7] Weak localization effects were neglected for the above mentioned reasons. The first order screened interaction model predicts the following values for the interaction constants: $g_1 = 1$, $g_2 = g_3 = g_4 = F/2$. The factor F depends on the Fermi wave vector k_F and the inverse screening length K and has its usual meaning. For our 10° samples F is about 0.5. For a sample with parallel dislocations the following values[20] were deduced: $g_1 = 2.2$, $g_2 = 0.14$, $g_3 = 0.62$, $g_4 = 0.26$. It is obvious that there is a substantial difference between the derived interaction constants and their theoretical estimate, especially g_1 is much larger than predicted. The fitting of the magneto resistance data in the parallel field configuration gives a Landé factor g = 5 and F = 1.76. The large g-factor is consistent with the observation of asymptotic log B behaviour above fields of about 5 T at 4.2 K. The value of F, however, is too large compared with the calculated value F = 0.5. On the other hand, the experimentally determined coefficient of the logarithmic temperature dependence in zero magnetic field which amounts to 0.68 is in reasonable agreement with F = 0.5.

If the interaction constants g_i are calculated for the the 10° specimens with perpendicular dislocations, one obtains $g_1+g_2 = 0.91$, $g_3 = 0.27$, $g_4 = 0.08$. If the possible errors are taken into account the parameters for the two kinds of samples differ by a factor of about 2.5.

The equations on which the previous analysis is based are valid in the first order of F. They should be applicable to systems with a high density of carriers such as thin metallic films for which F should be much smaller than 1. For the inversion layers in p-type germanium studied by us F is about 0.5 which makes it necessary to take interactions of higher order into account. This has been done recently by Fukuyama et al.[26] A re-evaluation of the interaction constants with $F^* = 0.86$ leads to the following values: $g_1 = 1.19$, $g_3 = 0.38$, $g_2 = g_4 = 0.12$. Now the agreement between theory and experiment can be considered as rather satisfactory.

For the samples with perpendicular dislocations, however, no reasonable set of parameters is obtained. This seems to be caused by the strong linear component of the magneto resistance when the field is perpendicular to the grain boundary. We attribute the difference in the deduced g_i-values to the different dislocation orientation in the two kinds of specimens investigated. The theory which we used for analyzing the data assumes that the disorder in the system is random. Bicrystal specimens, however, show a pronounced anisotropy both in conductivity and magneto resistance. This anisotropy can be attributed to the linear array of dislocations which should be present in our specimens. Even if no clean cut dislocation lattice were present, one should expect that linear disorder shows up at the grain boundary. Regularly spaced dislocations should set up a superlattice potential which can interact with the free carriers. The minigap to appear in the energy vs wave vector relation in a 10° bicrystal specimen is expected at a wave vector roughly two times larger than the Fermi wave vector. Therefore a rather intense interaction of the free holes with the superlattice seems feasible. We shall discuss these matters in detail in a forthcoming publication.

So far we have analyzed our data in terms of a one carrier theory. The subband calculations have suggested, however, that we are dealing with two kinds of holes. In order to obtain more accurate parameters for our two-dimensional system we need a theory which takes into account two kinds of holes. At present the substantial field dependence of the Hall coefficient for our 10° samples[18] does not allow a sufficiently precise deduction of the total hole density.

CONCLUSION

The study of the galvano magnetic properties of p-type inversion layers adjacent to grain boundaries in germanium bicrystals at helium temperatures has revealed that we are dealing with a two-dimensional electronic system with disorder. The dislocations which stabilize the grain boundary introduce random as well as a periodic disorder. In specimens in which the dislocations are oriented parallel to the current, the random disorder seems to dominate. In samples in which the dislocation lattice is perpendicular to the current, a very pronounced linear magneto resistance is observed, which is not compatible with the theory of two-dimensional disordered systems. Both resistivity and Hall effect of the samples studied increased logarithmically with decreasing temperature. The samples with parallel dislocations showed a logarithmic increase of resistance in high magnetic fields and no negative magneto resistance was observed. This indicates that weak localization effects are of minor importance. The data can be explained in a rather satisfying fashion by recent interaction theories.

ACKNOWLEDGEMENTS

The experimental work which is reported here was done when the authors were members of the Max-Planck-Institut fuer Festkoerperforschung, Hochfeld-magnetlabor Grenoble. We would like to thank Mr. A. Koehler of the MPI fuer Festkoerperforschung, Stuttgart for growing the bicrystals and Mrs. S. Jahn for assistance in preparing the manuscript.

REFERENCES

1. W. Shockley, Phys. Rev. 91:228 (1953).
2. See, e.g., H.F. Mataré, Defect Electronics in Semiconductors, Wiley Interscience (1971).
3. G. Landwehr and P. Handler, J. Phys. Chem. Solids 23:891 (1962).
4. E. Abrahams, P. W. Anderson, D. C. Licciardello and T. V. Ramakrishnan, Phys. Rev. Lett. 42:672 (1979).
5. B. L. Altshuler, A. G. Aronov and P. A. Lee, Phys. Rev. Lett. 44:1288 (1980).
6. B. L. Altshuler, D. Khemel'nitzkii, A. L. Larkin and P. A. Lee, Phys. Rev. B 22:5142 (1980).
7. H. Fukuyama, J. Phys. Soc. Japan 50:3562 (1981).
8. A. Kawabata, Surface Science 113:527 (1982).
9. G. Landwehr, phys. stat. sol. 3:440 (1963).
10. See, e.g.: a) G. Landwehr and E. Bangert in: "Two Dimensional Systems, Heterostructures and Superlattices", G.Bauer, F.Kuchar and H.Heinrich Eds., Springer Verlag (1984), p.40;
    b) G. Landwehr, in: "Festkoerperprobleme XV", H.J. Queisser Ed., Vieweg (1975), p.49.
11. E. Bangert, G. Landwehr and S. Uchida, Solid State Comm. 45:869 (1983).
12. H. F. Mataré and H. A. R. Wegener, Z. Physik 148:631 (1957).

13. B. M. Vul and E. I. Zavaritskaya, Proc. of the 14th Int. Conf. on Semiconductor Physics, Edinburgh 1978, Inst. of Physics Conference Series No.43, London (1979), p. 421.
14. J. Hornstra, J. Phys. Chem. Solids 5:129 (1958).
15. H. J. Moeller, Phil. Mag. A 43:1045 (1981).
16. F. J. Ohkawa and Y. Uemura, Suppl. Progr. Theor. Phys. 57:164 (1975).
17. See, e.g., T. Ando; A. B. Fowler and F. Stern, Rev. of Modern Physics 54:437 (1982).
18. G. Landwehr, E. Bangert and S. Uchida, Solid State Electronics, in print.
19. T. Ando, Phys. Rev. B 13:3468 (1976).
20. S. Uchida and G. Landwehr, in: Application of High Magnetic Fields in Semiconductor Physics, Grenoble 1982, G. Landwehr Ed., Springer Lecture Notes in Physics 177 (1983), p.65.
21. R. C. Dynes, Surf. Science 113:540 (1982).
22. G. Remenyi, S. Uchida, G. Landwehr, A. Briggs and E. Bangert, Surf. Science 142:43 (1984).
23. "Anderson Localization", Y. Nagaoka and H. Fukuyama Eds., Springer Solid State Sciences 39 (1982).
24. A. M. Finkelstein, Z. Phys. B 56:189 (1983).
25. B. L. Altshuler and A. G. Aronov, Solid State Comm. 46:429 (1983).
26. H. Fukuyama, Y. Isawa and H. Yasuhara, J. Phys. Soc. Japan 52:16 (1983).

STRUCTURAL PROPERTIES OF TWO-DIMENSIONAL METAL-AMMONIA LIQUIDS IN GRAPHITE

S. A. Solin

Department of Physics and Astronomy
Michigan State University
East Lansing, MI 48824-1116

I. INTRODUCTION

The vast literature on three-dimensional (3D) bulk metal-ammonia solutions[1-3] is testimony to the importance of these novel fluids as arenas for the study of the metal-insulator (M-I) problem.[4] Since phase transitions such as the M-I transition exhibit marked dependence of the spatial dimensionality of the systems being explored,[5] it is obvious that the prospect of studying metal-ammonia solutions in a two-dimensional (2D) environment is an intriguing one.

As a result of our work[6,7] on graphite intercalation compounds (GIC's) and the early chemical studies carried out by Rudörff and coworkers[8] my colleagues and I realized that one might be able to synthesize 2D metal-ammonia solutions by simultaneous chemical insertion of an alkali metal M = Na, K, Rb, Cs, etc., and ammonia into the galleries between graphite layers. The resultant compounds are called alkali-ammonia ternary GIC's. The synthesis of such compounds is in fact straightforward for all the alkali metals[8,9] even though, for example, neither sodium nor ammonia will separately intercalate graphite to form binary GIC's.

As in the case of 3D metal-ammonia fluids a knowledge of the structural properties of the 2D fluid is prerequisite to any succinct understanding of its physical properties. Therefore, we have to date carried out a series of studies of the structures of heavy alkali (K, Rb, Cs) metal-ammonia GIC's and the dependence of these structures on the chemical techniques used for preparation.[10] The purpose of this paper is to review briefly some results of the above-mentioned studies. We will focus on the potassium-based compound $K(NH_3)_xC_{24}$, $0 \leq x \leq 4.33$ and show that the saturated material ($x = 4.33$) does indeed constitute the structural analogue of the 3D $K\text{-}NH_3$ solution.

II. SAMPLE PREPARATION AND THE ABSORPTION ISOTHERM

Samples of $K(NH_3)_xC_{24}$ were prepared by sequential intercalation of highly-oriented pyrolytic graphite (HOPG)[11] first with potassium[12] to form KC_{24} and then by exposing this binary compound to ammonia vapor, the pressure of which could be controlled to vary the composition x. The KC_{24} material is a stage-2 binary GIC (each pair of K layers is separated by a pair of C layers in a regularly stacked configuration) with a layer stoichiometry of MC_{12}.[6]

The composition (x) dependence of $K(NH_3)_xC_{24}$ on NH_3 vapor pressure is shown in Figure 1, the results of which were obtained by both gravimetric (McBain balance[13]) and volumetric (gas handling[14]) techniques. Note that both the abcissa and ordinate span five decades in Figure 1, a dynamic range that is atypical of absorption isotherms.

On the basis of x-ray diffraction studies we have discovered that in the ammonia pressure, P_A, in the range 10^{-4}atm $< P_A < 2 \times 10^{-3}$atm., the samples retain their stage-2 character but the K^+ layers contain small amounts of NH_3 defects. The rapid rise in x at $P_A \approx 2 \times 10^{-3}$atm. corresponds to a large ingestion of ammonia to form (still) a stage-2 material in which the galleries contain a liquid $K\text{-}NH_3$ alloy of relatively low NH_3 concentration. At $P_A \approx 1$ atm there is another rapid rise in NH_3 content, but this increase is also characterized by a staging phase transition[7,10] to a pure (highly-ordered) stage one compound in which every interlayer gallery contains both K^+ and NH_3 species. The composition at $P_A \approx 10$ atms. is $K(NH_3)_{4.33}C_{24}$ and, because no potassium is expelled from the specimen during ammonia uptake, the K/C layer stoichiometry is now KC_{24} in contrast to the KC_{12} value for the binary stage-2 compound.

When ammonia is desorbed from the saturated compound, the specimen does not reversably return to its pristine KC_{24} state, but rather forms a residue compound with $1.5 \leq x_{min} \leq 2.1$ at $P_A \approx 10^{-3}$atm. However, the sample properties exhibit complete reversability upon subsequent repeated cycling between 10^{-4} and 10 atm. Most noteworthy in this cycling process

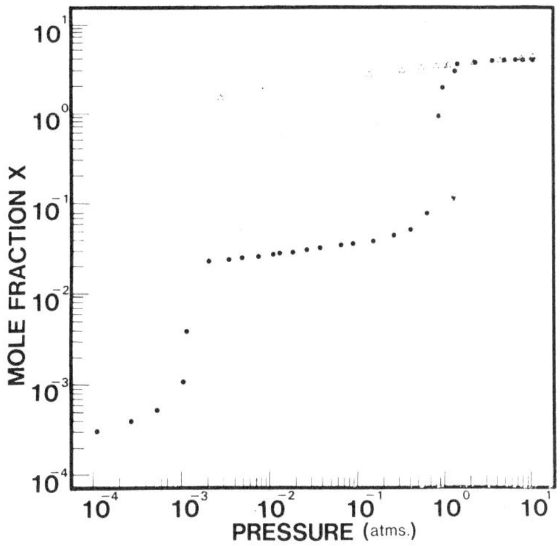

Fig. 1. The variation of the mole fraction x with $P(NH_3)$ for the ternary GIC $K(NH_3)_xC_{24}$ as measured by a "gas-handling" technique (see text). Experimental points indicated with a solid dot (●) represent the initial interacalation of NH_3, those with open circles (O) are for deintercalation. The McBain balance data from Figure 1 has been replotted as open triangles.

is the reproducible staging phase transition at $P_A \approx 3$ atms. In fact, by studying this transition we have obtained important clues to the fundamental physics of the staging phenomena in GIC's[7,10] but that work is peripheral to the topic we are focusing on here. The remainder of this paper will be concerned with the structural properties of the saturated $K(NH_3)_{4.33}C_{24}$ compound and in particular the intraplanar structure of the K-NH_3 layers.

III. INPLANE DIFFRACTION STUDIES

To study the structure of the K-NH_3 layers we measured the inplane ($\vec{q} \perp \vec{c}$) diffuse x-ray diffraction from $K(NH_3)_{4.33}C_{24}$.[15] Because the

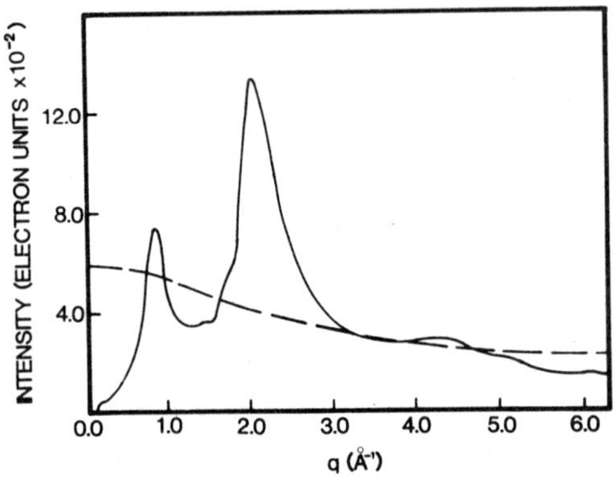

Fig. 2. The diffuse in-plane ($\vec{q}\perp\vec{c}$) scattering I'(q) from $K(NH_3)_{4.33}C_{24}$ (solid line) and the incoherent contribution (dashed line) as described in the text.

diffracted intensity was weak and the pyrex ampules in which highly reactive GIC's are normally housed yield an intense diffuse contribution which masks the scattering from our sample, it was necessary to design an x-ray transparent container which was nevertheless capable of sustaining pressures of the order of 10 atms. To accomplish this, the samples were housed in pyrex tubes, the ends of which were epoxied to a thin-walled (<0.005") aluminum can which contained the specimen.

The diffraction data reported here were acquired with incident MoKα radiation from a Rigaku 12 kw rotating anode source coupled to a vertically bent graphite monochromator and a computer-controlled Huber 4-circle diffractometer. A NaI scintillation detector was also used.

In Figure 2 is shown the corrected inplane diffraction pattern ($\vec{q}\perp\vec{c}$) for $K(NH_3)_{4.33}C_{24}$ acquired at room temperature. The solid line of Figure 2 results from applying several correction factors to the as recorded data. Let $I_{exp}(q)$ be the observed (hk0) inplane diffraction pattern. This contains Bragg reflections associated with the powder pattern of the aluminum sample can and the ordered carbon layers. A reference (hk0) pattern acquired from a sample free region of the aluminum can was recorded, appropriately scaled to and subsequently subtracted from $I_{exp}(q)$. The Bragg peaks associated with the carbon layers were removed (for clarity) from the resulting pattern to yield $I'_{exp}(q)$ which is the observed diffuse scattering. The pattern $I'_{exp}(q)$ was further corrected for absorption, and the Lorentz polarization factor to yield I'(q) as follows:

$$I'(q) = I'_{exp}(q)[(T/\cos\theta)\exp(-\mu T/\cos\theta)]^{-1} \times$$

$$[(1 + \cos^2 2\theta' \cos^2 2\theta)/(1 + \cos^2 2\theta')]^{-1} . \qquad (1)$$

Here T is the sample thickness, μ is the effective absorption coefficient, θ is the diffraction angle, and θ' is the graphite monochrometer diffraction angle for the (004) MoKα reflection. The corrected diffuse scattering function I'(q) which is shown in Figure 2 was scaled to oscillate about the incoherent scattering contribution

$$I_{Inc}(q) = \sum_{uc} f_m^{\,2} + i(m) \qquad (2)$$

(shown as a dashed line in Figure 2) at high q and from this scaling the ordinate scale of Figure 2 (in electron units) was established.[16] In Eq. (2), f_m is the q-dependent atomic scattering factor of the m'th atom, uc → unit of composition which in our case is $K(NH_3)_{4.33}$ and i(m) is the Compton modified scattering from the K, N, H, and C atoms.

We have deduced the two-dimensional pair distribution function $2\pi r\rho(r)$ of the K-NH_3 liquid using a Bessel function transform of I'(q),[17] namely

$$2\pi r\rho(r) = 2\pi r\rho_o + r\int_0^{q_m}[I'(q) - I_{Inc}(q)](q/f_e^2)\, J_o(q,r)e^{-\alpha^2 q^2}dq \qquad (3)$$

Here ρ_o is the average areal density, f_e is the q-dependent scattering factor per electron, and α is a damping factor that minimizes cut-off errors associated with the finite range, q_m, of the integral of Eq. (3).[18] The pair correlation function that results from the application of Eq. (3) to the data of Figure 1 is shown in Figure 3. Note that $2\pi r\rho(r)$ represents the number of atoms in an annular ring between r and r+dr as measured from an arbitrary reference atom. Therefore $2\pi r\rho(r)$ approaches the linear function $2\pi r\rho_o$ at large r where the correlations

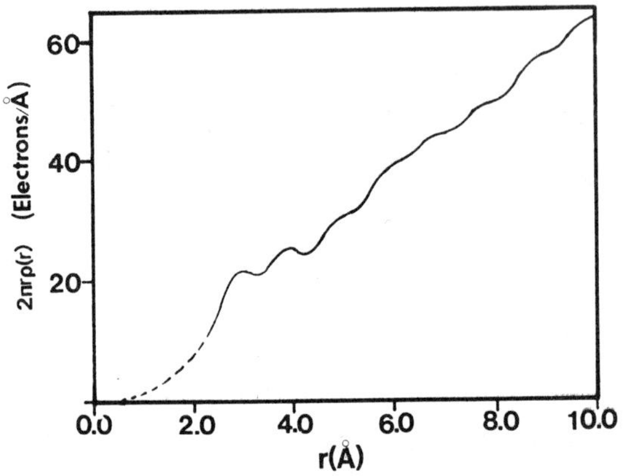

Fig. 3. The pair distribution function $2\pi r\rho(r)$ deduced by applying Eqs. 2 and 3 (see text) to the data of Figure 2 with $\mu = 5.0\ mm^{-1}$, $\alpha = 0.05$ Å, and T = 1.0 mm. The dashed line is an extrapolation in the region where finite q cut-off errors (see text) are important.

die out. This behavior is analogous to the pair correlation function $4\pi r^2\rho(r)$ approaching the parabolic form $4\pi r^2\rho_o$ for a 3D liquid or glass since $4\pi r^2\rho(r)$ is the number of atoms in a spherical shell between r and r+dr.

From a knowledge of the sizes of the NH_3 molecule and the K^+ ion one can speculate upon the origin of the peaks in the pair correlation function. For instance the peaks at ≈3.0 and 4.0Å can be associated with K-NH_3 and NH_3-NH_3 correlations respectively while the weak peak at r ≈ 7.0Å may correspond to K^+-K^+ near neighbor correlations. These weak peaks derive roughly, but not exactly from the peaks in the diffraction pattern at 2.10$Å^{-1}$ and 0.85$Å^{-1}$ respectively.

Speculations about the origins of the peaks in the pair-correlation function are especially dangerous when applied to a binary (K-NH_3) liquid or glass because the Bessel function back transform of the diffraction pattern does not have a direct (1 to 1) correspondence to the real pair correlation function[18] which is the convolution of the K^+-K^+, K^+-NH_3, and NH_3-NH_3 pair functions. Thus, the most appropriate method for deducing the structure of the K-NH_3 layers in $K(NH_3)_{4.33}C_{24}$ is to calculate the diffraction pattern from a structural model and compare the calculated pattern with that observed experimentally. We have followed just such a procedure and describe our results herewith.

IV. STRUCTURAL MODELS

The following structural model semiquantitatively accounts for the in plane scattering intensity of the K-NH_3 2D liquid intercalant in $K(NH_3)_{4.33}C_{24}$. We assume that the NH_3 molecules and the K^+ ions can be represented as coplanar hard discs. If $\vec{r}_m$ is the position of the center of the mth particle (i.e. NH_3 or K^+ disc) and f_m its q-dependent scattering factor, where $f_{NH_3}(0) = 10$ and $f_{K^+}(0) = 18$, then the scattered intensity for any distribution of discs will be[18]

$$I(q) = \sum_m f_m e^{i\vec{q}.\vec{r}_m} \sum_n f_n^* e^{-i\vec{q}\cdot\vec{r}_n} \qquad (4)$$

Let there be L layers of K-NH_3 in the scattering volume with N particles (discs) in each layer. Then if a_z is the C-axis repeat distance and $\vec{r}_m = \vec{R}_{mS} + Sa_z\hat{k}$ for the Sth layer, Eq. (1) can be rewritten as

$$I(q) = L \sum_{m,n}^{N} f_{mS} f_{nS}^* e^{i\vec{q}.(\vec{R}_{mS}-\vec{R}_{nS})} +$$

$$\sum_{m,n}^{N} \sum_{S\neq S'}^{L} f_{mS} f_{nS'}^* e^{i\vec{q}.(\vec{R}_{mS}-\vec{R}_{nS'})} e^{i\vec{q}.(S-S')a_z\hat{k}} \qquad (5)$$

which for in-plane scattering with $\vec{q} = q\hat{x}$ becomes

$$I(q) = L \sum_{m,n}^{N} f_{mS} f_{nS}^* e^{iq(x_{mS}-x_{nS})} + \sum_{m,n}^{N} \sum_{S\neq S'}^{L} f_{mS} f_{nS'}^* e^{iq(x_{mS}-x_{nS'})} \qquad (6)$$

If there is no interlayer correlation, i.e. x_{mS} and $x_{nS'}$ are uncorrelated, the second term in Eq. (6) is equal to zero. We have verified the lack of interlayer correlation in $K(NH_3)_{4.33}C_{24}$ by recording "flat"[2] (hk = const, ℓ) c^* x-ray scans.[19] For a sufficiently large system which is amenable to cylindrical averaging, the first term in Eq. (6) can be recast in the familiar[6] form

$$I(q) = L \sum_{m,n}^{N} f_m f_n^* J_o(qR_{mn}) \tag{7}$$

where $R_{mn} = |\vec{R}_m - \vec{R}_n|$, J_o is the zeroth order Bessel function and we have dropped the subscript S. The second term in Eq. (6), which has been omitted in previous treatments of in-plane diffuse scattering,[17,20] can significantly influence both the calculated $I_\perp(q)$[20] and the pair correlation function deduced from the Bessel transform of the observed $I_\perp(q)$[17] if interlayer correlations are nonnegligible. Such correlations may be present in a lattice gas model[20] of GIC structure and in a system that is about to undergo a 2D-3D order-disorder transition.[6]

In order to account for the observed diffraction pattern $I_\perp(q)$ shown in Figure 2 we have applied Eq. (7) to several computer generated structural distributions of K and NH_3 discs. Because K^+ ions can effectively polarize and bind up to six NH_3 molecules[1] we concentrated on distributions of symmetric planar n-fold coordinated clusters. For reasons which will be made clear below, $n \leq 4$. In order to satisfy the stoichiometry of our specimens, each distribution contained exactly 500 K^+ discs and 2165 NH_3 discs. In an n-fold coordinated distribution each K^+ disc was in contact with n NH_3 discs such that the lines of centers between the K^+ and NH_3 discs make equal angles of $\frac{2\pi}{n}$.

An n-fold coordinated distribution was generated as follows: A seed cluster was nucleated around a K^+ by symmetrically attaching n NH_3 discs with different radii randomly selected from a gaussian distribution (see below). Then a K^+ or NH_3 disc was randomly selected from the remaining collection of 499 K^+ discs and (2165-n) NH_3 discs. If the selection yielded a K^+ disc, n NH_3 discs were attached to it and the rigid cluster was moved, in a series of small translational and rotational steps, towards the seed cluster until the distance between the K^+ centers of the seed and the added cluster was minimized. If the next randomly selected particle was an NH_3 disc it was moved in a series of translational steps until it was as close as possible to the center of the seed cluster. The above described process was repeated until all of the particles had been assembled.

It is known from NMR measurements[21] that at room temperature the NH_3 molecules spin rapidly about their C_3 axes which in turn are tilted with respect to the graphite C-axis. The tilt angle varies rapidly in time such that at any instant there is a distribution in the orientations of the NH_3 molecules. Thus the cylindrical projections, of the surfaces of revolution about the C_3 axes onto the graphite basal plane, have a distribution in radii which we have modeled on the basis of NMR measurements[21] as

$$P(r) \alpha \exp[-(r - r_{min})^2/(\delta r)^2] - \exp[-(r_{max} - r_{min})^2/(\delta r)^2] \qquad (8)$$

where $r_{max} = 2.00Å$, $r_{min} = 1.30Å$, $r_{min} \leq r \leq r_{max}$, and $0 \leq \delta r \leq 0.3$ Å. The radii of all K^+ ions in a distribution were fixed at a value r_{K^+} approximately equal to the ionic radius of potassium. For each n-fold coordinated distribution the parameters δr and r_{K^+} were varied to achieve the best fit to the data of Figure 1 using Eq. 7. The fits obtained were quite sensitive to r_{K^+} and n, but were relatively insensitive to the other parameters and were size independent, i.e. macroscopic, for distributions containing more than 50 K^+ discs.

The 4-fold coordinated distribution shown in the inset of Figure 4 yielded the best fit to the data of Figure 2 as can be seen from Figure 4. Notice that the amplitudes, positions and widths of the features of the measured $I_\perp(q)$ are reasonably accounted for, the only exceptions being the shoulder on the high q side of the $2.07Å^{-1}$ peak and the peak at $\approx 4.24Å^{-1}$. Also note that the theoretical areal density calculated from the distribution shown in the inset of Figure 1 is equal, within experimental error, to the experimentally determined density. This is not the case, for example, for a zero-fold coordination (see below) even though the same stoichiometry obtains. Thus, the areal density is sensitive to the ammonia tilt angle distribution and is not simply fixed by the layer stoichiometry.

The total calculated in-plane scattering intensity is the sum of the pair scattering profiles, i.e. $I_\perp(q) = i_\perp^{K^+-K^+}(q) + i_\perp^{NH_3-NH_3}(q) + i_\perp^{K-NH_3}(q)$. These individual pair contributions are quite revealing as can be seen from Figure 5. The sharp peak in I (q) at $0.88Å^{-1}$ results from the dominance of the K^+-K^+ contribution over the K^+-NH_3 contribution of opposite sign at that q value. The former results from the fact that the K^+-K^+ nearest-neighbor distances show little variation in the 4-fold coordinated distribution as is evident from the inset of Figure 4. The sharp negative K^+-NH_3 contribution at $0.88Å^{-1}$ results from the fact that the ammonia and potassium discs have roughly the same diameter. Also note from Figure 5c that the K^+-NH_3 contribution consists of correlations superposed on $J_0[q \times (R_{mn} = 2.89Å)]$. This Bessel function background results from the presence of many K^+-NH_3 near-neighbor pairs separated by

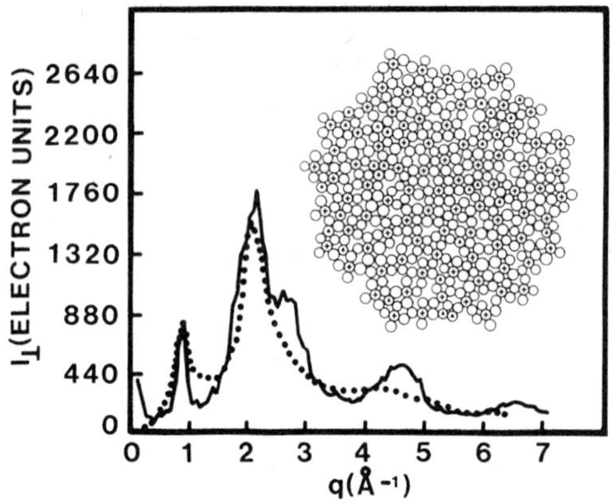

Fig. 4. The calculated (———) in-plane diffraction patterns of the K-NH_3 liquid in $K(NH_3)_{4.33}C_{24}$ scaled at q = 0.88Å to the measured room temperature (••••) pattern of Fig. 2. The calculated pattern was deduced by applying Eq. (7) of the text to the 4-fold coordinated distribution shown in the inset which depicts K^+ ions (⊕) bound to 4 NH_3 ions (○) and unbound NH_3 ions (Ⓐ). The parameters used were r_{max} = 2.00Å, r_{min} = 1.30Å, δr = 0.23Å, and r_{K^+} = 1.46Å.

2.89Å in the 4-fold coordinated distribution. It is evident from Figure 5c that the Bessel function contribution is responsible for the enhanced amplitude of $I_\perp(q)$ at the high q shoulder of the 2.07Å^{-1} peak and at the 4.24Å^{-1} peak. These enhancements represent a minor deficiency in our model which we associate with its oversimplified structural character.

The physical size of the NH_3 molecules[22] limited to 4 the number which with reasonable probability could be symmetrically bound to potassium in a planar configuration and still satisfy the stoichiometry/areal density requirements when selected from the appropriate tilt angle distribution. Therefore, as noted above we tried to fit the data of Figure 4 with distributions having $n < 4$. Two such distributions with

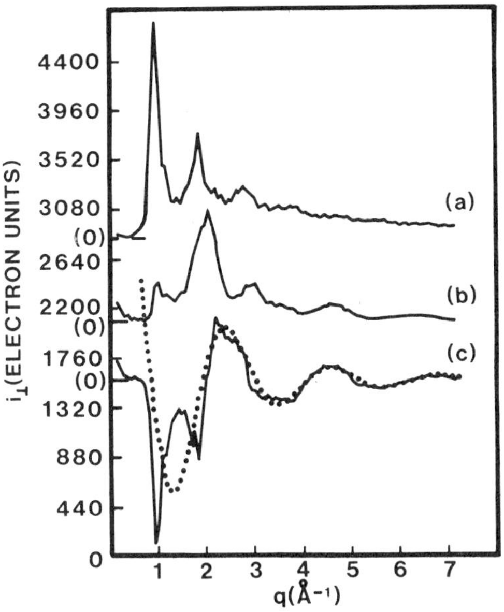

Fig. 5. The K^+-K^+ (a), NH_3-NH_3(b), and K^+-NH_3 (c) in-plane pair scattering functions calculated using the parameter set given in the caption of Figure 4, Eq. (7) and the distribution shown in the inset of Figure 4. The dots (●) are a plot of $J_o[q \times (R_{mn} = 2.89Å)]$ (see text).

$n = 3$ and $n = 0$ are shown together with their corresponding best fits in Figures 6 and 7 respectively. It is obvious from Figs. 6 and 7 that these other distributions yield fits inferior to the $n = 4$ fit, particularly with regard to the sharp peak at $0.88Å^{-1}$, and that the

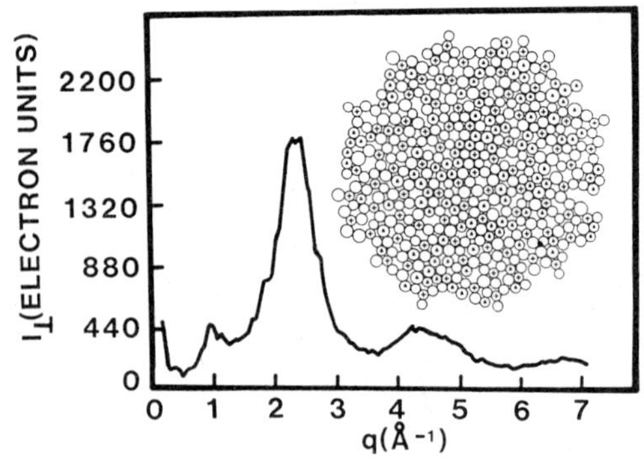

Fig. 6. The calculated in-plane diffraction pattern for the 3-fold coordinated K-NH_3 distributions [inset] with parameters r_{max} = 2.00Å, r_{min} = 1.30Å, δr = 0.30Å, and r_{K^+} = 1.33Å.

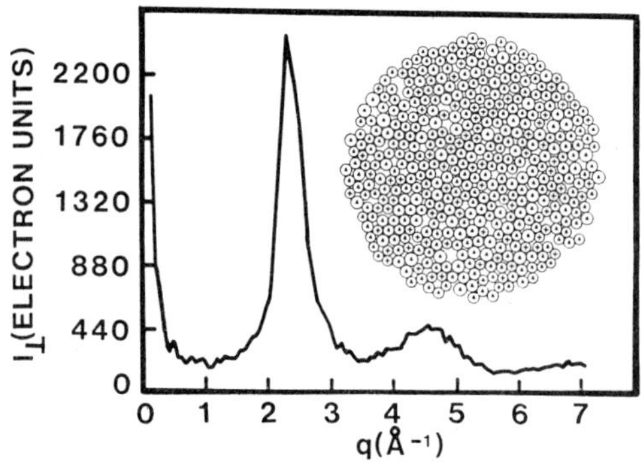

Fig. 7. The calculated in-plane diffraction pattern for the 0-fold coordinated, free ammonia, distribution [inset] with parameters r_{max} = 2.00Å, r_{min} = 1.30Å, δr = 0.30Å, and r_{K^+} = 1.33Å.

observed diffraction pattern is quite sensitive to the cluster coordination.

The real 2D K-NH_3 liquid is dynamic and almost certainly contains K-NH_3 clusters with differing coordination number. We could have accounted for this effect by introducing several additional fitting parameters to characterize the probability P(n) of finding an n-fold coordinated cluster in the liquid. Even more parameters could have been introduced to model asymmetries in the angular distributions of the NH_3 molecules about the K^+ cores. While such procedures would improve the fit to the data of Figure 4, they would also tend to mask the essential structual features of the 2D K-NH_3 liquid. These are the predominance of the symmetric 4-fold coordinated cluster and the distribution in tilt angles for the C_3 axes of the NH_3 molecules. It is not surprising that the simple model we have employed does not yield a fully quantitative fit to the data of Figure 4. But it is significant that this model produces as good a semiquantitative fit as is observed.

The success of the n = 4 distribution is in part a manifestation of the structure of 3D metal-ammonia solutions.[1] For bulk K-NH_3 liquids, the K^+ ion is octahedrally (6-fold) coordinated to NH_3. But the gallery height in $K(NH_3)_{4.33}C_{24}$ is such that only a monolayer of NH_3 molecules can occupy the interlayer space.[7,10] The resultant 4-fold coordinated planar K^+-NH_3 cluster is derived from the octahedron by the removal of two apical ammonia molecules. Thus it is evident that the K-NH_3 layers in $K(NH_3)_{4.33}C_{24}$ do indeed constitute the 2D structural analogue of a bulk K-NH_3 solution.

Note that to date the structures of all 2D "liquids" intercalated into graphite have been dominated by the host potential.[23] The 2D liquid-like character was a minor feature in comparison to the role played by substrate-induced registry, modulations, etc. In contrast, the K-NH_3 layers intercalated into graphite are essentially simple 2D liquids at room temperature. Substrate effects, though no doubt present, are of secondary importance.

Finally, the fact that the structure of the K-NH_3 layers in $K(NH_3)_{4.33}C_{24}$ is the analogue of the structure of the 3D bulk K-NH_3 fluid does not mean the physical properties of the 2D and 3D fluids will be related by dimensionality effects. Nevertheless, there is sufficient likelihood of dimensionally-derived differences that an investigation of the physical properties of the 2D fluid is clearly warranted. Of particular interest in this regard will be measurements of the composition dependence of the contribution of the K-NH_3 layers to the basal plane conductivity of $K(NH_3)_xC_{24}$ for $x \geq 3$, the region in which the sample is structurally a single phase stage-1 compound. Moreover, it will be important to expose the samples to much higher (≈100 atm) ammonia pressures than 10 atm so that values of x higher than 4.33 can be achieved and the likelihood of an insulating (more dilute in K^+ ions) layer is enhanced.

V. ACKNOWLEDGEMENTS

The work reviewed in this paper was carried out with several collaborators who made major conceptual and intellectual contributions to its content. These collaborators include Y.B. Fan, S.K. Hark, X.W. Qian, D. Stump, P. Vora, and B.R. York. I am grateful for the opportunity to interact with them. This research was supported by the NSF under grant DMR 82-11554.

VI. REFERENCES

1. J.C. Thompson, Electrons in Liquid Ammonia (Clarendon Press, Oxford, 1976).
2. J.C. Thompson, Rev. Mod. Phys. 40, 704 (1968).
3. J. Jortner and N.R. Kester, editors, Electrons in Fluids: The Nature of Metal-Ammonia Solutions (Springer-Verlag, New York, 1973).
4. N.F. Mott and E.A. Davis, Electronic Processes in Non-Crystalline Materials (Clarendon Press, Oxford, 1971).
5. S.K. Sinha, editor, Ordering in Two Dimensions (North-Holland, New York, 1980).
6. S.A. Solin, Adv. Chem. Phys. 49, 455 (1982).
7. B.R. York, S.K. Hark, and S.A. Solin, Solid State Comm. 50, 595 (1984).
8. W. Rudorff and E. Schultze, Angew. Chem. 66, 305 (1954).
9. X.W. Qian, Y.B. Fan, and S.A. Solin, to be published.
10. B.R. York and S.A. Solin, Phys. Rev., in press.
11. A.W. Moore in Chemistry and Physics of Carbon (Dekker, New York, 1973), p. 69.
12. A. Herold, Bull. Soc. Chim. Fr. 999 (1955).
13. J.W. McBain and A.M. Baker, J. Am. Chem. Soc. 48, 690 (1926).
14. See reference 10 and references therein.
15. S.A. Solin, Y.B. Fan, and B.R. York, Proceedings of the Materials Research Society, Boston, 1984, in press.
16. H.P. Klug and L.E. Alexander, X-Ray Diffraction Procedures for Polycrystalline and Amorphous Materials (Wiley, New York, 1974).
17. R. Clarke, N. Caswell, S.A. Solin, and P.M. Horn, Phys. Rev. Lett. 43, 2018 (1979).
18. B.E. Warren, X-Ray Diffraction (Addison-Wesley, Reading, Mass., 1969), pp. 264-275.
19. D. Stump, X.W. Qian, B.R. York, and S.A. Solin, to be published.

20. H. Zabel, Y.M. Yan, and S.C. Moss, Physica 99B, 453 (1980).
21. H. Resing, B.R. York, and S.A. Solin, Bull Am. Phys. Soc. 29, 294 (1984) and to be published.
22. A.H. Norten, J. Chem. Phys. 49, 1692 (1968).
23. R. Clarke, J.N. Gray, H. Homma, and M.J. Winokur, Phys. Rev. Lett. 47, 1407 (1981).

PHYSICAL PROPERTIES OF THE QUASI-TWO-DIMENSIONAL COMPOUND La_2NiO_4

J. M. Honig and
D. J. Buttrey*

Purdue University
Department of Chemistry
Chemistry Building
West Lafayette, Indiana
47907

*Department of Physical Chemistry
University of Cambridge
Lensfield Road
Cambridge CB2 1EP England

INTRODUCTION

The physical properties of quasi one- or two-dimensional (1D or 2D) systems have long held considerable facination for theorists. As detailed in a recent review article[1], the theoretical analysis of 2D systems is much simpler than that of its 3D counterpart, thus permitting several problems to be solved exactly that are intractable in the three-dimensional world. The 1D analogue displays unique features, that are at the forefront of a large world-wide research effort. Research on 2D materials has perhaps been equally intense; it spans a wide area involving surface films; inversion layers; planar materials such as graphite and clays; layered dichalcogenides with intercalates that pry apart the weakly bonded layers; and compounds in which layers are stacked in a regular arrangement with distinct periodicities. In general, layered materials display unusual properties reflecting the special characteristics of their lower dimensionality.

The continuing interest in the properties of lower-dimensional systems is attested to by the many conference proceedings and review articles that have been published over the last eight years; for a representative listing of recent publications see Refs. 2-7. The two-dimensional phase transitions and the critical behavior of systems near the transition point have received special attention in this period.

Of particular relevance to the present publication are prior studies on quasi-two-dimensional materials that crystallize in the K_2NiF_4 configuration. As is well established, such structures may be viewed as two-dimensional sheets of NiF_2 separated by a bi-layer of KF. Alternatively, the structure may be considered to be formed of perovskite ($KNiF_3$) layers in a staggered fashion without face-sharing (See Fig. 1a and b). The magnetic properties

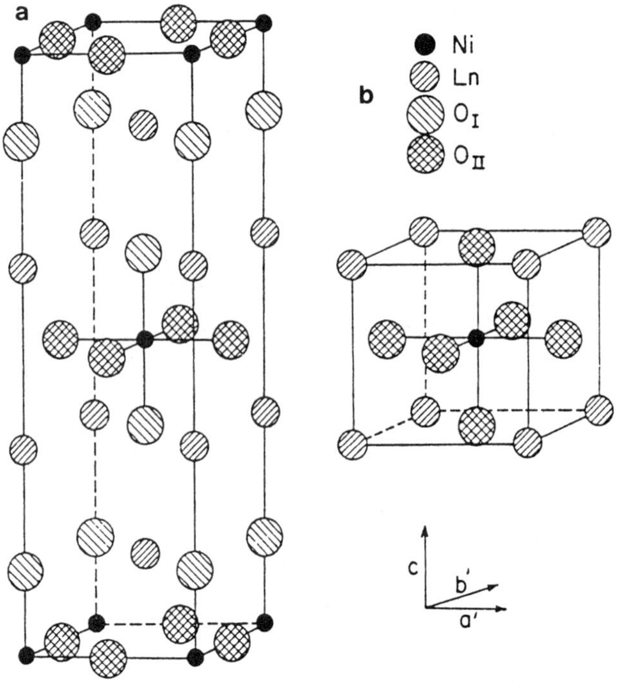

Fig. 1(a) and (b) The Body Centered Tetragonal K_2NiF_4-type Structure of Ln_2NiO_4 (left), Neglecting Distortions, and the Closely Related Perovskite Structure of $LnNiO_3$ (right).

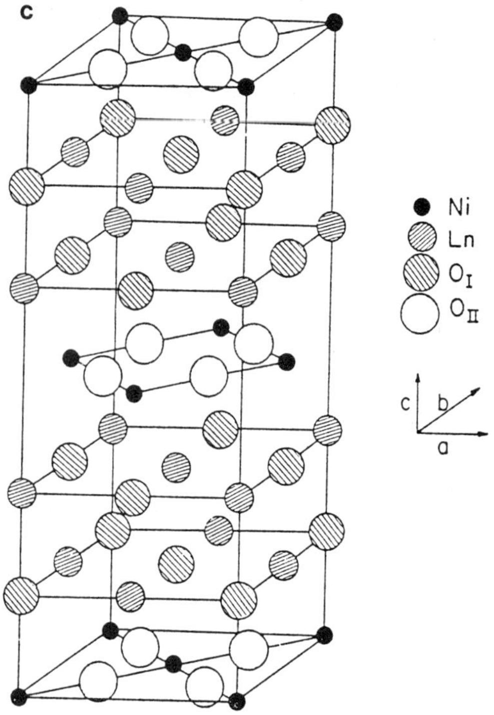

Fig. 1(c) Orthographic Projection of the Face Centered Ln_2NiO_4 Structure Neglecting Distortions. Relative to Fig. 1(a), $a' = \sqrt{2}a$, $b' = \sqrt{2}b$.

of such layered components are detailed in a somewhat dated but still very useful review by de Jongh and Miedema[8], who carefully discuss the novel features seen in lower dimensional materials and the theoretical framework within which these may be understood.

Whereas considerable effort has been devoted to the study of physical properties of K_2NiF_4 - type compounds of primarily ionic character much less work has been carried on covalently bonded 2D crystalline materials. Representative among the latter is the compound La_2NiO_4; in zero order approximation this material may be viewed as consisting of metallic NiO_2 sheets separated from each other by insulating LaO bi-layers. One can thereby largely suppress the interactions among electrons in adjacent conducting sheets. Furthermore, the NiO distance within the basal planes, 1.94 Å, is considerably shorter than the corresponding distance, 2.09 Å, in NiO. The c/a ratio at room temperature is in the range 3.23 - 3.27, depending on oxygen stoichiometry; this demonstrates the very significant separation of adjacent conducting planes achieved in this structure. One thus anticipates that La_2NiO_4 might exhibit physical properties characteristic of quasi-2D materials.

Until very recently large single crystals of La_2NiO_4 were unavailable. All of the prior electrical and magnetic measurements were thus carried out on polycrystalline specimens which yield only orientation-averaged results. Published investigations in this category are listed in References 9-22; this work dealt primarily with preparative techniques, characterization by x-ray crystallography or electron microscopy, electrical conductivity, and magnetic susceptibility measurements. The first susceptibility studies were carried out by Smolenskii et al.[12] in 1962; the initial conductivity investigations were reported by Foëx[10] in 1961. For recent reviews of various aspects of the above measurements the reader is referred to articles by Singh et al.[23] and by Ganguly and Rao[24]. In retrospect it now appears that in many cases the importance of oxygen stoichiometry was not given adequate attention.

Small single crystals of La_2NiO_4 were first grown by Foëx and coworkers[11] who melted or calcined the starting materials. Recently, large crystals were obtained in the authors' laboratory[25-27]. These are now of sufficient size for the anisotropy studies discussed in this publication.

GROWTH AND CHARACTERIZATION OF SINGLE CRYSTALS

The procedure for growth of single crystals has been described elsewhere in detail[25,27]. Radio-frequency induction was used to heat appropriately prereacted La_2O_3 and NiO powder in air above the melting point; material adhering to the water-cooled crucible remained relatively cool, serving as a thin skull. The melt was then gradually lowered through the stationary work coil, so as to initiate growth of crystals in a Bridgman-like process. Single crystal specimens were selected from the boule core for further studies. All samples used for measurements were verified to be single crystals as indicated by x-ray Laue back-reflection photographs. The La/Ni ratio of selected crystals was found to be 2.00 ± 0.01 by use of standard wet chemical methods. High resolution electron microscopy gave no indication of intergrowth of the Ruddlesden-Popper phases such as $La_3Ni_2O_7$ or higher homologs[22]. The oxygen/metal ratio was determined by iodometric titration of Ni^{3+} formed by incorporation of excess oxygen.

In prior structural investigations La_2NiO_4 was classified as belonging to the ideal tetragonal K_2NiF_4 structure I4/mmm[9,11,15-17, 20,21]. However, electron diffraction patterns from single crystal specimens grown as described above, revealed the presence of weak superstructure spots[26], indicating that the symmetry is lower than tetragonal. The intensity of the superlattice spots was greatly enhanced by heating the specimens under reduced oxygen fugacities, thereby substantially decreasing the excess oxygen content; evidently, the presence of the smaller Ni^{3+} ions and of cation vacancies in nonstoichiometric samples stabilizes the K_2NiF_4 structure. The lower symmetry may be attributed to the bending[27] of the Ni-O-Ni angles away from 180°, which becomes particularly pronounced in the most stoichiometric specimens.

No evidence of this distortion was apparent in x-ray powder patterns of unannealed samples with large Ni^{3+} content. An investigation of the La_2NiO_4 structure by single crystal x-ray techniques[28] was unsuccessful in providing acceptable levels of statistical agreement, suggesting the existence of microscopic inhomogeneities. X-ray powder diffraction patterns taken in conjunction with the above study[27] are exhibited in Fig. 2 for unannealed specimens and for samples annealed at an oxygen fugacity of $\log f_{O_2} = -8.5$ at 1470 K for one week. The principal peaks are indexed on the larger face-centered unit cell shown in Fig. 1c; here $a' \cong \sqrt{2}a$. The splittings observed for the annealed sample clearly show the existence of an orthorhombically distorted unit cell with lattice parameters $a_o' = 5.520$ Å, $b_o' = 5.456$ Å, $c_o = 12.55$ Å.

For the work cited below it is important to note that on heating annealed samples the distortion weakened and then disappeared at approximately 650 K; above this temperature only the electron diffraction pattern corresponding to the ideal tetragonal K_2NiF_4 structure remained. On cooling, the superstructure spots reappeared and ultimately regained their original intensity at room temperature. These findings were reproducible in subsequent cyclings.

Magnetic Properties

Magnetic susceptibility (χ) studies were carried out on oriented unannealed specimens. A sensitive microbalance was used to determine the magnetic susceptibility by the Faraday method; the field intensity was determined by Hall voltage measurements. These studies were restricted to the temperature range of 65 to 300 K; a variable temperature controller maintained a given setting to within 0.1 K. The system was calibrated against $CoHg(SCN)_4$. Typical results are shown in Fig. 3. The data for these unannealed specimens are in reasonable agreement with those of Ganguly et al.[19] except for two features: A weak dependence of χ on the applied magnetic field was observed in the range of 5 to 6.5 kG; also, a definite cusp was encountered at ~150 K for $H||c$; this had been missed in earlier studies, in which the temperature interval between successive measurements was much greater and the results were necessarily orientation-averaged.

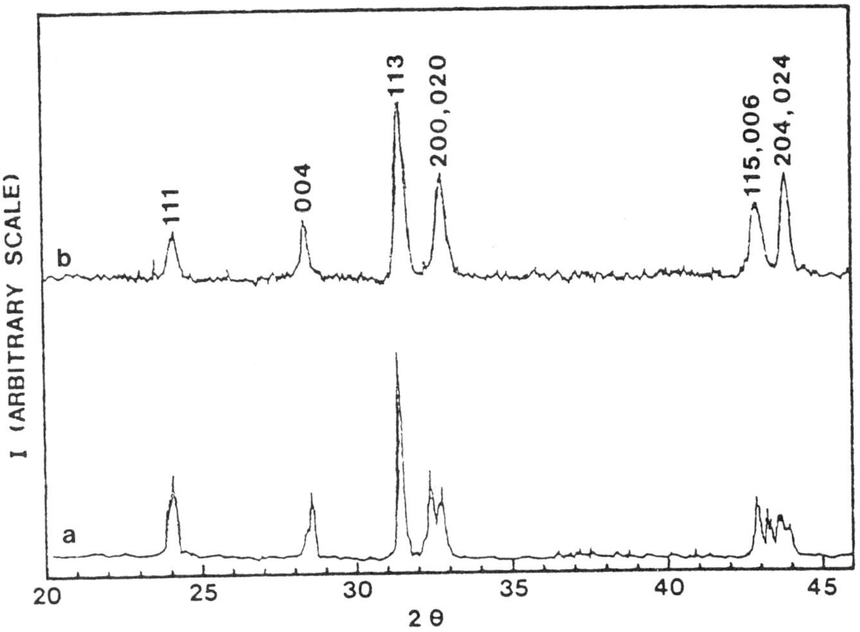

Fig. 2 A comparison of X-ray Power Diffraction Patterns (Cu K_{α}) of La_2NiO_4. (a) unannealed and (b) annealed at log f_{O2} = -8.5 and 1470 K, indexed after the face centered unit cell showin in Fig. 1(c).

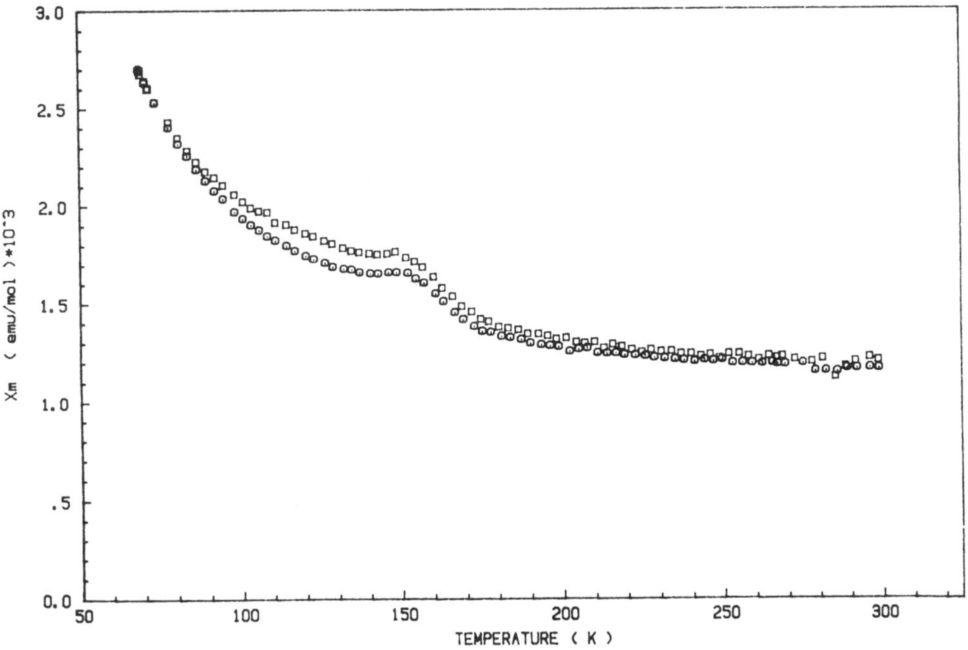

Fig. 3 Molar Magnetic Susceptibility for La_2NiO_4 for Unannealed La_2NiO_4,H||c. Squares, data taken at 6500 G; circles, data taken at 5500 G.

These magnetic field effects suggest the presence of long range antiferromagnetic ordering with a small net ferromagnetic component, as manifested by the weak magnetic image force and torque (the easy axis is parallel to c). One should note that the weak ferromagnetism may be correlated with the observed distortion and is manifested by a slight canting of the magnetic moments away from the basal plane. These effects should become increasingly pronounced as the extent of the deviation from ideal stoichiometry is reduced and a more distorted structure is attained. The results of a detailed study along these lines will be reported in a future publication.

Conductivity and Seebeck Measurements

Four-probe conductivity measurements were carried out in two cells which spanned the range 78-1230 K, including the region in which La_2NiO_4 undergoes the transition from the distorted to the undistorted structure. Provision was made for automatic polarity reversals and for null reading subtraction. A current pulse though a small heater provided a temperature gradient to establish Seebeck voltages. Voltage and temperature gradients were measured in alternation during the decay of the pulse, and linear least squares fits of each were used to evaluate the Seebeck coefficient.

The electrical resistivity for two representative specimens is shown in Fig. 4 as plots of log ρ vs. $10^3/T$. The curves labelled U or A represent data for specimens that were unannealed (Ni^{3+}/Ni^{2+} ratio $\approx$ 0.1) or annealed at log f_{O_2} = -4.6 (Ni^{3+}/Ni^{2+} < 0.005) at 1470 K for approximately one week. Close inspection reveals the following features: (i) Below 350 K the two sets of curves are qualitatively similar: the resistivity for specimen U is somewhat greater than that for specimen A; however, the resistivity for current flow along the pseudo tetragonal c-axis is larger by 1.5 to 2 orders of magnitude than that for current flow within the basal plane. (ii) The conductivity activation energies for specimens U and A are in the neighborhood of ε_ρ = 0.067 and 0.050 eV respectively. These values are at most several multiples of kT; thus, even if the ε_ρ were considered exclusively as a mobility activation energy ε_u, the energy barrier to charge transport is only slightly larger than the thermal energy of the carriers. Thermoelectric measurements cited below indicate that ε_u is actually considerably lower. In these circumstances electron behavior in La_2NiO_4 very likely is a regime intermediate between itinerant and strictly localized. As a point of departure we therefore use a modification of Goodenough's energy diagram[20], as shown in Fig. 5(a). Transport along the basal plane is believed to occur by carriers in a narrow $\sigma(x^2-y^2)$ 'valence' band, separated by a gap of no more than 0.14 eV from a $\sigma^*(x^2-y^2)$ 'conduction' band. Transport in the orthogonal direction involves carriers in a very narrow d_{z^2} 'valence band' whose edge is energetically close to that of the $\sigma(x^2-y^2)$ band. The gap separating the d_{z^2} 'valence' and $d^*_{z^2}$ 'conduction bands' is nearly the same as the $\sigma(x^2-y^2)$ band gap. In the model proposed here and sketched in Fig. 5, the charge carrier densities $n_{||}$ and $n_\perp$ for transport in the basal plane and along the c axis are nearly the same. The difference in resistivity for current flow perpendicular and parallel to the c-axis would then be largely ascribed to differences in mobility factors, as is consistent

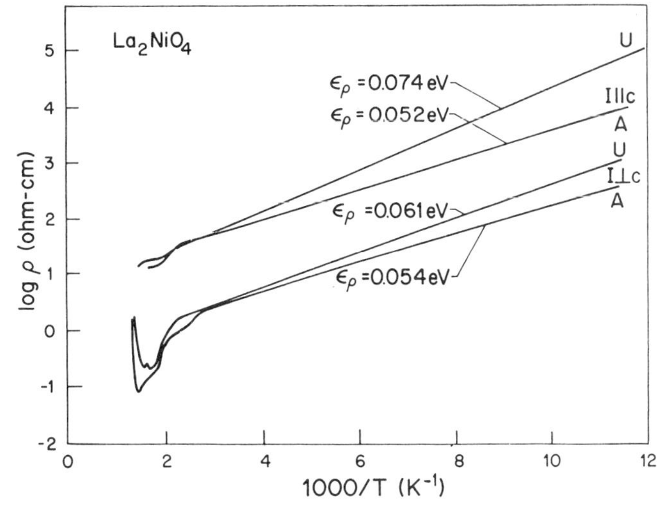

Fig. 4 Plots of Log ρ vs. $10^3/T$ for Unannealed (U) and Annealed (A) Single Crystal Specimjens of La_2NiO_4. ε_ρ is the observed resistivity activation energy. Upper curves: current directed along c axis; lower curves: current flow alon- basal plane. Sample anneal at 1400°C in pure CO_2 for one week.

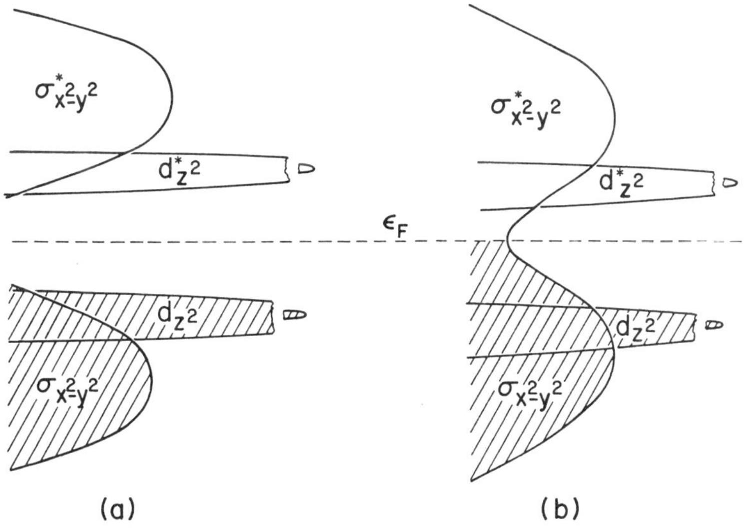

Fig. 5 Modified Gooenough Energy Band Diagram for La_2NiO_4 in the Semiconducting (a) and Metallic State (b). See text.

with the large interplanar separation. (iii) One of the most interesting features of Fig. 4 occurs in the neighborhood of 650 K, where the changeover from the orthorhombic to the tetragonal phase is encountered. Namely, whereas both sample U and sample A exhibit a sudden drop in resistivity at the indicated temperature, for current flow in the basal plane, no unusual features are encountered when current flows along the c-axis. (iv) La_2NiO_4 undergoes a two-dimensional phase transition for current flow in the basal plane. This is documented in Fig. 6 which shows the variation in resistivity of unannealed La_2NiO_4 in the high temperature range from 650 to 1230 K. The metallic characteristics of the sample above 700 K are readily apparent. On the basis of the resistivity at high temperature of $\sim 10^{-1}$-10^{-2} ohm-cm, La_2NiO_4 should be considered as a poor metal, which is in conformity with the DOS diagram of Fig. 5(b), in which the density of states at the Fermi level is small and the collapsed $\sigma(x^2-y^2)$ band is still narrow. The disappearance of the band gap is undoubtedly linked to the changeover from the orthorhombic to the tetragonal phase. (v) The resistivity change in the region from approximately 400 to 650 K is very irreproducible; it depends sensitively on the history of the thermal cycling of the samples. The very steep rise in resistivity with temperature depicted in Fig. 4 apparently is an artifact due to the formation of domains and the development of strained regions in the course of the structural change. This tended to be a serious problem; for example, slow cooling of a specimen through the critical temperature range caused crystals to crack, whereas rapid quenching past the range enabled the data to be collected which are displayed in Fig. 4. Thus, consistent and reliable data could be taken only below or above the transition region. There can be no doubt, however, about the changeover with temperature of the sample from semiconducting characteristics with a low conductivity activation energy to metallic properties, provided current flow was directed along the basal planes. No comparable changes were encountered for current flow along the orthogonal direction. This fact clearly illustrates the quasi-two-dimensional nature of La_2NiO_4.

A set of typical Seebeck measurements is displayed in Fig. 7. For technical reasons these studies were confined to below room temperature and thus encompass only the range where La_2NiO_4 displays semiconducting characteristics; also, the temperature gradient was constrained to lie in the basal plane. The following additional points are noteworthy: (vi) only for the unannealed specimen was the plot of Seebeck coefficient α vs. $1/T$ reasonably linear over a relatively large $1/T$ interval. For samples annealed at $\log f_{O_2} = -4.6$ or -8.5 a sigmoidal-type plot was obtained. This observed behavior could be attributed to either the more pronounced magnetic ordering effects encountered in the more nearly stoichiometric specimen or to changes in carrier localization in samples which exhibit greater distortions from the ideal K_2NiF_4 structure. (vii) In the range between 250 and 300 K all three curves are nearly linear and have parallel slopes with an average Seebeck activation energy near $\varepsilon_\alpha = 0.038$ eV. This should be compared with ε_σ values of 0.054 and 0.061 eV. While scant attention should be paid to the exact numerical values of ε_α because of the very limited linearity of the plots in Fig. 7, it is gratifying that ε_α is comparable to but does not exceed ε_ρ. This is an essential requirement for any simple theory of single or multicarrier conduction phenomena.[29] If taken literally, it would yield conductivity activation energies of the order of 0.015 to 0.040 eV.

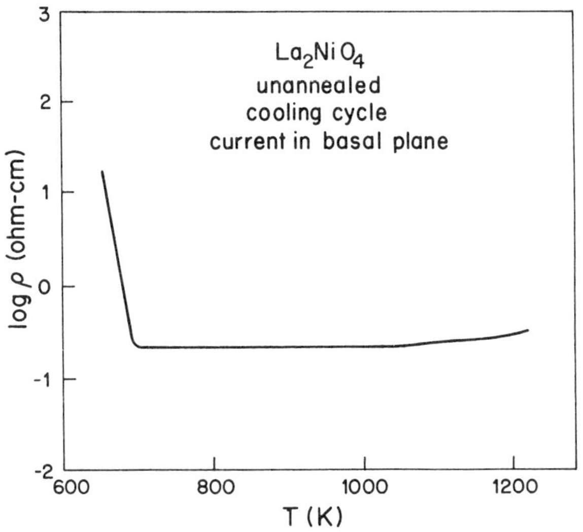

Fig. 6 Plot of Log ρ vs. T in the High Temperature Range for Unannealed La_2NiO_4 Single Crystals. Current flow in basal plane.

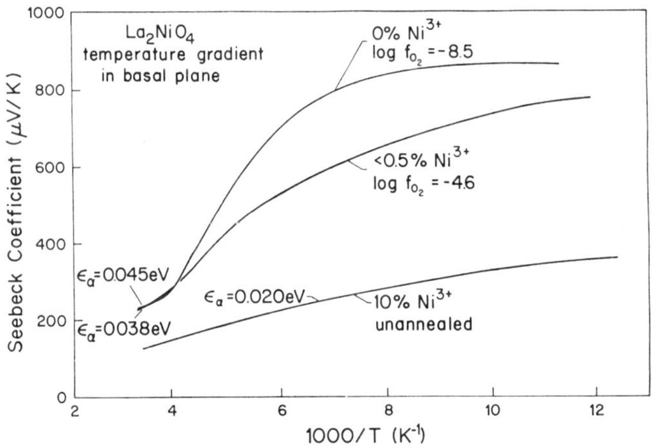

Fig. 7 Seebeck Coefficients for Several Single Crystal La_2NiO_4 Specimens. Temperature gradient directed along basal plane. ε_α is the number activation energy.

SUMMARY AND CONCLUSIONS

Several types of measurements have been summarized to show that La_2NiO_4 single crystals at low temperatures are distorted orthorhombically from the ideal K_2NiF_4 structure, and that they undergo a distortive transition in the neighborhood of 650 K. Electrical resistivity studies show that such crystals exhibit a two-dimensional metal-semiconductor transformation in that same region. Magnetic investigations show that long-range antiferromagnetic order exists in this material below $\sim$150 K. A more detailed study of the effect of oxygen nonstoichiometry on electrical and magnetic properties of La_2NiO_4 and other lanthanide nickelates is in progress.

References

1. J. M. Kosterlitz and D. J. Thouless in "Progress in Low Temperature Physics", D. F. Brewer, ed. (North Holland, Amsterdam, 1978) Vol. VII, p. 373 ff.
2. M. N. Barber, Physics Reports 59, 375 (1980).
3. S. K. Sinha, ed. "Ordering in Two Dimensions", Proc. Int. Conf., Lake Geneva, Wisconsin, May 1980 (North Holland, Amsterdam, New York, 1980).
4. J. M. Kosterlitz in "Phase Transitions in Surface Films" J. G. Dash and J. Ruvalds, eds.; NATO Advanced Study Institute Series B-51 (Plenum, New York, 1980), p. 193 ff.
5. T. D. Schultz in "The Physics and Chemistry of Low Dimensional Solids", L. Alcácer, ed., NATO Advanced Study Institute, 56 (Reidel, Dordrecht, Netherlands, 1980) p. 1 ff.
6. G. A. Thomas, ibid. p. 31 ff.
7. S. K. Sinha, Comm. Solid State Phys. 10, 212 (1982).
8. L. J. de Jongh and A. R. Miedema, Adv. Phys. 23, 1 (1974).
9. A. Rabenau and P. Eckerlin, Acta Cryst. 11, 304 (1958).
10. M. Foëx, Bull. soc. chim. France 1961, 109.
11. M. Foëx, A. Mancheron, and M. Line, Comp. rend. 250, 3027 (1960).
12. G. A. Smolenskii, V. M. Yudin, and E. S. Sher, Soviet Physics - Solid State 4, 2452 (1963) [Fiz. Tverd. Tela 4, 3550 (1962)].
13. G. A. Smolenskii, V. A. Bokov, S. A. Kizaev, E. J. Maltzer, G. M. Nedhir, V. P. Plakhty, A. G. Tutov, and V. M. Yudin, Proc. Int. Conf. Magn. Nottingham 1964; (Inst. Phys. London, 1965), p. 354 ff.
14. P. Ganguly and C. N. R. Rao, Mat. Res. Bull. 8, 405 (1973).
15. B. Willer and M. Daire, Compt. rend. Acad. Sci. Paris, 267, 1482 (1968).
16. B. Grande and H. K. Müller-Buschbaum, Z. anorg. allg. Chem. 433, 152 (1977).
17. H. K. Müller-Buschbaum and U. Lehmann, Z. anorg. allg. Chem. 447, 47 (1978).
18. M. Seppanen, Scand. J. Met. 8, 191 (1979).
19. G. Ganguly, S. Kollali, C. N. R. Rao and S. Kern, Magn. Lett. 1, 107 (1980).
20. J. B. Goodenough and S. Ramasesha, Mat. Res. Bull. 17, 383 (1982).
21. K. K. Singh, P. Ganguly, and C. N. R. Rao, Mat. Res. Bull. 17, 493 (1982).
22. J. Drennan, C. P. Tavares, and B. C. Steele, Mat. Res. Bull. 17, 621 (1982).
23. K. K. Singh, P. Ganguly, and J. B. Goodenough, J. Solid State Chem. 52, 254 (1984).
24. P. Ganguly and C. N. R. Rao, J. Solid State Chem. 53, 193 (1984).
25. D. J. Buttrey, H. R. Harrison, J. M. Honig, and R. R. Schartman, J. Solid State Chem. 54, 407 (1984).
26. C. N. R. Rao, D. J. Buttrey, N. Otsuka, P. Ganguly, H. R. Harrison, C. J. Sandberg, and J. M. Honig, J. Solid State Chem. 51, 266 (1984).
27. D. J. Buttrey, Ph.D. Thesis, Purdue University, 1984.
28. W. R. Robinson and J. Tuley, unpublished results.
29. M. Pai and J. M. Honig, Physica Stat. Sol. (b) 108, K79 (1981).

ONE ELECTRON BAND STRUCTURE OF A COLLECTION OF RESONANT STATES

J. Friedel and C. Noguera

Physique des Solides
Université Paris Sud
Laboratoire Associé au CNRS
91405 Orsay, France

SUMMARY

Successive elastic scatterings on resonant states broaden the core of the corresponding virtual bound levels. If these states have all the same energy and equal coupling strength with the delocalised states, the peak of the density of states at the core is split into two peaks separated by a gap. This is a real gap if the resonant states form a periodic lattice, but a pseudogap if they are distributed at random in three dimensions. This last result is shown using an approximation equivalent to the coherent potential approximation. In one or two dimensions, the CPA appoach leads to inconsistencies coherent with the fact that all states are localised.

INTRODUCTION

Our purpose is to discuss the one electron band structure when the electron considered is elastically scattered successfully by a number of identical resonant states distributed at random in space.

This problem is related to a number of physical situations - resonant elastic scattering of electrons by atomic vapours, by valence fluctuations in rare earths metals and alloys - where however other complications arise from exchange and correlation effects.

The interference effects we have in mind have been studied for a pair of resonant states[1], and also when such states form a periodic lattice[2,3]. In the lattice case, an energy gap appears in the centre of the resonance. As far as we know, the case of randomly distributed resonating centres has not yet been studied, although the more complex cases with direct hopping between resonant states have also been considered both for a pair of states[4] and, to some extend, for solid crystalline solutions[5].

1 - The model

We use here the simplest possible model. A band of N_k delocalised $|k\rangle$ states, of energy E_k between E_1 and E_2, with a density of states $n_o(E)$ per unit energy, interact with N localised states $|i\rangle$, distributed on sites R_i, and with equal energy E_o. States $|k\rangle$ and $|i\rangle$ are assumed normalised and all orthogonal. The local coupling potentials between states $|k\rangle$ and $|i\rangle$ are assumed of strength v all equal and independent of $|k\rangle$ and $|i\rangle$. If $\mathcal{V}$ is the large volume considered, N_k and n_o are proportional to $\mathcal{V}$, while v is proportional to $\mathcal{V}^{-1/2}$. The one electron hamiltonian then writes

$$H = H_o + V$$

$$H_o = \sum_k |k\rangle E_k \langle k| + \sum_i |i\rangle E_o \langle i|$$

$$V = \sum_{i,k} \left(|k\rangle\, v\, e^{-ikR_i} \langle i| + |i\rangle\, v^+ e^{+ikR_i} \langle k| \right)$$

In this paper, we shall only consider the density of states n(E) of the perturbed problem, and its difference $\delta n(E) = n(E) - n_o(E)$ due to the resonant scatterings.

n(E) is classically related to the trace of the resolvant G(z) :

$$G = G_o + G_o V G_o + G_o V G_o V G_o + \dots \qquad (4)$$

Thus

$$\mathrm{Tr}\, G(z) = A + B \qquad (5)$$

where

$$A = \sum_i \langle i|G(z)|i\rangle = \frac{N}{Z-E_o} + \frac{N|v|^2}{(Z-E_o)^2} \sum_k \frac{1}{Z-E_k} + \frac{|v|^4}{(Z-E_o)^3} \sum_{k,k'} \frac{e^{i(k'-k)(R_j-R_i)}}{(Z-E_k)(Z-E_{k'})}$$

$$+ \frac{|v|^6}{(Z-E_o)^4} \sum_{\substack{k,k',k''\\ i,j,\ell}} \frac{e^{i[(k''-k')(R_\ell-R_i)+(k'-k)(R_j-R_i)]}}{(Z-E_k)(Z-E_{k'})(Z-E_{k''})} + \dots \qquad (6)$$

$$B = \sum_k \langle k|G(z)|k\rangle = \sum_k \frac{1}{Z-E_k} + \frac{N|v|^2}{Z-E_o} \sum_k \frac{1}{(Z-E_k)^2} + \frac{|v|^4}{(Z-E_o)^2} \sum_{\substack{k,k'\\ i,j}} \frac{e^{i(k'-k)(R_j-R_i)}}{(Z-E_k)^2 (Z-E_{k'})}$$

$$+ \frac{|v|^6}{(Z-E_o)^3} \sum_{\substack{k,k',k''\\ i,j,\ell}} \frac{e^{i(k''-k')(R_\ell-R_i)+(k'-k)(R_j-R_i)}}{(Z-E_k)^2(Z-E_{k'})(Z-E_{k''})} + \dots \qquad (7)$$

Because of the non diagonal form assumed for V, only the terms in (4) which contain even powers of V appear in the trace of G. Indeed B is just the trace over $|k\rangle$ states of an effective hamiltonian $H' = H_o + V'$ where

$$V' = V\,G_o\,V = \sum_{k,k'} |k> \frac{|v|^2}{Z-E_o} \sum_i e^{i(k'-k)R_i} <k'| \qquad (8)$$

The contribution of B to the density of states is the same as that due to an effective potential

$$w(z) = \frac{|v|^2}{Z-E_o} \qquad (9)$$

scattering delocalised states $|k>$ on each site i. However the contribution of A has a more complex meaning.

2 - The isolated impurity[6]

For an isolated impurity on site O, N=1 and (6) and (7) reduce to

$$A = <O|G(z)|O> = \frac{1}{Z-E_o} + \frac{|v|^2}{(Z-E_o)^2} \sum_k \frac{1}{Z-E_k} + \frac{|v|^4}{(Z-E_o)^3} \sum_{k,k'} \frac{1}{(Z-E_k)(Z-E_{k'})} + \ldots$$

$$B = \sum_k <k|G(z)|k> = \sum_k \frac{1}{Z-E_k} + \frac{|v|^2}{(Z-E_o)} \sum_k \frac{1}{(Z-E_k)^2} + \frac{|v|^4}{(Z-E_o)^2} \sum_{k,k'} \frac{1}{(Z-E_k)^2(Z-E_{k'})} + \ldots$$

Putting

$$F_o(Z) = \sum_k \frac{1}{Z-E_k} \qquad (10)$$

and noting that

$$\frac{d\,F_o(z)}{dz} = -\sum_k \frac{1}{(Z-E_k)^2} \qquad (11)$$

we obtain

$$\mathrm{Tr}\,G(z) = F_o(z) + \left[\frac{1}{Z-E_o} - w\,\frac{dF_o(z)}{dz}\right] \frac{1}{1-w\,F_o(z)} \qquad (12)$$

where $w(z)$ is given by (9).

The density of states can then be written

$$n(E) = n_o(E) + \delta n(E) = -\frac{1}{\pi}\,\mathrm{Im}\,\mathrm{Tr}\,G(E+i\varepsilon)$$

with δn related to the second term in (12) :

$$\delta n = \frac{(1-|v|^2 \frac{dF_o}{dE})\,n_o(E)\,|v|^2 - [E-E_o\,|v|^2\,F_o(E)]\,\frac{dn_o}{dE}\,|v|^2}{[E-E_o-|v|^2\,F_o(E)]^2 + \pi^2\,n_o^2(E)|v|^4} \qquad (13)$$

We used the fact that $F_o(E+i\varepsilon) = F_o(E) - i\,\pi\,n_o(E)$, where $F_o(E)$ is the principal part of the integral. In the limit $v \to 0$, this gives a resonant state centred on $E'_o = E_o + |v|^2 F_o(E_o)$

and of width $\pi\, n_o(E'_o)\,|v|^2$. A lorentzian shape

$$\delta n \cong \frac{n_o(E'_o)\,|v|^2}{(E-E'_o)^2+\pi^2\, n_o{}^2(E'_o)\,|v|^4} \tag{14}$$

is well followed in this limit over an energy range small compared with the total band width E_2-E_1 but large compared with the energy width of the resonance, if $n_o(E)$ varies slowly enough in that range (Figure 1).

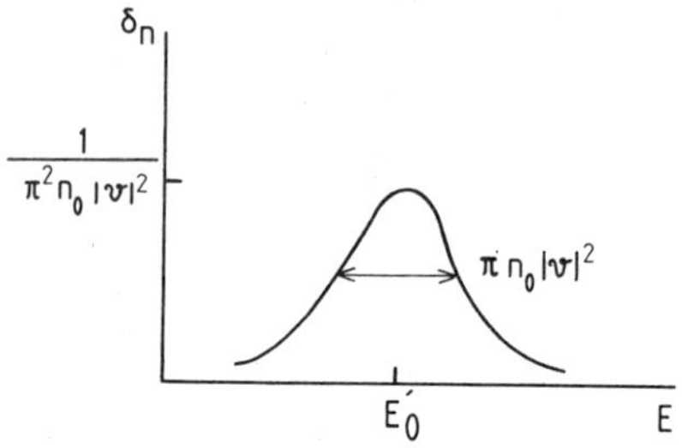

Figure 1. Virtual bound level for an isolated impurity.

As $n_o \propto \nu$ and $|v|^2 \propto \nu^{-1}$, n is independent of ν : in the limit of large volumes ν, it is an infinitesimal correction on n_o.

3 - Lattice of resonant states[2].

If now N resonant states $|i\rangle$ build a periodic lattice, of sites $\underset{\sim}{R}_i$,

$$\sum_j e^{i\underset{\sim}{k}(\underset{\sim}{R}_j-\underset{\sim}{R}_i)} = N \sum_{\underset{\sim}{K}} \delta(\underset{\sim}{k}-\underset{\sim}{K}) \tag{15}$$

where $\underset{\sim}{K}$ are the periods of the associated reciprocal lattice. Equations (6), (7) then give

$$A = \frac{N}{Z-E_o} + \frac{N|v|^2}{(Z-E_o)^2}\sum_{\underset{\sim}{k}}\frac{1}{Z-E_{\underset{\sim}{k}}} + \frac{N^2|v|^4}{(Z-E_o)^3}\sum_{\underset{\sim}{k},\underset{\sim}{K}}\frac{1}{(Z-E_{\underset{\sim}{k}})(Z-E_{\underset{\sim}{k}+\underset{\sim}{K}})}$$

$$+ \frac{N^3|v|^6}{(Z-E_o)^4}\sum_{\underset{\sim}{k},\underset{\sim}{K},\underset{\sim}{K}'}\frac{1}{(Z-E_{\underset{\sim}{k}})(Z-E_{\underset{\sim}{k}+\underset{\sim}{K}})(Z-E_{\underset{\sim}{k}+\underset{\sim}{K}'})} + \ldots$$

$$B = \sum_{\underset{\sim}{k}} \frac{1}{Z-E_{\underset{\sim}{k}}} + \frac{N|v|^2}{Z-E_o} \sum_{k} \frac{1}{(Z-E_{\underset{\sim}{k}})^2} + \frac{N^2|v|^4}{(Z-E_o)^2} \sum_{\underset{\sim}{k},\underset{\sim}{K}} \frac{1}{(Z-E_{\underset{\sim}{k}})^2 (Z-E_{\underset{\sim}{k}+\underset{\sim}{K}})}$$

$$+ \frac{N^3|v|^6}{(Z-E_o)^3} \sum_{\underset{\sim}{k},\underset{\sim}{K},\underset{\sim}{K}'} \frac{1}{(Z-E_{\underset{\sim}{k}})^2 (Z-E_{\underset{\sim}{k}+\underset{\sim}{K}})(Z-E_{\underset{\sim}{k}+\underset{\sim}{K}'})} + \ldots$$

Noting that

$$\frac{d^n F_o(z)}{dz^n} = (-1)^n \, n! \sum_{\underset{\sim}{k}} \frac{1}{(Z-E_{\underset{\sim}{k}})^{n+1}}$$

we obtain

$$\mathrm{Tr}\, G(z) = \frac{N}{Z-E_o} + \frac{N|v|^2}{(Z-E_o)^2} F_o(z) - \frac{N^2|v|^4}{(Z-E_o)^3} \frac{d\, F_o(z)}{dz} + \ldots$$

$$+ F_o(z) - \frac{N|v|^2}{Z-E_o} \frac{d\, F_o(z)}{dz} + \frac{N^2|v|^4}{(Z-E_o)^2} \frac{1}{2!} \frac{d^2 F_o(z)}{dz^2} + \ldots + UT$$

UT represents umklapp terms :

$$UT = \sum_{\underset{\sim}{k},\underset{\sim}{K}} \frac{N^2|v|^4}{(Z-E_o)^2} \left(\frac{1}{Z-E_o} + \frac{1}{Z-E_{\underset{\sim}{k}}}\right) \frac{1}{(Z-E_{\underset{\sim}{k}})(Z-E_{\underset{\sim}{k}+\underset{\sim}{K}})} \chi$$

with

$$\chi = 1 + \sum_{\underset{\sim}{K}'} \frac{N|v|^2}{Z-E_o} \frac{1}{Z-E_{\underset{\sim}{k}+\underset{\sim}{K}'}} + \sum_{\underset{\sim}{K}',\underset{\sim}{K}''} \frac{N^2|v|^4}{(Z-E_o)^2} \frac{1}{(Z-E_{\underset{\sim}{k}+\underset{\sim}{K}'})(Z-E_{\underset{\sim}{k}+\underset{\sim}{K}''})} + \ldots$$

Summing obvious series gives

$$\mathrm{Tr}\, G(z) = \frac{N}{Z-E_o} + \left[1 + \frac{N v^2}{(Z-E_o)^2}\right] F_o\left(Z - \frac{N|v|^2}{Z-E_o}\right) + UT \qquad (16)$$

with

$$UT = \sum_{\underset{\sim}{k},\underset{\sim}{K}} \frac{N^2|v|^4}{(Z-E_o)^2} \left(\frac{1}{Z-E_o} + \frac{1}{Z-E_{\underset{\sim}{k}}}\right) \frac{1}{(Z-E_{\underset{\sim}{k}})(Z-E_{\underset{\sim}{k}+\underset{\sim}{K}})} \frac{1}{1 - \sum_{\underset{\sim}{K}'} \frac{N\, w}{Z-E_{\underset{\sim}{k}+\underset{\sim}{K}'}}}$$

In the spirit of the analysis for a single impurity, we assume that E_o falls within the band (E_1, E_2) of lowest energy, and neglect the influence of bands of higher energy. This means that we neglect the UT terms.

Except for the special case where $E = E_o$ to which we

return later, the perturbed density of states n(E) is then non zero only where $F_o(E-\frac{N|v|^2}{Z-E_o})$ has an imaginary part, for $Z=E+i\varepsilon$. This means

$$E_1 < E - \frac{N|v|^2}{E-E_o} < E_2.$$

The four roots of

$$E(E-E_o) - N|v|^2 = E_i(E-E_o) \quad (i = 1, 2)$$

then separate the E axis in two allowed energy bands, pictured figure 2. In the limit $v \to 0$, they read

$$\begin{aligned} E_1 - \frac{N|v|^2}{E_o-E_1} &< E < E_o - \frac{N|v|^2}{E_2-E_o} \\ E_o + \frac{N v^2}{E_o-E_1} &< E < E_2 + \frac{N|v|^2}{E_2-E_o} \end{aligned} \qquad (17)$$

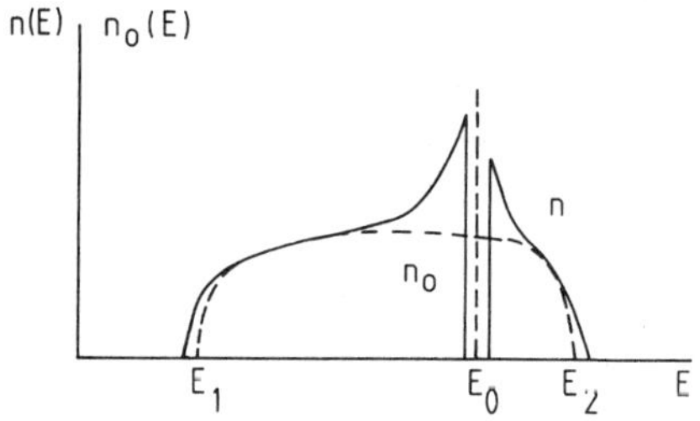

Figure 2. Perturbed and unperturbed densities of states n(E) and $n_o(E)$ for a lattice of resonant states.

Equation (16) then leads to

$$n(E) = N\delta(E-E_o) - N|v|^2 F_o(E-\frac{N|v|^2}{E-E_o}) \frac{d\delta(E-E_o)}{dE}$$

$$+ \left|1 + \frac{N|v|^2}{(E-E_o)^2}\right| n_o(E- \frac{N|v|^2}{E-E_o}) \qquad (18)$$

The third term in (18), presently discussed, leads to a slight broadening of the band (E_1, E_2) and to a resonant feature near E_o, with a small gap pictured figure 2. Thus for $v \to 0$, it gives, outside the gap,

$$n(E) \cong n_o(E) + \frac{N|v|^2}{(E-E_o)^2} n_o(E) - \frac{N|v|^2}{E-E_o} \frac{dn_o(E)}{dE} + O(|v|^4 \frac{d^2 n_o}{dE^2}). \qquad (19)$$

For a band where n_o does not vary too rapidly with energy near E_o, the first corrective term gives the wings of the lorentzian (14).

Thus the coherence of successive scatterings opens a gap in the centre of the resonance peak, but does not alter the long range lorentzian shape of the peak, where the various sites simply add their densities of states. The total correction δn increases then as n_o proportionally to $\mathcal{V}$ for a given lattice structure.

The density of states at $E = E_o$ remains to be discussed. If we write

$$E = E_o + \eta ,$$

the two first terms in (18) give

$$N\delta(\eta) - N|v|^2 \frac{d\delta(\eta)}{d\eta} F_o(E_o + \eta - \frac{N|v|^2}{\eta}).$$

When $E \to E_o$ and thus $\eta \to 0$, the argument of F_o tends to infinity. A development in moments, valid in that case, gives

$$F_o(x) = \sum_{n=0}^{\infty} \frac{M_n}{x^{n+1}}$$

where $M_n = \int n_o(E') E'^n dE'$ and $M_o = N_k$. Using the fact that $\eta \frac{d\delta(\eta)}{d\eta} = - \delta(\eta)$, the two first terms in (18) give finally

$$(N - N_{\underset{\sim}{k}}) \; \delta(E-E_o) \qquad (20)$$

In the simplest case, there is only one state $|i>$ on each site i, and there is one site i per lattice cell. This is for instance the case if the $|i>$ states are s states and the $|k>$ states free particles states. Then $N = N_k$ and contribution (18) vanishes. But contribution (20) differs from zero if states $|i>$ on each site i are degenerate (for instance f states), in which case $N > N_k$.

Finally result (19) could of course have been obtained directly by a reasoning in reciprocal space. The crossing of the broad band $E_{\underset{\sim}{k}}$ with the flat band E_o is lifted by the coupling potential $\mathcal{V}$. If we assume again that E_o falls within the first band $E_{\underset{\sim}{k}}$, and if we neglect mixing of higher order bands (Figure 3), we can write the wave functions in terms of Bloch functions of the boad and flat bands. For s states, with one resonant state per lattice site,

$$|\psi_{\underset{\sim}{k}}> = a_{\underset{\sim}{k}} |\underset{\sim}{k}> + b_{\underset{\sim}{k}} \frac{1}{\sqrt{N}} \sum_i e^{i\underset{\sim}{k}\underset{\sim}{R}_i} |i>.$$

The corresponding energy is given by

$$\begin{vmatrix} E_k - E & \sqrt{N}\, v^+ \\ \sqrt{N}\, v & E_o - E \end{vmatrix} = 0$$

or

$$E = \frac{1}{2} \left[E_o + E_k \pm \sqrt{(E_k - E_o)^2 + 4N\, |v|^2} \right] .$$

The perturbed density of states n(E) is then related to $n_o(E)$ by

$$n(E) = n_o(E_k) \frac{dE_k}{dE} = \left[1 + \frac{N\, v^2}{(E-E_o)^2} \right] n_o(E - \frac{N|v|^2}{E-E_o}) .$$

This is indeed the third term in equation (18). A supplementary flat band, leading to a delta function at $E = E_o$, would appear, as easily shown that way, if the |i> states are degenerate.

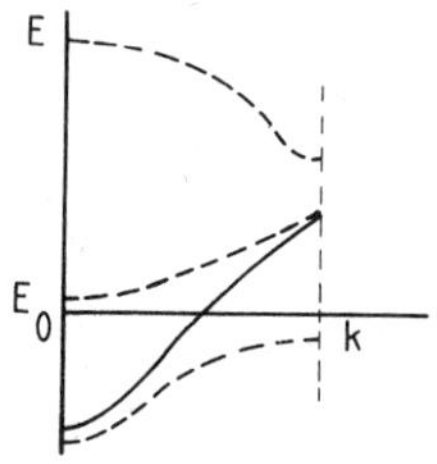

Figure 3. Lifting of crossing of flat and boad bands.

4 - The completely disordered case

We consider now N resonating sites put completely at random in a volume $\mathcal{V}$. This means that we neglect any short range of the sites i, any possible direct hopping between neighbouring resonating states and the possibility that sites i could be on the sites of a periodic lattice. We shall come back on these assumptions in the final discussion.

For complete randomness, the quantity

$e^{i \underset{\sim}{K}(\underset{\sim}{R}_i - \underset{\sim}{R}_n)} = 1$ if $\underset{\sim}{R}_i = \underset{\sim}{R}_n$ or, for $\underset{\sim}{R}_i \neq \underset{\sim}{R}_n$, if $\underset{\sim}{K} = 0$;

it has an average value equal to zero otherwise when summed over sites i and n. There are N cases such that $\underset{\sim}{R}_i = \underset{\sim}{R}_n$, and N(N-1) cases where $\underset{\sim}{R}_i \neq \underset{\sim}{R}_n$. Thus

$$\sum_{i,n} e^{i\underset{\sim}{K}(\underset{\sim}{R}_i - \underset{\sim}{R}_n)} = N \left[1 + (N - 1)\, \delta(\underset{\sim}{K}) \right] \qquad (21)$$

Similarly

$$\sum_{i,n,m} e^{i(\mathbf{k}_1\mathbf{R}_i+\mathbf{k}_2\mathbf{R}_m+\mathbf{k}_3\mathbf{R}_n)} = N \Big[\delta(\mathbf{k}_1+\mathbf{k}_2+\mathbf{k}_3)$$

$$+(N-1)\{\delta(\mathbf{k}_1+\mathbf{k}_2)\,\delta(\mathbf{k}_3)+\delta(\mathbf{k}_1+\mathbf{k}_3)\,\delta(\mathbf{k}_2)+\delta(\mathbf{k}_2+\mathbf{k}_3)\,\delta(\mathbf{k}_1)\}$$

$$+(N-1)(N-2)\,\delta(\mathbf{k}_1)\,\delta(\mathbf{k}_2)\,\delta(\mathbf{k}_3)\Big] \qquad (22)$$

and so on for expressions involving higher summations.

In analogy with the previous cases, the successive terms in the developments (6), (7) can then be regrouped in terms of a function $F_o(Z-\Sigma)$ with a suitable choice of $\Sigma(z)$. These terms are those which contain simple products of factors of the type $\sum_{\mathbf{k}} \frac{1}{(Z-E_{\mathbf{k}})^m}$. In contrast, one cannot sum up in this way contributions of the type

$$\sum_{\mathbf{k},\mathbf{k}',\mathbf{q}} \frac{1}{(Z-E_{\mathbf{k}})(Z-E_{\mathbf{k}+\mathbf{q}})(Z-E_{\mathbf{k}'})(Z-E_{\mathbf{k}'-\mathbf{q}})} \, . \qquad (23)$$

We shall neglect such terms, thus use an approximation very analogous to the standard CPA.

More precisely, the terms of the multiple scattering developments (6), (7) will be represented by diagrams, using the following developments :

- each line represents a propagator $\frac{1}{Z - E_{\mathbf{k}}}$.
- each dot represents an interaction with a resonant state : the value of the interaction is w (equation (9)), and the sum of all wave vectors $\mathbf{k}$ must be zero at the dot.

Defining $G_n = \sum_{\mathbf{k}} \frac{1}{(Z-E_{\mathbf{k}})^n}$, a diagram such as that pictured in the Table I for 2 sites and 4th order of perturbation, which appears in A, will read $w^4G_1{}^2G_2$, and its degeneracy is 3.

We will neglect all diagrams involving convolutions of propagators such as (Figure 4)

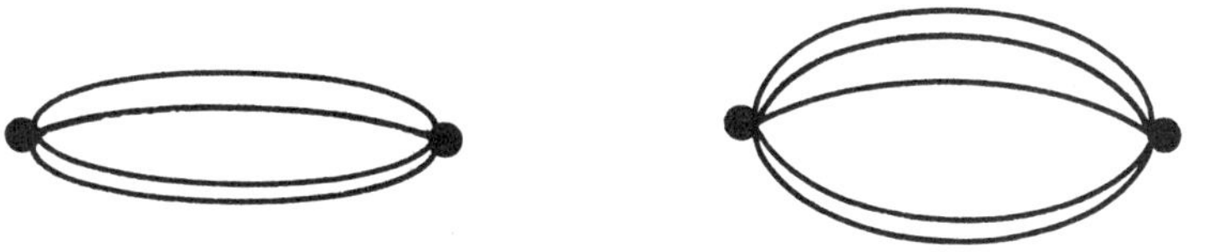

H_4 $\qquad$ I_5

Figure 4. First neglected diagrams.

TABLE I - Degeneracy g of diagrams.

m \ n	1 site	2 sites	3 sites
1	 g=1		
2	 g=1	 g=1	
3	 g=1	 g=3	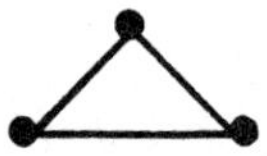 g=1
4	 g=1	 g=6	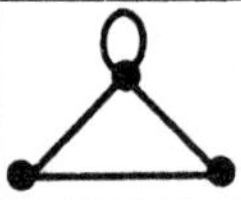 g=4  g=2
5			 
g	1	C_m^2	C_m^3 $2C_m^4$

$$H_4 = \sum_{\underset{\sim}{q}} \left[\sum_{\underset{\sim}{k}} \frac{1}{(E-E_{\underset{\sim}{k}})(E-E_{\underset{\sim}{k}+\underset{\sim}{q}})} \right]^2 \tag{24}$$

$$I_5 = \sum_{\underset{\sim}{k},\underset{\sim}{k}',\underset{\sim}{q}} \frac{1}{(E-E_{\underset{\sim}{k}})^2(E-E_{\underset{\sim}{k}+\underset{\sim}{q}})(E-E_{\underset{\sim}{k}'})(E-E_{\underset{\sim}{k}'+\underset{\sim}{q}})} \tag{25}$$

which appear respectively in the developments of A and B.

Table I indicate the types of diagrams that have been kept for computing A and B, together with their degeneracy g. They are classified as a function of the number (n) of sites and of the order (m) of the interaction. One should remember that each n site diagram is weighted by a factor N(N-1)(N-2) ... (N-n+1).

The summation involving 1 site gives respectively

$$A = \frac{N}{Z-E_o-|v|^2F_o(z)} \quad \text{and } B = F_o\left(Z - \frac{N|v|^2}{Z-E_o-|v|^2F_o(z)}\right) .$$

When taking into account 1 site and 2 sites diagrams (<u>excluding H_4 and I_5</u>),

$$A = \frac{N}{Z-E_o-|v|^2F_o(z)} - \frac{N(N-1)|v|^4 \ \partial F_o/dz}{[Z-E_o-|v|^2 \ F_o(z)]^3}$$

and

$$B=F_o(z) - \frac{N|v|^2 \ \partial F_o/\partial z}{Z-E_o-|v|^2 \ F_o(z)} + N(N-1)|v|^4\{ \ \frac{1}{2} \ \frac{\partial^2 F_o/\partial z^2}{[Z-E_o-|v|^2 \ F_o(z)]^2}$$

$$+ \frac{|v|^2(\partial F_o/\partial z)^2}{[Z-E_o-|v|^2 \ F_o(z)]^3} \quad .$$

In the limit $N \to \infty$, $N(N-1)$ can be replaced by N^2. Then to this order of approximation

$$A = \frac{N}{Z-E_o-|v|^2 \ F_o(Z- \frac{N|v|^2}{Z-E_o-|v|^2 \ F_o(z)})}$$

and $B = F_o\{ \ Z - \dfrac{N|v|^2}{Z-E_o-|v|^2 \ F_o(z- \dfrac{N|v|^2}{Z-E_o-|v|^2 \ F_o(z)})} \}$

Within the same types of approximations, such a development in continued fractions can be extended to include 3 sites diagrams. It leads for $n \to \infty$ to

$$A = \frac{N}{Z-E_o-|v|^2 \ F_o(Z-\Sigma)}$$

$$B = F_o(Z-\Sigma) \tag{26}$$

with

$$\Sigma = \frac{N|v|^2}{Z-E_o-|v|^2 \ F_o(Z-\Sigma)} \tag{27}$$

and thus also

$$A = \frac{\Sigma}{|v|^2} \tag{28}$$

From equations (26)-(28), it is clear that the perturbed density of states $n(E)$ is zero where $\Sigma(E+i\varepsilon)$ is real and $F_o \ [E+i\varepsilon - \Sigma(E+i\varepsilon)]$ is also real. Putting then

$$\Sigma(E+i\varepsilon) = \sigma \ \text{ real,}$$

$n(E) = 0$ if

$$E - \sigma(E) < E_1 \text{ or } > E_2 \ .$$

where E_i (i = 1, 2) are the two band edges (Figure 2). The limits of the perturbed band n(E) are then given by

$$E = E_i + \sigma \tag{29}$$

where, from (27),

$$\sigma = \frac{N|v|^2}{E-E_o-|v|^2 F_o(E_i)}$$

The four limiting values of E are then given by

$$E=\frac{1}{2}\left[E_o+E_i+|v|^2 F_o(E_i) \pm \sqrt{[E_o-E_i+|v|^2 F_o(E_i)]^2 + 4N|v|^2}\right] \tag{30}$$

In the usual limit $|v| \to 0$, the perturbed bands are thus

$$E_1 - \frac{N|v|^2}{E_o-E_1+|v|^2 F_o(E_1)} < E < E_o + |v|^2 F_o(E'_i) + \frac{N|v|^2}{E_o-E'_i+|v|^2 F_o(E'_i)}$$

$$E_o+|v|^2 F_o(E''_i) + \frac{N|v|^2}{E_o-E_i''+|v|^2 F_o(E''_i)} < E < E_2 - \frac{N|v|^2}{E_o-E_2+|v|^2 F_o(E_2)}$$

where $E'_i = E_1$ and $E''_i = E_2$ or $E'_i = E_2$ and $E''_i = E_1$ depending on the form of the band $n_o(E)$ and the position of E_o. The band structure is thus qualitatively similar to that pictured figure 2 for a lattice, with again a small gap around E_o.

From (26), (28) one deduces the perturbed density of states

$$n(E) = -\frac{1}{\pi} \mathfrak{Im} \{ \Sigma(E+i\varepsilon) + F_o [E+i\varepsilon - \Sigma(E+i\varepsilon)] \} .$$

Taking into account that $F_o(z)$ has poles only on the real axis, this can be written

$$n(E) = \frac{N|v|^4 n_o(E-\mathfrak{Re}\Sigma)}{[E-E_o-|v|^2 F_o(E-\mathfrak{Re}\Sigma)]^2 + \pi^2|v|^4 n_o^2(E-\mathfrak{Re}\Sigma)} + n_o(E-\mathfrak{Re}\Sigma) \tag{31}$$

where $\mathfrak{Re}\Sigma$ is the real part of $\Sigma(E+i\varepsilon)$. Near the limits of the gap, $\mathfrak{Re}\Sigma$ can be replaced by its limit σ at the gap which, from (29) and (30), reads

$$\sigma \cong \frac{N|v|^2}{E_o-E_i+|v|^2 F_o(E_i)} .$$

The first term in (31) then gives rise to a double peak of n(E) with lorentzian wings, on either side of the energy gap, again as in a lattice.

Discussion

The approximation developped here for the disordered case is similar to a CPA type of approach of multiple scattering by effective potentials w' (obviously related to w, equation (9)) on the resonant sites.

Indeed if N such sites were distributed at random on a lattice (Figure 5) with N_L sites, one knows that the selfenergy

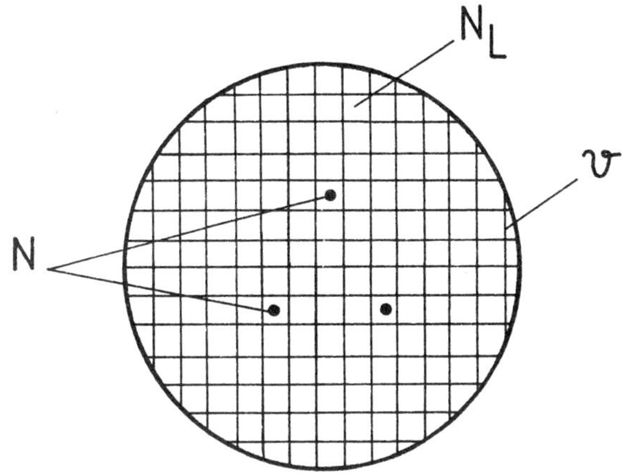

Figure 5. N resonant states at random on a lattice with N_L sites.

in the CPA approximation reads, with our definition for F_o

$$\Sigma(z) = cw' + \Sigma(z)\left[w' - \Sigma(z)\right] N_L F_o \left[z - \Sigma(z)\right]$$

where $c = N/N_L$. Hence

$$\Sigma(z) = \frac{N\ w}{1-(w-N_L^{-1}\Sigma)\ F_o(z-\Sigma)}$$

if $w' = N_L\ w$. This equation reduces to (27) in the limit of continuous disorder where $N_L \to \infty$ while N and ϑ are kept constant.

With the folding of energies such that 4 band edges appear, the approximation developped must therefore be expected to hold well in the perturbed band n(E) away from E_1 and E_2, but also not too near the band gap at E_o : the lorentzian tails of the resonant states are preserved. By analogy with the discussion of the CPA, one can expect that the multiple scattering terms neglected in this approach would alter somewhat n(E) near to E_o, by replacing the real gap obtained by a pseudogap filled with a small density of localised states.

It is worth pointing out that if $F_o(E_i)$ is usually finite in three dimensions, because $n_o(E)$ varies parabolically with $(E-E_i)$ near the band edges E_i, this is not true in two or one dimensions, where $F_o(E)$ tends to infinity for $E \rightarrow E_i$. The discussion leading to equations (30) then suggests that the central gap becomes equal to the band width : all states are localised ; they can't be represented in a CPA type of approximation.

Strictly speaking, the same sort of difficulty arises if n_o is the density of free particles. However in that case it is not reasonable to assume v independent of E_k in the limit of large energies. One must introduce a progressive cut E_2, which will keep $F_o(E)$ finite.

This paper arises from a discussion with Nevill Mott in 1983 on coherency effects between resonant states. It is dedicated to his eightieth birthday. A discussion with Ducastelle on relation to CPA is gratefully aknowledged.

REFERENCES

1. Caroli B. (1967), J. Phys. Chem. Solids, 28, 1427.
2. Friedel J. (1967), Pure Metals and Solid Solutions, McGraw Hill, New York.
3. Coqblin B. and Blandin A. (1968), Adv. Phys. 17, 281.
4. Alexander S. and Anderson P.W. (1964) Phys. Rev., 133, A 1594.
5. Brouers F. and Vedyayev A.V., Phys. Rev. B., 1972, 5, 348.
6. Anderson P.W. (1961), Phys. Rev. 124, 41.

INELASTIC SCATTERING AND LOCALIZATION IN TWO DIMENSIONS

Elihu Abrahams

Serin Physics Laboratory
Rutgers University
Piscataway, NJ 08854

It is an honor to be able to contribute to a volume in honor of Sir Nevill Mott. His many contributions and profound insights in the physics of disordered systems have laid the groundwork for a variety of recent developments. Thus it is very appropriate to describe some current research in the area of electron localization.

The scaling theory of localization[1-4] and its interesting predictions, especially in two dimensions (2D), have stimulated a variety of experimental investigations. Among those which have been particularly instructive are magnetoresistance measurements in 2D metal films[5] which have provided remarkable confirmation of the physical picture and indeed the details of the theory in the weakly-localized regime.

An important physical quantity which can be determined from the magnetoresistance measurements is related to the inelastic scattering rate of electrons in the disordered metal. Therefore, this quantity has been studied extensively, both theoretically and experimentally. The measurements give a direct determination of the length scale[1] for the coherent backscattering[5] responsible for weak localization effects.

The argument which relates the electron inelastic scattering rate $1/\tau_i$ to localization was first given by Thouless. In a time τ_i, an electron which is diffusing in the static disorder goes a distance

$$L_T = \sqrt{D\tau_i} \qquad (1)$$

where D is the diffusion constant. This is the length scale beyond which coherent static localization effects will be destroyed by incoherent energy transfers due to inelastic events. The same argument was used[7] in the scaling theory to show that the relevant length scale for the scaling was just this inelastic length L_T. Thus in 2D, the logarithmic conductivity decrease with scale is given by

$$\Delta\sigma = -(e^2/\hbar\pi^2)\ln L/v_F\tau \quad . \tag{2}$$

Here, we identify L with the inelastic length L_T. In (2), τ is the elastic scattering time and v_F is the Fermi velocity. Since inelastic rates usually vary as a power of T, say $1/\tau_i \propto T^p$, a logarithmic temperature dependence of σ is predicted:

$$\Delta\sigma = -(e^2/\pi^2\hbar)(p/2)\ln(T_o/T) \quad . \tag{3}$$

This behavior is confirmed by experiment. Of course, it is well known that essentially the same result (with p=1) is found from conductivity corrections caused by electron-electron interaction effects.[8] Therefore, conductivity measurements alone cannot unambiguously confirm the localization theory nor can they determine either the inelastic scattering rate or its exponent p. The separation of the localization and interaction effects is achieved by magnetoconductance measurements.[4,5,9] This is possible because the electron-electron interaction conductivity correction is not affected in weak to moderate magnetic fields[4] while the coherence of the localization-inducing backscattering is destroyed. Thus a positive magnetoconductance is expected when the magnetic length $(\hbar c/eH)^{1/2}$ becomes less than the inelastic length L_T. From magnetoconductance data, therefore, the details of the localization theory in 2D have been confirmed and inelastic rates have been determined.

From the simple argument[6,7] mentioned above, it is natural to assume that the inelastic length L_T entering the conductivity (or magnetoconductivity) correction (3) is determined by the quasiparticle lifetime in the disordered system. A variety of processes can contribute to the lifetime. Among them are electron-electron,[10,11] electron-phonon,[12] electron-paramagnon, electron-superconducting fluctuation[13] scattering.

The electron-electron scattering inelastic effect has been a subject of a number of theoretical and experimental investigations. In the pure metal, the well-known "Fermi-liquid" result[14,7] is a $1/\tau_i$ of order $E_F(T/E_F)^2$ (we use $k_B = \hbar = 1$). Some time ago it was recognized[10,11,15] that the diffusive character of electron propagation causes important modifications to the influence of electron-electron interactions on a variety of thermodynamic and transport quantities. In particular, the electron-electron scattering contribution to the inelastic rate gives a contribution proportional to $T^{d/2}$ where d is the dimensionality. In 1981 this quantity was discussed in detail[15] by Abrahams et al for d=2 with the intent of determining the τ_i to be used [via (1)] in the conductivity correction. They pointed out that the decay rate of the electron eigenstates ("exact eigenstates") in the presence of impurities is the relevant quantity, not the decay of electron states of the pure system. They found[15,16] that

$$1/\tau_i = \frac{1}{2\tau}\left(\frac{T}{E_F}\right)\ell n(T_1/T) \quad . \tag{4}$$

The scale T_1 was rather large so that the existence of the logarithmic correction was in principle experimentally verifiable. The physical origin of the divergent logarithm will be discussed below.

At about the same time, a different method was used by Altshuler et al[17] (AAK). They evaluated directly the localization corrections to the conductivity including the scattering of electrons from the fluctuations of the electromagnetic field. This is a way of treating electron-electron collisions with the screened Coulomb interaction. The result found by AAK is an inelastic cutoff for $\Delta\sigma$:

$$1/\tau_\phi = \frac{1}{2\tau}\frac{T}{E_F}\,\ell n E_F\tau \tag{5}$$

The difference between (4)[15] and (5)[17] led Fukuyama and Abrahams[18] (FA) to examine the question of whether the quasiparticle decay rate $1/\tau_i$ is in fact the correct inelastic time to be used for the conductivity correction. They calculated the influence of the screened electron-electron interaction on the zero-wave number particle-particle diffusion propagator ("Cooperon") which is the mathematical manifestation of the coherent backscattering responsible for localization. For the q=0 Cooperon cutoff, or mass, they found [compare (5)]

$$1/\tau_c = \frac{1}{2\tau}\frac{T}{E_F}\,\ell n(T_1/T) \tag{6}$$

precisely the same as the result (4) of Ref. 15 for the quasiparticle lifetime.

The question of the existence of the singular logarithm predicted in Ref. 15 and by FA[18] but not by AAK[17] was a source of some discussion, especially since an overview of the experimental results was at best inconclusive. The origin of the singularity can be traced back to the fact that in 2D the screening, inhibited as it is by the static disorder, is comparatively ineffective at large distances. Then the contribution to the electron-electron interaction of density fluctuations of long wavelength ($> L_T$) eventually contributes a $\ln T_1/T$ factor to both the quasiparticle lifetime and, up to a certain order, the Cooperon mass. In 3D, an extra factor q in the density fluctuation phase space removes the singularity.

The result of FA was reexamined by Fukuyama.[19] He showed that the result (5) of AAK could be obtained within the method of FA by ignoring everywhere the contribution of density fluctuations over distances greater than L_T. He justified this procedure by invoking the momentum dependence of the Cooperon mass $1/\tau_c(q)$. Fukuyama pointed out that the conductivity is determined by the $q \approx L_T^{-1}$ Cooperon, not the $q = 0$ one used in the discussion of FA. He then neglected the very long wavelength contributions which are the source of the singular logarithm. However, it can be shown[20] that the long wavelength ($> L_T$) density fluctuations contribute a singular $\ln(T_1/T)$ factor to $1/\tau_c(q \sim L_T^{-1})$ just as they do to $1/\tau_c(0)$. Thus it appears that a small wave number cutoff L_T^{-1} is not the correct source of the discrepancy between FA and AAK, in spite of the suggestion of Ref. 19 and an earlier, similar, one of AAK.[17]

During the past year, Aronov[21] and Eiler[22] have correctly identified the origin of the difference between FA[18] and AAK.[17] It turns out that there is a contribution to the Cooperon mass which was neglected by FA and in Ref. 19. It is essentially a scattering-in process for the particle-particle correlation and is of the form of Fig. 2c of FA (see also Fig. 1 of Ref. 21). It can be shown explicity[20] that when this contribution is included, the singular logarithmic factor of FA turns into the non-singular one of AAK. However, it has not been verified that no other contributions of this order are present. All of the calculations we have been discussing are approximate self-consistent perturbative ones in that the inelastic cutoffs appear in the graphs used to evaluate the inelastic cutoffs. This is necessary to avoid divergences in the calculations. Although the recent results suggest that the divergent logarithm of FA is cancelled in higher order, it has not been verified that all terms of the same order have been taken into account. Furthermore, it has not been verified that the approximations used so far are conserving[23] and derivable from a single contribution to the free energy.

We emphasize, as discussed thoroughly by Eiler,[22] that the methods of FA and AAK have the same physical content. Therefore, the question of the nature of the logarithm is one to be answered by a consistent calculation in either approach. The magnetoconductance observations at very low temperature consistently yield a linear T dependence for $1/\tau_\phi$ thereby confirming the theoretical suggestions[15-17] that the diffusively modified electron-electron scattering determines the length scale for the coherent localization effects.

Other contributions to the inelastic cutoff $1/\tau_\phi$ do not suffer from the delicate behavior of the screened Coulomb interaction at large distances in 2D and are expected to be the same as the contribution to the quasiparticle scattering rate $1/\tau_i$. The electron-phonon scattering contribution is not expected to be particularly enhanced by disorder although the temperature dependence changes. It was worked out some time ago by Schmid[12] in 3D. The extension to 2D is straightforward and yields a T^3 dependence for $1/\tau_i$ instead of T^2 as in the clean case. At low temperature, this process will be masked by the electron-electron scattering which, as we have seen, is linear in T.

Recently, the contribution of electron-paramagnon scattering was worked out by Chang and Abrahams.[24] Here the effect of a strong short-ranged opposite-spin electron-electron interaction is considered. Then the electrons undergo inelastic scattering processes in which spin fluctuations (paramagnons) indicate the electron-electron interaction. The result for $1/\tau_\phi$ is essentially

$$1/\tau_\phi = \frac{3}{4\tau}\frac{T}{E_F\tau}\frac{I^2}{1-I}\ln E_F\tau \quad , \qquad (7)$$

a result similar to the AAK result for the electron-electron contribution, Eq. (5). In Eq. (7), $(1-I)^{-1}$ is the Stoner enhancement factor. Since this process has a linear T dependence it will be hard to distinguish it from the electron-electron scattering contribution.

The electron inelastic scattering rate due to electron-superconducting fluctuation scattering was investigated[25] some time ago since it is an essential ingredient in fluctuation phenomena just above the superconducting critical temperature T_c. The results of these early calculations

$$1/\tau_i = \frac{\ln 2}{E_F\tau}[T/\ln(T/T_c)] \quad . \tag{8}$$

were valid in zero magnetic field and temperatures not too close to T_c. In this problem, the inelastic scattering acts as a pair-breaker and shifts T_c downward so the singularity in (8) at $T = T_c$ is removed in a better calculation. This has been discussed in a recent reexamination of the problem.[26] Another important effect which is pointed out in Ref. 26 is that the magnetic field which is applied for magnetoconductivity measurements strongly affects the superconducting fluctuations. The details of the full T, H behavior of the contribution to $1/\tau_\phi$ are too complicated to be discussed here; the interested reader is referred to Ref. 26..

I have given here a brief recapitulation of recent work on inelastic scattering effects in the localization problem. The confusion surrounding the presence or not of the singular $\ln T_1/T$ factor in the electron-electron scattering rate has been discussed and the suggestion made that it is not of fundamental importance for the theory. In fact, the technique of magnetoconductivity measurements and the resulting determination of various scattering rates has provided an impressive confirmation of the predictions of the scaling theory of localization in the two-dimensional weakly localized regime.

I would like to thank many experimental and theoretical colleagues for discussions on various topics in localization theory. On the particular problem of the inelastic effects, I am especially indebted to J.M.B. Lopes dos Santos and H. Fukuyama.

References

1. E. Abrahams, P.W. Anderson, D.C. Licciardello and T.V. Ramakrishnan, Phys. Rev. Lett. 42, 673 (1979).
2. L.P. Gorkov, A.I. Larkin and D.E. Khmel'nitskii, JETP Lett. 32, 248 (1979).
3. F.J. Wegner, Z. Physik 35, 207 (1979).
4. S. Hikami, A.I. Larkin and Y. Nagaoka, Prog. Theor. Phys. 63, 707 (1980).
5. G. Bergmann, Physics Rep. 107, 1 (1984).
6. D.J. Thouless, Phys. Rev. Lett. 39, 1167 (1977).
7. P.W. Anderson, E. Abrahams and T.V. Ramakrishnan, Phys. Rev. Lett. 43, 718 (1970).
8. B.L. Altshuler, A.G. Aronov and P.A. Lee, Phys. Rev. Lett. 44, 1288 (1980).
9. B.L. Altshuler, D.E. Khmel'nitskii, A.I. Larkin and P.A. Lee, Phys. Rev. B 22, 5142 (1984).
10. A. Schmid, Z. Physik 271, 251 (1974).
11. B.L. Altshuler, A.G. Aronov, Zh. Eksp. Teor. Fiz. 77, 2028 (1979) [Sov. Phys. JETP 50, 968 (1979)].

12. A. Schmid, Z. Physik 259, 421 (1973).
13. E. Abrahams, M. Redi and J.W.F. Woo, Phys. Rev. B 1, 208 (1970).
14. See for example, P. Morel and P. Nozières, Phys. Rev. 126, 1909 (1962).
15. E. Abrahams, P.W. Anderson, P.A. Lee and T.V. Ramakrishnan, Phys. Rev. B 24, 6783 (1981).
16. J.M.B. Lopes dos Santos, Phys. Rev. B 27, 1189 (1983).
17. B.L. Altshuler, A.G. Aronov and D.E. Khmel'nitskii, Solid State Commun. 39, 619 (1981); J. Phys. C 15, 7367 (1982).
18. H. Fukuyama and E. Abrahams, Phys. Rev. B 27, 5976 (1983).
19. H. Fukuyama, J. Phys. Soc. of Jpn. 53, 3299 (1984).
20. E. Abrahams and J.M.B. Lopes dos Santos, unpublished.
21. A.G. Aronov, Physica 126B, 314 (1984).
22. H. Eiler, J. Low Temp. Phys. 56, 481 (1984).
23. G. Baym, Phys. Rev. 127, 1391 (1962).
24. M.-c. Chang and E. Abrahams, Phys. Rev. B (1985).
25. B.R. Patton, Phys. Rev. Lett. 27, 1273 (1971); J. Keller and V. Korenman, Phys. Rev. B 5, 4367 (1972).
26. W. Brenig, M.-c. Chang, E. Abrahams and P. Wölfle, Phys. Rev. B 31 (1985).

EXISTENCE OF A SHARP ANDERSON TRANSITION IN DISORDERED TWO-DIMENSIONAL SYSTEMS

G. M. Scher

Hewlett-Packard Company
Fort Collins Integrated Circuits Division
3404 East Harmony Road
Fort Collins, CO 80525, U.S.A.

D. Adler

Center for Materials Science and Engineering
Massachusetts Institute of Technology
Cambridge, MA 02139, U.S.A.

ABSTRACT

We derive the quantum-mechanical energy-dependent diffusivity directly from the Kubo-Greenwood formula for electrical conductivity. The resulting expression suggests a numerical method for the solution of the Schrodinger equation without the introduction of sample boundaries. Instead, we simply analyze the diffusion of a particle outwards from an initial bulk site. This method is applied to calculate the conductivity of a two-dimensional system in the presence of various degrees of time-independent disorder. In the case of the Anderson model for non-interacting electrons, we find a sharp transition from localized to diffusive behavior, contrary to the predictions of scaling theory.

Photographs of the energy-resolved electronic probability waveform during its evolution in time indicate a phenomenon similar to classical percolation, but with the distinct quantum feature of a minimum metallic conductivity, as predicted by Mott. An explanation of the numerical results is based on first principles and utilizes only the essential physical ingredients of the problem: disorder, diffusion, and quantum-mechanical probability amplitude.

INTRODUCTION

There has been a great deal of effort expended recently on the problem of disorder-induced localization [1]. The macroscopic question is whether or not there is a transition from insulating to conducting behavior as the degree of disorder is reduced. Microscopically (if we exclude the effects of electron-electron interactions) we can restate the problem in terms of the extent of the diffusion of an electronic wavepacket outward from a certain initial confinement. The existence of a transition from finite to infinite diffusion lengths after an infinite amount of time has been called an Anderson transition [2].

Current one-parameter scaling theory claims the absence of an Anderson transition and a minimum metallic conductivity [3] in disordered two-dimensional systems [4]. However, in this theory, the one parameter is taken to be the conductivity $\sigma(L)$ of a square sample of area L^2. Since conductivity is a bulk property, the concept of a $\sigma(L)$ depending on sample shape, size, and boundary conditions is inadequate. Thus far, neither theoretical analysis nor low-temperature experiments on MOSFET devices has confirmed or refuted this result. Numerical calculations have met with difficulty in two dimensions due to the prohibitive amounts of computer time and storage necessary to solve some version of the Schrodinger Equation on a lattice large enough to observe, or rule out the possibility of observing, a transition induced by varying disorder. Unfortunately, efforts to overcome this limitation have involved use of arbitrary boundary conditions, smoothing functions, finite imaginary energies in Green's functions, etc. The conclusions obtained from such calculations have been conclusions more about the effect of the various boundary conditions and inserted parameters than about the influence of disorder on the motion of the electron. Calculation of eigenstates and eigenfunctions implies infinite energy resolution, hence knowledge of the electron's motion for an infinite period of time.

In this paper, we take a different approach [5]. We show that a change in the diffusive behavior of the electron as the degree of disorder is varied can be observed at finite times, in finite amounts of space, and with finite computer time and storage. (However, we do use longer times and larger computational lattices than previous investigators.) We do this by applying the Kubo-Greenwood formula for the dc conductivity to obtain an expression for the energy-dependent diffusion coefficient. Using this result, we numerically solve the Schrodinger equation in the presence of increasing degrees of time-independent disorder for a two-dimensional lattice. We find a sharp Anderson transition at a specific ratio of disorder to bandwidth. We discuss the results in detail, including the discrepancy with the results of scaling theory.

MODEL AND RESULTS

The Kubo-Greenwood formula for the dissipative part of the dc electrical conductivity can be written:

$$\sigma(E_F,T) = \lim_{u \to o} \mathrm{Re}\, \frac{2e^2}{\Omega} \frac{iu}{\hbar} \sum_{n,n'} |\langle n|\hat{x}|n'\rangle|^2 [(f(E_{n'})-f(E_n)]\ (E_n-E_{n'}-iu)^{-1} \tag{1}$$

where n and n' are states of the system having volume Ω, $\hat{x}$ is the displacement operator in the x-direction, and f(E) is the Fermi-Dirac distribution function. This leads to the following expression for the energy-dependent diffusion coefficient:

$$D(E_F,T) = \lim_{u \to o} \frac{u^2}{2\hbar^2} \int_0^\infty dt\ e^{-ut/\hbar} \underbrace{\left\langle \sum_{\underline{r}_i} \langle (x_i-x_o)^2 |K(\underline{r}_i,\underline{r}_o;t)|^2 \right\rangle}_{\equiv L_x^2(t) = \frac{1}{2} L^2(t)} \tag{2}$$

(the outer brackets ⟨ ⟩ denote the configuration average)

where $K(\underline{r}_i,\underline{r}_o;t) = \sum_n q(E_n)\Psi_n(\underline{r}_i)\Psi_n^*(\underline{r}_o)e^{-iE_nt/\hbar}$ is the energy-restricted propagator, $q(E)$ being a weighting function that is zero except in the energy range near E_F. To make connection between diffusivity and conductivity, $\sigma = e^2 2g(E_F)D$, with $q(E) = \eta\,(-df/dE)^{1/2}$; η is a normalization factor, $\eta = [\delta^3 \int_{-\infty}^{\infty} dE \frac{-df}{dE} \; g(E)]^{-1/2}$, with the lattice spacing = δ (we consider discrete rather than continuous space), and $g(E)$ is density of states per unit energy per unit volume.

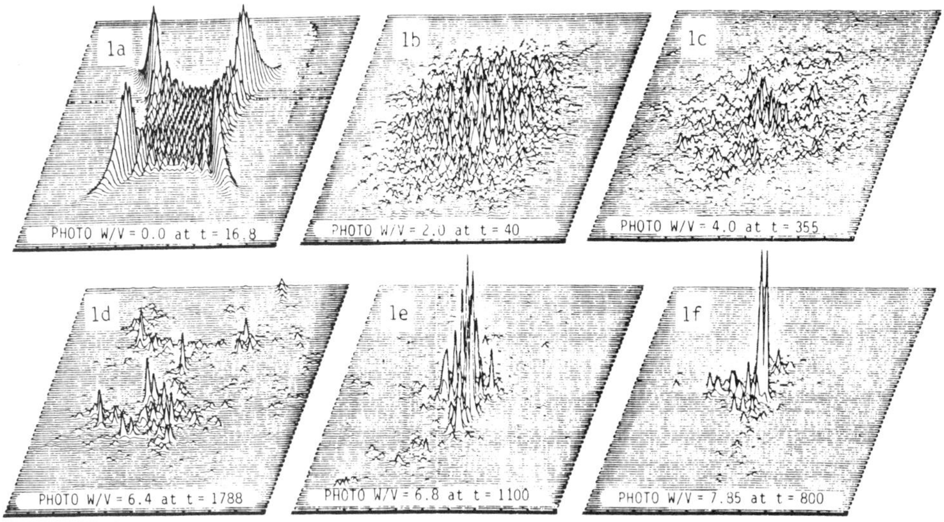

Fig. 1. These computer photos show electronic probability at particular stages of diffusion for various degrees of disorder. The initial wavepacket of the electron is localized in both space and energy--at time t=0, roughly 90% of the probability is contained on the 100 lattice sites at the center of the lattice; the electron's energy is at the center of the energy band, with a resolution equal to 1% of the bandwidth.

The Hamiltonian describing the system is (Anderson model for static disorder--non-interacting electrons):
$H = W\sum_i \varepsilon_i \, |i\rangle\langle i| + V\sum_{\{i,j:\text{ nearest neighbors}\}} |i\rangle\langle j|$, where ε_i are random numbers between 0 and 1.

The ratio W/V measures the degree of disorder: large W/V ⇒ strong disorder implies localization of the electron, small W/V ⇒ weak disorder implies the possibility of electron diffusion in two dimensions.

We observe a transition of <u>some</u> sort at W/V≈6.4. The 101 x 101 site photos show diffusion for the cases W/V=0.0 (free particle), 2.0, 4.0, 6.4, 6.8, and 7.85. (The size of the computational lattice is enlarged to 141 x 141 and 180 x 180 so that the expanding wavepacket never sees any boundaries.) The random numbers E_i are chosen to be identical in all of these cases. What appears facinating is that such a small decrease in disorder can give rise to such a marked difference in diffusive behavior. On the side of the transition where diffusion occurs, we observed a minimum metallic conductivity $\sigma=0.12\ e^2/\hbar$; on the insulating side of the transition, $\sigma=0$.

To use Eq. (2) for calculating diffusivity requires evaluation of $\langle \underline{r}_i | q(H) e^{-iHt/\hbar} | \underline{r}_o \rangle$. We may write this quantity in the form $\int_{-\infty}^{\infty} dt' \, Q(t') \langle \underline{r}_i | e^{-iH(t-t')/\hbar} | \underline{r}_o \rangle$, where $Q(t')$ is the Fourier transform of $q(E)$, $Q(t') = \frac{1}{2\pi\hbar} \int_{-\infty}^{\infty} dE \, q(E) \, e^{-iEt'/\hbar}$ and $\langle \underline{r}_i | e^{-iH(t-t')/\hbar} | \underline{r}_o \rangle$ is the ordinary propagator. We then simply watch the evolution of the electronic wavepacket $\tilde{K}(\underline{r}_i, \underline{r}_o; t)$ in time by iteratively solving the time-independent Schrodinger equation. A similarly motivated approach was used by Prelovsek [6]. Our observations differ from his, however, because we formed the initial wavepacket to reduce transient diffusive behavior, used longer times, examined the same random number configuration at several degrees of disorder, and had the benefit of revealing probability photographs at various stages of the diffusion of the electron. Figures 1 and 2 present a sampling of our numerical results. The computer generated photographs in Fig. 1a-1f are snapshots of electronic probability at various times and degrees of disorder for the same configuration of random numbers. The $L^2(t)$ curves in Fig. 2 involve two different random number configurations.

DISCUSSION

Our numerical data are useful in answering the question, "What does electronic motion look like in a static disordered lattice?" The purpose of an analytic theory is to explain why electronic motion looks like it does. A theory will be totally satisfactory only after explaining the various particular wiggles we see in the $L^2(t)$ curves and wavefunction photographs, as well as any general trends or averaged behavior that may be discernable. If the theory ends up predicting a sharp transition, a universal curve or number, or something else aesthetically pleasing, that would be nice. But the desire to see such a prediction must be subordinated to being able to explain, in as sharp detail as possible, the actual solution of the Schrodinger equation. Specifically, the theory must:

-predict the observed transition at $W/V \approx 6.5$ from strongly localized to at least moderate percolative behavior. (It is not so important to say whether the transition is to really extended states--we haven't observed a spatially infinite sample, we haven't observed infinitely extended states. We have observed a transition of some sort in the finite time and effectively finite space of our calculation samples.)

-explain why the transition is so abrupt. Why does such a small decrease in disorder produce such a large increase in localization length (or possibly, states delocalized altogether)?

-explain why the long-sustained linear slope in $L_X^2(t)$ near the transition is the observed value of approximately 0.33.

-explain why $L^2(t)$ ceases linear behavior and flattens out at the observed value of t, dependent on W/V (W/V>6.5).

-explain why diffusive behavior is so different in two dimensions and one dimension. Why is there never any $L^2(t) \sim t$ behavior in one dimension, no matter how weak the disorder, while in two dimensions, very stable linearity in $L^2(t)$ is observed? [Again, since we have watched the electron for only a finite amount of time, we do not know whether such linear behavior will persist

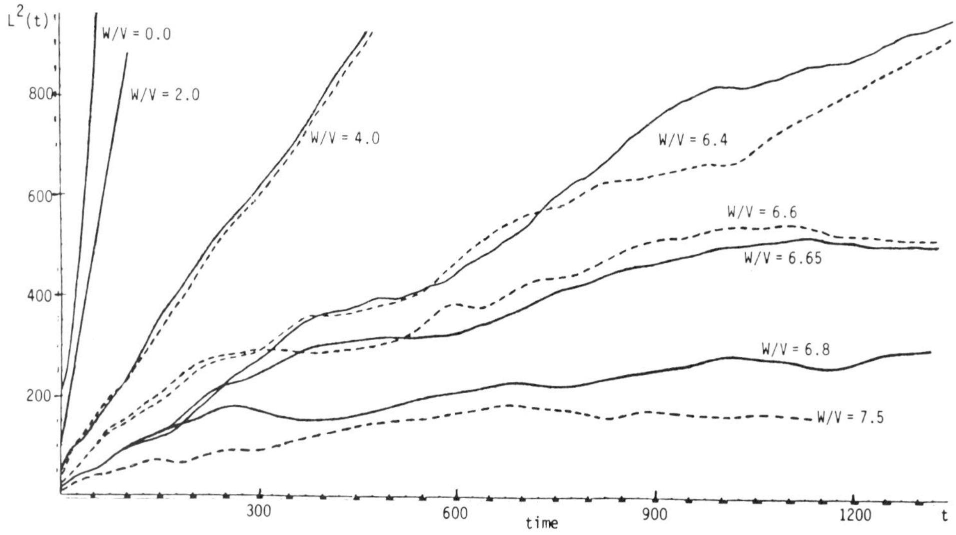

Fig. 2. $L^2(t)$ for different values W/V of disorder. The solid curves result for one configuration of random numbers, the dashed curves for a second configuration. For weak disorder, the statistical variation in $L^2(t)$ is seen by fluctuations in a given $L^2(t)$ curve, which is just about the same as the difference in $L^2(t)$ between configurations. At W/V=4.0 there is no dropoff in the slope of $L^2(t)$ over the observed time period, contrary to the 30% or greater dropoff predicted by scaling theory [7]. At W/V=6.4 we see a transition; a percolating web just barely is capable of forming (see Fig. 1d). We have carried out the calculation to t=2200 for W/V=6.4 and and $L^2(t=2200)=1325$, diffusion continuing far beyond the localization length L^2_{max} = 729 claimed by MacKinnon and Kramer [8]. At W/V = 6.6, 6.65, 6.8, and 7.5, $L^2(t)$ flattens out, localization occuring when the time is around 300 to 600 near the transition, and earlier for the greater disorder.

indefinitely. But why is linearity (i.e. the behavior at finite times is indistinguishable from linear) present in two dimensions and totally absent in one dimension?]

-explain why the diffusivity, $D = dL_x^2(t)/dt$ (in the linear $L^2(t)$ region), is greater away from the band center than it is at the band center for weak disorder in two dimensions. Comparing the diffusion of a delta-function (composed of states throughout the energy band) with that of a wavepacket resolved in energy to within 1% of the band center, we find that at W/V=2.0 our calculations yield D (delta-function)=7.5, $D(E_F=0.0)=4.0$. For greater disorder the opposite is true--at W/V=6.4 the $L^2(t)$ at the band center is linear, while the delta-function $L^2(t)$ steadily declines in slope.

While these questions are asked with respect to the Anderson model, we expect by the nature of the ingredients which go into our answers, to be able to discern which numerical observations are specific to the Anderson model and which would arise more generally for other disordered $V(\underline{r})$.

We structure an argument as follows: Assume the diffusivity $D(E_F,T)$ is a certain non-zero number $\overline{D}$. Are the two relevant equations, $\lim_{t\to\infty} L^2(t)/2t=\overline{D}$ and $H\Psi(\underline{r},t)=i\hbar\, d\Psi(\underline{r},t)/dt$ incompatible? If they are, we know D cannot take on the value we gave it. Repeat the question, giving D a different non-zero value. Thus we determine which values of D are possible and which are not. A necessary condition for D to be non-zero does not serve as a sufficient condition for showing that D is non-zero. However, we will see that studying whether the necessary condition is satisfied is sufficient to explain all of the aspects of the electronic motion during the finite time of our observation.

Anderson localization involves solution of the one-electron Schrodinger equation for a complicated quantum-well potential. So it makes sense in creating a theory to look at a simple quantum well problem and ask the same questions about it that we wish to ask about the more complicated problem. The simple potential we choose is a slightly asymmetric double well, as sketched in Fig. 3. Consider the propagator $\tilde{K}(x_i,x_o;t)$, with x_i and x_o in different wells: $\tilde{K}(x_i,x_o;t) = \sum_n q(E_n)\Psi_n(x_i)\Psi_n(x_o)e^{-iE_nt/\hbar}$. Choose $q(E_n)$ so that only states of energy E_1 and E_2 are included in $\tilde{K}(x_i,x_o;t)$. Then $\tilde{K}(x_i,x_o;t) = \Psi_1(x_i)\Psi_1(x_o)e^{-iE_1t/\hbar} + \Psi_2(x_i)\Psi_2(x_o)e^{-iE_2t/\hbar}$. Now, what we would like

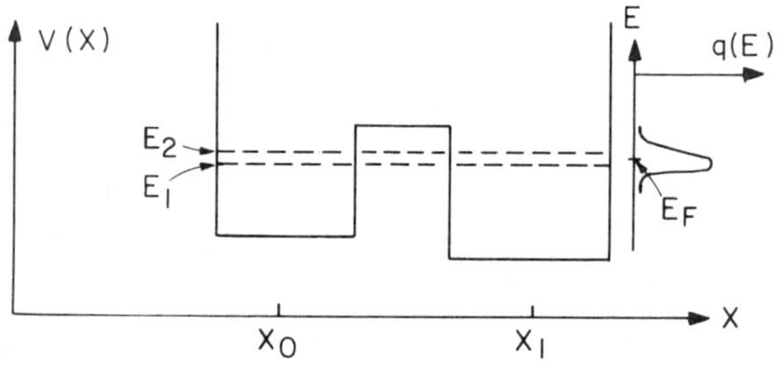

Figure 3. Asymmetric double-well potential used as a simple model for the analysis of Anderson localization (see text).

to find out is whether states $\Psi_1(x)$ and $\Psi_2(x)$ have significant magnitude of probability in both wells. If we know $V(x)$ (as we easily do here), then this question can be answered straightforwardly by determining E_1, E_2, $\Psi_1(x)$, and $\Psi_2(x)$. But since we are making an analogy with the case we are really interested in, i.e. $V(r)$ for a random potential in two (or three--or one) dimensions, where it is not trivial to find E_1, E_2, etc.. let us try to analyze the double-well problem with only the following two items of information:

(1) The difference in energy between the two eigenstates is known to be ΔE.

(2) The evolution of the propagator $\tilde{K}(x_i, x_o; t)$ in time is known out to $t = \hbar\pi/\Delta E$.

Then if $|\tilde{K}(x_i, x_o; t)|^2$ over this time period (from $t=0$ to $t=\hbar\pi/\Delta E$) remains less than or equal to some maximum value, say K_{max}, we can conclude that $|\tilde{K}(x_i, x_o; t)|^2$ will never become greater than K_{max} for all time. (Between $t=0$ and $t=\hbar\pi/\Delta E$, all possible phase differences of $e^{-iE_1 t/\hbar}$ and $e^{-iE_2 t/\hbar}$ are seen.)

In the more complicated case of interest, the random potential, we are not given piece of information (1) in terms of a precise, single value ΔE, as in the double-well problem, but rather are given the average ΔE between states constituting the propagator. If we are to talk about the energy spacing between states, however, it is necessary to define a specific system having a certain number of sites, thus a certain number of states, and hence an identifiable ΔE_{ave}. How do we decide where to place these boundaries? We could place them anywhere, as long as they have not been reached by the electron during the time t that we have been watching the evolution of the propagator. While this is intuitive, it is imprecise in that some very small amount of probability is on all sites at all times. Note however, that a placement of boundaries to include only sites covered (at times $\leq t$) with probability greater than some small finite value captures the physical qualitative differences observed in our numerical photographs--at low disorder ($W/V = 2.0$) all sites out to a radius $\approx (L^2(t))^{1/2}$ are covered; as disorder is increased to $W/V = 6.4$, the webbing of probability covered sites gets thinner and thinner.

Thinking of boundaries in this way, we fundamentally differ from pre-deciding on a square of some length L, as Thouless does in defining his ratio $\delta E(L)/\Delta E(L)$. The only place to put the boundary that is not arbitrary, and makes maximum use of the information available, is around only those sites covered by electronic probability prior to time t. This may be a very irregular-looking boundary, possibly with interior gaps. (Consider placing such a boundary in the case of $W/V = 6.4$--see photos in Fig. 1.) It will be interesting in future calculations to store in computer memory the time t, and magnitude $|\tilde{K}(\underline{r}_i, \underline{r}_o; t)|^2$ for which the probability acquires its maximum value at each site r_i. Such a distribution of times and magnitudes should lend analytical precision to these ideas.

The main difference between classical and quantum percolation is the following: Classically, if a particle has an energy just barely above the percolation threshold, it randomly bumps along through the cluster of available sites, often wandering into dead ends before finally managing to find its way out. The particle can take arbitrarily long to do this, hence the conductivity at the threshold can become arbitrarily small.

In the quantum case, however, the possibility of a discrete energy spectrum puts a stopwatch on the electron. If the electron is diffusing too slowly, $\hbar/t$ becomes narrower than ΔE_{ave}, diffusion ceases (percolation is choked off) and $\sigma=0$. Thus the quantum D and σ cannot take on arbitrarily small values in two dimensions. Counting the number of sites N_{sites} covered by probability out to time t in our calculations near W/V = 6.4, we find that $\Delta E_{ave}/(\hbar/t) = (1/N_{sites})\times[1/g(E_F)]\times(t/\hbar)\approx 1$.

In one dimension, Thouless and Kirkpatrick [9] demonstrated a certain scaled curve of conductivity as a function of time (<u>not</u> of spatial length L) applicable over a wide range of W/V. The curve gives $\lim_{t\to\infty} \sigma(t)\to 0$. This can be understood in terms of our condition $\Delta E_{ave} t/\hbar < \pi$. Since linear diffusion in one dimension would imply an effective sample length $[L^2(t)]^{1/2}\approx(Dt)^{1/2}$, we see that no matter what D is, $\Delta E_{ave} t/\hbar \sim (Dt)^{-1/2} t/\hbar > \pi$, thus contradicting the possibility of diffusion. In three dimensions, the condition is $\Delta E_{ave} t/\hbar \sim [Dt+C(t)]^{-3/2} t/\hbar < \pi$, where C(t) could be any function growing less than linear in time. Then there are two possibilities. If $C(t)\sim t^n$, $n<2/3$, a $\sigma_{min} \neq 0$ could exist, whereas if $2/3 \leq n < 1$, the conductivity could take on arbitrarily small values.

The condition for diffusion involves making a connection between boundary conditions and the average energy spacing ΔE_{ave}. This initially was pursued by Thouless in studying the ratio $\delta E(L)/\Delta E(L)$. But there are somewhat subtle differences between $\delta E(L)/\Delta E(L)$ and our analogous ratio $(\Delta E_{ave} t/\hbar)^{-1}$. (Indeed there are subtle differences between Thouless's own two explanations [10] of precisely what is meant by $\delta E(L)/\Delta E(L)$.) The average energy spacing $\Delta E(L)$ for Thouless is the level spacing in a square block of length L. For us, ΔE_{ave} is the level spacing in a system enclosing the wavepacket up to time t. $\delta E(L)$ for Thouless is the mean change in energy of a level due to a change from periodic to anti-periodic boundary conditions. Our analogous quantity is $\hbar/t$, which determines the phase variation $\exp(-iE_n t/\hbar)$ among terms $\Psi_n(\underline{r}_i)\Psi_n^*(\underline{r}_o)$ in the propagator.

We wish to put ourselves in the shoes of the electron, and figure out why we get stopped so much more effectively from diffusion at W/V= 6.6 than at W/V = 6.4. As the electron, we certainly have no awareness of any periodic or anti-periodic boundary conditions. Such mathematical conveniences cannot answer <u>why</u> we get stopped differently as the disorder changes. The only information that is relevant to our motion between time=0 and time=t is contained in $V(\underline{r})$ at those sites we have made it out to by time t.

We see that an important quantity is the ratio between the energy resolution of a level and the energy spacing between levels. This is not, however, the conductivity or the conductance. As we see in our photographs, this ratio is not sufficient to predict any sort of universal behavior--for each value of W/V (or for the different parameters describing a different model of disorder) the appearance of the wavepacket is distinctly and interestingly different. Also, wavepackets located in energy about various positions E_F in the spectrum vary in their diffusive motion.

The problem one can recognize in a local approach such as ours is that we cannot predict what happens after the time t of our observation of the wavepacket for weak disorder. We have discovered one significant ratio, which explains what is seen around $W/V = 6.4$. What if some other phenomenon, of a totally different nature, should come into importance, say at $W/V = 1.0$? Maybe at very large times and sample sizes, the apparently linear $L^2(t)$ curves would change for a reason different than that discussed near $W/V = 6.4$. The only way to rule out this possibility with complete assurance would be to assume $D=0$ at weak disorder, and find some incompatibility with the Schrodinger equation.

While we cannot be completely certain about diffusion continuing in the limit of infinite time, if we assume that it does we can say something about the eigenstates of the disordered system. Kaveh and Mott have shown that linear diffusion of a Gaussian wavepacket means the states fall off as $\Psi(\underline{r}) \sim 1/r$ in two dimensions and $\Psi(\underline{r}) \sim 1/r^2$ in three dimensions [11]. Their conclusion is based on examining the diffusion in k-space of an electron initially localized in momentum for weak disorder. This behavior has been called "power law localization", but such use of the word "localization" can be misleading since power law decay of the eigenstates does not rule out linear diffusion unless the drop off is steeper than $1/r^2$ in two dimensions or $1/r^{5/2}$ in three dimensions. To see this, we define the center or the wavefunction $\Psi_n(\underline{r})$ as $\underline{r}_{cen} = \langle n|\hat{\underline{r}}|n\rangle$ and express the squared width of the wavefunction as $w_n^2 = \langle n|(\hat{\underline{r}} - \underline{r}_{cen})|n\rangle = \langle n|\hat{\underline{r}}^2|n\rangle - |\underline{r}_{cen}|^2 = \sum_{n'=n} |\langle n|\hat{\underline{r}}|n'\rangle|^2$.

If the expectation value of w_n^2 is finite for states at the Fermi level, then the dc conductivity and $D(E_F,T)$ vanish [12]. Therefore, to have a non-zero D requires $w_n^2 = \int d\underline{r}\, |\Psi_n(\underline{r})|^2 r^2$ not be finite. It is also worth pointing out that eigenstates for the free particle, which has infinite diffusivity, can be chosen of the form $e^{ikr}/r^{1/2}$ (in two dimensions) and e^{ikr}/r (in three dimensions). We summarize these observations in the table below:

	One Dimension	Two Dimensions	Three Dimensions
CRYSTAL V(r) is periodic	$L^2(t) \sim t^2$ $\sigma = \infty$ $\Psi(x) \sim$ constant	$L^2(t) \sim t^2$ $\sigma = \infty$ $\Psi(\underline{r}) \sim 1/r^{1/2}$	$L^2(t) \sim t$ $\sigma = \infty$ $\Psi(\underline{r}) \sim 1/r$
WEAK DISORDER		$L^2(t) \sim t$ $\sigma \neq 0$ $\Psi(\underline{r}) \sim 1/r$	$L^2(t) \sim t$ $\sigma \neq 0$ $\Psi(\underline{r}) \sim 1/r^2$
STRONG DISORDER	$L^2(t) \sim$ const $\sigma = 0$ $\Psi(x) \sim e^{-\alpha x}$	$L^2(t) \sim$ const $\sigma = 0$ $\Psi(\underline{r}) \sim e^{-\alpha r}$	$L^2(t) \sim$ const $\sigma = 0$ $\Psi(\underline{r}) \sim e^{-\alpha r}$

Scaling theory asserts that for a two-dimensional time-independent disordered system, in the limits of low disorder and $L \to \infty$, Ohm's Law applies and σ approaches a constant. We do not see the validity of this assertion. If the disorder $\to 0$ limit is done before the $L \to \infty$. limit, the assumption is not true, since $\sigma \sim L^2(t)/t \sim t^2/t \to \infty$. [$\beta(L) = d \ln \sigma / d \ln L$ is therefore positive.] And if the reverse is done, the $L \to \infty$ limit is taken before the disorder $\to 0$ limit, the assumption is also faulty since in this case L is no longer a parameter which can be scaled from L to 2L to 4L, and so forth. (Electron-phonon interactions, which produce finite conductivity in crystals, also serve to inhibit zero conductivity in disordered systems.)

CONCLUSIONS

We have shown that a sharp transition exists in disordered two-dimensional systems near a critical value, W/V = 6.4. We analyzed the implications for current theories of localization and conclude that the conflict between our results and the deductions from scaling-theory arguments is due to a faulty assumption in the latter.

ACKNOWLEDGEMENTS

We are very grateful to Michel Baranger for his inspiration and help. This paper is an outgrowth of a Ph.D. thesis submitted by one of us (GMS) to Harvard University. The author appreciates the input he received from B. I. Halperin, P. A. Lee, and William Paul, who provided friendly and instructive reviews of the research while serving on his Doctoral Thesis Committee. Both of us have benefitted enormously from both personal contact and insightful correspondence with Nevill Mott.

REFERENCES

1. See, for example, the proceedings of the Bar-Ilan Conference, Phil. Mag. B 50, No. 2 (1984).
2. P. W. Anderson, Phys. Rev. 109, 1492 (1958).
3. N. F. Mott, Phil. Mag. B 43, 941 (1981).
4. See, for example, E. Abrahams, P. W. Anderson, D. C. Licciardello, and P. V. Ramakrishnan, Phys. Rev. Lett. 42, 693 (1979).
5. G. M. Scher, Ph.D. Thesis, Harvard University, 1983 (unpublished); see also G. M. Scher, J. Non-Cryst. Solids 59-60, 33 (1983).
6. P. Prelovsek, Phys. Rev. B 18, 3657 (1978).
7. P. A. Lee, Numerical Studies of Localization in Two Dimensions, in: Proc. of the 4th Taniguchi International Symposium, ed. by Y. Nagaoka and H. Fukuyama (Springer-Verlag, Berlin, 1982), p. 62.
8. A. MacKinnon and B. Kramer, Phys. Rev. Lett. 47, 1546 (1981); Phys. Rev. Lett. 49, 695 (1982).
9. D. J. Thouless and S. Kirkpatrick, J. Phys. C 14, 235 (1981).
10. D. J. Thouless, The Anderson Model for Disordered Solids, in: Les Houches Lectures on Ill Condensed Matter, ed. by R. Balian, R. Maynard and G. Toulouse (North Holland Publishing Company, Amsterdam, 1979) p. 61; J. Non-Cryst. Solids 35-36, 3 (1980).
11. M. Kaveh and N. F. Mott, J. Phys. C 14, L177 (1981).
12. B. I. Halperin, Phys. Fennica 8, 215 (1973).

FLUCTUATION KINETICS AND THE MOTT HOPPING

M. Ya. Azbel'

School of Physics and Astronomy, Raymond and Beverly Sackler Faculty of Exact Sciences, Tel-Aviv University
Ramat Aviv, Tel Aviv 69978, Israel

ABSTRACT

This paper discusses the properties of systems whose transport is dominated by a single quenched fluctuation. The Lee oscillations in the Mott hopping are specifically considered.

FLUCTUATION KINETICS

During the last years there was an extensive experimental and theoretical study of transport in small low-dimensional (low-D) systems at low temperatures. Their most remarkable features were the supersensitivity of resistivity to the Fermi energy and the instability to voltages ∿1mV in MOSFET channels[2] and to currents ∿100 nA in extra-thin (300Å x 10Å x 10Å) Nb wires.[3]

All such phenomena are presumably related to "fluctuation kinetics", when transport is dominated by a single quenched fluctuation ("flucton"). This paper demonstrates that fluctuation kinetics is characteristic of systems where eigenstates are strongly localized. This occurs[4] in 1D, 2D and sufficiently dirty 3D cases. The paper discusses examples and some manifestations of fluctuation kinetics in such systems. In more detail it considers the Lee oscillations in the Mott hopping conductivity.

First I demonstrate that strong localization may lead to fluctuation kinetics. Then I review strong narrow resonances and oscillations of 1D conductivity with the Fermi energy ε_F (experimentally ε_F in FET channels is tuned by the gate voltage), and discuss current instabilities, long temperature relaxation time and other manifestations of fluctuation kinetics in small systems. Section 2 describes a microscopic picture of the Mott hopping and the Lee oscillations in the Mott hopping conductivity.

Start with zero temperature T=0, then strong localization yields[4] the exponentially large resistance R and exponentially small conductance G:

$$R \propto \exp(L/L_o); \quad G \propto \exp(-L/L_o). \tag{1.1}$$

Here L is the length of a system and L_o is the localization length, which depends on the impurity concentration per electron c and the Fermi energy ε_F.

The very relation (1.1) implies that the role of fluctuations is enormously blown up. Indeed, the relative mean quadratic fluctuation of c is $\alpha\ 1/\sqrt{LSNc}$, where S is the sample cross-section area and n is the electron density. When $S<L/L_o^2nc$, then even the mean c fluctuation leads to large fluctuations in resistance (ΔR) and conductance (ΔG): $\Delta R >> R$, $\Delta G >> G$. Larger fluctuations are exponentially improbable, but they yield $\Delta L_o \sim L_o$ and thus an exponential increase in R and G. They dominate[5] the average R_{av} and G_{av}. Since exponentially rare fluctuations, and thus R_{av}, G_{av}, cannot be observed in practice, typical (representative) R_r and G_r do not equal R_{av} and G_{av}, as they do in a conventional case. Rather, R_r and G_r are[4]

$$R_r \sim \exp\ (\ell nR)_{av};\quad G_r \sim \exp\ (\ell nG)_{av};\quad (\ell nR)_{av} = L(1/L_o)_{av}. \tag{1.2}$$

Here $(1/L_o)_{av}$ is a "well-behaved" quantity, whose average and representative values are the same.

However, even a very low temperature T radically changes the situation.[6] Then the contribution into G comes from all energies in the interval T, i.e. G is averaged over the energies in this interval. As a result, it may be dominated by the single highest conductance localized eigenstate ("flucton"). When the Fermi energy changes, the flucton changes too, and conductance exhibits exponentially large fluctuations.[6-8] Even its temperature dependence yields giant fluctuations.[7]

Another important implication of finite temperature is inelastic scattering. A phenomenological-type approach[9] reduces it to a kind of "uncertainty principle" for strong conductivity peaks:

$$t_r \sim \delta\varepsilon_e/\delta\varepsilon_r, \tag{1.3}$$

$$\delta\varepsilon_e \sim t_{nr}^{1/2}\Delta\varepsilon_o. \tag{1.4}$$

Here t_r and t_{nr} are the "resonant" and "non-resonant" transmission coefficients; $t \sim G\pi h/e^2$; and $\delta\varepsilon_r$ and $\delta\varepsilon_e$ are the resonant and eigenstate total (inelastic included) width; $\Delta\varepsilon_o$ is the interlevel spacing in a given localization well (whose size is the wave function localization length $2L_o$). When $T > \delta\varepsilon_e$, then the resonance width is $\delta\varepsilon_r \sim T$. In a small sample 1D non-resonant G_{nr} is described (see Section 2) by the Mott formula[11,29] $G_{nr} \propto \exp[-(T_1/T)^{1/2}]$, where T_1 is the Mott temperature. So, Eqs. (1.3, 1.4) yield:

$$G_r \propto t_r \propto \exp\ [-(T^*/T)^{1/2}], \tag{1.5}$$

$$(T^*/T_1)^{1/2} \underset{\sim}{\sim} 0.5. \tag{1.6}$$

Resonances exponentially die out when $\delta\varepsilon_e$ exceeds the interlevel (in the whole system) spacing $\Delta\varepsilon$. By Eq. (1.4), this happens at a temperature when

$$\Delta\varepsilon/\Delta\varepsilon_o \sim t_{nr}^{1/2} \sim \exp\ [-(T^*/T)^{1/2}]. \tag{1.7}$$

In a multichannel case Eqs. (1.6, 1.7) may be presented in the following form. Introduce $\alpha = (T^*/T_1)^{1/2}$, where T^* and T_1 are the Mott temperatures for adjacent G maximum and minimum. Then α's are randomly scattered and, by Eq. (1.6), $\alpha_{av} = 0.5$ [with relative fluctuations, as usual, $\propto$ (total number of strong resonances)$^{-1/2}$]. Temperature T_v, where a resonance vanishes, yields, by Eq. (1.7),

$$\Gamma = (T^*/T_v)^{1/2} = \ell n\ (L/2L_o), \tag{1.8}$$

where in 1D (cf. Section 2)

$$L_o \sim \begin{cases} hv/T_1, \text{ if } k_F d/2<1, \\ \dfrac{4\pi\hbar^2}{dm^* T_1}, \text{ if } k_F d/2>1, \end{cases} \tag{1.9}$$

k_F is the Fermi wave vector; d is the sample width; the thickness yields one channel; 1D assumes $d<L_o$. Inelastic scattering becomes unimportant when[9,10] $T<T_e \sim T^*(L_o/L)^2$. A rough estimate for T_1, when ε_F is above the scatter potential barrier, is[21] $T_1 \sim (2\pi\hbar^2/dm^*\lambda_s)\exp(-\pi k_F a_i)$, where λ_s is the interscatterer distance, and a_i is the effective size of a scatterer. Relations for α, α_{av}, Γ, T_1, agree with experiments[1,12]; temperature T_e has not yet been reached experimentally.

Equations (1.5-1.7) agree with experiments[1,12], but do not elucidate physics related to resonant scattering. This new physics was presented by P. Lee[13]. He argued that at sufficiently high temperatures, tunneling resonances are replaced by oscillations in the Mott hopping conductivity. The resistances of consequent non-coherent Mott hops add up and yield the average resistance R_{av} (rather than the average conductance, as resonant tunneling). In small systems the main contribution into R_{av} comes from the "bottle-neck" flucton with the largest Mott hop. This flucton depends on and changes with the Fermi energy ε_F, thus leading to the Lee oscillations. Their specific feature is the temperature dependence of the maxima and minima energies (while resonance tunneling energies are T-independent) - see Section 2. These energies shift by $\sim (T_1 T)^{\frac{1}{2}}$.

A conventional theory of the Mott hopping assumes that the probability of the best hops is current-independent (but see refs. 13, 18). A current I implies that during the Mott 1D hopping time $\tau \sim (L_o/v)\exp[(T_1/T)^{\frac{1}{2}}]$ (where L_o/v is a "single attempt time" for the Fermi velocity v, and the exponent accounts for the probability of a hop) $I\tau$ electrons pass through every cross-section. If $I\tau > 1$, i.e. if

$$I > (v/L_o)\exp[-(T_1/T)^{\frac{1}{2}}], \tag{1.10}$$

then the "best" localized states are "overcrowded". So, from time to time an electron has to switch to the "second best" choice of a hop with exponentially higher hopping time and resistance. Then a current instability develops. If, e.g., $v \sim 10^6$ cm/sec, $L_o \sim 1000$ Å, this happens at $I > 20\exp[-(T_1/T)^{\frac{1}{2}}]$(nA).

Large τ may also make it almost impossible to reach a very low equilibrium temperature in a reasonable time. Suppose the number of localized states is N. Then the equilibrium thermal population of a given state with the energy ε_n is $N\rho^o(\varepsilon_n)$ where ρ^o is the Fermi function. If N is relatively small, then, say, $N\rho^o(\varepsilon_n) = 33/77$, and thermal equilibrium corresponds to the time average over the time $t \gg 77\tau$, which may be macroscopically large. Cooling implies the change in the thermal population of all states (not too far from ε_F), which may include simultaneous hops. The corresponding relaxation time may be extraordinarily large.

Current instabilities and temperature relaxation time in small systems call for a detailed theory. In general, fluctuation kinetics is far from being completed. The problems include the quantitative theory of time development in resonant tunneling (simple models were studied in ref. 14), microscopic theory of resonant tunneling and Lee oscillations at finite temperatures (for first attempts see refs. 13, 15), and accounting for electron-electron interaction.

Fluctuation kinetics may be expected in every case of localization. These include conductivity oscillations[1,9,12,13,16] current instabilities,[2,3] 1/f noise,[17,2] "junction" resonances,[18a] electromagnetic waves,[21,22] electric tunneling and breakdown,[19] thermal conductivity,[20] acoustic waves,[21] shallow hydrodynamic waves on a random bottom,[23] quantized Hall effect breakdown[24] and possibly irregular oscillations in magnetic field.[24a] The promising study of all these phenomena is just beginning.

MOTT HOPPING AND LEE OSCILLATIONS

The main goal of this Section is the analytical theory of the Lee oscillations, which were suggested and numerically studied in ref. 13. First I derive a general formula for the Mott hopping conductivity, then I present a "conventional" Mott formula, explore the role of "bottle-neck" hops, and finally determine the characteristics of the Lee oscillations.

Consider the Mott hopping between the n-th and (n+1)-st localized states. The resulting current j_n is

$$j_n = \rho_n (1-\rho_{n+1}) w_{n,n+1} - \rho_{n+1} (1-\rho_n) w_{n+1,n}, \tag{2.1}$$

where ρ_n is the distribution function and $w_{n,n+1}$ is the probability of a hop from n to (n+1). In the equilibrium $\rho_n = \rho_n^o$; ρ_n^o is the Fermi function, $j_n = 0$, so, in agreement with the detailed balance principle,

$$\frac{w_{n,n+1}}{w_{n+1,n}} = \exp\left(- \frac{\varepsilon_{n+1}-\varepsilon_n}{T} \right), \tag{2.2}$$

ε_n being the energy of the n-th state.

When, e.g. $\varepsilon_{n+1} - \varepsilon_n >> T$, then $w_{n,n+1} = w_{n+1,n} \exp[-(\varepsilon_{n+1} - \varepsilon_n)/T]$, and $w_{n+1,n}$ is the temperature-independent overlap of probability densities. If the space distance between the states is Δr_n, then

$$w_{n+1,n} \sim \exp(-\Delta r_n/L_o), \tag{2.2a}$$

Substitute into Eq. (2.1)

$$\rho_n = \rho_n^o + \rho_n^o (1-\rho_n^o) \xi_n. \tag{2.3}$$

Then the linear (over ξ_n) approximation yields

$$j_n = \tilde{w}_n (\xi_n - \xi_{n+1}), \tag{2.4}$$

$$\tilde{w}_n = (1-\rho_n^o) \rho_{n+1}^o w_{n+1,n} = (1-\rho_{n+1}^o) \rho_n^o w_{n,n+1} \sim \exp(-\Delta r_n/L_o) w_n^*. \tag{2.5}$$

When, e.g., $\varepsilon_{n+1} > \varepsilon_n$, then (cf. ref. 25)

$$w_n^* \sim \begin{cases} \exp(-|\varepsilon_{n+1} - \varepsilon_n|/T), & \text{if } \varepsilon_{n+1} > \varepsilon_F > \varepsilon_n \\ \exp(-|\varepsilon_{n+1} - \varepsilon_F|/T), & \text{if } \varepsilon_n > \varepsilon_F \\ \exp(-|\varepsilon_F - \varepsilon_n|/T), & \text{if } \varepsilon_F > \varepsilon_{n+1}, \end{cases} \tag{2.6}$$

i.e. only the states in the vicinity of ε_F are important.

The Mott hopping typically proceeds in a unique way, without "branching". Then, as in a 1D case, the current is conserved, $j_n = j$, and, by Eq. (2.4),

$$\xi_{n+1} = - j \sum^{n} \tilde{w}_m^{-1}. \tag{2.7}$$

The ratio of ξ's at the outgoing and incoming ends determines the transmission coefficient and hence the Landauer-type resistance:[26]

$$R \propto (\pi\hbar/e^2) \sum \tilde{w}_m^{-1}, \tag{2.8}$$

i.e. indeed R is the sum of the resistances $\tilde{w}_m^{-1}$ of consequent hops.[27] The number of terms in Eq. (2.8) is proportional to the sample length $L/\Lambda L_o$. The number of different parallel channels is proportional to the cross-section area $S/\Lambda^2 L_o^2$. In a conventional Mott case L, S are large, and

$$R \sim (\pi\hbar/e^2)(L_o \Lambda L/S)(\tilde{w}_m^{-1})_{av}. \tag{2.9}$$

To determine the average $\tilde{w}_m^{-1}$ by Eqs. (2.5, 2.6) it is convenient to plot localized eigenstates in the dimensionless coordinates ε/T, $\vec{r}/L_o$. Each state corresponds to $\Delta\varepsilon/T \sim 1/V\rho_D T$ (ρ_D is the D-dimensional density of states per unit energy and volume) and is randomly situated at $\vec{r}/L_o$. So, a (D+1)-dimensional volume per state is $\sim (1/V\rho_D T)(V/L_o^D) \sim 1/L_o^D \rho_D$, and an average dimensionless distance Λ between randomly located states is

$$\Lambda \sim (L_o^D \rho_D T)^{-\frac{1}{D+1}} \sim (T_D/T)^{\frac{1}{D+1}}. \tag{2.10}$$

(If L_o strongly depends on ε, this estimate is easily amended - cf. ref. 7). By Eqs. (2.5) and (2.6), one obtains the Mott formula:[11]

$$R \sim (\pi\hbar/e^2)(L_o L \Lambda/S)(\tilde{w}_m^{-1})_{av}; \quad (\tilde{w}_m^{-1})_{av} \sim \exp(-\Lambda) \sim \exp[-(T_D/T)^{\frac{1}{D+1}}], \tag{2.11}$$

where T_D is the Mott temperature.

Equation (2.11) assumes that in Eq. (2.8) there exists a continuous system of hops which does not include an anomalously high resistance hop. This assumption may be invalid. Consider, for instance, 1D (one- or poly-channel) sample $L \times d \times \tilde{d}$, where

$$L/L_o \gg \Lambda > d/L_o,\ \tilde{d}/L_o. \tag{2.12}$$

(Note that this condition of one dimensionality depends on temperature and is always valid at sufficiently low temperatures). An anomalously large hopping distance $\Lambda^* > \Lambda$ corresponds to a $\Lambda^* \times \Lambda^*$ void in the vicinity of ε_F/T somewhere at L/L_o. The average number of such voids is

$$N_v \sim (L\Lambda^*/\Lambda^2 L_o) \exp(-a\Lambda^{*2}/\Lambda^2); \quad a \sim 1 \tag{2.13}$$

It is $\gtrsim 1$, if

$$\Lambda^* < \Lambda_m = \Lambda \sqrt{\ln(L/\Lambda L_o)/a}. \tag{2.14}$$

The contribution of these voids into the resistance is R_v,

$$(e^2/\pi\hbar) R_v \sim (L\Lambda^*/\Lambda^2 L_o) \exp(-a\Lambda^{*2}/\Lambda^2) \exp(\Lambda^*). \tag{2.15}$$

It is maximal, when

$$\Lambda^* = \Lambda_o = \Lambda^2/2a. \qquad (2.16)$$

Suppose $\Lambda_o < \Lambda_m$, (and $\Lambda_o < L/L_o$), i.e. L is exponentially large (or the temperature is sufficiently high):

$$L > L^* \equiv \Lambda L_o \exp(\Lambda^2/4a) \sim L_o (T_1/T)^{1/2} \exp(T_1/4aT), \qquad (2.17)$$

where I accounted for a 1D Eq. (2.10). Then Λ_o dominates the resistance, and by Eqs. (2.15) and (2.16),

$$(e^2/\pi\hbar)R \sim (L/2aL_o) \exp(\Lambda^2/4a) \sim (L/L_o) \exp(T_1/4aT), \qquad (2.18)$$

which is the Kurkijarvi result[28]; the condition (2.17) agrees with ref. 29. When $\Lambda_o > \Lambda_m$, i.e. Eq. (2.17) is violated, then R is dominated by a single term with $\Lambda^* \sim \Lambda_m$, and

$$(e^2/\pi\hbar)R \sim \exp(\Lambda_m) \sim \exp\left(\Lambda\sqrt{\ln(L/\Lambda L_o)/a}\right)$$
$$\sim \exp\{(T_1/Ta)^{1/2}\sqrt{\ln[(L/L_o)(T/T_1)^{1/2}]}\} \qquad (2.19)$$

This brings us back to a 1D Eq. (2.11), but with slightly quicker temperature dependence and the renormalized Mott temperature

$$T_1^* \approx (T_1/a) \ln(L/L_o). \qquad (2.20)$$

(cf. ref. 29). So, when $L < L^*$ from Eq. (2.17), resistance is dominated by a single Mott hop[30] ("flucton").

Naturally, this reasoning is easily generalized to a 2D or 3D case, when the sample width d is $>\Lambda L_o$ or when d and thickness $\tilde{d}$ are $>\Lambda L_o$. Then, e.g., at the temperature, when $\tilde{d} < L_o\Lambda$, but $d \sim L_o\Lambda$, dimensionality changes from 2 to 1. Correspondingly, 2D $\ln R \propto T^{-1/3}$ changes to a 1D $\ln R \propto T^{-1}$, (and then to 1D $\ln R \propto T^{-1/2}$).

Equation (2.19) demonstrates the average behavior of a flucton. A specific flucton ($n=\nu$) yields $R \propto w_{\nu}^{-1}$, which, by Eqs. (2.5,2.6), is exponential with T^{-1}. The resistance is, e.g., independent of ε_F if $\varepsilon_{\nu+1} > \varepsilon_F > \varepsilon_\nu$; decreases $\propto \exp[(\varepsilon_{\nu+1} - \varepsilon_F)/T]$ with ε_F if $\varepsilon_{\nu+1} > \varepsilon_\nu > \varepsilon_F$; increases $\propto \exp[(\varepsilon_F - \varepsilon_\nu)/T]$ with ε_F, if $\varepsilon_F > \varepsilon_{\nu+1} > \varepsilon_\nu$. Thus, conductance G exhibits, with ε_F, a plateau, narrow ($\sim T$) exponentially high maxima (when the change in ε_F leads to the change from "G-increasing" flucton to "G-decreasing" flucton), narrow exponentially deep minima (when the change is from "G-decreasing" to "G-increasing" flucton). The energy ε_F, e.g., of a G maximum, corresponding to the switch from $n=\nu$ to $n=\mu$, is $\varepsilon_F = \frac{1}{2}(\varepsilon_\mu + \varepsilon_{\nu+1}) + \frac{1}{2}(T/L_o)(\Delta r_\nu - \Delta r_\mu)$, and on average shifts with T by $\sim T\Lambda_m \sim (T_1/T)^{1/2}[\ln(L/L_o) - \frac{1}{2}\ln(T_1/T)]^{1/2}$. The distance between adjacent minima and maxima is also $\Delta\varepsilon_F \sim T\Lambda_m$.

At very low temperatures $T < T_e$ the Lee oscillations are replaced by resonance tunneling. The resonance ε_F is T-independent.

Thus, one may construe a detailed theory of the Lee oscillations, which accounts only for incoherent inelastic Mott hops, and of the resonant tunneling, which accounts only for coherent elastic scattering. A complete picture still calls for its development. This type of situation may be characteristic of fluctuation kinetics.

REFERENCES

1. A.B. Fowler, A. Hartstein, R.A. Webb, Phys. Rev. Lett. 48:196 (1982).
2. K.S. Ralls, W.J. Skocpol, L.D. Jackel, R.E. Howard, L.A. Fetter, R.W. Epworth, D.M. Tennant, Phys. Rev. Lett. 52:228 (1984).
3. J.H. Claassen, S.A. Wolf, C. Leemann, J.H. Elliot, R. Orbach, Proc. Intern. Confer. on Localization, Interaction and Transport Phenomena in Impure Metals, Braunschweig, 1984, p. 11.
4. E. Abrahams, P.W. Anderson, D.C. Licciardello, T.V. Ramakrishnan, Phys. Rev. Lett. 42:673 (1979), and P.W. Anderson, D.J. Thouless, E. Abrahams, D.S. Fisher, Phys. Rev. B22:3519 (1980), and P.W. Anderson, Phys. Rev. B23:4828 (1981).
5. B.S. Anderek, E. Abrahams, J. Phys. C13:L383 (1980). See also E.N. Economou and C.M. Soukoulis, Phys. Rev. Lett. 46:618 (1981), C.M. Soukoulis and E.N. Economou, Sol. St. Comm. 37:409 (1981).
6. M.Ya. Azbel', Sol. St. Comm. 45:527 (1983).
7. M.Ya. Azbel', D.P. Di Vincenzo, Sol. St. Comm. 49:949 (1984), and Phys. Rev. B, to be published.
8. M.Ya. Azbel', P. Soven, Phys. Rev. B27:831 (1983), see also I.M. Lifshitz, V.Ya. Kirpichenkov, Sov. Phys. JETP 50:499 (1979).
9. M.Ya. Azbel', A. Hartstein, D.P. Di Vincenzo, Phys. Rev. Lett. 52:1641 (1984); see also R.F. Kwasnick, M.A. Kastner, J. Melngailis, P.A. Lee, Phys. Rev. Lett. 52:224 (1984).
10. M.Ya. Azbel', Topics in Solid State Physics, Springer-Verlag, 1985.
11. N.F. Mott, E.A. Davis, Electronic Processes in Non-Crystalline Solids, Oxford University Press, Oxford, 1971.
12. A.B. Fowler, A. Hartstein, R.A. Webb, Physica 117B-118B:661 (1983) and A. Hartstein, R.A. Webb, A.B. Fowler, J.J. Wainer, Surf. Science, 142, 1(1984), and R. A. Webb, A. Hartstein, J. J. Wainer and A. B. Fowler, to be published.
13. P. A. Lee, Phys. Rev. Lett. 53, 2042 (1984). The relation between the Lee oscillation and the Breit-Wigner formula, accounting for both elastic and inelastic scattering, was presented by P. Lee, A. D. Stone, preprint.
14. M. Buttiker, L. Landauer, Phys. Rev. Lett. 49:1739 (1982), and K.W.H. Stevens, J. Phys. C16:3649 (1983); C17:5735 (1984).
15. A.D. Stone, M.Ya. Azbel', P.A. Lee, to be published; see also Y. Gefen, G. Schön, to be published.
16. W.E. Howard, F.F. Fang, Solid State Electronics 8:82 (1965), and J.A. Pals, W.J.J.A. Van Heck, Appl. Phys. Letters 23:550 (1973), and J.M. Voschenko, J.N. Zemel, Phys. Rev. B9:4410 (1974), and R.J. Tidey, R.A. Stradling, M. Pepper, J. of Phys. C7:L353 (1974) and G. Voland, K. Pagnia, Appl. Phys. 8:211 (1979), and M. Pepper, J. Phys. C12:L617 (1979), and Phil. Mag. B42:947 (1980) and Surf. Phys. 98:L218 (1980), and M. Pepper, M.J. Wren, R.E. Oakley, J. Phys. C12:L897 (1979), and M. Pepper, M.J. Wren, J. Phys. C15:L617 (1982), and W.J. Skocpol, L.D. Jackel, E.L. Hu, R.E. Howard, L.A. Fetter, Phys. Rev. Lett. 49:951 (1982), and R.G. Wheeler, K.K. Choi, A. Goel, R. Wisnieff, D.E. Prober, Phys. Rev. Lett. 49:1674 (1982).
17. R.F. Voss, J. Phys. C11:L923 (1978).
18. M. Pollak, I. Riess, J. Phys. C9:2339 (1976), and B.I. Shklovskii, Sov. Phys. Semicond. 10:855 (1976).
18a. P.H. Woerler, G.A.M. Hurkx, Proc. Intern. Confer. on Localization, Interaction and Transport Phenomena in Impure Metals, Braunschweig, 1984, p. 34.
19. B. Ricco, M.Ya. Azbel', Phys. Rev. 51:1795 (1983) and B. Ricco, M.Ya. Azbel', Phys. Rev. B29:1970, 4356 (1984).
20. R. Maynard, E. Akkermans, Phys. Rev. Lett., to be published.
21. M.Ya. Azbel', Phys. Rev. B28:4106 (1983).

22. J.B. Pendry, J. Phys. C16:3493 (1983), and U. Frisch, J. -L. Gautero, to be published.
23. E. Guazzelli, E. Guyon, J. Physique-Lettres 44:L837 (1983).
24. M.E. Cage, R.F. Dziuba, B.F. Field, E.R. Williams, Phys. Rev. Lett. 51:1374 (1983).
24a. R.A. Webb, S. Washburg, C.P. Umbach, and R.B. Laibowitz, Proc. Intern. Confer. on Localization, Interaction and Transport Phenomena in Impure Metals, Braunschweig, 1984, p. 119.
25. A. Miller and E. Abrahams, Phys. Rev. 120:745 (1960) and V. Ambegaokar, B.I. Halperin, and J.S. Langer, Phys. Rev. B4:2612 (1971).
26. M.Ya. Azbel', Sol. St. Comm. 43:515 (1982).
27. This reasoning is immediately generalized to the hopping between the n-th and q adjacent states. Then

$$j_n = \sum_{m=1}^{q} \sum_{p=0}^{m-1} (\rho_{n-p} w_{n-p,n+m-p} - \rho_{n+m-p} w_{n+m-p,n-p}).$$

28. J. Kurkijarvi, Phys. Rev. B8:922 (1973).
29. W. Brenig, G.H. Döhler adn H. Heyszenau, Phil. Mag. 27:1093 (1973).
30. At very low temperatures $T < T_e = T_1 (L_o/L)^2$, when $L < \Lambda$, one hop covers the whole length, and again the Mott $\ell n R \propto T^{-1}$. However, at such temperatures, inelastic scattering becomes unimportant, and resonant tunneling dominates.[7,13]

ASPECTS OF 2D AND 3D CONDUCTION IN DOPED SEMICONDUCTORS

A.P. Long,[a] D.J. Newson,[b] and M. Pepper[a] [b]

a) GEC Hirst Research Centre, East Lane, Wembley
Middlesex, U.K.

b) Cavendish Laboratory, Cambridge CB3 0HE, U.K.

ABSTRACT

In this article we present some recent results on conduction in doped semiconductors. There are two topics considered here. The first is the use of the GaAs MESFET (Schottky gated FET) as a system for the investigation of transitions between 3D and 2D impurity band conduction. A new temperature dependent magnetoresistance has been found and a model is presented. The second part of the paper is concerned with the magnetic field induced transition in InP and GaAs and differences in the characteristics; for example, a sample of InP with doping approximately twice the Mott criterion shows a clear σ_{min} whereas GaAs of lower doping clearly does not. We also present results on Sb-doped Si. Here a new effect has been observed at low temperatures which is thought to be a transition from the conduction band to the impurity band which is enhanced by the application of a magnetic field.

DIMENSIONALITY TRANSITION IN GaAs FET's

Introduction

We have previously shown that the GaAs Schottky gate FET can be used to investigate the transition from 3D to 2D electronic conduction [1,2]. The Fermi energy, E_F, obtained from the periodicity of Shubnikov-de Haas oscillations has been obtained as a function of the thickness of the conducting channel. The results were compared to the calculations of Berggren [1], based on the progressive reduction in the number of quantised sub-bands, and good agreement has been found. Shrinkage of wave-functions with a strong magnetic field results in a metal insulator transition and by varying the thickness of the conducting layer (between $\simeq 1\mu m$ and zero) can be observed in 2D or 3D.

In this article we present results on the correction to quantum interference and the electron-electron interaction. Here 2D effects occur when the inelastic diffusion length or interaction length exceed the thickness of the conducting channel. In particular, we present results on the magnetoresistance in the presence of a magnetic field in the plane of the channel. A new temperature dependent magnetoresistance is found.

Experimental

A cross-section through one of the devices is shown in Fig. 1. The device consists of 3 layers of GaAs grown by M.O.C.V.D. and a Schottky gate:

(1) a p+ substrate doped $5 \times 10^{18} cm^{-3}$;
(2) a nominally undoped spacer layer of thickness $2\mu m$ to act as an insulator;
(3) a $0.4\mu m$ thick active epilayer of n-GaAs;
(4) a Ti-Au Schottky gate.

There are two depletion layers in the device:

(1) below the Schottky barrier; and
(2) at the back interface.

The back depletion region ensures that the channel is contained in a region of constant doping and negates interface scattering. The devices were defined in a Hall bar geometry, $750\mu m \times 100\mu m$, with an inter-probe spacing of $150\mu m$. The uncertainty in channel thickness due to growth techniques is estimated to be $< 10Å$ over $150\mu m$.

The thickness of the conducting channel may be calculated by assuming the depletion approximation.

$$t = \left[\frac{-2\epsilon (V_p - \Phi)}{eN_d}\right]^{1/2} - \left[\frac{-2\epsilon (V_g - \Phi)}{eN_d}\right]^{1/2} \quad (1)$$

where Φ is the Schottky barrier potential, ϵ the dielectric constant, N_d the doping level of active layer, V_p the pinch-off voltage (namely, that voltage at which the channel has zero thickness) and V_g the voltage on the gate relative to the conducting channel.

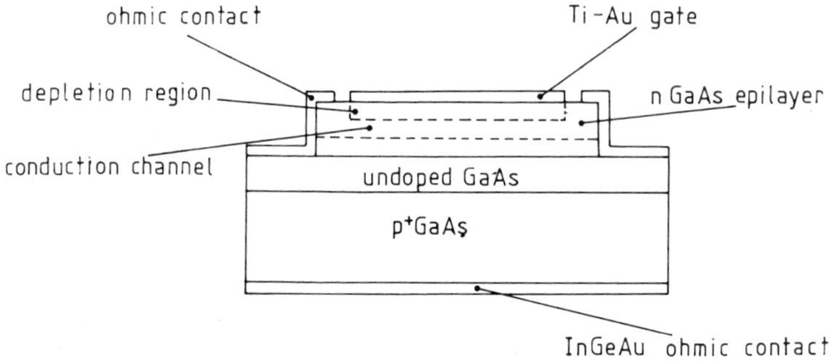

Fig. 1 A cross-section through one of the special GaAs MESFETS.

The samples used in this work were doped slightly above the Mott critical criterion ($n_c \simeq 2\text{-}3 \times 10^{16} cm^{-3}$ in GaAs) and so the low temperature electronic properties are influenced by localisation and interaction effects. The sub-band energies and Fermi energy as a function of channel thickness were calculated previously [1] for various doping levels. As the channel thickness is reduced the 3D density of states splits up until only the ground sub-band is occupied. Once in the ground sub-band, the Fermi energy falls linearly with thickness.

When many electronic sub-bands are occupied and the channel is thick, the conduction appears 3D. The low-field magnetoresistance goes as B^2 at low fields and $B^{1/2}$ at higher B, in agreement with the theory of quantum interference. The low field data allow the extraction of inelastic length and enable us to verify that $L_{in} << a$ as expected. The magnetoresistance was observed to be isotropic. The higher field data is plotted in Fig. 2 against $B^{1/2}$ and is nearly temperature independent, again showing the localisation effects to be 3D. It should be noted that there is no evidence for spin-orbit effects at fields down to 0.002 Tesla, giving a lower estimate of a spin-orbit time of 1×10^{-9} seconds.

As the channel thickness is reduced, the electrons become confined to a narrower potential. Once $L_{in} < a$ the localisation effects become 2D as inelastic scattering is not possible orthogonal to the plane of the channel. Similarly, once $L_{int} < a$ interaction effects become 2D.

The magnetoresistance for perpendicular (dashed) and parallel (solid) fields at different temperatures is plotted in Fig. 3 for a channel thickness of 440Å. At 1.3K, L_{in} = 875Å and L_{int} = 727Å and hence both localisation and interaction effects are 2D, by the criteria outlined above. The perpen-

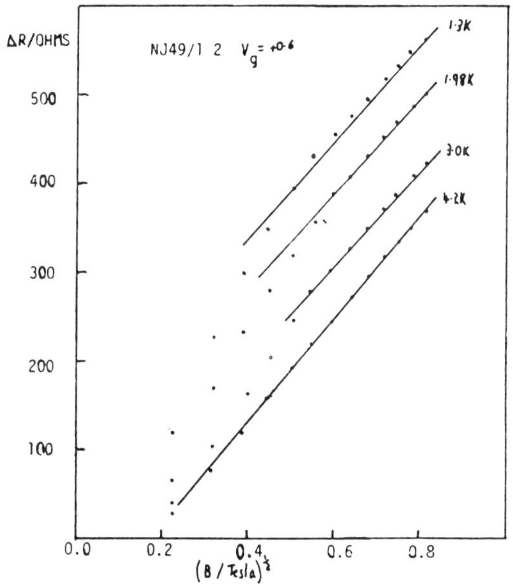

Fig. 2 Magnetoresistance plotted against $B^{1/2}$ for NJ49/1#2. V_g = +0.6v. a = 0.31μm. This shows 3D quantum interference effects.

dicular data gives good fits to the theoretical expression of Hikami [3], where the suppression of quantum interference was calculated. The parallel data are now not identical to the perpendicular data, as was the case in 3D. The parallel magnetoresistance is much smaller than the perpendicular magnetoresistance at low fields, but becomes of comparable size at larger values of field [4].

The effect of a parallel magnetic field on quantum interference in a thin slab was treated by Altshuler and Aronov [5] who found that

$$\delta\sigma = \frac{e^2}{2\pi^2\hbar} \ln\left[\frac{a^2 L_{in}^2 B^2 e^2}{3\hbar^2} + 1\right] \tag{2}$$

This provides an alternative means of calculating L_{in}. The results from this method together with the results obtained from perpendicular field are plotted in Fig. 4. The excellent agreement shows the validity of the theories. For a parallel field, the crossover from weak-to-strong magnetic field behaviour occurs when the flux through an area $L_{in}a$ is equal to one flux quantum, i.e., $L_{in}a \simeq L_c^2$.

At higher magnetic fields, we observe a negative magnetoresistance $\simeq B^{1/2}$. However, 3D quantum interference magnetoresistance is temperature independent, implying that a different mechanism is responsible for this behaviour. The temperature dependence is in the wrong direction for it to be accounted for by 3D interaction effects. In Fig. 5 we plot resistance against Ln T for different values of channel thickness in a fixed magnetic field, indicating that the behaviour takes the form $\delta\sigma \simeq B^{1/2}$ Ln T. It should be stressed that this behaviour cannot be accounted for by a mis-

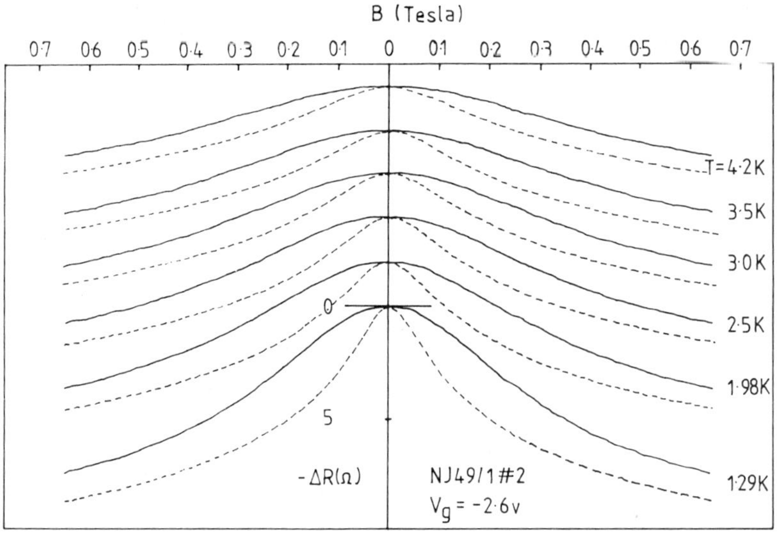

Fig. 3 Low-field perpendicular and parallel magnetoresistance (dashed and solid lines, respectively) for $V_g = -2.6$v. NJ49/1#2. $a = 0.044\mu$m.

alignment of the sample with the magnetic field: the parallel magnetoresistance is nearly as large as the perpendicular magnetoresistance. A misalignment by an angle θ would cause an effective perpendicular field of $B\sin\theta$, and hence a much smaller magnetoresistance.

We now underline the reason for this behaviour: a fuller discussion will be found elsewhere [6].

In a disordered "metallic" system, electrons diffuse with a diffusion constant $D = 1/d\, V_F^2 \tau_{el}$, where V_F is the Fermi energy velocity, τ_{el} the elastic

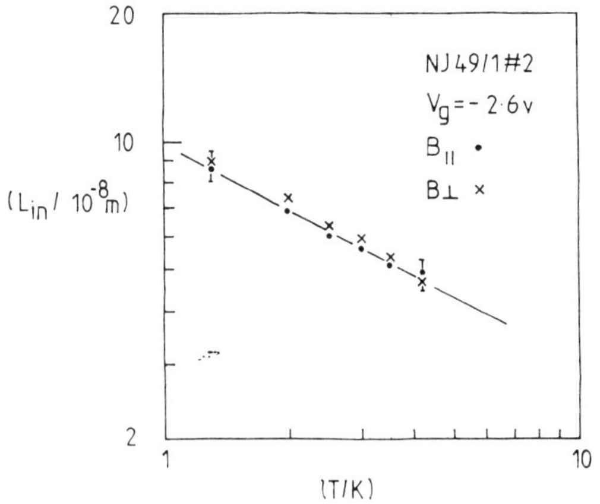

Fig. 4 Inelastic length obtained from perpendicular and parallel magnetoresistance. $V_g = -2.6v$. NJ49/1#2. $a = 0.044\mu m$.

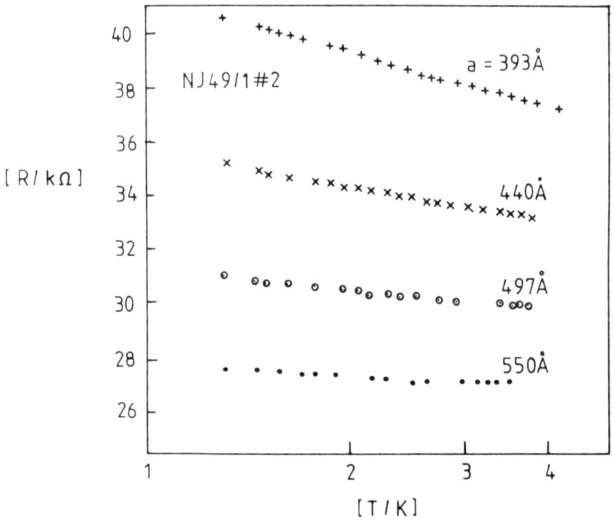

Fig. 5 Resistance against ln T for different values of channel thickness in a fixed parallel magnetic field of 0.51 Tesla.NJ49/1#2.

mean-free-time, and d is the dimensionality. Following Bergmann [7], phase coherence is maintained until the electron suffers an inelastic scattering event: usually an electron-electron collision at the temperatures considered here. Such collisions occur every τ_{in}, the electron diffusing a mean distance of $\sqrt{(D\tau_{in})}$ between such scattering events. Due to such diffusion, an incident electron-wave with wave-vector $\underline{K}$ may give rise to a coherently back-scattered spot at $-\underline{K}$, which decreases in size with time. This effectively reduces the electron diffusion constant and hence the conductivity. Integration of the size of the spot over time between the appropriate limits (τ_{el} and τ_{in} in the case of zero magnetic field) gives the size of the conductivity correction.

$$\text{In 2D} \qquad \delta\sigma = e^2/2\pi^2\hbar \, \mathrm{Ln}\tau_{in} \tag{3}$$

$$\text{In 3D} \qquad \delta\sigma = e^2/(\pi^2\hbar\sqrt{(D\tau_{in})}) \tag{4}$$

A magnetic field introduces an additional phase between identical paths in opposite direction. We may thus introduce an effective magnetic length, $L_c = \sqrt{\hbar/eB}$, as the distance in real space over which the electron loses phase coherence due to the magnetic field. If L_c is shorter than L_{in}, then the magnetic field determines the conductivity correction: in a 2D system of zero thickness and a 3D system the conductivity corrections are temperature independent and take the form for $L_c << L_{in}$:

$$\text{3D} \qquad \delta\sigma = \frac{e^2}{2\pi^2\hbar} \, 0.605\sqrt{(eB/\hbar)} \tag{5}$$

$$\text{2D} \qquad \delta\sigma = \frac{e^2}{2\pi^2\hbar} \, \ln B \tag{6}$$

In a conducting channel, there is a momentum uncertainty h/a orthogonal to the channel. We take account of this uncertainty by employing the method of Kaveh and Mott [8]. They found that the probability, $|a(q)|^2$ of finding an electron in a state shifted by q from its original state

$$\exp(ik.r) \text{ is } |a(q)|^2 = 1/N(E_F)\pi\hbar Dq^2) \tag{7}$$

So the probability of diffusing to all states other than the original is

$$R = \Sigma \, |a(q)|^2 \tag{8}$$

They showed that

$$\delta\sigma \simeq (1-2R)\sigma_b \tag{9}$$

In a strong magnetic field, i.e., $L_c < a$, the electron is localised with a decay constant $\simeq L_c$. So, $\Delta q_z \simeq 2\pi/L_c$

$$\text{Hence,} \quad R = \frac{L^2 a}{N(E_F)\pi D(2\pi)^2} \ln T \frac{1}{L_c}$$

Thus, from Eq. (9),

$$\delta\sigma \simeq aB^{1/2} \ln T \tag{10}$$

We may also obtain this result by the theory of Bergmann [7] outlined above.

Under certain conditions, the momentum uncertainty perpendicular to the channel affects the size of the coherently back-scattered spot at $-\underline{K}$ (discussed earlier) and hence enters into the localisation problem by changing the integration limits of the contribution to the conductivity. In zero magnetic field, the spot, which in 3D is a sphere of radius $1/L_{in}$, becomes an ellipsoid of axes $1/L_{in} \times 1/L_{in} \times 1/a$. The last term in the product is from uncertainty orthogonal to the channel. Let us now introduce a magnetic field. This may be visualized in 3D as a sphere of radius $1/L_c$ (it is a sphere because the electron diffuses in real space, and hence all paths will have a phase difference introduced). When $1/L_c$ is smaller than $1/a$, the sphere becomes an ellipsoid $1/L_c \times 1/L_c \times 1/a$.

The intersection between the magnetic sphere/ellipsoid and the inelastic ellipsoid govern the integration limits and hence the conductivity correction as a function of temperature and magnetic field: the limits are between $1/\ell$ (the reciprocal of the elastic length) and the exterior surface provided by the sphere and ellipse. There are essentially three distinct cases:

(1) $1/L_{in} < 1/a < 1/L_c$. In this case, the sphere lies totally outside the ellipsoid and the magnetic field only provides the integration limit. The conductivity is temperature independent, and the quantum interference correction behaves in a 3D manner.

(2) $1/L_{in} < 1/L_c < 1/a$. In this case, the limits of integration are provided by a complex surface between the sphere and the ellipsoid.

However, part of the limit is temperature dependent, being defined by L_{in} which is itself temperature dependent, thus the conductivity correction is temperature dependent (Fig. 6).

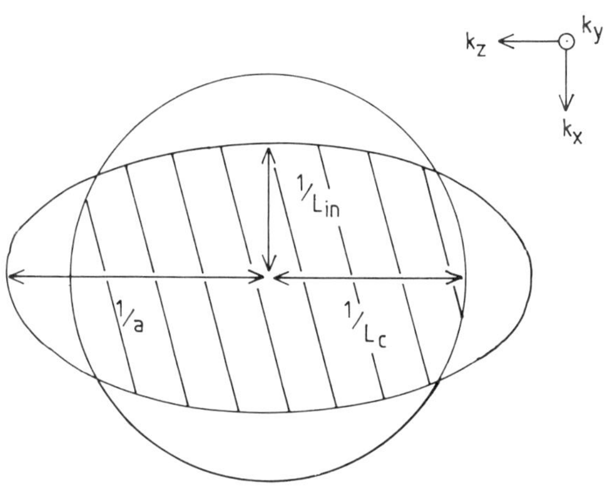

Fig. 6 The coherently back-scattered spot at $-\underline{K}$, in the case $1/L_{in} < 1/L_c < 1/a$. The integration limits are temperature dependent and hence the conductivity is temperature dependent.

The volume in k-space is roughly equal to:

$$\simeq \pi(1/a-1/L_c)^2 a/L_{in}$$

Following Bergmann [7], this gives a proportion of coherently interfering states

$$I_{coh} = \frac{\ell}{2\pi^2 k_F}\,(1-2a/L_c)1/(D\tau) \tag{11}$$

Upon integration, this gives a conductivity correction

$$\delta\sigma = \frac{e^2}{2\pi^2\hbar}\,(1-2a/L_c)\ \ln T \tag{12}$$

Thus, the temperature dependent part of the conductivity has a maximum value when the channel has zero thickness, corresponding to an inversion layer or a heterojunction, when a parallel field has no effect on quantum interference as no flux is enclosed.

(3) $1/L_c < 1/L_{in} < 1/a$. In this case, the magnetic field has only a small effect on localisation. This effect may be calculated by noting that the system behaves like an interferometer, and to a first approximation we must add intensities rather than amplitudes.

This results in a negative magnetoresistance

$$\delta\sigma = \frac{e^2}{2\pi^2\hbar}\, a^2 L_{in}{}^2 B^2 \tag{13}$$

This expression, (13), is in agreement with the low field limit of the expression of Altshuler and Aronov [9] to within a numerical factor.

In summary, we predict that as the magnetic field increases, we will pass through B^2 into $B^{1/2}$ lnT behaviour, and once L_c is shorter than the channel thickness we will lose the temperature dependence of quantum interference. Thus, in a parallel field the electron gas will behave as though it were 3D, whereas in a perpendicular field it behaves in a 2D manner. We note that as $a \to 0$, as is the case for Si-MOS or GaAs-AlGaAs heterojunctions, then

$$\delta\sigma = \frac{e^2}{2\pi^2\hbar}\ \ln T \tag{14}$$

This agrees with standard theory that a parallel field has no delocalising action on a 2DEG of zero thickness. We note that the observed effects are not explicable by the coherent interference of scattered waves reflecting off the surfaces of the confining potential. Volkov [10] showed that such surface scattering would produce a $B^{1/2}$ magnetoresistance once $L_c < a$ in the perpendicular orientation: we are observing a Ln B dependence on magnetic field in this orientation.

In addition to the quantum interference (commonly known as weak localisation) discussed above, electron-electron interactions [9,11] in the presence of disorder are also important. The exchange and Hartree energies cause the energy of a state $|\underline{k}\rangle$ to be broadened. However, the electron-electron correlation also leads to $|\underline{k}\rangle$ becoming a wave packet including states $|\underline{k}+\underline{q}\rangle$. This corresponds to an energy broadening $\Delta = \hbar D q^2$. These effects lead to repulsion of states away from the Fermi energy and a reduction in $N(E_F)$. They give rise to a conduction correction

$$\delta\sigma = e^2/2\pi^2\hbar\,(2-2F)\ln T \qquad \text{in 2D} \qquad (15)$$

$$\delta\sigma = e^2/2\pi^2\hbar\,(4/3-2F)\sqrt{kT/D\hbar} \qquad \text{in 3D} \qquad (16)$$

where F is the normalised mean of the potential averaged over the Fermi surface.

There is also a positive mangetoresistance due to interaction effects. This can be shown to take the form

$$\delta\sigma \simeq -\ln B \qquad \text{in 2D} \qquad (17)$$

$$\delta\sigma \simeq -\sqrt{B} \qquad \text{in 3D} \qquad (18)$$

in fields such that $g\mu B >> kT$. This interaction magnetoresistance will not be discussed below because of the small fields used. We cite the good fits to the theory of Hikami [3] for the 2D quantum interference magnetoresistance as evidence for this.

The dimensionality of interaction effects is governed by the interaction length $L_{int} = \sqrt{(D\hbar/kT)}$ and may thus be obtained from conductivity measurements. We have observed such a crossover between 3D and 2D interaction behaviour of temperature dependence as the thickness is reduced. This is discussed elsewhere [12].

MAGNETIC LOCALISATION IN 3D SEMICONDUCTORS

Introduction

Magnetic shrinkage of wavefunctions has for some time been a valuable technique for the investigation of the metal-insulator transition. The most pronounced effects are found in the light mass semiconductors with the longest Bohr orbit, and the transition can be induced in a sample which is doped well above the transition. Application of a strong field also sharpens the distinction between the different conduction processes (E_2 and E_3) when states at the Fermi energy are localised. Earlier work on this topic is reviewed elsewhere [13].

Here we present some preliminary results on the effects of a magnetic field on transport in the III-V semiconductors GaAs and InP and a new effect in Sb-doped Si. Our motivation for investigating the III-V's was to determine if a minimum metallic conductivity σ_{min} is absent when the transition is induced by a strong field as appears the case in the absence of the field [14]. Different results were obtained, the GaAs sample showed an enhancement of the interaction correction, with arbitrarily small values of metallic conduction, at high fields instead or a transition to strong localisation. On the other hand, InP showed a sharp transition with a well-defined value of σ_{min}. Sb-doped Si showed a conductivity anomaly which we interpret as an electronic phase transition between conduction and impurity band states. Possibly, the differences between the samples of GaAs and InP used here reflect whether the electrons are in a conduction band or impurity band state. This will be very sensitive to doping and compensation, a change in $N(E_F)$ producing a large effect on the energies and the possibility of a transition.

InP and GaAs

We have investigated transport in n type InP of doping $7 \times 10^{16} cm^{-3}$, just over twice the critical value for the metal-insulator transition. In the absence of the magnetic field, the conductivity decreased linearly with temperature, T, between 50K and 10K indicating a correction due to quantum interference. The magnitude of the electron-electron scattering time required to fit the data was in good agreement with theory [15]. Below 4K increased with decreasing T linearly as $T^{1/2}$, indicating the existence of the interaction contribution to σ [16]. An application of the magnetic field at temperatures below 4.2K resulted in a large positive magnetoresistance at fields above 5 Tesla and negative below, Fig. 7. The negative magnetoresistance arose from the suppression of quantum interference and is gradually reduced as the temperature drops and the interaction positive magnetoresistance extends down to lower temperatures. Analysis of these corrections is presented elsewhere [15].

The large increase in resistance for B> 5 Tesla may be due to a Shubnikov-de Haas oscillation but may also indicate a transition to an impurity band state as the system is driven towards the metal-insulator transition at ≃ 9.8 Tesla. Unexpectedly the gradient m in the $T^{1/2}$ interaction correction ($\delta\sigma \propto m\, T^{1/2}$) was found to decrease with increasing field

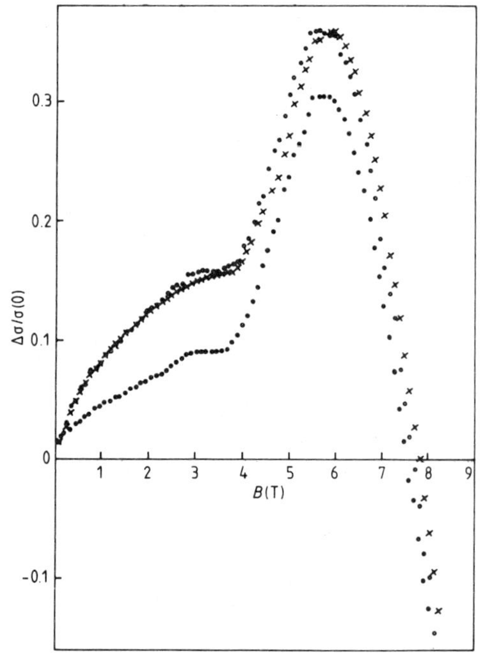

Fig. 7 Conductivity of n type InP ($N_d = 7 \times 10^{16} cm^{-3}$) versus magnetic field for a range of temperatures. . 52mK ◦ 740mK x 1.3K

and was zero near and at the transition, Figs. 8 and 9. This resulted in temperature independent conduction down to $\simeq$ 30mK at and close to the transition and clear observation of σ_{min} (as has been reported in compensated InP) [17] which is close to the value expected in the presence of a strong field ($15\Omega^{-1}cm^{-1}$) $0.05\ e^2/\hbar L_c$, where L_c is the cyclotron orbit. (Here we note that in the presence of the strong field the effective Bohr orbit is $\simeq$ 50Å, i.e., within a factor of two of silicon.) On the insulating side of the transition, the conductivity varies as σ_{min} exp $(-W/kT)$ and the intercept value of σ_{min} changes within a factor of two as the field increases, which may be due to wavefunction distortion or a change in the Anderson localisation criterion. The behaviour of the activation energy W as a function of magnetic field is shown in Fig. 10. As seen a linear relationship is obtained and the energy varies as $\simeq 3 \times 10^{-5}$ eV/Tesla which is $1/2g\mu_B B$, where g is the Lande g value, μ_B is the Bohr magneton and B is the magnetic field in Tesla. This strongly suggests that the excitation process is to a mobility edge in an upper Hubbard band which is slowly increasing in energy with increasing field. The factor 1/2 implies that the strongly overlapping Hubbard bands have similar values of density of states.

The conclusion of this work on InP, which is in accord with other work on compensated material [17], is that in the presence of a strong magnetic field a well-defined σ_{min} is found if interaction effects are absent or do not result in a temperature dependent conductivity correction.

Measurements on a sample of GaAs of doping $4 \times 10^{16}cm^{-3}$ show the appropriate behaviour, Fig. 11. Here for B = O is above the appropriate value of $\sigma_{min} = 4\Omega^{-1}cm^{-1}$; application of the field results in an initial nega-

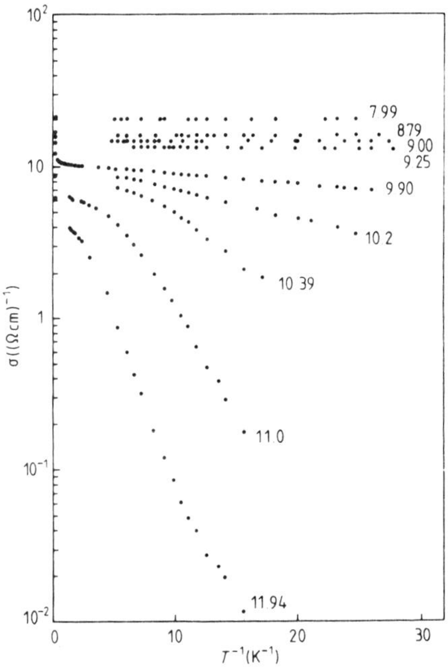

Fig. 8 Conductivity versus reciprocal temperature for different values of magnetic field in Tesla.

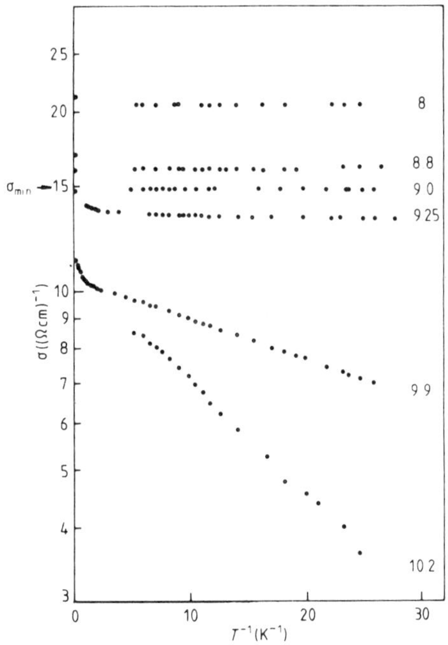

Fig. 9 The results shown in Fig. 8, illustrating the transition region.

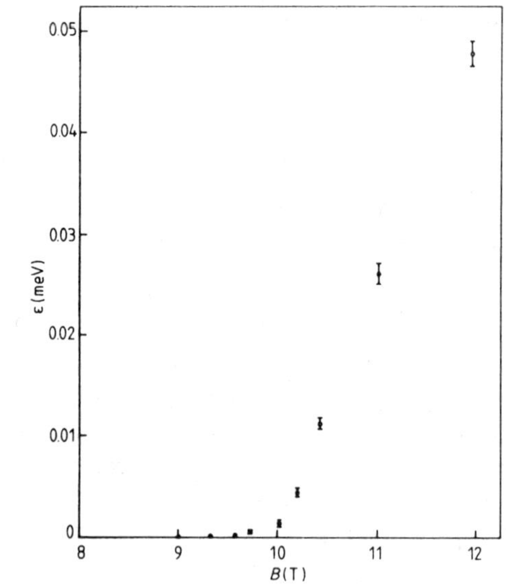

Fig. 10 Excitation activation energy obtained from the results of Fig. 5 is plotted against magnetic field B in Tesla.

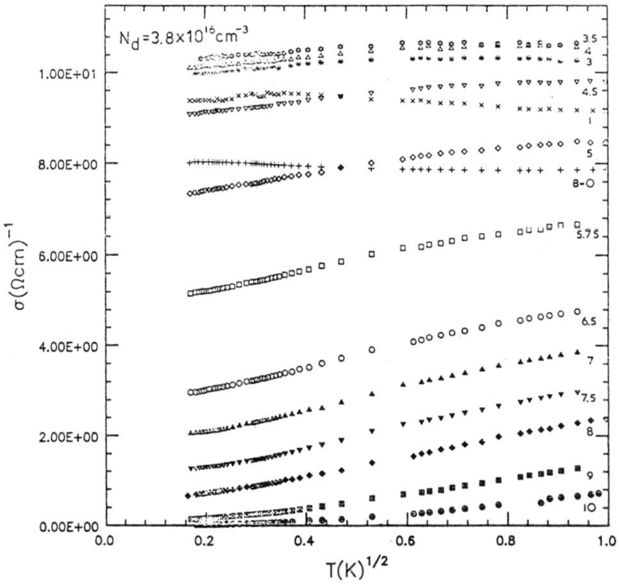

Fig. 11 Conductivity of GaAs, doped to $4x10^{16}cm^{-3}$, is plotted against $T^{1/2}$ as a function of magnetic field. Metallic behaviour is found at values of considerably below $\sigma_{min} = 4\Omega^{-1}cm^{-1}$.

tive magnetoresistance, but with further increase of B this becomes positive and a clear $T^{1/2}$ correction is found. The decrease in conductivity is marked and taking the existence of this correction as indicating metallic behaviour values of metallic conduction can be found up to a factor of 50 below σ_{min}. It is significant that other epitaxial GaAs layers of higher doping, and possibly more compensated, did not show this effect as the magnetic field caused strong localisation with activated conduction.

In conclusion, there appear to be considerable differences in behaviour which are not understood. Measurements on a wider range of specimens of different doping and compensation are required.

Sb-Doped Si

We have investigated the properties of Sb-doped Si at low temperatures and present a summary of our earlier work [17]. The samples used were in the doping range 2.7 x 10^{18} to 4.9 x 10^{18} and spanned the metal-insulator transition. Here, we will be discussing results on samples very close to the transition. For metallic samples the temperature dependence and magnetoresistance were dominated by the interaction effect and there was a correction which varied with temperature as m $T^{1/2}$. Comparison with theory is difficult because of the importance of intervalley scattering, but, as expected, m was found to change sign with doping. Figure 12 shows the $T^{1/2}$ correction for a sample of doping 4.9 x $10^{18}cm^{-3}$. When B exceeds 1 Tesla, the sign of the temperature dependence changes due to the conversion of the screening term to (4/3-F) from (4/3-3F) were F = 0.8. As expected, further increase of the field did not result in a change in the value of m. Decreasing the doping of the silicon resulted in the observation of a series of effects not previously observed, and these will now be discussed in order of decreasing doping. It is to be noted that the doping, N_D, was obtained from Hall measurements and so is accurate relatively but not absolutely.

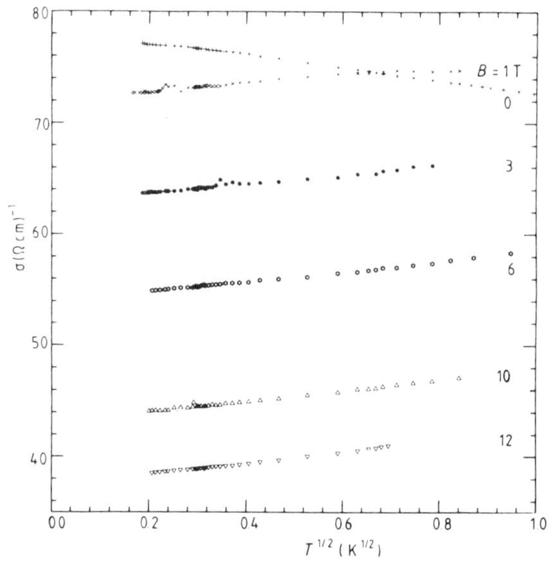

Fig. 12 $T^{1/2}$ dependence of σ at a doping level of 4.9 x $10^{18}cm^{-3}$, the magnetic field in Tesla is indicated.

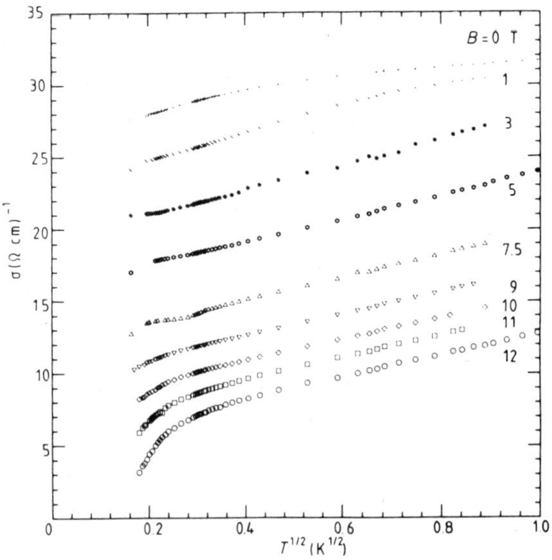

Fig. 13 $T^{1/2}$ dependence of σ at a doping level of $3.2 \times 10^{18} cm^{-3}$.

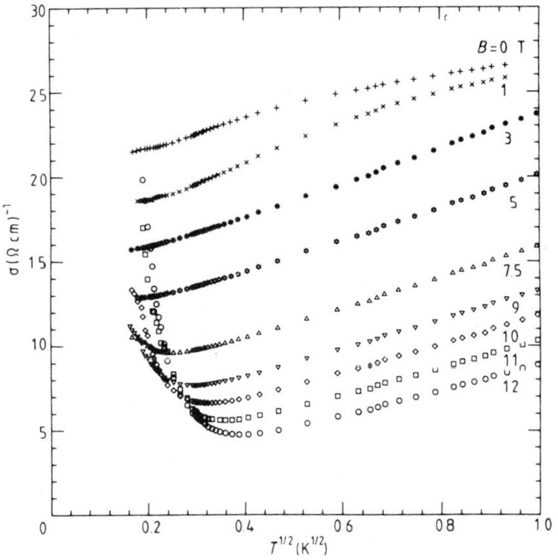

Fig. 14 σ versus $T^{1/2}$ for a doping level of $3.1 \times 10^{18} cm^{-3}$.

1. Figure 13 shows the results obtained on a sample of doping $3.2 \times 10^{18} cm^{-3}$ for different values of magnetic field B in Tesla (T). We note that the gradient of the $T^{1/2}$ plots are similar and the change in σ is large, up to a factor of 2. At the lowest temperature and highest field conditions, σ drops faster than $T^{1/2}$.

2. A slight decrease in the doping to 3.1×10^{18} gives similar results to those in Fig. 13 at higher temperatures, but at low temperatures now shows a rapid increase with decreasing temperature, Fig. 14. The temperature at which this rise starts is raised by the magnetic field, the increase in σ is considerable and at the lowest temperatures the values of σ for B = 0, and B = 12 Tesla are very close.

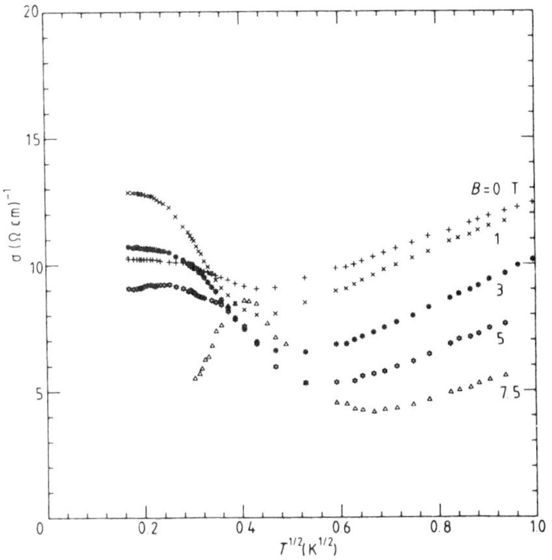

Fig. 15 σ versus $T^{1/2}$ for a value of doping of $2.95 \times 10^{18} cm^{-3}$.

3. It is possible to observe both these departures from the $T^{1/2}$ law, the rapid drop and the rise. The observation of the drop before the rise is often found but appears not to behave in a systematic manner with doping. Figure 15 shows results obtained at a doping level of $2.95 \times 10^{18} cm^{-3}$. Here, the rise in at the lowest temperatures is clear, but at the highest value of magnetic field (7.5 Tesla) the rise cannot be sustained and starts to fall rapidly.

We suggest that these changes are the result of an electronic phase transition. This is from the high temperature phase (A) to the low temperature phase (B) and is enhanced by the magnetic field. Phase A corresponds to the conduction band and B to the impurity band. Mott [18] has pointed out that the transition, occurring when the Free Energy in the impurity band is lower, is aided by the Brinkman-Rice mass enhancement in the impurity band [19]. It is also helped by the reduction in the density of states in phase A caused by the interaction, as $\delta\sigma/\sigma = \delta N(F_F)/N(E_F)$ the large values of $\delta\sigma/\sigma$ imply a considerable reduction in $N(E_F)$. In this sense, the transition is interaction driven and follows on from the $T^{1/2}$ correction, the large interaction correction may give rise to a valley instability and a change in occupation leading to the transition.

The conduction proceeds as follows. Applying the magnetic field at higher temperatures enhances the interaction $T^{1/2}$ correction and this is observed as the temperature decreases. The reduction in the density of states can result in either Anderson localisation at E_F in the conduction band prior to the transition occurring and electrons transferring to the impurity band. As the first electrons to enter the impurity band may be localised, it is difficult to distinguish between these two possibilities. The conductivity now falls rapidly with decreasing temperature in an activated manner until appreciable transfer to the impurity band has occurred. At this state, σ increases with decreasing temperature and is then flat, i.e., metallic behaviour. The temperature at which the transition occurs is enhanced by the magnetic field but this can also result in a metal-insulator transition in phase B. Figures 16a and 16b show these effects

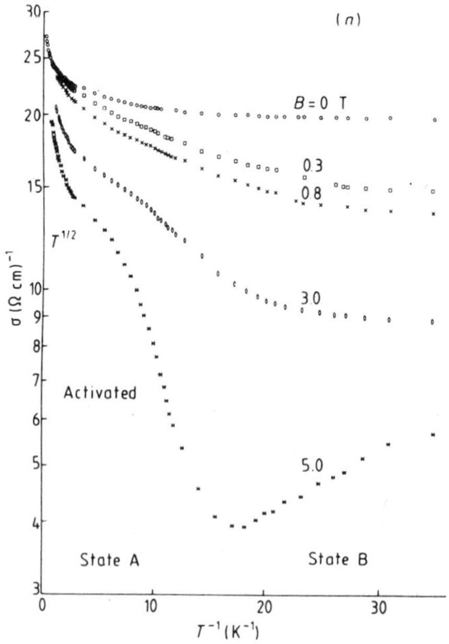

Fig. 16a σ plotted against T^{-1} for different values of magnetic field illustrating the $T^{1/2}$ and activated regions of transport, the sample doping is 3.08 x 10^{18}cm^{-3}.

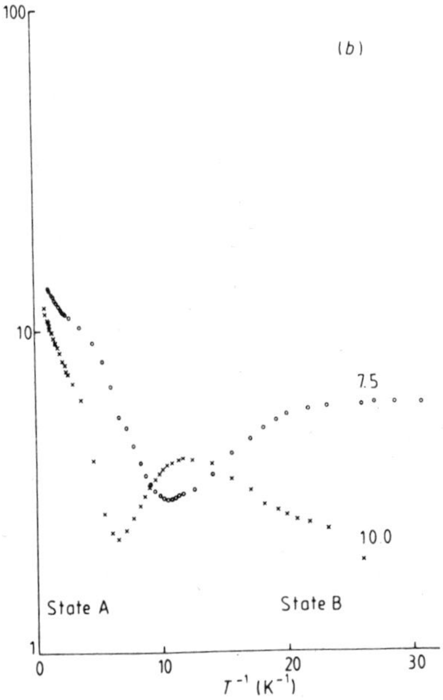

Fig. 16b Conductivity of the sample used in 16a at higher values of magnetic field illustrating the metal insulator transition in Phase B between 7.5 and 10 Tesla.

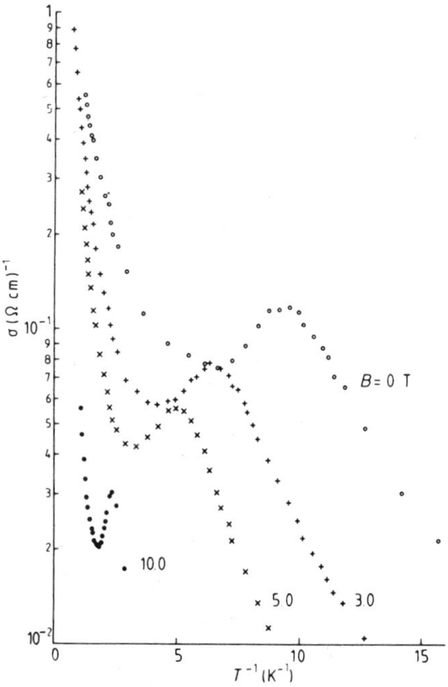

Fig. 17 σ plotted against T^{-1} for a sample of doping level 2.7 x 10^{18}cm^{-3} for different values of magnetic field.

clearly; the transition from $T^{1/2}$ to activated behaviour in phase A, the rise in σ in phase B and finally the introduction of metal-insulator transition in B by increasing the magnetic field to 10 Tesla from 7.5 Tesla. Reduction of the doping to $2.7 \times 10^{18}cm^{-3}$ results in activated conduction for B = 0 in phase A. However, the transition to phase B is still found and is enhanced by a magnetic field, Fig. 17. In this figure, it is seen that both phases are localised but B is less localised than A.

The results of Fig. 16a indicate the extreme of a σ_{min} in phase B. This would be about $8\Omega^{-1}cm^{-1}$ in rough agreement with Mott's formula $\sigma_{min} = 0.05e^2/\hbar L_c$, where L_c is the cyclotron radius. Phase A does not possess a σ_{min}, the presence of the strong $T^{1/2}$ correction reducing the value of the conductivity to arbitrarily small values. Extrapolating the zero magnetic field conductivity to zero temperature fits the same empirical relation as phosphorus doped silicon [20] $\sigma(0) = \sigma_c(\frac{n}{n_c} - 1)^{1/2}$, where we take the critical doping n_c as $2.9 \times 10^{18}cm^{-3}$, although near n_c the index which gives the best fit is unity.

In summary, we suggest that an electronic phase transition has occurred between the conduction band and an impurity band, the two phases possessing very different characteristics. The dependence of the temperature where σ is a minimum (which we take as the critical temperature T_c) has been plotted as a function of magnetic field. If σ goes from the $T^{1/2}$ law into phase B, then T_c varies as B^2; on the other hand, if there is an intervening activated region, T_c varies as $B^{1/2}$.

ACKNOWLEDGEMENTS

During the course of the investigations described in this paper, we have enjoyed many stimulating discussions with Professor Sir Nevill Mott and Professor M. Kaveh.

REFERENCES

1. D.A. Poole, M. Pepper, K.F. Berggren, G. Hill, and H.W. Myron, J. Phys. C 15, L21 (1982).
2. D.A. Poole, M. Pepper, and H.W. Myron, Physica 117B+118B, 687 (1983).
3. S. Hikami, A.I. Larkin, and Y. Nagaoka, Prog. Theor. Phys. 67, 707 (1980).
4. A. Kawabata, J. Phys. Soc. Japan 49, 628 (1980).
5. B.L. Altshuler and A.G. Aronov, JETP Lett. 33, 499 (1981).
6. C. McFadden, D.J. Newson, M. Pepper, and N.J. Mason, J. Phys. C, to be published.
7. G. Bergmann, Phys. Rev. B 28, 2914 (1983).
8. M. Kaveh and N.F. Mott, J. Phys. C. 14, L177 (1981).
9. B.L. Altshuler and A.G. Aronov, Solid State Comm. 30, 115 (1979).
10. V.A. Volkov, JETP Lett. 11, 394 (1982).
11. M. Kaveh and N.F. Mott, J. Phys. C 15, L707 (1982).
12. D.J. Newson, C. McFadden, and M. Pepper, Phil. Mag. (1985), to be published.
13. M. Pepper, J. Non-Cryst. Solids 32, 161 (1979).
14. E. Abrahams, P.W. Anderson, D.C. Licciardello, and T.V. Ramakrishnan, Phys. Rev. Lett. 42, 673 (1979).

15. A.P. Long and M. Pepper, J. Phys. C:Solid State Physics C17, 3391 (1984).
16. A. Kawabata, Solid State Comm. 34, 431 (1980).
17. G. Biskupski, H. Dubois, J.L. Wojkiewicz, A. Briggs, and G. Remenyi, J. Phys. C:Solid State Physics C17, L411 (1984).
18. N.F. Mott, Metal-Insulator Transitions, (Taylor and Francis, London, 1974).
19. W.F. Brinkman and T.M. Rice, Phys. Rev. B 7, 1508 (1973).
20. T.F. Rosenbaum, R.F. Milligan, M.A. Paalanen, G.A. Thomas, R.N. Bhatt, and W. Lin, Phys. Rev. B 27, 7009 (1983).

LOCALIZATION PHENOMENA AND AC CONDUCTIVITY IN WEAKLY DISORDERED QUASI-ONE-DIMENSIONAL AND LAYERED MATERIALS AND IN ANISOTROPIC LOW DIMENSIONAL SYSTEMS

Yu. A. Firsov

A.F. Ioffe Physico-Technical Institute
Academy of Sciences of the USSR
194021 Leningrad, USSR

ABSTRACT

For quasi-one-dimensional (quasi-1d) and layered materials, for highly anisotropic two-dimensional systems and wires the influence of weak disorder due to randomly distributed impurities is studied in a diagrammatic self-consistent approximation (SCDT). It is shown that for quasi-1d systems there is a threshold value, w_C, for the interchain exchange integral. For $w < w_C$ a quasi-1d system is in a localized state and for $w > w_C$ it is in a metallic state. In contrast with a quasi-1d case, for a layered material extended states in an upper part of energy band appear almost immediately when a weak interlayer tunneling is switched on. The mobility edge is determined as a function of the strength of this tunneling. An expression for the AC conductivity $\sigma(\omega)$ is found in both cases. For highly anisotropic 2d systems localization radii $r_{\parallel}$ and $r_{\perp}$ are determined and $\sigma(\omega)$ is found. A possible relevance of this model to a system of dislocations on the grain boundaries of a germanium bicrystal is noted. A model of a wire consisting of N parallel, weakly coupled chains is considered. It is shown that with an increase in interchain coupling a rather sharp transition from a localized state with a small localization radius ($r_C \sim \ell$, where ℓ is the mean free-path) to a localized state with a large localization radius ($r_C \simeq N\ell$) should take place. A very important conclusion is that for any anisotropic system, the effect of anisotropy can be completely absorbed into anisotropic diffusion coefficients $D_{\parallel}(\omega)$ and $D_{\perp}(\omega)$ and that the ratio $D_{\perp}(\omega)/D_{\parallel}(\omega)$ remains constant for $0 < \omega\tau < 1$. The accuracy of SCDT is discussed. A short review of the present state of the theory of the metal-insulator transition in disordered systems is given.

1. INTRODUCTION

Anderson was the first to investigate the problem of localization [1]. This work brought a scientific precision into the whole field of disordered systems. Ioffe and Regel [2] realized the importance of this concept for analyzing experimental data. Since then many theorists made valuable contributions to the field. However, it would not be an exaggeration to say that the generation of scientists to which the present author also belongs began to derive basic knowledge of disordered systems and was able to get to the core of the matter, thanks to original papers of Sir Nevill Mott published by him during the last 25 years and to his comprehensive review articles and books [3-8]. In them we can find most of the concepts which have brought us to where we are. Already in 1961 Mott and Twose [3] pointed out the possibility of weak localization in a one-dimensional (1d) system in the presence of weak disorder (see also refs. 5 and 8). In 1979 this concept was generalized by Abrahams, Anderson, Licciardello and Ramakrishnan for the case of isotropic two-dimensional (2d) systems.

In this paper, I shall consider the localization induced by weak disorder in highly anisotropic low-dimensional systems, in quasi-one-dimensional (quasi 1d) and in layered (quasi 2d) materials. The strength of disorder in disordered materials is characterized by the value of the dimensionless parameter $k\ell \sim \epsilon\tau/\hbar$, where k is the wave vector, ϵ is the energy of quasi-particle, ℓ is the mean free path, τ is the relaxation time. The disorder is weak if the following inequality holds

$$\epsilon\tau/\hbar >> 1 \text{ or } k\ell >> 1 \tag{1}$$

For ordinary three-dimensional (3d) systems, the inequality (1) coincides with the condition of applicability of the Boltzman equation which predicts a finite value $\sigma_{3d}^{(0)} = e^2 n\tau/m$ for conductivity as $T \to 0$ (T is the temperature) and thereby rules out the possibility of localization due to weak disorder for d = 3. For a tight-binding model of strongly disordered 3d system, Anderson formulated a criterion showing when localization should take place [1]. Ioffe and Regel [2] have shown that in a sense this reduces to the condition $k_F\ell \simeq 1$, where k_F is the Fermi wave vector. On the other hand, in 1d metallic chain all states are localized even at arbitrary small disorder, $k_F\ell >> 1$ [3-8]. In Russia, we call this phenomenon a Mott localization. Its mechanism is very distinct from the "classical" mechanism of strong localization considered by Anderson in 1958 [1]. As we comprehend now it is wholly due to the quantum interference effect which is also responsible for even weaker localization for d = 2. The mathematical description of localization phenomena reduces to an investigation of the evolution in time for an electron wave packet initially (for t = 0) localized on a lattice site (let is be x = 0). If for $t \to \infty$ it spreads all over the chain we say that diffusion takes place. On the other hand, if it will turn out to be restricted to a region of finite length, we say that diffusion is forbidden (diffusion constant $D \equiv 0$) and localization takes place. Of course, we assumed that averages over impurity distribution have been taken.

The necessary information about all these processes can be obtained from the density correlator $\rho(0,x,t) \equiv \rho_t(x)$. This quantity was rigorously calculated with the help of Berezinskii's technique [11,12]. For $t \to \infty$ it takes the form

$$\rho_\infty(x) = \frac{1}{4\pi^{1/2}}\left(\frac{\pi^2}{8}\right)^2\left(\frac{4\ell}{|x|}\right)^{3/2} e^{-x/4\ell} \qquad \text{for } x \geq 4\ell \tag{2}$$

The width of this distribution is called localization radius r_c, where

$r_c \simeq 4\ell$ since $k_F \sim \pi/a$, for weak disorder ($k_F\ell >> 1$) we get $r_c >> a$, where a is the lattice spacing. This is the reason why this phenomenon may be called a weak localization, in contrast to a strong Anderson localization for which $r_c \simeq a$. However, the difference between both mechanisms of localization is much deeper and due to their different physical nature. It has been shown [11-14] that the Mott localization described by (2) is wholly due to quantum interference of electron waves during a process of multiple backward scattering caused not by one but by many impurities ($r_c >> n_{imp}^{-1}$). As we shall see later, such weak localization is not limited to 1d systems. The classical conductivity obtained from the Boltzman equation is

$$\sigma_{1d}^{(0)} = \frac{e^2 n_{1d}\tau}{m} = \frac{e^2}{\pi}\frac{k_F}{m}\tau = 2\frac{e^2\ell}{h} \tag{3}$$

Quantum corrections to $\sigma_{1d}^{(o)}$ show an obvious tendency to localization (see § 2). For d = 1 all these corrections can be summed up with the help of the Berezinskii technique [14]. For $\omega\tau << 1$ the AC conductivity is equal to

$$\sigma_{1d}(\omega) = \sigma_{1d}^{(o)}\left\{4(\omega\tau)^2\left[\ell n^2(2\omega\tau)+(2C-3)\ell n 2\omega\tau+\text{const}\right]-i4\xi(3)\omega\tau\right\} \tag{4}$$

The result $\mathrm{Re}\sigma(\omega) \sim \omega^2\ell n^2\omega$ was first obtained by Mott [3,4]. Rigorous calculations show that due to numerical reasons the range of the applicability of the Mott law ($\mathrm{Re}\sigma(\omega) \sim \omega^2\ell n^2\omega$) is rather narrow ($\omega\tau \leq 0.01$). Nevertheless, this result has an important meaning. For $\omega\tau > 0.01$ numerical calculations [14] show that $\mathrm{Re}\sigma(\omega)$ has a maximum near $\omega\tau \simeq 0.6$. The static dielectric constant $\epsilon'(0)$ may be determined from the relation

$$\epsilon'(0) = 1 - 4\pi \lim_{\omega\to 0} \mathrm{Im}\sigma(\omega)/\omega \tag{5}$$

However, the dimension of σ_{1d} is [$cm^2 sec^{-1}$] (instead of [sec^{-1}] as for σ_{3d}) and $\epsilon'(0)$, determined from (4) and (5) will not be dimensionless. Usually, one has in mind a simplified model of quasi-1d systems consisting of uncoupled 1d chains (tunneling or hopping between chains is absent). Denoting by s the area of transverse cross-section per chain and dividing (3) and (4) by s instead of n_{1d} [cm^{-1}], we get $n = n_{1d}/s$ for the concentration per unit volume, as usual. Then, from (4) and (5) it follows

$$\lim_{\omega\to 0} \epsilon'(\omega) = 1 + 4\xi(3)\frac{4\pi\sigma_{1d}^{(o)}\tau}{s} = 1 + 4\xi(3)(\omega_p\tau)^2 >> 1 \tag{6}$$

where ω_p is the "plasma frequency," $\omega_p^2 = 4\pi ne^2/m$. The value of $\epsilon'(\omega)$ changes its sign near $\omega\tau \simeq 0.6$ where $\mathrm{Re}\sigma(\omega)$ has a maximum [14]. For $\omega\tau >> 1$ one gets [14]

$$\sigma(\omega) = \frac{\sigma_{1d}^{(o)}}{s}\left[\frac{2}{(\omega\tau)^2} + i\frac{1}{\omega\tau}\right] \tag{7}$$

It is interesting to note that all these results will change very drastically when we take into account tunneling between chains (see § 4).

In 1977 Thouless [15] argued that for a 3d system in the shape of a long wire (made of isotropic metal) all states should be exponentially localized with a localization length

$$r_c \sim \ell\ (k_F d)^2 \sim \ell N \tag{8}$$

where d is the radius of a wire, k_F is the 3d Fermi wave vector, N is the number of atoms in a cross-section. Result (8) was also obtained from the microscopic theory of Weller, Prigodin and by the present author [16]. In Section 8 we shall consider the special model of a highly anisotropic wire consisting of N weakly coupled chains and investigate the crossover from Mott's localized regime ($r_c \simeq 4\ell$) to Thouless's localized regime ($r_c \sim N\ell$) which takes place with increasing coupling strength.

In Section 4 the theory is generalized to a more realistic model of a quasi-1d-system with interchain tunneling taken into account, its strength being characterized by the value of the transverse overlap integral w. The following principal question arises. Will the quantum interference effects (leading to Mott localization for d = 1) immediately disappear when arbitrary weak tunneling between chains is switched on? It was stated in reference 13 that $\sigma(0) \sim w^2$, which means that localization should immediately disappear for arbitary small w. The present author and Prigodin [17,18] have shown that in reality a threshold value of w exists, $w_c \simeq \hbar/\tau$. For $w < w_c$ quantum interference is effective and the system is in a localized regime, for $w > w_c$ it is in a metallic state. The range of applicability of (4) appears to be very limited, even for $w < w_c$. Formula (6) for $\epsilon'(0)$ is also very different from (6) (see (51) in § 4). All this means that real quasi-1d systems cannot be fully described on the basis of strictly 1d models [19]. Results obtained [20] for weakly disordered layered materials also differ from those for strictly 2d systems. An investigation of isotropic 2d system in the presence of weak disorder turned out to be a more difficult problem in as much as rigorous mathematical method, whose accuracy for d = 2 may be comparable to the precision of Berezinskii technique for d = 1 has not yet been developed. The scaling theory by Abrahams et al. [9,10] based on ideas developed earlier by Thouless [21] leads to the result that, due to the quantum interference effects, weak localization for d = 2 takes place even at arbitrary small disorder, in contradiction to Mott [4]. Many objections have been raised to the scaling theory of localization but most of them are refuted now. Moreover, the scaling theory assumptions for d = 2 are justified to some extent by mapping of the original hamiltonian onto field theoretical models like the n-orbital per site model [22], the nonlinear σ model of interaction matrices [23], and by the supersymmetry theory [24]. In this paper, I adhere to a diagrammatic analysis. The first quantum correction to the conductivity $\sigma_{2d}^{(o)} = \frac{e^2 n_{2d}}{m}\tau = \frac{2e^2}{h} k_F \ell$ for $\omega \neq 0$ is given by [25]

$$\sigma_{2d}(\omega) = \sigma_{2d}^{(o)}\left[1 - \lambda\ \ell n\left(\frac{1}{-i\omega\tau}\right)\right] \quad \text{where } \lambda = \frac{\hbar}{2\pi\epsilon_F\tau} \tag{9}$$

For $\omega = 0$, quantum corrections depend on the characteristic size L of the sample [9,26]

$$\sigma_{2d}(L) = \sigma_{2d}^{(o)}\left(1 - \lambda\ \ell n \frac{L}{\ell}\right) \tag{9a}$$

A very simple derivation of (9) and (9a) has been presented by Kaveh and

Mott [27]. According to (9a), $\sigma_{2d}(L)$ tends to zero when $L \to r_c$, where

$$r_c^{(2d)} = \ell \exp\left(\frac{1}{2\lambda}\right) = \ell \exp\left(\pi \frac{\epsilon_F \tau}{\hbar}\right) >> \ell \qquad (10)$$

It may be shown (see Section 2, Eq. (23)) that r_c is the localization radius for $d = 2$. It is obvious that for $L > r_c$ formula (9a) is inapplicable, since it gives $\sigma < 0$. $\sigma_{2d}(L)$ for $L > r_c$ is given by (11a). According to (9), $\mathrm{Re}\sigma(\omega)$ diminishes with decreasing $\omega\tau$ which also shows the tendency toward localization ($\mathrm{Re}\sigma_{(o)} \equiv 0$). The next order corrections (see chapter 2) also maintain this tendency. Self-consistent diagrammatic method (SCDT) for selective summation of infinite sets of such "dangerous" diagrams for $\lambda << 1$ was proposed by Wölfle and Vollhardt [28,29]. They have obtained the following formula for $\sigma_{2d}(\omega)$

$$\sigma_{2d}(\omega) = \sigma_{2d}(o) \left[\frac{1}{\lambda}\left(\frac{\omega}{\omega_o}\right)^2 - i\frac{\omega}{\omega_o}\right] \quad \text{if } \omega < \omega_o = \frac{1}{\tau} e^{-\frac{1}{\lambda}} \qquad (11)$$

For $\omega \to 0$ we get $\mathrm{Re}\sigma(\omega) \to 0$, which manifests localization (compare with (4)). The dimensionless conductance $g(L)$ as function of the sample characteristic size L was also obtained by SCDT [28,29]. For $L > r_c$ (when (9a) is inapplicable) they got

$$g_{2d}(L) = \sigma_{2d}(L)\frac{\hbar}{e^2} = \frac{1}{2\pi} \ell^{-L/r_c} \left(1 + \frac{L}{r_c}\right) \ell n \left[1 + \left(\frac{r_c}{L}\right)^2\right] \qquad (11a)$$

which shows that $g_{2d}(L) \to 0$ for $L \to \infty$, in agreement with conclusions of the scaling theory [9,10]. From (10), (11) and (11a), it follows that at $T = 0$ for $d = 2$ localization should take place. For $d = 2$ localization is even weaker than for $d = 1$ ($r_{c,2d} >> r_{c,1d}$). However, its physical origin is the same as for $d = 1$, precisely the quantum interference effect. This will be discussed in more detail in Section 2 where a short review of SCDT will be given. In Section 3 SCDT will be generalized for the case of highly anisotropic systems. The reader who is not interested in all these theoretical details may safely skip Sections 2 and 3, with the exception of those places in Section 3 where the models for quasi-1d, quasi-2d and anisotropic 2d systems are formulated. Physical results obtained for these systems can be found in Sections 4-8. In Section 9 I give a comparative analysis of the various predictions concerning the metal-insulator transition for noninteracting electrons obtained by different methods. In Section 10 the role of e-e interactions in the localization process is shortly discussed.

2. SELF-CONSISTENT DIAGRAMMATIC TREATMENT OF LOCALIZATION DUE TO WEAK DISORDER IN 1D- AND ISOTROPIC 2D SYSTEMS. A SHORT REVIEW

First, I shall discuss the most important results for quantum corrections to $\sigma^{(0)}$ obtained in the case of weak disorder, $\lambda = \hbar/2\pi\epsilon_F\tau << 1$. These corrections are due to quantum interference effects, which are very important for $d = 1$ [11,12] and $d = 2$ [9,25,26]. The present author and V.N. Prigodin [17,18,20] recently have shown them to be also important for highly anisotropic 3d systems. These corrections may be evaluated from the Kubo formula for $\sigma(\omega)$ by means of successive renormalization of diffuson lines by cooperons (cooperon insets in diffuson self-energy part) and vice versa [18,28-30]. Results of such iterative renormalization procedure are as follows

$$\sigma(\omega)/\sigma_o = D(\omega)/D_o = \delta(\omega) = 1 + \delta^{(1)}(\omega) + \delta^{(2)}(\omega) + \ldots \qquad (12)$$

In the case d = 1 for $\tilde{\omega} = \omega\tau << 1$ one gets

$$\delta^{(1)}(\omega) = -\frac{1}{2(-i\tilde{\omega})^{1/2}}, \; \delta^{(2)}(\omega) = \frac{a_2^{(2)}}{-i\tilde{\omega}} + \frac{a_1^{(2)}}{(-i\tilde{\omega})^{1/2}} + a_o^{(2)} \qquad (12a)$$

Numerical values of the constants in (12a) depend on the upper cut-off K_o for a wave vector (usually, $K_o = 1/\ell$). The first term $\delta^{(1)}(\omega)$ corresponds to the well-known result for a wire [25] (in the limit of only one atom in cross-section). Such series are not tractable because a higher-order term $\delta^{(2)}$ also contains a contribution $(-i\tilde{\omega})^{-1/2}$ (of course, it is "small" compared to the main term $(-i\tilde{\omega})^{-1}$in $\delta^{(2)}$) which is comparable to $\delta^{(1)}$. We know the rigorous answer (4) for d = 1 from which it follows that a very delicate mutual cancellation of all terms containing ω in negative powers should take place. To see this more sophisticated methods should be applied and one of them is SCDT [28,29]. In the case of d = 2 ($\tilde{\omega} << 1$) quantum corrections up to λ^4 may be obtained from the general expression given in Section II in Ting's paper [30]

$$\delta^{(1)} = -\lambda \ell n\left(\frac{1}{-i\tilde{\omega}}\right); \; \delta^{(2)} = \lambda^2 \ell n\left(\frac{1}{-i\tilde{\omega}}\right); \; \delta^{(3)} = \lambda^3\left[\ell n^2\left(\frac{1}{-i\tilde{\omega}}\right) - \ell n\left(\frac{1}{-i\tilde{\omega}}\right)\right]$$

$$\delta^{(4)} = -\lambda^4\left[\frac{4}{3}\ell n^3\left(\frac{1}{-i\tilde{\omega}}\right) + \ell n^2\left(\frac{1}{-i\tilde{\omega}}\right) - \ell n\left(\frac{1}{-i\tilde{\omega}}\right)\right] \qquad (12b)$$

The general structure of the quantum corrections is given by the formula

$$\delta^{(n)}(\tilde{\omega}) = \lambda^n \sum_{K=1}^{n-1} a_K^{(n)}\left[\ell n\left(\frac{1}{-i\tilde{\omega}}\right)\right]^K \qquad (13)$$

Therefore, expansion (12) may be expressed as series in powers of $\ell n(1/-i\tilde{\omega})$

$$\delta(\omega) = 1 - \sum_{n=1}^{\infty} f_n(\lambda)\,\ell n^n\left(\frac{1}{-i\tilde{\omega}}\right) \qquad (14)$$

here $f_n(\lambda)$ are determined as series in powers of λ, $f_n(\lambda) = \sum_{K=1}^{\infty} b_K^{(n)} \lambda^{n+K}$.

In principle, numerical coefficients $a_K^{(n)}$ in (13) for $K < n-2$ (for instance, in the last term for $\delta^{(4)}$ in (12b)) may be determined rigorously only if we take into account less divergent diagrams not incorporated in the iterative procedure [28-30] mentioned above. The localization criterion ($\ell im_{\omega\to 0} \mathrm{Re}\,\sigma(\omega) \to 0$) may be expressed as follows

$$\ell im_{\omega\to 0} \mathrm{Re}\,\delta(\omega) \to 0 \qquad (15)$$

Since every term in (14) diverges as $\tilde{\omega} \to 0$, it is necessary first to sum the series in (14) and only then take the limit $\tilde{\omega} \to 0$. This program may be realized within the framework of the self-consistent diagrammatic treatment (SCDT) developed by Wölfle and Vollhardt [26,27] on the basis of self-consistent current-relaxation theory by Götze et al. [29,30]. The most important consequence of this approach is the conclusion that a self-consistency condition may be written directly for the diffusion coefficient $D(\omega)$ connected with the conductivity $\sigma(\omega)$ by Einstein relation

$$\sigma(\omega) = e^2 N(\epsilon_F) D(\omega) \tag{16}$$

where $N(\epsilon_F) = dn/d\epsilon_F$ is the density of states at the Fermi energy. For isotropic d-dimensional systems the corresponding SCDT equation takes the form

$$D_d(\omega) = D_d^{(o)} - \frac{1}{\hbar\pi N_d(\epsilon_F)} \int \frac{d^d q}{(2\pi)^d} \frac{D_d(\omega)}{-i\omega + D_d(\omega) q^2} \tag{17}$$

here $D_d^{(o)} = V_F \ell/d$ is the nonrenormalized diffusion coefficient which follows from Boltzman equation. The general definition for $N_d(\epsilon_F)$ is

$$N_d(\epsilon_F) = \int \frac{d^d K}{(2\pi)^d} \, \delta(\epsilon(\vec{K}) - \epsilon_F) \tag{18a}$$

For an isotropic energy spectrum, $\epsilon(K)$, one gets from (18a)

$$N_1(\epsilon_F) = \frac{1}{\pi\hbar V_F} = \frac{m}{\pi\hbar^2 K_F}; \; N_2(\epsilon_F) = \frac{m}{2\pi\hbar^2}; \; N_3(\epsilon_F) = \frac{2}{(2\pi)^2} \frac{m}{\hbar^2} K_F \tag{18}$$

It is very important that in this approximation the dependence of D on q appears to be inessential. As $D_d(q,\omega) \to D_d(\omega)$, with (18) taken into account, Eq. (17) may be rewritten as [29]

$$\frac{D_d(\omega)}{D_d^{(o)}} = \delta(\omega) = 1 - d\,\lambda\, K_F^{2-d} \int_0^{q_o} \frac{q^{d-1} dq}{-i\omega/D_d(\omega) + q^2} \tag{19}$$

The upper cut-off q_o is usually set equal to $1/\ell$. After evaluation of the integral over q we come to the following SCDT equations

$$\delta(\tilde{\omega}) = 1 - \frac{1}{\pi} \left(\frac{\delta(\tilde{\omega})}{-i\tilde{\omega}}\right) \operatorname{arctg} \left(\frac{\delta(\tilde{\omega})}{-i\tilde{\omega}}\right) \qquad \text{for } d = 1 \tag{20a}$$

$$\delta(\tilde{\omega}) = 1 - \lambda \ell n \left(\frac{\delta(\tilde{\omega})}{-i\tilde{\omega}}\right) \qquad \text{for } d = 2 \tag{20b}$$

Successive iteration of these equations leads to (12a) and (12b). For $\tilde{\omega} = 0$ the only possible solution of (20a) and (20b) is $\delta(\omega) = 0$, which means that due to quantum interference effects in the presence of weak disorder ($K_F\ell \gg 1$) localization takes place. Solutions of (20a) and (20b) for $\tilde{\omega} \equiv \omega\tau \ll 1$ can be easily found. Let us put

$$\delta(\tilde{\omega}) = -i(\tilde{\omega}\delta_1^{(1)} + \tilde{\omega}^3 \delta_1^{(2)}) + \tilde{\omega}^2 \delta_2^{(1)} + \tilde{\omega}^4 \delta_2^{(2)} \tag{21}$$

and substitute (21) in (20a) or (20b). We can show that Eqs. (20a) and (20b) are satisfied by (21) and get

$$\delta(\tilde{\omega}) = -i\tilde{\omega} + 8\tilde{\omega} \qquad \text{for } d = 1 \tag{21a}$$

$$\delta(\tilde{\omega}) = -i \frac{\tilde{\omega}}{\tilde{\omega}_o} + \left(\frac{\tilde{\omega}}{\tilde{\omega}_o}\right)^2 \frac{1}{\lambda} \qquad \text{for } d = 2 \tag{21b}$$

here $\tilde{\omega} = \omega_o\tau = e^{-1/\lambda}$. Equation (21b) applies for $\omega < \omega_o$. These results may be compared with (4) and (11) quoted in the Introduction. The level repulsion effect that gives rise to a weak $(\ell n\tilde{\omega})^{d+1}$ factor [5] is not obtained within SCDT. However the most important conclusion for us about the inevitability of localization stipulated by weak disorder for d = 1 and d = 2 seems to be convincing. In SCDT the density response function $\chi(q,\omega)$ (see ref. 29), rather than the function $\rho_\infty(x)$ (see Introduction), is calculated directly. According to ref. 29 it may be expressed as

$$\chi(q,\omega) = \chi^T \frac{D(q,\omega)q^2}{-i\omega + D(q,\omega)q^2} \tag{22}$$

where $\chi^T = N(\epsilon_F)$ is the isothermal compressibility. From (22) it follows that $D(\omega) = -i\omega r_c^2$, where r_c is the characteristic length

$$r_c = \begin{cases} \ell & \text{for } d = 1 \\ \left(\frac{D_o}{\omega_o}\right)^{1/2} \simeq \ell\, e^{1/2\lambda} & \text{for } d = 2 \end{cases} \tag{23}$$

As we shall see below r_c should be interpreted as the localization radius. Equation (22) can be rewritten as

$$\chi(q,\omega) = N(\epsilon_F) \frac{r_c^2 q^2}{1+q^2 r_c^2} \tag{22a}$$

More convenient for physical interpretation is the density relaxation function $\emptyset(q,\omega)$ as introduced in mode-coupling theories [31,32]

$$\emptyset(q,\omega) = \frac{\chi - \chi^T}{r_c^2 \omega} = -\frac{N(\epsilon_F)}{\omega} \frac{1}{1+q^2 r_c^2} \tag{24}$$

From (24) it follows that $\lim_{t\to\infty} \emptyset(q,t) \neq 0$. Indeed,

$$\lim_{t\to\infty} \emptyset(q,t) = \int_{-\infty}^{\infty} \emptyset(q,\omega) e^{-i\omega t} d\omega \sim N(\epsilon_F) \frac{1}{1+q^2 r_c^2} \tag{25}$$

Equation (25) gives a Fourier transform for the function $\emptyset_\infty(r)$ describing an exponential localization ($\exp(-r/r_c)$) in r-space with a characteristic length r_c, which is usually called localization radius. At this point, I finish the review of the results obtained for 1d- and isotropic 2d-systems and come to highly anisotropic systems.

3. SELF-CONSISTENT DIAGRAMMATIC TREATMENT OF LOCALIZATION IN HIGHLY ANISOTROPIC SYSTEMS IN THE PRESENCE OF WEAK DISORDER

In this paragraph a short description of SCDT equations obtained by the present author and V.N. Prigodin for quasi-1d systems [17,18], layered materials [20] and anisotropic 2d planes [33] is given. The SCDT approach for systems limited in size [34] will be explained in Sections 7 and 8.

1. Quasi-1d systems. Assuming the 1d metallic chains (which are parallel to each other) to be packed into a square lattice with lattice spacing $a_\perp$ and considering the nearest-neighbor coupling only, the energy spectrum is given by

$$\epsilon(\vec{p}) = \epsilon_F + V_F(|p_\parallel| - \epsilon_F) - w\psi(\vec{p}_\perp) \tag{26}$$

where $\psi(\vec{p}_\perp) = \cos(p_x a_\perp/\hbar) + \cos(p_y a_\perp/\hbar)$, p_x and p_y being perpendicular components of momentum, V_F and $p_\parallel$ are the Fermi velocity and momentum of electron along the chain, w is the interchain exchange integral, whose value characterizes the strength of interchain tunneling.

For $w/\epsilon_F \ll 1$, the Fermi surface consists of two open corrugated planes, w/ϵ_F being the measure of this corrugation. With increasing of parameter w/ϵ_F we move from the system of uncoupled 1d chains ($\tilde{w} \equiv 0$) to anisotropic 3d system. To make a transition to a 1d case to be more obvious we suppose impurities to be disposed on chains, their potential being effective only within a given chain. Impurities on different chains are assumed to be independent and act incoherently. Nevertheless, for $w \neq 0$, a short range scattering appears which influences the transverse motion too. Scattering processes accounting for the moment transfer $\Delta p_\parallel$ for a motion along the chain may be represented in a simplified form as "backward" ($\Delta p_\parallel \simeq 2p_F$) and "forward" ($\Delta p_\parallel \simeq 0$) scatterings. Corresponding relaxation times are τ_1 and τ_2, respectively. The total relaxation time τ and the corresponding total mean free path ℓ are defined as

$$\frac{1}{\tau} = \frac{1}{\tau_1} + \frac{1}{\tau_2} \quad , \quad \ell = V_F\tau \tag{27}$$

In our case, the parameter $w\tau$ is arbitrary, but $\epsilon_F\tau/\hbar \gg 1$. In refs. 17 and 18 a diagrammatic analysis was performed. It was shown [18] that the vertex function describing the scattering of two particles with close energies possesses singularities in two channels: cooper channel (electron-electron scattering) and diffusion channel (electron-hole scattering). In the first approximation they may be considered as independent and in a ladder approximation we got for cooperon propagator γ and diffusion propagator λ similar expressions (for details see ref. 18)

$$\gamma = \lambda = \frac{V_F}{2\tau}\left\{\frac{1}{\tau}\frac{1}{[-i\omega + D_\parallel(o)\, q_\parallel^{\,2} + \tilde{D}_\perp(o)(2-\psi(\vec{q}_\perp))]}\right\} \tag{28}$$

where $D_\parallel(o) = V_F^2\tau_2/2$ is the longitudinal diffusion constant corresponding to a result following from the 1d Boltzman equation. It is quite natural that it is proportional to τ_2 accounting only for backward scattering. Quantity $\tilde{D}_\perp(o) = w^2\tau/\hbar$ may be called a transverse diffusion constant. In usual units it is equal to $a_\perp^2 w^2\tau/2\hbar = D_\perp(o)$. For the e-h channel ($\gamma$), $\vec{q}$ is the difference of the e- and h- moments. For an e-e channel (λ), $\vec{q}$ is the total moment of two particles. If

$$\tilde{\omega} \equiv \omega\tau \ll 1, \; \tau D_\parallel(o) q_\parallel^{\,2} \simeq \ell\, \ell_2 q_\parallel^{\,2} \ll 1, \; (w\tau)^2 \ll 1 \tag{29}$$

the denominator in (28) is small, showing "infrared divergency" in both channels, but with a different momentum variable. For $w\tau > 1$, the quantity $2 - \psi(\vec{q}_\perp)$ should be expanded in series and instead of $\tilde{D}_\perp(o)(2 - \psi(\vec{q}_\perp))$, we get $\tilde{D}_\perp(o) q_\perp^2 a_\perp^2/2 = D_\perp(o) q_\perp^2$. In this case we get the usual pole

$$\frac{1}{\tau}\frac{1}{(-i\omega + D_\parallel(o)\, q_\parallel^{\,2} + D_\perp(o) q_\perp^{\,2})} = \frac{1}{-i\omega + \frac{1}{2}\ell_\parallel^{\,2} q_\parallel^{\,2} + \frac{1}{2}\ell_\perp^{\,2} q_\perp^{\,2}} \tag{30}$$

with anisotropic diffusion constants.

As it was pointed out [35] localization may be obtained only if interaction between both modes (cooperon and diffuson) is taken into account. The influence of diffuson on cooperon in the first approximation is accounted for by inserting diffuson lines into cooperon bubbles (see ref. 18) and vice versa. As a result, instead of (28) we get

$$\gamma^{(1)} = \lambda^{(1)} = \frac{V_F}{2\tau}\left\{\frac{1}{\tau}\cdot\frac{1}{[-i\omega + D_{\parallel}^{(1)}(\omega)q_{\parallel}^{2} + \tilde{D}_{\perp}^{(1)}(\omega)(2-\psi(\vec{q}_{\perp}))]}\right\} \quad (31)$$

where instead of $D_{\parallel}^{(o)}(\omega)$ and $D_{\perp}^{(o)}(\omega)$ in denominator there appear diffusion coefficients $D_{\parallel}^{(1)}(\omega)$ and $\tilde{D}_{\perp}^{(1)}(\omega)$ including first order quantum corrections

$$D_{\parallel,\perp}^{(1)}(\omega) = D_{\parallel,\perp}^{(o)} - \frac{1}{\hbar\pi N_1(\epsilon_F)}\int\frac{d^3q}{(2\pi)^3}\,\frac{D_{\parallel,\perp}^{(o)}}{-i\omega + D_{\parallel}^{(o)}q_{\parallel}^{2} + \tilde{D}_{\perp}^{(o)}(2-\psi(\vec{q}_{\perp}))} \quad (32)$$

where $N_1(\epsilon_F) = (\hbar\pi V_F)^{-1}$ is the 1d density of states on a Fermi surface. $N_1(\epsilon_F)$ appears in (32) only if the Fermi surface is open. In fact, the second term in (32) is the first order quantum correction due to the quantum interference effect.

Investigation of the higher-order quantum corrections to γ and λ along the lines given above leads us to the conclusion that for small $w\tau$, $q_{\parallel}\ell$ and $\omega\tau$ the pole structure for diffuson and cooperon (see eqs. (28) and (31)) remains unchanged (if we do not bother about additional factors in the numerator which are very close to unity if inequalities (29) are fulfilled). In any order n diffusion coefficients $D_{\parallel,\perp}^{(n)}(\omega)$ appearing in poles of $\gamma^{(n)}$, $\lambda^{(n)}$ may be obtained by iteration procedure of the following SCDT equations

$$D_{\parallel,\perp}(\omega) = D_{\parallel,\perp}^{(o)} - \frac{1}{\pi\hbar N_1(\epsilon_F)}\int\frac{d^3q}{(2\pi)^3}\,\frac{D_{\parallel,\perp}(\omega)}{-i\omega + D_{\parallel}(\omega)q_{\parallel}^{2} + D_{\perp}(\omega)(2-p(\vec{q}_{\perp}))} \quad (33)$$

which may be compared to (17). Equation (32) is easily obtained from (33) on the first step of iteration. As in the isotropic case [29], it turned out to be crucial to keep only the frequency dependence of $D(q,\omega)$. Therefore, q-dependence in $D(q,\omega)$ is neglected and we write $D(\omega)$ in (33).

If $w\tau < 1$, integration over $\vec{q}_{\perp}$ should be performed over all the transverse parts of a Brillouin zone $(-\pi/a_{\perp} \le q_x, q_y \le \pi/a_{\perp})$. The limits of the integral over $q_{\parallel}$ should be restricted*) to the values of $|q_{\parallel}| < q_o \sim 1/\ell_2$ where $\ell_2 = V_F\tau_2/2$. It is very important that Eq. (33) preserve the initial value of anisotropy of the diffusion coefficients for all ω $(\omega\tau < 1)$. Indeed, after introducing the dimensionless functions $\delta_{\parallel}(\omega)$ and $\delta_{\perp}(\omega)$

$$\delta_{\parallel}(\omega) = \frac{D_{\parallel}(\omega)}{D_{\parallel}^{(o)}} = \frac{\sigma_{\parallel}(\omega)}{\sigma_{\parallel}(o)};\quad \delta_{\perp}(\omega) = \frac{D_{\perp}(\omega)}{D_{\perp}^{(o)}} = \frac{\sigma_{\perp}(\omega)}{\sigma_{\perp}(o)} \quad (34)$$

We see from (33) that they both satisfy the same equation, that is $\delta_{\parallel}(\omega) = \delta_{\perp}(\omega) = \delta(\omega)$, and the equation for $\delta(\omega)$ is as given below.

---

* This problem is not solved completely within SCDT, but in our case a change of cut-offs influences only a numerical factor in the final formula (44) which varies very slightly even if we extend the limits of integration over $q_{\parallel}$ from $-\infty$ to ∞.

$$\delta(\omega) = 1 - \frac{1}{\hbar\pi N_1(\epsilon_F)}\int_0^{q_o}\frac{dq_\parallel}{\pi}\int_{-\pi}^{\pi}\frac{dx}{2\pi}\int_{-\pi}^{\pi}\frac{dy}{2\pi}\Big[-i\frac{\omega}{\delta(\omega)} + D_\parallel^{(o)}q_\parallel^2 + \tilde{D}_\perp^{(o)}(2-\cos x - \cos y)\Big]^{-1} \qquad (35)$$

where $D_\parallel^{(o)} = \ell_2 V_F$, $\tilde{D}_\parallel^{(o)} = w^2\tau/\hbar$. This is the SCDT equation for quasi-1d systems with open Fermi surfaces [17,18]. The very important conclusion that the effect of anisotropy can be completely absorbed into anisotropic diffusion coefficients and that the ratio $D_\parallel(\omega)/D_\perp(\omega)$ remains constant for $0 < \omega\tau < 1$ was first obtained in 1983 by the present author and V.N. Prigodin [17]. In 1984 Bhatt and Wölfle [36] treating an example of an anisotropic 2d system came to a similar conclusion. As we see below this is a general property of anisotropic systems. Recently it was confirmed experimentally [37].

2. Layered Materials. Quasi-2d system may be considered as stacks of metallic planes coupled together by an interplane exchange integral w, whose value characterizes now the strength of the interplane coupling. It is assumed that $w << \epsilon_F$. Then the energy spectrum of such a system is given by

$$\epsilon(\vec{p}) - \epsilon_F = V_F\,(|p_\parallel| - p_F) - w\cos(p_\perp a_\perp/\hbar) \qquad (36)$$

where $\vec{p}_\parallel$ is the momentum (2d vector) and V_F is the Fermi velocity characterizing electron motion in a layer plane, $p_\perp$ is the momentum component in direction perpendicular to layer plane $(-\pi < p_\perp a_\perp < \pi)$. For $w/\epsilon_F << 1$ the Fermi surface is open and represents a periodically corrugated cylinder. The parameter w/ϵ_F characterizes the measure of this corrugation. For $w/\epsilon_F \to 1$ the Fermi surface becomes closed. Changing the value of the parameter w/ϵ_F one may come from a highly anisotropic quasi-2d system to an anisotropic 3d layered material.

To make the transition to a 2d system when $w \to 0$ more obvious we assume impurities to be distributed on planes, their potential being effective only within a given layer. The random potentials created by the impurities located on different planes are statistically independent and act incoherently. As a result this potential may be considered as short ranged with respect to $p_\perp$.

We assume scattering to be sufficiently weak, so that $\epsilon_F\tau >> 1$ but the parameter $w\tau$ may be arbitrary. I shall not reproduce here the diagrammatical analysis which resembles the reasoning given aboe in section 12 of this paragraph. All necessary details may be found in [20]. Again, we come to the conclusion that the initial anisotropy for transport coefficients remains. Therefore, we have only one unknown function

$$\delta(\omega) = \frac{D_j(\omega)}{D_j(o)} = \frac{\sigma_j(\omega)}{\sigma_j(o)}$$

Here $\sigma_j^{(o)} = \ell^2 D_j^{(o)} N_2(\epsilon_F)/a_\perp$ is the conductivity of a layered system, $a_\perp$ is the distance between layers. The SCDT equation for $\alpha(\omega)$ may be written as follows with $j = \parallel, \perp$:

$$\delta(\omega) = 1 - \frac{1}{\hbar\pi N_2(\epsilon_F)} \int_{-\pi}^{\pi} \frac{dq_\perp}{2\pi} \int \frac{d(q_\parallel^2)}{2\pi} \left[-i\omega/\delta(\alpha) + D_\parallel^{(o)} q_\parallel^2 + \tilde{D}_\perp^{(o)}(1-\cos q_\perp)\right]^{-1} \tag{37}$$

This is the main SCDT equation for layered (quasi-2d) systems with open Fermi surfaces (corrugated cylinders). Here indices $\parallel$ and $\perp$ denote parallel or perpendicular to layer plane.

$$D_\parallel^{(o)} = V_F \ell/2; \quad \tilde{D}_\perp^{(o)} = w^2 \tau/2; \quad \ell = V_F \tau$$

As may easily be seen, for $w \to 0$ Eqs. (35) and (37) transform to Eqs. (20a) and (20b) for 1d- and 2d- systems.

3. Highly Anisotropic 2d System. Let us consider the system of weakly coupled chains on an infinite plane. The energy spectrum of a system is given by

$$\epsilon(p) = \epsilon_F + V_F(|p_\parallel| - p_F) - w\cos(p_\perp q_\perp/\hbar) \tag{38}$$

where $p_\parallel$ is the moment and V_F is the Fermi velocity along the chain. Here, in distinction with a quasi-2d case, $p_\parallel$ is a 1d vector, $p_\perp$ is the momentum component in direction perpendicular to chains, $-\pi < p_\perp a_\perp < \pi$; $a_\perp$ being the interchain spacing. We suppose, for simplicity, that scattering is characterized by only one parameter τ, accounting for a backward scattering along the chain. As it was mentioned in Section 1, a short range scattering for a transverse motion will inevitably arise, if $w \neq 0$.

It may be shown again [31] that anisotropy is retained and we need only one function δ. The corresponding SCDT equation takes the form [31] (compare with (35) and (37))

$$\delta(\omega) = 1 - \frac{1}{\hbar\pi N_1(\epsilon_F)} \int_0^{q_\perp^o} \frac{d(q_\perp a_\perp)}{\pi} \int_0^{1/\ell_\parallel} \frac{dq_\parallel}{\pi} \left[-i\omega/\delta(\omega) + \tilde{D}_\parallel^{(o)} q_\parallel^2 + \tilde{D}_\perp^{(o)}(1-\cos q_\perp a_\perp)\right]^{-1} \tag{39}$$

It is essential that $N_1(\epsilon_F)$ rather than $N_2(\epsilon_F)$ comes into (39) which corresponds to the case when the Fermi surface is supposed to be open ($w/\epsilon_F << 1$). For $w\tau/\hbar << 1$ the upper cut-off is $q_\perp^{(o)} = \pi/a_\perp$, as above.. However, for $\tilde{w} >> 1$ (but $w/\epsilon_F << 1$) it should be put equal to $1/\ell_\perp$, where

$$\ell_\perp = \langle V_\perp^2 \rangle^{1/2} \tau \simeq \frac{w}{\hbar} a_\perp \tau = a_\perp \tilde{w} \tag{40}$$

and the term $D_\perp^{(o)}(1-\cos a_\perp a_\perp)$ in square brackets of (39) should be replaced by $D_\perp^{(o)} a_\perp^2 q_\perp^2 = D_\perp^{(o)} a_\perp^2$, as in Section 1. Then we get for $w > 1$ (but $w/\epsilon_F << 1$)

$$\delta(\omega) = 1 - \frac{a_\perp}{\hbar\pi N_1(\epsilon_F)} \int_0^{1/\ell_\perp} \frac{dq_\perp}{\pi} \int_0^{1/\ell_\parallel} \frac{dq_\parallel}{\pi} \left[-i\omega/\delta(\omega) + q_\parallel^2 D_\parallel^{(o)} + q_\perp^2 D_\perp^{(o)}\right]^{-1} \tag{41}$$

4. <u>Concluding Remarks</u>. There exist several methods for deriving the SCDT equation. Some of them were mentioned in Section 2. In Section 1 of this paragraph I have tried to summarize the key points of the approach proposed by me and V.N. Prigodin [17,18]. However, all derivations are based on the same approximations. Only the limited number of the most divergent terms in every step of the iteration procedure is included (for instance, see the discussion of the case d = 2 in Section 2). Radical changes of these methods are needed to go beyond this approximation. The most remarkable feature of all SCDT equations for the case of weak disorder $(K_F\ell)^{-1} \ll 1$) is the very simple form of the frequency dependence of the integrand, ω appears only in its denominator. This makes the analysis of SCDT equations to be very simple and excludes purely mathematical errors. In all known cases when the results obtained by other methods (analytical or numerical) may be considered rigorous, the predictions of the SCD treatment are in reasonable agreement with them [28,38]. However, we should not forget that the procedure of evaluating SCA equations is based on selective summation of infinite sets of certain classes of diagrams containing infrared divergencies in cooper and diffusion channels. To make the calculations of higher order quantum corrections more tractable it is usually assumed that not all but some of the functions (occurring on the intermediate stages of the calculations) that change slowly in an interesting interval of the variables ω, $\vec{q}$ may be considered to be approximately constant. Of course, these mathematical simplifications may be clarified to some extent by employing arguments based on the mode-coupling approach [28,29,31,32]; however, it does not solve all the problems. Therefore, I hesitate to declare blankly that further improvements of this procedure would not expose small frequency dependent corrections in the denominator of the integrand in the SCDT equation which may affect subtle details in the behavior of $\delta(\omega)$ especially in cases when $\delta(\omega)$ becomes very small. This may appear to be essential in the vicinity of the metal-insulator transition, especially with respect to the problem of the minimum metallic conductivity. I shall return to this point in Section 9.

4. LOCALIZATION THRESHOLD IN A QUASI 1D SYSTEMS AND AC CONDUCTIVITY

As we know, for 3d isotropic materials, the Anderson localization takes place only for sufficiently strong disorder, when $K_F\ell \leq 1$. In this case, the localization length $r_c \sim a$, where a is the lattice constant. On the other hand, for d = 1 all states are localized already for arbitrarily small disorder (Mott localization) and the localization length r_c may be very large ($r_c \sim b \sim a(K_F\ell) \gg a$) if $K_F\ell \gg 1$. How does this change take place, continuously or sharply? One possible way of approaching this problem is to take a system of N weakly coupled chains in the presence of weak disorder and then let $N \to \infty$ which corresponds to a real quasi-one-dimensional (quasi 1d) system. The degree of "quasi-one-dimensionality" may then be characterized by the value of the dimensionless parameter w/ϵ_F, where ϵ_F is the Fermi energy for a single metallic chain, w is the interchain exchange integral characterizing the strength of interchain tunneling or the width of the energy spectrum in the transverse direction (see below). For quasi-1d systems $w/\epsilon_F \ll 1$. The strength of weak disorder is characterized by the parameter $K_F \ell \sim \epsilon_F \tau/\hbar$, which is supposed to be large.

As we see later, the single important parameter, which characterizes both the degree of quasi-one-dimensionality and the strength of disorder in our model is the product $K_F \ell w/\epsilon_F = w\tau$. However, the problem of N chains (for finite N) appears to be rather difficult (see Sections 5-7 below). Therefore, I shall treat here the infinite quasi-1d-system by means of the SCDT equation (35) which is applicable if the limit $N \to \infty$ is taken from the very beginning.

Our main conclusion is that Mott's localization (for weak disorder, $K_F \ell >> 1$) found until now only for 1d systems does not disappear abruptly as very weak interchain tunneling is "switched on." It disappears only if w becomes larger than the threshold value $w_c \sim 0.3\tau^{-1}$. Hence, weak disorder influences very strongly on physical properties of quasi-1d systems for $w \lesssim w_c$. This result was obtained by the present author and V.N. Prigodin in 1983 [17] for a special case $\tau_1 = \tau_2$. Independently, but somewhat later, Apel and Rice [39] came to similar results for w_c on the basis of a one-parameter scaling approach. Introducing the following dimensionless variables $\tilde{\omega} = \frac{1}{2}\omega\tau_2$, $\tilde{w} \le w\left(\frac{1}{2}\tau\tau_2\right)^{1/2}$ we rewrite Eq. (35) for $\delta(\omega)$ in the form

$$\delta = 1 - \int_0^1 \frac{dz}{\pi} \int_{-\pi}^{\pi} \frac{dx}{2\pi} \int_{-\pi}^{\pi} \frac{dy}{2\pi} \left[z^2 + \tilde{w}^2 (2 - \cos x - \cos y) - i\tilde{\omega}/\delta \right]^{-1} \tag{42}$$

For $\tilde{w} > 0$ system is in the localized regime. Therefore, for small $\tilde{w}$ we may put $\delta = -i\tilde{\omega}\xi^2$ (see Section 2) and in the limit $\tilde{\omega} \to 0$ we come to the following equation for dimensionless localization length ξ

$$1 = \int_0^1 \frac{dz}{\pi} \int_{-\pi}^{\pi} \frac{dx}{2\pi} \int_{-\pi}^{\pi} \frac{dy}{2\pi} \left[\xi^{-2} + z^2 + \tilde{w}^2 (2 - \cos x - \cos y) \right]^{-1} \tag{43}$$

The value of $\tilde{w}$ for which $\xi \to \infty$ (we denote it as $\tilde{w}_c$) determines the localization threshold. From (43) it immediately follows that $\tilde{w}_c \approx 0.3$ or

$$w_c \approx 0.3\,(\tau\tau_2/2)^{-1/2} \tag{44}$$

For $w < w_c$ quasi-1d system is in a localized state and for $|\epsilon| = (w_c - w)/w_c << 1$ from (43) one gets

$$1 = -a_1 + \frac{1}{\tilde{w}} a_2 - a_3 \frac{1}{\tilde{w}^2 \xi} \tag{45}$$

where a_1, a_2 and a_3 are the numerical factors of the order of unity. From (45) it follows that $\xi \simeq b/|\epsilon|$, where b is of the order of unity. Hence, for parallel and perpendicular localization radii one has

$$r_\parallel \simeq \ell/|\epsilon|; \quad r_\perp \simeq a_\perp \tilde{w}/|\epsilon| \quad \text{for } |\epsilon| = (w_c - w)/w_c << 1 \tag{46}$$

Such critical behavior is typical for the system which is near the metal-insulator transition. The SCDT approach gives critical indices for $r_\parallel$ and $r_\perp$ equal to 1. For $\omega = 0$ and $w > w_c$ from (43) one gets

$$\delta(0,w) \sim 1.25\,(w-w_c)/w_c;\quad \sigma_{\parallel,\perp}(o) = \sigma_{\parallel,\perp}^{(o)}\,1.25(w-w_c)/w_c \tag{47}$$

Thus, the conductivity increases monotonously from its zero value at $w = w_c$.

This means that within the SCD treatment for a quasi-1d system, we did not obtain a minimum metallic conductivity. Perhaps this is due to the deficiency of SCDT (see the end of Section 3). Nevertheless, I am sure that the insulator-metal transition should occur near w_c. This can be explained on the basis of the following very simple physical argument. What was really done? With the help of the SCDT equation we have investigated for a quasi-1d system the role of disorder, which is "weak" ($\epsilon_F\tau \gg 1$) for an electron propagating along the chain but appears to be sufficiently strong ($\tilde{w}\tau < 1$) for a transverse motion of an electron. At first glance, the following picture for an electron motion may be imagined. Being initially located on chain No. 1, the electron becomes localized (Mott localization) and remains within some segment whose length is of the order of $r_{\ell a}^{(1)}$. In as much as $w \neq 0$, it may tunnel (the probability of tunneling being small) to one of the neighboring chains (let this be chain No. 2) and thus "overcome the difficulties" which prevented its movement along the first chain. Now it moves to and from along chain No. 2 (again within some piece whose length is of the order of $r_{\ell a}^{(1)}$) and again it may tunnel to the next chain, and so on. If these considerations turn out to be true there would be no reason for the electron to be localized. From the SCD treatment it follows that for sufficiently small w ($w < w_c$) such a simple-minded picture is invalid since due to intrachain scattering there exists an uncertainty in energy $\sim 1/\tau$. If this is larger than w - the band width with respect to $p_\perp$ - such simple picture could not be valid. Only for $w > 1/\tau$ it makes sense.

Our conclusion that for T = 0 the dc conductivity in quasi-1d system is equal to zero for $w < w_c$ is in obvious conflict with results obtained by Abrikosov and Ryzhkin [13,40]. Approximations used in [40] are equivalent to the neglect of renormalization for $D_\perp$ in (33). In this case, the localization threshold indeed does not appear. However, we have shown that even the simplest way of taking into account the w-corrections to $D_\perp^{(o)}$ brings about finite values for w_c. Of course, the numerical factor 0.31 in (44) depends on the quality of the approximation and may be improved by a more rigorous approach.

From (44) it follows that the value of w_c depends not only on τ_2 (backward scattering) but on τ_1 (forward scattering) as well. An increase of w_c due to including forward scattering may easily be understood on qualitative grounds because it leads to an additional disorder which should enhance the tendency towards localization. Near the threshold ($|\epsilon| \ll 1$ and for $|\tilde{\omega}/\delta| \ll 1$) Eq. (42) can be simplified (see ref. 17) and the following results for ac conductivity are easily obtained. For w/w_c (localized regime)

$$\frac{\sigma_{\parallel,\perp}(\omega)}{\sigma_{\parallel,\perp}^{(o)}} = \begin{cases} -i\dfrac{\tilde{\omega}}{|\epsilon|^2} + 2\dfrac{\tilde{\omega}^2}{|\epsilon|^5} & \text{for } \tilde{\omega} \ll |\epsilon|^3 \ll 1 \quad (48a) \\ (\sqrt{3} - i)\,|\tilde{\omega}|^{1/3} & \text{for } |\epsilon|^3 \ll \tilde{\omega} \ll 1 \quad (48b) \end{cases}$$

Thus, a quasi-1d system for $w < w_c$ behaves like a disordered 3d insulator near the mobility edge (compare with refs. 39 and 40) if $|\epsilon| << 1$. For $w > w_c$ (extended regime)

$$\frac{\sigma_{\parallel,\perp}(\omega)}{\sigma_{\parallel,\perp}(o)} \sim \begin{cases} (\sqrt{3} - i)|\tilde{\omega}|^{-1/3} & \text{for } |\epsilon|^3 << \tilde{\omega} << 1 \quad (49a) \\ 1.25\epsilon + b\,(1-i)\left(\frac{\tilde{\omega}}{|\epsilon|}\right)^{1/2} & \text{for } \tilde{\omega} << |\epsilon|^3 << 1 \quad (49b) \end{cases}$$

Here b is a constant of order unity. This behavior is typical for a 3d metal near the mobility edge [39,40]. The specific d = 1 frequency dependence of $\sigma(\omega)$ restores only for $w << w_c$

$$\sigma_{\parallel}(\omega)/\sigma_{\parallel}(o) \sim -i\tilde{\omega} + 8\,\tilde{\omega}^2 \quad (50)$$

As mentioned above, the SCD treatment is unable to give a factor of the form $\ell n^2\tilde{\omega}$ as a second term in (50). For $w > w_c$ a frequency dependence of that type may appear only in the vicinity of $\omega\tau \simeq 1$ (since $\tilde{w}_c \simeq 0.3$) where SCDT does not work well enough. In conclusion, I give the expression for both components of the dielectric permeability tensor near the threshold (in the localized regime)

$$\chi_{\parallel} = \frac{\omega_p^2\tau^2}{4\pi}\cdot\frac{1}{\epsilon^2}\,;\quad \chi_{\perp} = \frac{ne^2w^2\tau^2a_{\perp}^2}{\epsilon_F\hbar^2} \approx \frac{e^2K_F}{\epsilon_F}\left(\frac{w\tau}{\hbar}\right)^2\frac{1}{\epsilon^2} \quad (51)$$

where $\omega_p^2 = 4\pi N_1(\epsilon_F)/a_{\perp}^2$ is the "plasma frequency." This differs from the result (6) for $\chi_{\parallel}$ obtained for a strictly 1d system by the large factor $|\epsilon|^{-2} = w_c^2/(w_c-w)^2 >> 1$. The value of $\chi_{\perp}$ for 1d system should be equal to zero.

5. MOBILITY EDGE AND AC CONDUCTIVITY FOR WEAKLY DISORDERED QUASI-2D MATERIALS

In this paragraph, I shall shortly describe the results for layered systems which may be obtained on the basis of SCDT Eq. (37) (for a more detailed description, see ref. 20).

If $\tilde{w} = w\tau < 1$ and $\left|-\frac{i\tilde{\omega}}{\delta}\right| \leq 1$

both integrals in (37) can easily be evaluated and Eq. (37) takes the form

$$\delta(\tilde{\omega}) = 1 + \lambda\ell n\left[-\frac{i\tilde{\omega}}{\delta(\tilde{\omega})} + \tilde{w}^2 + \left[-\frac{i\tilde{\omega}}{\delta(\tilde{\omega})}\left(-\frac{i\omega}{\delta} + 2\tilde{w}^2\right)\right]^{1/2}\right] \quad (37a)$$

Here $\lambda = \hbar/2\pi\epsilon_F\tau = 1/\pi k_F\ell << 1$. In a limit $\tilde{w} \to 0$ Eq. (37a) transforms into Eq. (20b) for strictly 2d system. To find a localized state for which $\lim_{\omega\to 0}\mathrm{Re}\,\delta(\omega) \to 0$ as earlier (see (21)), we put for $\tilde{\omega} \to 0$:

$$\delta(\tilde{\omega}) = -i\tilde{\omega}\,\xi^2 + a\omega^2$$

where ξ is the dimensionless parameter that characterizes the scale for the localization radii (see (55) below). Then from (37a) it follows that

$$1 = \lambda\,\ell n\,2 - \lambda\,\ell n\left(\xi^{-2} + \tilde{w}^2 + \left[\xi^{-2}\left(\xi^2 + 2\tilde{w}^2\right)\right]^{1/2}\right) \quad (52)$$

In contrast to the quasi-1d case we have here two free parameters. I shall fix λ (or $\tilde{\epsilon}_F = \epsilon_F\tau/\hbar$) and vary $\tilde{w}$. Then from (52) it follows that all states at the Fermi level are localized, if $\tilde{w} < \tilde{w}_c$ where $\tilde{w}_c$ is the solution of (52) for $\xi \to \infty$

$$\tilde{w}_c = \sqrt{2}\, \exp(-\pi\tilde{\epsilon}_F) \quad \text{or} \quad w_c = (2\hbar/\tau)\, \exp(-1/2\lambda) \tag{53}$$

In localized regime ($w < w_c$) from (52) one gets for ξ

$$\xi = 1/\tilde{w}_c\,[1-(\tilde{\omega}/\omega_c)^2] \tag{54}$$

which leads to the following results for both localization radii

$$r_{\|} = \ell_o\,\xi;\ r_{\perp} = a_{\perp}\tilde{\omega}\xi. \tag{55}$$

$\mathrm{Re}\,\delta(\omega, \tilde{w}) \neq 0$ for $w > w_c$ and a layered system is in a metallic state

$$\sigma_{\|,\perp}\,/\sigma_{\|,\perp}^{(o)} = \delta(0,\tilde{w}) = 1-2\lambda\ell n(\sqrt{2}/\tilde{w}) = (\epsilon_F - \epsilon_c)/\epsilon_F \tag{56}$$

Here $\epsilon_c = (\hbar/\pi\tau)\ell n(\sqrt{2}/\tilde{w})$ may be identified with a mobility edge. For $\tilde{\omega} > \tilde{\omega}_c$ this edge is below the Fermi level ($\epsilon_c < \epsilon_F$) and the system is in the extended regime (metallic state). On the other hand, $\epsilon_c > \epsilon_F$ if $\tilde{w} < \tilde{w}$, and the system is in a localized regime (insulator state). Since disorder is supposed to be weak ($\epsilon_F\tau/\hbar >> 1$), ω_c is exponentially small (see (53)). Therefore, the SCDT treatment claims that even a very weak interlayer coupling completely destroys the weak localization predicted for purely 2d systems by SCDT [28,29] and by one-parameter scaling theories [9,10]. It is quite obvious that the influence of different weak perturbations lending 3d features to 2d systems should be important in thin films too. For instance, it may be due to weak tunneling of electrons from the film to a metallic substrate (for instance, see ref. 41). The most important conclusion is that a weak localization in 2d and quasi-2d systems is a very subtle phenomenon that can easily be destroyed.

Now I shortly summarize the result for the ac conductivity which can be obtained from SCA Eq. (37a) for the metallic state.

I. $\tilde{w}_c < \tilde{w} \ll 1$, but $\Delta = (\epsilon_F - \epsilon_c)/\epsilon_F >> \lambda$ or $(w - w_c)/w_c >> 1)$

$$\frac{\sigma_{\|,\perp}(\omega)}{\sigma_{\|,\perp}^{(o)}} = \begin{cases} \Delta + (\lambda/\tilde{w})\ (2i\tilde{\omega}/\Delta)^{1/2} & \text{for } \tilde{\omega} << \tilde{w}^2 \\ 1 - \lambda\ell n(1/(-i\tilde{\omega})) & \text{for } \tilde{w}^2 << \tilde{\omega} << 1 \end{cases} \tag{57}$$

Thus, for $\tilde{\omega} >> \tilde{w}^2$ the corrugation of the Fermi surface becomes unimportant and the frequency dependence is the same as for the 2d case (compare with (9)). For $\tilde{\omega} << \tilde{w}^2$ a layered system shows 3d behavior [42,43] but the scale is different.

II. $\Delta << \lambda << 1$. The system is near the Mott transition.

$$\frac{\sigma_{\|,\perp}(\omega)}{\sigma_{\|,\perp}^{(o)}} = \begin{cases} \Delta + (\lambda/\tilde{w}_c)(-2i\tilde{\omega}/\Delta)^{1/2} & \text{for } \tilde{\omega} << \Delta^3(\tilde{w}_c/\lambda)^2 \\ (\lambda/\tilde{w}_c)^{2/3}(-i\tilde{\omega})^{1/3} & \text{for } \Delta^3(\tilde{w}_c/\lambda)^2 << \tilde{\omega} << \tilde{w}_c^2 \\ 1 - \lambda\ell n(1/(-i\tilde{\omega})) & \text{for } \tilde{w}_c^2 << \tilde{\omega} << 1 \end{cases} \tag{58}$$

In the first two frequency regions we have scaling behavior typical for an

Anderson 3d metal-insulator transition [42,43], but all scales are different. In the dielectric phase ($w < w_c$) the following results from Eq. (37a) may be obtained.

I. Near the Mott transition, $\Delta << \lambda$

$$\frac{\sigma_{\|,\perp}(\omega)}{\sigma_{\|,\perp}(0)} = \begin{cases} -2i\tilde{\omega}\xi^2 + 4\frac{\tilde{w}_c}{\lambda}\xi^5\tilde{\omega}^2 & \text{for } \tilde{\omega} << \Delta^3(\tilde{w}_c/\lambda)^2 \\ (\tilde{w}_c/\lambda)^{-2/3}(-i\tilde{\omega})^{1/3} & \text{for } \Delta^3(\tilde{w}_c/\lambda)^2 << \tilde{\omega} << \tilde{w}_c^2 \\ 1 - \lambda \ln(1/-i\tilde{\omega}) & \text{for } \tilde{w}_c^2 << \tilde{\omega} << 1 \end{cases} \tag{59}$$

The low frequency behavior is typical for a dielectric near the transition [43].

II. Far from the Mott transition, $\Delta >> \lambda$

$$\frac{\sigma_{\|,\perp}(\omega)}{\sigma_{\|,\perp}(0)} = \begin{cases} -2i\tilde{\omega}\xi^2 + 4(\tilde{w}_c/\lambda)\xi^5\tilde{\omega}^2 & \text{for } \tilde{\omega} << \tilde{w}_c^2 \\ 1 - \lambda \ln(1/-i\tilde{\omega}) & \text{for } \tilde{w}_c^2 << \tilde{\omega} << 1 \end{cases} \tag{60}$$

Thus, for small frequencies $\omega << \tilde{w}_c^2 \tau/\hbar \doteq (2\hbar/\tau)\exp(-2\tau\epsilon_F\tau/\hbar)$ we get a behavior typical for 3d insulators, but all scales are quite different, they are proportional to $[\exp(-2\pi\epsilon_F\tau/\hbar) << 1]$ which is the characteristic feature of weak localization for d = 2.

It is interesting that in contrast to the quasi-1d case a layered system exhibits purely 2d behavior in both phases (for certain frequency intervals). In the end I quote the results for the dielectric permeability tensor χ. Using relation (5) and formulae (59) and (60) we find in the localized regime

$$\chi_\| \simeq 8\pi\xi^2\tau\sigma_\|(0) = 2\xi^2(\omega_p\tau)^2 = \frac{\omega_p^2\tau^2}{1-(w/w_c)^2} \cdot \exp(2\pi\epsilon_F\tau/\hbar)$$

$$\chi_\perp \simeq 8\pi\xi^2\tau\sigma_\perp(0) = e^2N_2(\epsilon_F)a_\perp\left(\frac{w\tau}{\hbar}\right)^2 \frac{1}{1-(w/w_c)^2}\exp(2\pi\epsilon_F\tau/\hbar) \tag{61}$$

Here ω_p is the plasma frequency $\omega_p^2 = 2\pi ne^2/m_\|$. The dimensionless parameter $e^2N_2a_\perp$ is assumed to be small.

6. HIGHLY ANISOTROPIC 2D SYSTEM--CHAINS ON INFINITE PLANE

Let us consider the 2d system consisting of an infinite number ($N \to \infty$) of weakly coupled chains, parallel to each other and periodically arranged on a plane. This model may have some relevance to a system of dislocations on the grain boundaries of a germanium bicrystal [44-46]*. I shall return to this point in the next paragraph in which the case of finite number N (but $N >> 1$) will be considered. Now we turn to Eq. (39) obtained in paragraph 3, Section 3. After a substitution of $\delta = -i\tilde{\omega}\xi$ in (39) and taking the limit $\tilde{\omega} \to 0$ we come to two different equations for the definition of ξ for $w < 1$ and for $w > 1$. In the first case ($w < 1$) integration over $q_\perp$ within the limits $-\pi < q_\perp a_\perp < \pi$ can easily be performed and we transpose SCDT equation (39) into the form

*I am very grateful to Dr. Yu. G. Shreter who drew my attention to such a possibility.

$$1 = (\sqrt{2}/\pi)\xi \int_0^{\xi} du\,[(u^2+1)(u^2+1+2\tilde{w}^2\xi^2]^{-1/2} \tag{62}$$

from which it follows that ξ is of the order of unity for $\tilde{w} \leq 1$. This means that the localization radius along the chain $r_{\|}$ is of order ℓ. More interesting is the second case ($\tilde{w} > 1$), for which (after multiplying of integrand in (41) by τ/τ and introducing of dimensionless variables $z = \ell_{\|} q_{\|}$, $y = q_{\perp}\ell_{\perp}$) SCDT equation (41) takes the form

$$1 = \frac{a_{\perp}}{\ell_{\perp}} \frac{2}{\pi^2} \int_0^1 dz \int_0^1 dy \frac{1}{z^2+y^2+\xi^{-2}} = \frac{1}{2\pi\tilde{w}} \ell n \left(\frac{\xi}{\tilde{\xi}}\right) = \gamma \ell n \left(\frac{\xi}{\tilde{\xi}}\right) \tag{63}$$

where $\tilde{\xi} \approx 1$ and $\gamma = (2\pi\tilde{w})^{-1}$ is a new dimensionless coupling constant ($\gamma < 1$). From (63) it follows that

$$\xi = \xi_{\infty} = \tilde{\xi} \exp(\pi\tilde{w}) = \tilde{\xi} \exp(1/2\gamma) >> 1 \tag{64}$$

and for the localization radii we get for $\tilde{w} = w\tau/\hbar > 1$ (but $w/\epsilon_F << 1$)

$$r_{\|} \simeq \ell_{\|} \exp(\pi w\tau/\hbar); \; r_{\perp} = a_{\perp} (\hbar/w\tau) \exp(\pi w\tau/\hbar) \tag{65}$$

This should be compared to the answer $r_c \sim \ell \exp(\hbar\epsilon_F\tau/\hbar)$ for the isotropic case (para 2). One may ask how can we go from (65) to the result $r_{\|} \to r_{\perp} \to r_c$ when $w/\epsilon_F \to 1$? To see this we should consider the case of anisotropic but closed Fermi surfaces, and substitute $N_2(\epsilon_F) = (m_{\perp} m_{\|})^{1/2}/2\pi\hbar^2$ instead of $N_1(\epsilon_F)/a_{\perp}$ in (41) and put $a_{\perp}^{o} \to 1/\ell_{\perp}$. Then instead of the factor $(2\pi\tilde{w})^{-1}$ in (63) there arises another factor

$$\frac{1}{\hbar\pi N_2(\epsilon_F)} \frac{1}{\ell_{\perp}\ell_{\|}} \tau \to \frac{1}{\pi} \frac{\hbar}{\tau} \frac{1}{V_{\|}^{F} V_{\perp}^{F} (m_{\perp} m_{\|})^{1/2}} \simeq \frac{\hbar}{2\pi\tau\sqrt{w_{\|} w_{\perp}}} \tag{66}$$

which obviously tends to $\lambda = \hbar/2\pi\epsilon_F\tau$ for the isotropic case and yields $\xi = \exp(1/2\lambda)$ or $r_{\|} = r_{\perp} = r_c$.

Both components of the ac conductivity tensor $\sigma_{\|,\perp}(\omega) = \sigma_{\|,\perp}^{o}\delta(\omega)$ can easily be obtained from SCDT equation (41) (for $\tilde{w} > 1$)

$$\delta(\tilde{\omega}) = 1 - \gamma \ell n \left(\frac{\delta(\tilde{\omega})}{-i\omega}\right) = 1 - \frac{1}{2\pi\tilde{w}} \ell n \left(\frac{\delta(\tilde{\omega})}{-i\tilde{\omega}}\right) \tag{67}$$

which resembles Eq. (20b) for the isotropic case. The only difference is due to the presence of $\gamma = (2\pi w\tau/\hbar)^{-1}$ instead of $\lambda = (2\pi\epsilon_F\tau/\hbar)^{-1}$. Introducing now $\Omega_o = \frac{1}{\tau} e^{(-1/\gamma)} = \frac{1}{\tau} e^{-2\pi\tilde{w}}$ instead of $\omega_o = \frac{1}{\tau}\exp(-\pi\epsilon_F\tau/\hbar)$ (in para 2) one gets for $\omega << \Omega_o$ (compare with (21b)):

$$\frac{\sigma_{\|}(\omega)}{\sigma_{\|}^{o}} = \frac{\sigma_{\perp}(\omega)}{\sigma_{\perp}^{o}} = \left(\frac{\omega}{\Omega_o}\right)^2 \frac{1}{\gamma} - i\frac{\omega}{\Omega_o} \tag{68}$$

For $\tilde{\Omega}_o < \tilde{\omega} < 1$ (here $\tilde{\Omega}_o = \Omega_o\tau$) we put $\delta(\omega) = -i\delta_1 + \delta_2$ and get from (67) two following transcendental equations

$$\delta_1 = \pi\gamma/2 - \gamma \operatorname{arctg} \delta_1/\delta_2 \tag{69a}$$

$$\delta_2 = 1 - \gamma \ln\left[(\delta_1^2 + \delta_2^2)^{1/2}/\tilde{\omega}\right] \tag{69b}$$

From (69a) it follows that $\delta_1 = \pi\gamma/4$ for $\delta_1 = \delta_2$. From (69b) it follows that δ_1 may be equal to δ_2 only for

$$\tilde{\omega} \equiv \tilde{\Omega}_1 = \frac{\pi}{2\sqrt{2}} \exp(\pi/4)\ \gamma \exp(-1/\gamma) < \tilde{\Omega}_o \tag{70}$$

For $\tilde{\omega} > \tilde{\Omega}_1$, δ_2 increases more rapidly than δ_1 and for $\tilde{\omega} = \omega\tau \to 1$ we get

$$\delta_1 \simeq \frac{\pi}{2}\gamma\ ,\ \delta_2 \to 1 + \frac{1}{2}\left(\frac{\pi}{2}\right)^2 \gamma^3 \simeq 1 \tag{71}$$

The dependence of $\mathrm{Re}\,\frac{\sigma_{\parallel,\perp}(\omega)}{\sigma_{\parallel,\perp}(o)} = \delta_2(\omega)$ and $\mathrm{Im}\,\frac{\sigma_{\parallel,\perp}(\omega)}{\sigma_{\parallel,\perp}(o)} = \delta_1(\omega)$ on $\tilde{\omega}$ is sketched in Fig. 1.

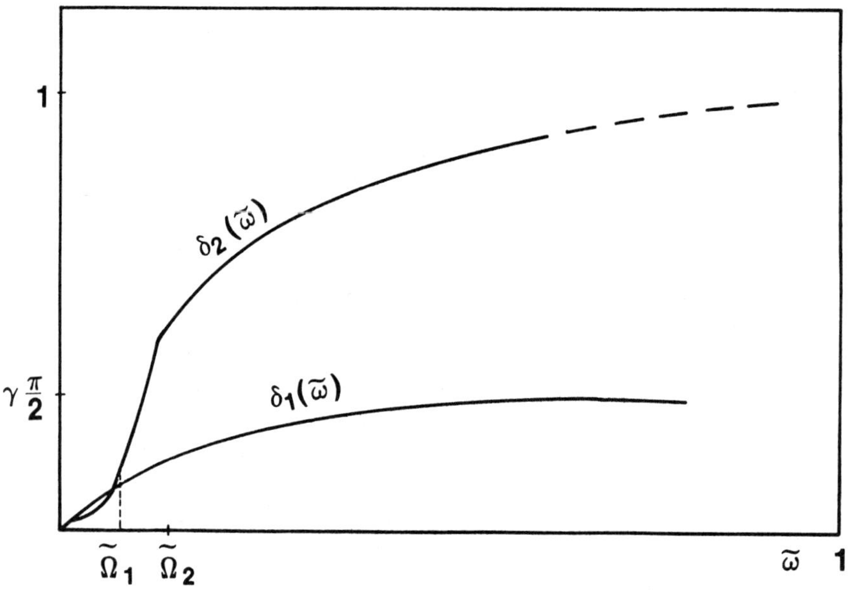

Fig. 1 The dependence of $\mathrm{Re}\,\frac{\sigma_{\parallel,\perp}(\omega)}{\sigma_{\parallel,\perp}(o)} = \delta_2(\tilde{\omega})$ and $\mathrm{Im}\,\frac{\sigma_{\parallel,\perp}(\omega)}{\sigma_{\parallel,\perp}(o)} = \delta_1(\tilde{\omega})$ on $\tilde{\omega} = \omega\tau$ for an anisotropic 2d system $\gamma = \frac{\hbar}{2\pi W\tau} < 1$.

The most important conclusions obtained in this paragraph can be summarized as follows:

1. Localization radii $r_{\parallel}$ and $r_{\perp}$ increase very rapidly with increasing $\tilde{w} = w\tau/\hbar$ (see (65)) attaining very large, albeit finite values.

2. Anisotropy of ac conductivity in a plane remains the same for all values of ω in the interval $0 < \omega\tau < 1$.

7. N WEAKLY COUPLED PARALLEL CHAINS ON A PLANE

Now we suppose the number of chains N to be finite, albeit very large ($N >> 1$). For what N_{max} this finite system behaves itself like the infinite one considered above in Section 6? How does N_{max} depend on $\tilde{w}$? The question is not of purely theoretical interest. For instance, for a system of dislocations on germanium bicrystal grain boundaries their number and distance $a_{\perp}$ between them (and, hence, $w(a_{\perp})$) can be varied experimentally by changing of the tilt angle between grains [44-46]. A consistent generalization of the diagrammatic SCA treatment [28,29] for the case of a discrete spectrum leads to the following prescriptions (for details see refs. 33 and 34). Now $q_{\perp}$ takes $2N + 1$ discrete values

$$q_{\perp}a_{\perp} \rightarrow \pi n/N \quad \text{for } -N < n < N \tag{72}$$

The integral over $q_{\perp}$ in (39) and in (41) should be replaced by the sum

$$\int_{-q_o}^{q_o} \frac{d(q_{\perp}a_{\perp})}{2\pi} \rightarrow \frac{1}{2N}\sum_{n=-N_{max}}^{N_{max}} \tag{73}$$

where $N_{max} = N$ for $\tilde{w} < 1$ and $N_{max} = \sqrt{2}N/\pi\tilde{w} = 2\sqrt{2}\gamma N$ for $\tilde{w} > 1$. In the last case it is necessary to expand the cosine function in the denominator of (39)

$$\tilde{w}^2(1-\cos q_{\perp}a_{\perp}) \rightarrow \tilde{w}^2(a_{\perp}q_{\perp}(n)/2)^2 \rightarrow (n/N_{max})^2 \tag{74}$$

Then, instead of (41) for $\tilde{w} > 1$ one gets (in the limit $\tilde{\omega} \rightarrow 1$) the following SCA equation

$$1 = \frac{\sqrt{2}}{\pi}\int_0^1 dz\, \frac{1}{2N}\sum_{n=-N_{max}}^{N_{max}} \frac{1}{z^2 + \xi^{-2} + (n/N_{max})^2} \tag{75}$$

The sum over n should be evaluated with the accuracy $O(1/N^2)$. This may be done with the use of the following approximate relation

$$\frac{N_{max}^2}{2N}\sum_{n=-N_{max}}^{N_{max}}\frac{1}{b^2+n^2}=\frac{N_{max}^2}{2N}\left[\frac{\pi}{b}\,\mathrm{cth}(\pi b)-\frac{2}{b}\,\mathrm{arctg}\left(\frac{b}{N_{max}+1}\right)-\frac{1}{b^2+(N_{max}+1)^2}+O\left(\frac{1}{N_{max}^3}\right)\right] \qquad (76)$$

In our case $b = N_{max}(z^2+\xi^{-2})^{1/2}$. The integral over z can easily be evaluated. The first term in (76) after integration over z gives for $\xi < N_{max}$

$$\frac{\sqrt{2}}{\pi}\frac{N_{max}}{2N}\,\pi\ell n\,(\xi+\sqrt{\xi^2+1}) \to 2\gamma\ell n 2\xi \qquad (77a)$$

and for $\xi > N_{max}$

$$\frac{\sqrt{2}}{\pi}\left\{\frac{1}{2N}\,\xi\,\mathrm{arctg}\left[\left(\frac{\xi}{N_{max}}\right)^2-1\right]^{1/2}+\sqrt{2}\pi\gamma\left\{\ell n\left(\xi+\sqrt{\xi^2+1}\right)-\ell n\left[\frac{\xi}{N_{max}}+\sqrt{\left(\frac{\xi}{N_{max}}\right)^2-1}\right]\right\}\right.$$
$$\left.+\gamma\left[1-\left(\frac{N_{max}}{\xi}\right)^2\right]^{1/2}\cdot f\left(\frac{N_{max}}{\xi}\right)\right\} \qquad (77b)$$

where $f(N_{max}/\xi)$ is a slowly varying function of order of unity in the whole interval $0 < N/\xi < 1$. The second term in (76) after integration over z for $\xi >> 1$, but for arbitrary ξ/N_{max} becomes

$$-\gamma K_1 + \gamma K_2\frac{1}{\xi^2} \qquad (78)$$

where the constants K_1 and K_2 are of order of unity $\left(K_1=\frac{4}{\pi}\sum_{n=0}^{\infty}\frac{(-1)^n}{(2n+1)^2}\right)$.

The last term in (76) gives an unimportant correction $\sim$ const/N and may be omitted. Thus for $1 << \xi < N_{max}$ the SCA equation takes the form

$$1 + K_1\gamma = 2\gamma\ell n 2\xi \qquad (79)$$

and we come to the result (for $N > \xi$)

$$\xi = \xi_\infty = \exp(1/2\gamma) \qquad (80)$$

Here ξ_∞ is the characteristic scale for the localization radii in an infinite anisotropic 2d system (see Section 6). For ξ tending to N_{max} from above ($\xi > N_{max}$), using (77b) and (78) we come to the same conclusion. However, for $\xi >> N_{max}$ we obtain from (77b) and (78) the following SCA equation

$$1 + \gamma K = \frac{1}{\sqrt{2}}\frac{\xi}{N} + 2\gamma\ell n(N_{max}) \qquad (81)$$

From (81) it follows (for $\xi_\infty > \xi > N_{max}$)

$$\xi = 2\sqrt{2}\gamma N\ell n(\xi_\infty/N_{max}) = N_{max}\,\ell n(\xi_\infty/N_{max}) > N_{max} \qquad (82)$$

The corresponding function $\xi(N)$ for $\tilde{w} > 1$ is shown in Fig. 2. Thus it may be said that for $\tilde{w} = \tilde{w}\tau/\hbar >> 1$ the system of N chains on a plane for $N > \xi_\infty = \exp(\pi\tilde{w})$ behaves like the infinite highly anisotropic 2d system considered in Section 6. For $w < 1$ a similar analysis shows that $r_\parallel \sim \ell_1$, $r_\perp \sim a_\perp$. Of course, to apply this model for describing real experiments [44–46] on germanium bicrystals (for small tilt angles) it is necessary to calculate quantum corrections for $T \neq 0$ with e–e interactions taken into account. However, formula $\sigma(\omega)$ for $T = 0$ obtained in Section 6 can be applied for finite temperatures when $1/\tau > \omega > 1/\tau_{in}$ where τ_{in} is the inelastic scattering time.

8. N COUPLED PARALLEL CHAINS PACKED IN A BUNDLE. A MICROSCOPIC MODEL OF ANISOTROPIC THIN WIRE

Theoretical investigations of thin wires began with the paper [15a] of Thouless. On the basis of qualitative physical arguments he came to the conclusion that for a 3d system in the shape of a long wire (made from

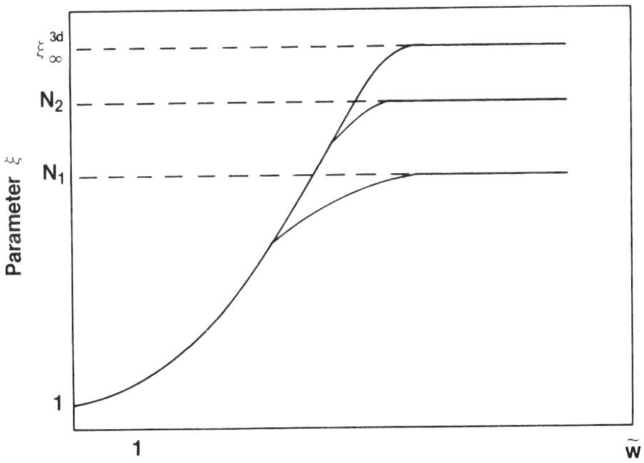

Fig. 2 The dependence of the characteristic scale $\xi(N)$ for the localization radii $r_\parallel$ and $r_\perp$ on the transverse coupling $\tilde{w} = \tilde{w}\tau/\hbar$.

$N_1 < N_2 < \xi_\infty^{3d} = \exp(\pi \frac{\epsilon_F \tau}{\hbar})$.

isotropic metal) all states are exponentially localized at $T = 0$ with a localization length

$$r_c\,(N) \sim \ell\,(k_F d)^2 \sim N\ell \tag{83}$$

where d is the radius of the wire, N is the number of atoms in cross-section. Later these considerations were generalized for non-zero temperature [15b] and a new length scale $L_{in} = (\ell . \ell_{in})^{1/2}$ (here ℓ_{in} is the inelastic mean-free-path) was introduced. This generalization is essential if we wish to apply the theory to real experiments on thin wires [47-49].

Here I shall consider the problem of an anisotropic thin wire at $T = 0$. What new can be said in this case? First of all, it is necessary to prove that the qualitative prediction (83) is rigorous from a mathematical point of view. For the system under consideration result (83) certainly fails to be applicable with an increase of the distance between chains, because tunneling between them weakens very drastically ($\tilde{w} \to 0$) and $r_c(N)$ will tend to ℓ instead of $N\ell$. How does this crossover between small and large localization lengths occur, smoothly or as an abrupt jump? What is the characteristic scale for $\tilde{w}$? Our model for the N coupled chains represents microscopically an N channel system, as discussed in the scaling theories [50-52] and it is interesting to investigate it without the additional hypotheses usually used in the scaling theories. Moreover, letting $N \to \infty$ we may gain some new insight into the problem of quasi-1d systems.

To treat this problem rigorously a modified Berezinskii's technique was developed in [16] by Weller, Prigodin and by the present author (see also the paper of the same three authors in book [16b]). The main physical result obtained in [16] for N large ($N \gg 1$) but finite and for $w\tau \gg 1$ reduces to the assertion that the localization length is $r_c^{(N)} = \alpha N \ell$, where the α is a numerical constant of order of unity. One can see that when $w\tau$ changes from 0 (set of N uncoupled chains) to $w\tau \gg 1$, the localization length $r_c^{(N)}$ increases very drastically (N times) and for $N \to \infty$ it tends to infinity (extended states). However, such a result based on a formal limit $N \to \infty$ is not conclusive because the main equation obtained is only applicable if "oscillating vertexes" are ignored, which limits the value of N from above.

Inclusion of these oscillating vertexes into the scheme and necessary generalization of the modified Berezinskii's technique developed in [16] were recently performed by the same authors and all basic equations were obtained (unpublished). However, the limiting case $N \to \infty$ is not solved at present. Here I shall present some new results obtained for this model within the SCD approach [34]. Following the prescriptions quoted in Section 7 I write the SCDT equation for $\tilde{w} < 1$ in the limit $\tilde{\omega} \to 0$ as follows (compare with (43)):

$$1 = \frac{\sqrt{2}}{\pi}\int_0^1 dz\,\frac{1}{N}\sum_{n_1=-N_{max}}^{N_{max}}\;\sum_{n_2=-N_{max}}^{N_{max}}\frac{1}{\xi^{-2}+z^2+2w^2\left[\sin^2\left(\frac{\pi}{2}\frac{n_1}{\sqrt{N}}\right)+\sin^2\left(\frac{\pi}{2}\frac{n_2}{\sqrt{N}}\right)\right]} \tag{84}$$

For a wire of square cross-section we have now (compare with (72)):

$$q_x a_\perp \to \pi n_1/\sqrt{N},\ q_y a_\perp \to \pi n_2/\sqrt{N};\ -\sqrt{N} \le n_1,\ n_2 < \sqrt{N} \quad (85)$$

For simplicity, I make the following approximation for the denominator in (84)

$$\sin^2(\pi n_1/2\sqrt{N}) + \sin^2(\pi n_2/2\sqrt{N}) \to n_1^2/N_{max}^2 + n_2^2/N_{max}^2;\ N_{max} = 2\sqrt{N}/\pi \quad (86)$$

Then instead of (84) we get

$$1 = \frac{\sqrt{2}}{\pi}\int_0^1 dz \frac{1}{\tilde{w}^2} \frac{N_{max}^2}{2N} \sum_{n_1=-N_{max}}^{N_{max}} \sum_{n_2=-N_{max}}^{N_{max}} \frac{1}{n_1^2 + n_2^2 + b^2} \quad (87)$$

where $b^2 = (N_{max}^2/2\tilde{w}^2)(z^2 + \xi^{-2})$. As in Section 7, we should evaluate both sums very carefully. The widely used procedure of introducing continuous variables (in our case they would be $x_1 = n_1/N_{max}$, $x_2 = n_2/N_{max}$) and transforming sums over n_1 and n_2 into integrals over x_1 and x_2 would be illegal if we want to get a correct dependence of ξ for $w\xi > N_{max}$. The first sum over n_1 may be obtained with the help of (76) if we substitute in (76) $\tilde{b} = \sqrt{b^2 + n_2^2}$ instead of b. The term corresponding to $n_2 = 0$ from the second sum (over n_2) may be considered along the lines given in Section 7 and after integration over z we get results similar to (77a) and (77b). For all other terms ($n_2 \neq 0$) of the second sum (over n_2) the formula (76) simplifies, because we can set

$$\mathrm{cth}(\pi b) \to 1,\ \mathrm{arctg}(\tilde{b}/N_{max}) \to \begin{cases} \pi/2 & \text{for } z > \tilde{w} \\ \tilde{b}/N_{max} & \text{for } z < \tilde{w} \end{cases} \quad (88)$$

Then (76) simplifies drastically, and we should only have to calculate the sum $\sum_{n_2=1}^{N_{max}} 1/\sqrt{b^2+n^2}$. This can be done if we use the following rigorous relation:

$$\sum_{n=1}^{N_{max}} \frac{1}{\sqrt{b^2 + n_2^2}} = \int_0^\infty J_0(bt)\exp(-t)\ \frac{1 - \exp(-tN_{max})}{1 - \exp(-t)} \quad (89)$$

The two-fold integral over z and t can easily be carried out. Now, with all key points of the treatment specified, I shall skip all tedious but straightforward calculations (for details, see ref. 34) and come to the results. Instead of Eq. (45) obtained in Section 4 for an infinite system $N \to \infty$ we get the following equations:

For $\sqrt{2}\ \xi\ \tilde{w} < N_{max} = 2\sqrt{N}/\pi$

$$1 = -a_1 + \frac{a_2}{\tilde{w}} - \frac{a_3}{\tilde{w}^2 \xi} + a_4 \frac{\xi}{N} \quad (90a)$$

For $\sqrt{2}\ \xi\ \tilde{w} > N_{max}$

$$1 = -a_1 + \frac{a_2}{\tilde{w}_1} + a_4' \frac{\xi}{N} - a_5 \frac{N^{1/2}}{\tilde{w}^3 \xi} \quad (90b)$$

All coefficients a_1, a_2, a_3, a_4, a_4', and a_5 are of order unity. The last term in (90a) and the two last terms in (90b) cannot be correctly obtained if we light-heartedly transform the sums over n_1, n_2 into a two-fold integral. However, these terms are very important as we shall see below. Let us rewrite Eqs. (90a) and (90b) in a more convenient form

$$a_3/\xi - a_4\tilde{\xi}/N = A\epsilon \tag{91a}$$

$$a_5 N^{1/2}/\tilde{\xi}^2 - a_4'\tilde{\xi}/N = B\epsilon \tag{91b}$$

where $\tilde{\xi} = \xi\tilde{w}$; $\epsilon = (w_1 - w)/w_c$, and A and B are constants of the order of unity.

From (91a) and (91b) we see that $\tilde{\xi} \sim \sqrt{N}$ for $\epsilon = 0$. From (91a) we obtain

$$\tilde{\xi} = A\frac{N\epsilon}{2a_4}\left[\sqrt{1 + 4a_4a_3\frac{1}{\epsilon^2 N}} - 1\right] \tag{92a}$$

From (91b) it follows that for $\epsilon < 0$ and $|\epsilon|\sqrt{N} > 1$ one gets $\tilde{\xi} \sim N/|\Delta| \sim N$. The behavior of ξ as function of w is shown in Fig. 3, where we see the

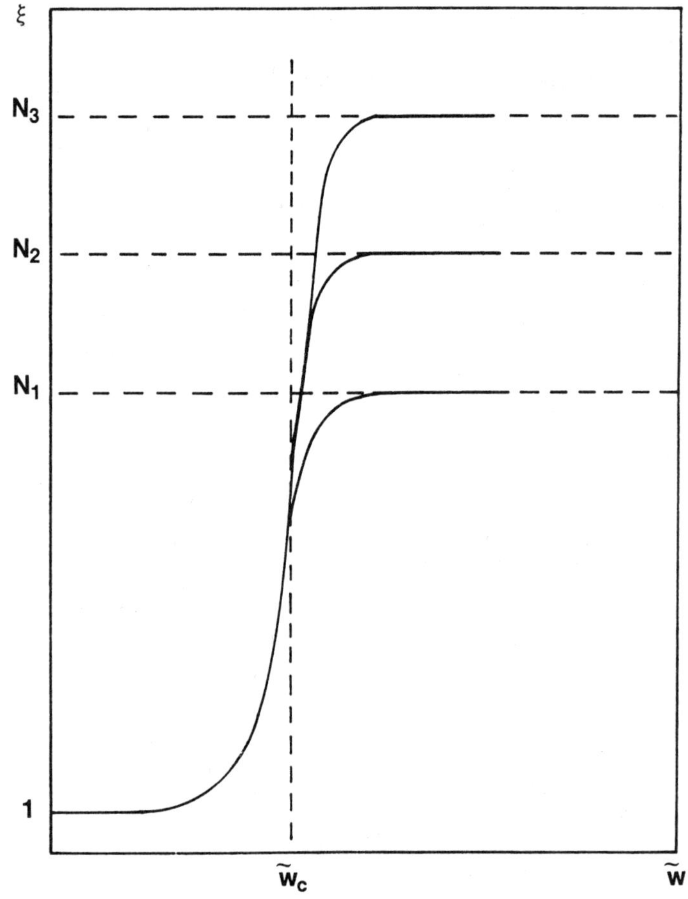

Fig. 3 The dependence on $\tilde{w}$ of the characteristic scale $\xi(\tilde{w})$ for a wire made of N parallel chains. $N_3 > N_2 > N_1$.

transition from the "small" localization length $\sim \ell$ (corresponding to independent chains) to the localization radius of a wire $\sim N\ell$. This "transition" takes place in the vicinity $w \simeq w_c$. Calculation of σ for $T \neq 0$ is in a progress now.

9. ON THE DESCRIPTION OF THE DISORDER-INDUCED METAL-INSULATOR TRANSITION IN 3D SYSTEMS

The reliability of any theoretical prediction depends on the accuracy of the method employed. All the results quoted above were obtained with the help of the self-consistent diagrammatic theory proposed earlier by Wölfle and Vollhardt and modified by the present author and Prigodin for the case of highly anisotropic systems. In Sections 2-3 I explained the basic assumptions and tried to form a true notion of the accuracy of this method. In all known cases when results were obtained independently by SCDT and by the other reliable methods there was a reasonable agreement between them (see Sections 1-2, further examples can be found in ref. 38). Therefore, I believe the prediction of localization in the presence of weak disorder in highly anisotropic systems to be true. However, more subtle details of behavior near the transition scarcely can be revealed by SCDT. In a sense, this approach is similar in spirit to the Van der Waals theory of critical points in fluids and the values of critical indices obtained in SCDT cannot be exact. For instance, from scaling theory near the mobility edge E_c one gets [42]

$$\sigma \sim \left(\frac{E_F - E_c}{E_F}\right)^{(d-2)\nu} ; \quad r_c \sim \left(\frac{E_F - E_c}{E_F}\right)^{-\nu} \tag{93}$$

SCDT yields the critical index $\nu = 1$, which certainly cannot be correct. The same can be said about the frequency dependence $\sigma(\omega) \sim \omega^{1/3}$ (or $\epsilon(\omega) \sim \omega^{-2/3}$).

A continuous decrease of σ near the transition point at which it smoothly vanishes does not contradict any fundamentals because it is unlikely that mechanism of transport near the localization threshold can be described by the Boltzman equation. Nevertheless, we also cannot reject a possibility of a discontinuous drop in σ at the transition point. To clarify this assertion I shall exploit an analogy with a usual phase transition. It is well known that critical fluctuations not taken properly into account in the mean field theory, after incorporating them into the scheme, change the value of the indices. Sometimes the critical fluctuations can even change the order of the phase transition (from the second order to the first order) causing an order parameter to drop discontinuously in the vicinity of T_c. We cannot exclude such a possibility for Anderson's localization transition too. To clear the matter up, it is necessary to go beyond approximations employed in SCDT (see Sections 2-3) or in one-parameter scaling theory which may be done with the help of renormalization group theory for which averaging over impurities is a straightforward procedure and the "Thouless conjecture," that conductance can be defined on a microscopic scale, is not employed. It seems to be a firmly established fact that if Coulomb effects are ignored, the conductance in the presence of weak disorder ($k_F\ell >> 1$) may be obtained with the help of renormalization group theory containing only one invariant coupling (one charge g). Thus, a one-parameter scaling description is noncontradictory, in principle. What is the form of $\beta(g)$? That is the question. Various approaches employed by different authors could not reveal up to now any discontinuities in the Gell-Mann Low function $\beta(g)$ or its derivative. However, some new interesting opportunities arise when e-e interactions are taken into account as shown below. Some other approaches are also worth mentioning here. It was established long ago [55] that the Anderson transition exists on a Cayley tree (Bethe lattice) which is believed to represent a space of infinite dimensionality ($d \to \infty$), and thus a transition with "mean field" exponents may be expected. However, present results contradict each other. For instance, in ref. 56 a minimum metallic conductivity at the transition is found for a tight-binding Anderson model. In ref. 57 with the help of an S-matrix approach it was found that the conductivity approaches zero smoothly. In ref. 58 with more mathematical rigor, authors came to a similar conclusion. However, Efetov [59] recently applied the supersymmetry method to a special model of a disordered granular system and found a discontinuous drop in σ at the transition point. I think it is not an exaggeration to say that the Bethe model is still far from being completely solved and many questions remain unanswered. Interesting arguments in favor of a minimum metallic conductivity can be found in the review article [60] by Bonch-Bruevich.

10. THE INFLUENCE OF e-e INTERACTIONS ON THE INSULATOR-METAL TRANSITION IN DISORDERED SYSTEMS

The interest in this problem sharply increased after 1974 when Mott's book "Metal-Insulator Transitions" appeared [7]. Since e-e interactions can by themselves precipitate a Mott-Hubbard type metal-insulator transition even in ordered systems it seems obvious that they should play an important role near the disorder-induced transition as well. In 1974 Schmid [61] has noticed that in an impure metal the energy relaxation time determined by inelastic e-e collisions is much smaller than in a pure metal, and that it may lead to a failure of the quasi-particle picture near the Fermi surface for $T \to 0$. In 1975, Efros and Shklovskii [62,63] pointed out that the Coulomb interactions can create a soft gap in the density of states near the Fermi level (see also [64]). Now it is quite clear that even in the presence of weak disorder, e-e interactions may seriously influence the single-particle density of states in metals also [65-69]. The following remark is very much to the point here. All these effects due to e-e interactions can be obtained within the SCD approach, too [70]. Altshuler and Aronov [65,66] have shown that in the presence of weak disorder a new set of logarithmic quantum corrections to the conductivity for $d = 2$ will arise if $T \neq 0$. A very illuminating interpretation of these effects was given by Mott and Kaveh [67]. These corrections were observed experimentally. Thus, it is obvious that a qualitatively new theory should be developed which can effectively sum up all logarithmic series. This program is in progress now [72-76]. Most of the basic conclusions concerning the impurity-driven metal-insulator transition that were obtained by ignoring the e-e interactions are not refuted in refs. 72-76. However, many important novel features of different stages of the localization process were discovered, especially for $d = 2$. For instance, Finkelstein [73c,d] argues that for $d = 2$ the system exhibits a tendency for spin density rearrangement and only after this happens, the Anderson instability develops and the localization transition takes place. However, the most important is the conclusion that the corresponding renormalization group theory [73] contains not one charge (conductance g) but several charges (five, in the general case) which seriously distinguishes it from the one-parameter scaling theory [9,10]. In principle, it makes it possible that smooth changes of one of the charges can trigger a discontinuous drop in g thus leading us back to the concept of a minimum metallic conductivity. In conclusion, I shall briefly describe how e-e interactions might modify the frequency dependence of the conductivity [73]. In the critical region of the metal-insulator transition the behavior of $\sigma(\omega)$ essentially depends on the magnitude of the charge z, because of the renormalization of the coefficients in the diffusion propagator

$$\frac{1}{Dq^2 - i z\omega} \tag{94}$$

According to ref. 73b, for the isotropic case we get

$$\sigma(\omega) \sim e^2 \left(\frac{\partial n}{\partial \mu} z\omega\right)^{(d-2)/d} \tag{95}$$

There are two possibilities: a) $z \to$ const (for instance, in the case of strong magnetic fields). Then for d = 3 we get $\sigma(\omega) \sim \omega^{1/3}$ as in the case of noninteracting electrons (when z = 1). b) $z \to 0$. This corresponds to the case of magnetic impurities

$$\sigma(\omega) \sim \omega^{(d-2)/(d-p)} \tag{96}$$

In the first order approximation the index is $p = \frac{\epsilon}{2}$, where $\epsilon = d - 2$. Then for d = 3 from (96) we get: $\sigma(\omega) \sim \omega^{2/5}$.

In the case of strong spin-orbital interactions, we get [74] $\sigma(\omega) \sim \omega^{1/3}$ which corresponds to the case a). The role of e-e interactions in highly anisotropic systems is not elucidated till now.

11. CONCLUSIONS

In this paper I surved the recent progress in the field of weak localization in highly anisotropic systems in the presence of weak disorder. The common feature of all the very different systems considered in Sections 4-8 is the crucial role of the quantum interference effect which manifested itself for the first time in the one-dimensional case considered by Mott and Twose in 1961. Of course, this effect shows itself in different ways in different systems, and the requirements on the accuracy of the mathematical methods for its exposure should not be the same (compare 1d and 2d cases). However, for weak disorder it is the only mechanism causing localization. The other general property of anisotropic weakly disordered systems is the observation (obtained within the SCD approximation) that the ratio $D_\perp(\omega)/D_\parallel(\omega)$ characterizing the anisotropy of the diffusion tensor, keeps its initial value $D_\perp(o)/D_\parallel(o)$ for $0 < \omega\tau < 1$. This property was discovered for the first time in 1983 [17] by the present author and Prigodin for a quasi-1d system and in 1984 [20] for a layered system. In 1984 Wölfle and Bhat [36] obtained the first quantum corrections for an anisotropic 2d system with a closed Fermi surface. They have shown that the anisotropy does not change with temperature (in a certain temperature interval). This prediction was observed experimentally [37]. It provided the first parameter-free test of the scaling theory of localization. The influence of weak disorder on a quasi-1d and a quasi-2d systems is very different. All states in a quasi-1d system turn out to be localized if weak interchain tunneling is smaller than the threshold value $w_c \simeq \tau^{-1}$. This localized regime differs from that for a single chain (see results for $\sigma(\omega)$ in Section 4). In a quasi-2d system extended states near the upper edge of a band immediately appear if a weak interlayer tunneling is switched on. With increasing w a borderline between extended and localized states (mobility edge) shifts down very rapidly. Already for very small w an insulator-metal transition takes place. The highly anisotropic 2d system of dislocations on the grain boundaries of a germanium bicrystal [44-46] may be a very fruitful study object for both theorists and experimentalists.

REFERENCES

1. P.W. Anderson, Phys. Rev. 109, 1492 (1958).
2. A.F. Ioffe and A.R. Regel, Progr. Semicond. 4, 237 (1960).
3. N.F. Mott and W.D. Twose, Adv. Phys. 10, 107 (1961).
4. N.F. Mott, Adv. Phys. 16, 49 (1967).
5. N.F. Mott, Phil. Mag. 17, 1259 (1968).
6. N.F. Mott, Adv. Phys. 26, 363 (1977).
7. N.F. Mott, Metal-Insulator Transitions, Taylor and Francis, London (1974).
8. N.F. Mott and E.A. Davis, Electron Processes in Non-Crystalline Materials, 2nd edition, Clarendon Press, Oxford (1979).
9. E. Abrahamas, P.W. Anderson, D.C. Licciardello and T.V. Ramakrishnan, Phys. Rev. Lett. 42, 673 (1979).
10. P.W. Anderson, Physica 117B & 118B, 30 (1983).
11. V.L. Berezinskii, Zh. Eksp. Teor. Fiz. 65, 1259 (1973). [Sov. Phys. JETP 38, 620 (1974).]
12. A.A. Gogolin, V.I. Mel'nikov and E.I. Rashba, Zh. Eksp. Teor. Fiz. 69, 327 (1975). [Sov. Phys. JETP 42, 168 (1976).]
13. A.A. Abrikosov and I.A. Ryzhkin, Adv. Phys. 27, 147 (1978).
14. A.A. Gogolin, Phys. Rept. 86, 2 (1982).
15. D.J. Thouless a) Phys. Rev. Lett. 39, 1167 (1977); b) Sclid State Comm. 34, 683 (1980).
16. W. Weller, V.N. Prigodin and Yu.A. Firsov, a) Phys. Stat. Sol. (b) 110, 143 (1982); b) Proc. of the Intern. Seminar on the Theory of Localization, Iohnsbach, DDR, December 5-9 (1983).
17. V.N. Prigodin and Yu.A. Firsov, Zh. Eksp. Teor. Fiz., Pis'ma 38, 241 (1983). [JETP Lett. 38, 284 (1984).]
18. Yu.A. Firsov and V.N. Prigodin, in Proc. of the Intern. Seminar on Theory of Localization Iohnsbach, DDR, December 5-9 (1983).
19. Yu.A. Firsov and V.N. Prigodin and Chr. Seidel, Phys. Rept., to be published in 1985.
20. V.N. Prigodin and Yu.A. Firsov, J. Phys. C 17, L979 (1984).
21. D.J. Thouless, Phys. Rept. 13C, 93 (1974).
22. F. Wegner, Zs. Phys. B 35, 207 (1979); Phys. Rev. B 19, 783 (1979).
23. S. Hikame, Phys. Rev. B 24, 2671 (1981).
24. K.B. Efetov, Adv. Phys. 32, 53 (1983).

25. L.P. Gor'kov, A.I. Larkin and D.E. Khmel'nizkii, Zh. Eksp. Teor. Fiz., Pis'ma 30, 248 (1979). [Sov. Phys. JETP, Lett. 30, 228 (1979).]
26. E. Abrahams and T.V. Ramakrishnan, J. Non-Cryst. Solids 35, 15 (1980).
27. M. Kaveh and N.F. Mott, J. Phys. C. 14, L177 (1981).
28. D. Vollhardt and P. Wölfle, Phys. Rev. 22B, 4666 (1980); Phys. Rev. Lett. 48, 699 (1982).
29. P. Wölfle and D. Vollhardt, in Anderson Localization, Y. Nagaoka, H. Fukuyama, eds., Springer-Verlag, Berlin-Heidelberg (1982) p. 26-43.
30. C.S. Ting, Phys. Rev. B26, 678 (1982).
31. W. Götze, Solid State Comm. 27, 1393 (1978); J. Phys. C. 12, 1297 (1979).
32. D. Belitz, A. Gold and W. Götze, Zs. Phys. B 44, 273 (1981).
33. Yu.A. Firsov and V.N. Prigodin, J. Phys. C. 18 (1985), to be published.
34. V.N. Prigodin, Yu.A. Firsov and W. Weller, Solid State Comm., to be published in 1985.
35. K.B. Efetov, A.I. Larkin and D.E. Khmel'nizkii, Zh. Eksp. Teor. Fiz. 79, 1120 (1980). [Sov. Phys. JETP 52, 568 (1980).]
36. P. Wölfle and R.N. Bhatt, Phys. Rev. B 30, 3542 (1984).
37. D.J. Bishop, R.C. Dynes, B.J. Lin and D.C. Tsui, Phys. Rev. B 30, 3539 (1984).
38. E.A. Kotov and M.V. Sadovskii, Zs. Phys. B 51, 17 (1983).
39. W. Apel and T.M. Rice, J. Phys. C. 16, L1151 (1983).
40. A.A. Abrikosov and I.A. Ryzhkin, Sol. St. Comm. 24, 317 (1977).
41. V.E. Kravtsov and I.V. Lerner, Zh. Eksp. Teor. Fiz. 86, 1332 (1984). [Sov. Phys. JETP 59 (1984).]
42. F. Wegner, Zs. Phys. B 25, 327 (1976).
43. B. Shapiro, Phys. Rev. B 25, 4266 (1982).
44. G. Landwehr and P. Handler, J. Phys. Chem. Solids 23, 891 (1962).
45. H. Matare, Defect Electronics in Semiconductors, New York (1971).
46. B.M. Vul and E.I. Zavaritskaya, a) Zh. Eksp. Teor. Fiz. 76, 1089 (1979). [Sov. Phys. JETP 49, 551 (1979).] b) Zh. Eksp. Teor. Fiz., Pis'ma 37, 571 (1983). [Sov. Phys. JETP, Lett. 37, 681 (1983).]
47. N. Giordano, Phys. Rev. B 22, 5635 (1960); Physica 107, 1 (1981).
48. J.T. Masden and N. Giordano, Phys. Rev. Lett. 49, 819 (1982).
49. A.E. White, M. Tinkham, W.J. Skocpol and D.C. Flunders, Phys. Rev. Lett. 48, 1752 (1982).
50. P.W. Anderson, D.J. Thouless, E. Abrahams and D.S. Fisher, Phys. Rev. B 22, 3519 (1980).
51. P.W. Anderson, Phys. Rev. B 23, 4828 (1981).
52. M. Ya. Azbel, Phys. Rev. Lett. 46, 675 (1981).
53. O.N. Dorokhov, Zh. Eksp. Teor. Fiz. 85, 1040 (1983). [Sov. Phys. JETP 58, 606 (1983).]
54. K.B. Efetov and A.I. Larkin, Zh. Eksp. Teor. Fiz. 85, 764 (1983). [Sov. Phys. JETP 58, 434 (1983).]

55. R. Abou-Chacra, P.W. Anderson and D.J. Thouless, J. Phys. C 6, 1734 (1973).
56. S.M. Girvin and M. Johnson, Phys. Rev. B 22, 3585 (1980).
57. B. Shapiro, Phys. Rev. Lett. 50, 747 (1983).
58. H. Kunz and B. Souillard, J. de Physique-Lett. 44, L411-L503 (1983).
59. K.B. Efetov, Zh. Eksp. Teor. Fiz., Pis'ma 40, 17 (1984).
60. V.L. Bonch-Bruevich, Usp. Fiz. Nauk. 140, 583 (1983).
61. A. Schmid, Zs. Phys. 271, 251 (1974).
62. A.L. Efros and B.I. Shklovskii, J. Phys. C 8, L49 (1975).
63. A.L. Efros, J. Phys. C. 9, 2021 (1976).
64. B.I. Shklovskii and A.L. Efros, Electronic Properties of Doped Semiconductors, Moscow, "Nauka," (1979).
65. B.L. Altshuler and A.G. Aronov, a) Zh. Eksp. Teor. Fiz., Pis'ma 30, 514 (1979). [Sov. Phys. JETP Lett. 30.] b) Zh. Eksp. Teor. Fiz. 77, 2028 (1979). [Sov. Phys. JETP 50, 968 (1979).] c) Solid State Comm. 38, 11 (1981).
66. B.L. Altshuler, A.G. Aronov and P.A. Lee, Phys. Rev. Lett. 44, 1288 (1980).
67. M. Kaveh and N.F. Mott, J. Phys. C 14, L183 (1981).
68. E.Abrahams, P.W. Anderson, P.A. Lee and T.V. Kamakrishnan, Phys. Rev. B 24, 6783 (1981).
69. C. Castellani, C. di Castro, G. Forgacs and E. Tabet, Nucl. Phys. B 225 [FS-3], 441 (1983).
70. M.I. Kaznel'son and V.M. Sadowskii, a) Fiz. Teor. Tela 25, 3371 (1983). [Sov. Phys. Solid State 25, 1942 (1983).] b) Zh. Eksp. Teor. Fiz. 87, 523 (1984). [Sov. Phys. JETP 60 (1984).]
71. B.I. Altshuler, D. Khmelnitzkii, A.I. Larkin and P.A. Lee, Phys. Rev. B 22, 5142 (1980).
72. W.L. McMillan, Phys. Rev. B 24, 2739 (1981).
73. A.M. Finkelstein, a) Zh. Eksp. Teor. Fiz. 84, 168 (1983). [Sov. Phys. JETP 57, 97 (1983).] b) Zh. Eksp. Teor. Fiz. 86, 367 (1984). [Sov. Phys. JETP 59, 212 (1984).] c) Zs. Phys. B 56, 189 (1984). d) Zh. Eksp. Teor. Fiz., Pis'ma 40, 63 (1984). [Sov. Phys. JETP Lett. 40, 796, 1984).]
74. B.L. Altshuler and A.G. Aronov, a) Sol. St. Comm. 46, 429 (1983). b) Zh. Eksp. Teor. Fiz., Pis'ma 37, 349 (1983). [Sov. Phys. K JETP Lett. 37, 410 (1983).]
75. C. Castellani, C. di Castro, P.A. Lee and M. Ma, Phys. Rev. B 30 (2), 527 (1984).
76. C. Castellani, C. di Castro, P.A. Lee, M. Ma, S. Sorella and E. Tabet, Phys. Rev. B 30, 1596 (1984).

CONTENTS OF COMPANION VOLUME:

TETRAHEDRALLY-BONDED AMORPHOUS SEMICONDUCTORS

Edited by
David Adler and Hellmut Fritzsche

PART ONE: GROWTH AND STRUCTURE

PART TWO: HYDROGEN INCORPORATION, IMPURITIES, AND DEFECTS

CONTENTS OF COMPANION VOLUME:

PHYSICS OF DISORDERED MATERIALS

Edited by
David Adler, Hellmut Fritzsche,
and Stanford R. Ovshinsky

PART ONE: GENERAL ASPECTS

PART TWO: ELEMENTS OF DISORDER

AUTHOR INDEX

SUBJECT INDEX